OPERATIONS WITH EQUATIONS

Addition Property of Equality

Subtraction Property of Equality

Multiplication Property of Equality   If $a = b$, then $a \cdot c = b \cdot c$.

Division Property of Equality   If $a = b$, then $\dfrac{a}{c} = \dfrac{b}{c}$, $c \neq 0$.

MW01503402

OPERATIONS WITH INEQUALITIES

If $a < b$, then $a + c < b + c$.   If $a < b$, then $\dfrac{a}{c} < \dfrac{b}{c}$ if $c > 0$;

If $a < b$, then $a - c < b - c$.

If $a < b$, then $a \cdot c < b \cdot c$ if $c > 0$; $\dfrac{a}{c} > \dfrac{b}{c}$ if $c < 0$.
$\qquad\qquad\quad a \cdot c > b \cdot c$ if $c < 0$;

If $P > 0$, then $|x| < P$ is equivalent to $-P < x < P$,

$\qquad$ and $|x| > P$ is equivalent to $x < -P$ or $x > P$.

DETERMINANTS

$$\begin{vmatrix} a & b \\ c & d \end{vmatrix} = ad - bc$$

$$\begin{vmatrix} a_1 & b_1 & c_1 \\ a_2 & b_2 & c_2 \\ a_3 & b_3 & c_3 \end{vmatrix} = a_1 b_2 c_3 - a_1 b_3 c_2 - a_2 b_1 c_3 + a_3 b_1 c_2 + a_2 b_3 c_1 - a_3 b_2 c_1$$

PROPERTIES OF EXPONENTS

If a and b are real numbers and m and n are integers, the following rules hold true.

Product Rule $b^m \cdot b^n = b^{m+n}$ **Power Rules** $(b^m)^n = b^{mn}$

$\qquad\qquad\qquad\qquad\qquad\qquad\qquad\qquad\qquad\qquad\quad (a \cdot b)^n = a^n \cdot b^n$

Quotient Rule $\dfrac{b^m}{b^n} = b^{m-n}$, $b \neq 0$ $\left(\dfrac{a}{b}\right)^n = \dfrac{a^n}{b^n}$, $b \neq 0$

Definition of a negative exponent $b^{-n} = \dfrac{1}{b^n}$, $b \neq 0$

Definition of a zero exponent $b^0 = 1$, $b \neq 0$

SPECIAL PRODUCTS

$(a + b)^2 = a^2 + 2ab + b^2$

$(a - b)^2 = a^2 - 2ab + b^2$

$(a + b)(a - b) = a^2 - b^2$

SPECIAL FACTORS

$a^2 + 2ab + b^2 = (a + b)^2$

$a^2 - 2ab + b^2 = (a - b)^2$

$a^2 - b^2 = (a + b)(a - b)$

$a^3 - b^3 = (a - b)(a^2 + ab + b^2)$

$a^3 + b^3 = (a + b)(a^2 - ab + b^2)$

Intermediate Algebra

Intermediate Algebra

C. Lee Welch

Gilbert M. Peter

Cuesta College

Scott, Foresman and Company

Glenview, Illinois

London, England

Available Supplement

Study Guide: includes chapter summaries and tests (with solutions) and additional programmed exercises.

Library of Congress Cataloging-in-Publication Data

Welch, C. Lee
    Intermediate algebra.

    Includes index.
    1. Algebra.   I. Peter, Gilbert M.   II. Title.
QA152.2.W433   1986   512.9   85-26279
ISBN 0-673-16587-6

    2   3   4   5   6—RRC—90   89   88   87   86

Preface

Intermediate Algebra is designed for students who need an algebra course that bridges the gap between elementary algebra and more advanced precalculus courses. A knowledge of elementary algebra is assumed.

The book was written to provide students and instructors with a meaningful and thorough text. The basic concepts of intermediate algebra are presented in a straightforward manner. The concepts are motivated by examples and reinforced by additional examples. Where it is considered appropriate, formal proofs are used to justify the ideas. Mathematical terms are carefully and clearly defined.

There is a common thread throughout the text: first, the development of an idea or skill; next, application of the idea or skill to problem solving; finally, review and expansion upon the idea or skill in a repetitive learning process.

Special Features

Set notation is introduced in Chapter 1 and used to show the relationships between the components of the number system. It is also used to provide a consistent means to identify the solution sets to equations throughout the text. Solution sets to inequalities are described using interval notation so as to prepare the student for more advanced courses.

A review of fractions is provided in Chapter 1 for those students who need it.

Word problems appear early in Chapter 1, and they are included in most exercise sets throughout the text and in a separate section at the end of most chapters. In an attempt to make the problems "meaningful," we have taken many problems from areas in which the units of measurement may not be familiar to the student. Consequently such units are not always included.

Linear equations and inequalities are introduced in Chapter 2 and expanded upon in Chapter 3. Many review exercises are scattered throughout the text to reinforce the techniques of graphing. This organization provides the student with the option of using systems of equations or inequalities to solve word problems throughout the remainder of the text. It also helps the student prepare for graphing nonlinear equations in Chapters 9 and 10.

Theory versus practice: A careful balance is maintained between the "how" and "why" of each mathematical concept. The discussion is intuitive but based on carefully stated mathematical definitions and theorems. Proofs are included in areas where they enhance the "why" of a concept.

Problems involving logarithms and exponents may be solved either with the use of scientific calculators or tables. We assume that most students will have calculators.

Chapter Organization

Clear format includes the pedagogical use of color within examples and for explanatory side comments. Definitions, rules, theorems, and procedures are highlighted to emphasize their importance and to make them easy to locate.

Caution symbols, ▶, are used to warn students about common errors.

Examples are numerous and detailed to help the student develop the skills and understanding necessary to be successful in solving the exercises. Approximately 520 examples are included. For clarity, the end of each example is indicated with a colored symbol, ■.

Exercises are carefully graded to provide a smooth transition from easier drill problems to more challenging applications. More than 5000 exercises are distributed throughout the book. Exercises are paired to ensure that all concepts are covered, regardless of whether odd or even problems are assigned. Answers to the odd-numbered exercises are included at the back of the text. The exercise sets are designed so that the average student can experience success, and the more capable student will be challenged.

Review problems are included in many of the exercise sets to help reinforce what the student has already learned. A set of review problems is also provided at the end of each chapter. The set of review problems at the end of each even-numbered chapter covers the two preceding chapters.

Calculator problems are included in most of the exercise sets.

Chapter summaries are provided at the end of each chapter to outline the important concepts. They are keyed to the appropriate sections for easy reference.

Study Guide

A self-paced **Study Guide** is available for students who would like more practice. The problems closely parallel those found in the text, section by section, and gradually lead the student from participating in a solution to actually supplying the entire solution to a problem. Many of the problems in Chapters 3–11 are taken directly from the text, and particular attention is paid to those in which the student is asked to "prove" or "show" that a concept is true.

The important concepts of each chapter are summarized in a manner similar to that in the text, and each chapter ends with a test for which complete solutions are supplied.

Instructor's Guide

The **Instructor's Guide** contains answers to all even-numbered exercises in the text. It also contains three alternative tests and a test bank for each chapter.

We wish to thank our reviewers, Ben P. Bockstege, Broward Community College; Bob Finnell, Portland Community College; M. Catharine Hudspeth,

California State Polytechnic; Robert Russell, West Valley College; Jean L. Sutherland, Evergreen Valley College; Wesley W. Tom, Chaffey College; and George Wales, Ferris State College. Their many helpful suggestions and comments have greatly improved the content and organization of the text.

In particular, we would like to thank our editors, Steve Quigley, Sarah Joseph, Kathleen Monahan, and Janet Tilden, for their excellent suggestions and encouragement. Last, but not least, we would like to thank our wives, Pauline and Marlene, for their patience and continued support.

C. Lee Welch
Gilbert M. Peter

Contents

11

Sequences, Series, and the Binomial Theorem 415

1

The Real Number System

Success in intermediate algebra may depend on how much is remembered from elementary algebra. Do you remember the terminology and the order for carrying out the operations of arithmetic and algebra? Do you remember what kind of numbers you would expect to find in the set of real numbers? Do you remember the properties of the real numbers well enough to use them in problem solving? If your answer to each of the above questions is yes, Chapter 1 should be easy. If your answer is no, Chapter 1 can provide the review that is essential for success. It is important that you master its contents, because it forms the basis for the work that will follow.

1.1 Sets and Set Notation

One of the basic concepts of all branches of mathematics is the *set*.

A **set** is a well-defined collection of objects.

The objects within a set are called **elements** or **members.** A set is well defined when one can state, without doubt, whether an element belongs to the set.

Sets are usually denoted by capital letters, and the elements of the set are placed within braces { }. For example, if S represents the seasons of the year, then

$S = \{$spring, summer, fall, winter$\}$.

The symbol $\in$ indicates that a particular object is an element of a set. For the set

$L = \{2, 4, 6, 8\}$,

$2 \in L$, $4 \in L$, $6 \in L$, and $8 \in L$. To indicate that 10 is not an element of L, use the symbol $\notin$: $10 \notin L$. This statement is read "10 is not an element of L." The slash negates a given symbol.

$2 + 3 = 5$ is read "2 plus 3 is equal to 5."

$2 + 3 \neq 9$ is read "2 plus 3 is not equal to 9."

$4 \in L$ is read "4 is an element of L."

$13 \notin L$ is read "13 is not an element of L."

There are two types of sets, *finite* and *infinite*. A set is a **finite set** if there is a number that describes exactly how many elements it contains. A set that is not finite is an **infinite set.**

Example 1   State whether the following sets are finite or infinite.

(a) $A = \{2, 4, 6\}$

   A is a finite set. It contains exactly 3 elements.

(b) $B = \{1, 2, 3, \ldots\}$

   B is an infinite set. The three dots mean that the set continues indefinitely in the same manner.

(c) The set of cards in a poker deck

   This set is finite, since it contains 52 members.   ■

The order in which the elements of a set are written is not important. Both $\{a, b, c\}$ and $\{b, a, c\}$ represent the same set. Also, the set $\{a, b, c, c\}$ is the same set as $\{a, b, c\}$. We will not repeat elements within a set. For example, let L be the set of letters in the word *mathematics*. Then

   $L = \{m, a, t, h, e, i, c, s\}.$

There are two ways to describe sets, the *listing method* and *set-builder notation*, which is also known as the rule method. With the **listing method,** each element in the set is named. With **set-builder notation,** the common characteristic of the elements in the set is described. For example,

   $P = \{$Mercury, Venus, Earth, Mars, Jupiter, Saturn, Uranus, Neptune, Pluto$\}$

identifies the set of planets in our solar system by listing all the elements. On the other hand,

   $P = \{x \mid x$ is a planet in our solar system$\}$

is read "P is the set of all x such that x is a planet in our solar system." (The vertical bar is read "such that.") This statement identifies the elements of the set by describing their shared characteristic. Notice that x could represent any one of the elements in the set. For this reason we say that x is a **variable.** A symbol that represents a fixed value throughout a discussion is called a **constant.**

Example 2   Write $A = \{x \mid x$ is a counting number less than 10$\}$ using the listing method.

   $A = \{1, 2, 3, 4, 5, 6, 7, 8, 9\}$ ■

Example 3   Write $V = \{a, e, i, o, u\}$ using set-builder notation.

   $V = \{x \mid x$ is a vowel$\}$ ■

A set that has no elements is called the **empty set** or **null set;** it is designated by the symbol $\emptyset$.

Two operations are used to combine sets. These are the *union* of sets and the *intersection* of sets.

The **union** of two sets A and B, written $A \cup B$, is the set of elements that are in set A, in set B, or in both sets A and B.

$$A \cup B = \{x \,|\, x \in A \quad \text{or} \quad x \in B\}$$

The **intersection** of two sets A and B, written $A \cap B$, is the set of elements that are common to both sets A and B.

$$A \cap B = \{x \,|\, x \in A \quad \text{and} \quad x \in B\}$$

Example 4   Let $A = \{1, 2, 3\}$, $B = \{3, 4, 5\}$ and $C = \{5, 6\}$. List the elements of the following.

(a) $A \cup B = \{1, 2, 3, 4, 5\}$

(b) $A \cup C = \{1, 2, 3, 5, 6\}$

(c) $A \cap B = \{3\}$

(d) $A \cap C = \emptyset$ A and C have no elements in common

(e) $(A \cup B) \cap C = \{1, 2, 3, 4, 5\} \cap \{5, 6\} = \{5\}$

(f) $(A \cap B) \cup C = \{3\} \cup \{5, 6\} = \{3, 5, 6\}$

(g) $(A \cup B) \cap (A \cup C) = \{1, 2, 3, 4, 5\} \cap \{1, 2, 3, 5, 6\}$

                 $= \{1, 2, 3, 5\}$ ■

The union and intersection of sets are often shown by using **Venn diagrams.** (See Figure 1.1.) In a Venn diagram, the operation under consideration is indicated by shading the region satisfied by it. The rectangle enclosing the sets A and B is used to define those areas, numbers or regions to be considered. It is called the **universal set.** Many other possible diagrams for the intersection and union of two sets exist; some will be considered after the concept of a *subset* is discussed.

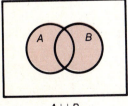

$A \cup B$

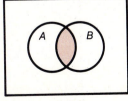

$A \cap B$

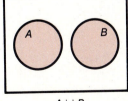

$A \cup B$

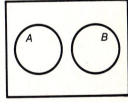

$A \cap B$

Figure 1.1

> A set A is a **subset** of a set B, written $A \subseteq B$, if and only if every element of A is also an element of B. Thus,
>
> $A \subseteq B$ if whenever $x \in A$, $x \in B$.

Example 5   Let $A = \{1, 2, 3, 4\}$, $B = \{2, 3, 4\}$, and $C = \{2, 3, 4, 5\}$.

(a) $B \subseteq A$ Every element of B is an element of A

(b) $B \subseteq C$ Every element of B is an element of C

(c) $A \nsubseteq C$ $1 \in A$ but $1 \notin C$

(d) $C \nsubseteq A$ $5 \in C$ but $5 \notin A$ ■

Every set that is not empty has more than one subset. For example, the subsets of $\{1, 2\}$ are $\{1\}$, $\{2\}$, $\{1, 2\}$, and $\varnothing$. Every set is a subset of itself, and the empty set is a subset of every set.

Example 6   List the subsets of the following sets.

(a) $\{a\}$

There are two subsets, $\{a\}$ and $\varnothing$.

(b) $\{a, b, c\}$

There are eight subsets; $\{a\}$, $\{b\}$, $\{c\}$, $\{a, b\}$, $\{a, c\}$, $\{b, c\}$, $\{a, b, c\}$ and $\varnothing$.   ■

How many subsets would you expect a set with four elements to have? With five elements? With six elements?

Example 7   Use Venn diagrams to indicate (a) $A \cup B$ if $B \subseteq A$ and (b) $A \cap B$ if $B \subseteq A$.

The regions are shown in Figures 1.2 and 1.3. Again, shading is used to indicate the portion under consideration.

(a) $A \cup B = A$ if $B \subseteq A$ (b) $A \cap B = B$ if $B \subseteq A$ ■

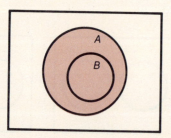

Figure 1.2

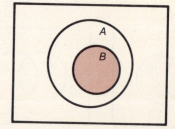

Figure 1.3

Example 8   Shade the region described as $\{x \mid x \in A \text{ and } x \notin B\}$.
   Vertical lines can be used to indicate all $x \in A$ and horizontal lines to indicate $x \notin B$. The portion with double shading is the solution set. (See Figure 1.4.) The key word is **and,** which means that both conditions must be met. ■

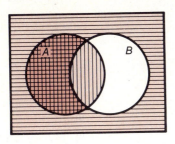

Figure 1.4

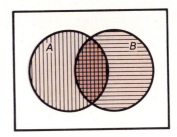

Figure 1.5

Example 9   Shade the region described as $\{x \mid x \in A \text{ or } x \in B\}$.
   Vertical lines are used to indicate when $x \in A$ and horizontal lines when $x \in B$. (See Figure 1.5.) The key word is **or,** which means either condition must be satisfied. The region with any shading forms the solution set. ■

▶ *Caution*   It is easy to use set terminology incorrectly. With respect to the set $A = \{1, 2, 3\}$,

Right	*Wrong*
$1 \in A$	$\{1\} \in A$
$\{1, 2\} \subseteq \{1, 2, 3\}$	$1 \subseteq A.$

Writing $\{1\}$ indicates a set, and sets are subsets of other sets, not elements of them. Similarly, 1 is an element of the set A and should not be called a subset of it.

Exercises 1.1

Write each of the following sets using the listing method. If the set is empty, write Ø.

1. $\{x \mid x$ is a letter in the word *add*$\}$
2. $\{t \mid t$ is a letter in the word *calculus*$\}$
3. $\{y \mid y$ is a day of the week starting with the letter A$\}$
4. $\{p \mid p$ was a woman president of the United States$\}$
5. $\{m \mid m - 4 = 0\}$
6. $\{x \mid x + 3 = 5\}$
7. $\{x \mid x$ is a counting number less than 5$\}$
8. $\{y \mid y$ is a counting number between 7 and 11$\}$
9. $\{2, 3, 4\} \cup \{1, 3, 5\}$
10. $\{2, 3, 4\} \cap \{1, 3, 5\}$

Indicate whether each statement is true *or* false.

11. $2 \in \{2, 3, 5\}$

12. $6 \in \{x \mid x$ is a counting number less than 9$\}$

13. $5 \subseteq \{4, 5, 9\}$

14. $-5 \in \{3, 4, 5\}$

15. $\emptyset \subseteq \{1\}$

16. $\{4, 1, 5\} \subseteq \{1, 4, 5\}$

17. If $C \cup D = D$ then $C \subseteq D$.

18. $\emptyset = \{0\}$

19. If C and D are any sets, then $C \cap D \subseteq D$.

20. $\{n \mid n$ is a counting number greater than 100$\}$ is an infinite set.

List the elements of the given set in each of the following problems if $A = \{1, 3, 5, 9\}$, $B = \{2, 4, 6, 8\}$ and $C = \{1, 2, 3, 6\}$.

21. $A \cup B$

22. $A \cap C$

23. $B \cap C$

24. $B \cup C$

25. $(A \cap B) \cup C$

26. $(A \cup C) \cap B$

27. $(A \cup B) \cup C$

28. $(A \cap B) \cap C$

29. $(A \cap B) \cap (A \cap C)$

30. $(A \cup B) \cap (A \cap C)$

31. $(B \cap C) \cap (B \cap A)$

32. $(A \cap B) \cap A$

33. $A \cup (B \cap C)$

34. $(A \cup C) \cap C$

Given two sets A and B with $A \cap B \neq \emptyset$, draw Venn diagrams and shade the appropriate region for each of the following. Use the diagram below.

35. $\{x \mid x \in A$ and $x \in B\}$

36. $\{x \mid x \notin A$ and $x \in B\}$

37. $\{x \mid x \in A$ or $x \notin B\}$

38. $\{x \mid x \in A$ or $x \in B\}$

39. $\{x \mid x \notin A$ and $x \notin B\}$

40. $\{x \mid x \notin A$ or $x \notin B\}$

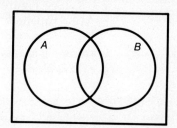

Given two sets A and B with $A \cap B = \emptyset$, draw Venn diagrams and shade the appropriate region for each of the following.

41. $\{x \mid x \notin A$ and $x \in B\}$

42. $\{x \mid x \in A$ and $x \in B\}$

43. $\{x \mid x \in A$ or $x \in B\}$

44. $\{x \mid x \in A$ or $x \in B\}$

45. $\{x \mid x \notin A$ or $x \notin B\}$

46. $\{x \mid x \notin A$ and $x \notin B\}$

Indicate whether each of the following sets is finite, infinite *or an* empty *set.*

47. $\{n \mid n$ is an even counting number greater than 1000$\}$

48. $\{n \mid n$ is a counting number and a multiple of 2$\}$

49. $\{n \mid n$ is an odd counting number and divisible by 4$\}$

50. $\{n \mid n$ is a counting number that is a multiple of 3 and is divisible by 6$\}$

List all of the subsets of each of the following sets.

51. $\{3\}$

52. $\{m\}$

53. $\{r, t\}$

54. $\{3, 5\}$

55. $\{0, 1, 2\}$

56. $\{a, b, c, d\}$

Write each of the following using set-builder notation.

57. $\{2, 4, 6, 8\}$

58. $\{3, 5, 7\}$

59. $\{a, e, i, o, u\}$

60. $\{w, x, y, z\}$

61. $\{7, 8, 9, \ldots\}$

62. $\{8, 10, 12, \ldots\}$

Given two sets C and D such that D ⊆ C, C ⊄ D, draw Venn diagrams and shade the appropriate region for the following.

63. $\{y \mid y \in C \text{ and } y \in D\}$ **64.** $\{y \mid y \notin C \text{ and } y \in D\}$

65. $\{y \mid y \in C \text{ and } y \notin D\}$ **66.** $\{y \mid y \notin C \text{ and } y \notin D\}$

The Eggleston family, consisting of Verle, Pam, Sherry, and Lisa, will be designated by {V, P, S, L}. In order to develop a cooperative family work schedule, they have agreed to set up work groups for house cleaning, yard cleanup, and other chores.

67. How many different three-person work groups could be formed?

68. How many different two-person work groups could be formed?

1.2   Real Numbers and the Number Line

Before beginning a study of the real numbers and the number line, we will review the symbols that indicate the relationship between two real numbers a and b. They are listed below.

Symbols	Interpretation
$a = b$	a is **equal to** b
$a > b$	a is **greater than** b
$a < b$	a is **less than** b
$a \geq b$	a is **greater than or equal to** b
$a \leq b$	a is **less than or equal to** b

Any of the above symbols can be negated by drawing a slash mark through it. The statement $a \not< b$ means "a is not less than b." This is the same as saying that a is greater than or equal to b.

The set of real numbers can be shown graphically on a number line. A **number line** is a straight line on which an arbitrary point labeled 0 (zero) is located. The point where 0 is located is called the **origin**. (See Figure 1.6.) The arrows at either end of the number line indicate that the line extends infinitely in both directions. The positive direction is to the right of zero and the negative direction to the left of zero.

Figure 1.6

Each point on the number line corresponds to a unique real number. The number line in Figure 1.7 shows a few of the points. The point labeled P corresponds to the number 3. The number 3 is called the **coordinate** of P. The coordinate of R is -2, and the coordinate of S is -1.

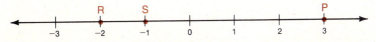

Figure 1.7

Example 1   Locate 2, $\frac{1}{2}$, $-2\frac{1}{2}$, -1.7 and $\frac{8}{5}$ on a number line.
The number line in Figure 1.8 shows the points. ■

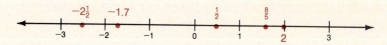

Figure 1.8

The size of two real numbers can be compared by looking at where they are located on a number line. The fact that numbers increase as you move from left to right on the number line leads to the following definition.

> For any two real numbers a and b, $a < b$ if a appears to the left of b on a number line.

Example 2   Use the number line in Figure 1.9 to tell whether the following statements are true.

(a) $5 < 6$ True, because 5 appears to the left of 6 on a number line.

(b) $-4 < -2$ True, because -4 appears to the left of -2 on a number line.

(c) $3 > -2$ True, because 3 appears to the right of -2 on a number line. ■

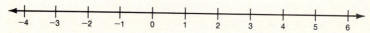

Figure 1.9

Every point to the right of zero corresponds to a point the same distance to the left of zero. The numbers represented by these points are called **additive inverses** or **opposites**.

5 and -5 are opposites or additive inverses.

2 and -2 are opposites or additive inverses.

$\frac{1}{5}$ and $-\frac{1}{5}$ are opposites or additive inverses.

We see that to find the opposite of a positive number a negative sign is placed before it. If the same operation is applied to a negative number, say -4, its opposite will be $-(-4)$. However, we know the opposite of -4 is 4.

For any real number a,

$-(-a) = a.$

The two real numbers 4 and -4 are said to be opposites. However, they do have a common property; on a number line they are the same distance from the origin. These distances are shown in Figure 1.10. The distance (number of units) on a number line between a real number and the origin is the **absolute value** of the number. The absolute value of a real number is shown by placing the number between two vertical bars.

$|a|$ is read "the absolute value of a."

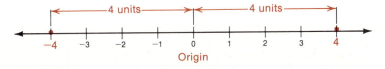

Figure 1.10

For example,

$|4| = 4$ 4 is 4 units from the origin

$|-5| = 5$ -5 is 5 units from the origin

$|-1.5| = 1.5$ -1.5 is 1.5 units from the origin.

See Figure 1.11.

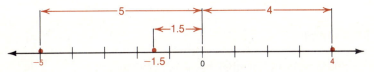

Figure 1.11

More formally, for any real number a the absolute value is defined as follows.

Absolute Value

$|a| = a$ if $a \geq 0$

$|a| = -a$ if $a < 0$ $a \in \{\text{real numbers}\}$

Example 3   Give the values of the following expressions.

(a) $|5| = 5$ since $5 \geq 0$

(b) $|-3| = -(-3) = 3$ since $-3 < 0$

(c) $|0| = 0$ ■

Up to this point the discussion has centered around the set of real numbers. The set of real numbers can be broken down into several subsets. The first is the set of **natural numbers,** which we will denote by N.

Natural numbers $= N = \{1, 2, 3, 4, \ldots\}$

Since the natural numbers are used for counting, they are also called the counting numbers.

When zero is placed within the set of natural numbers, a new set W, the **whole numbers,** is formed.

Whole numbers $= W = \{0, 1, 2, 3, \ldots\}$

The set formed by the union of the whole numbers and their additive inverses is known as the set of **integers** and is denoted by J.

Integers $= J = \{\ldots -3, -2, -1, 0, 1, 2, 3, \ldots\}$

Notice that

$$N \subseteq W \subseteq J.$$

The set of integers is a subset of a still broader subset of the real numbers, Q, the set of **rational numbers.** We will define it using set-builder notation:

Rational numbers $= Q = \left\{ \dfrac{a}{b} \,\middle|\, a \in J, \, b \in J, \, b \neq 0 \right\}.$

For example,

$\dfrac{3}{5}$ is a rational number because $3 \in J$ and $5 \in J$;

$\dfrac{-1}{8}$ is a rational number because $-1 \in J$ and $8 \in J$;

0.17 is a rational number because $0.17 = \dfrac{17}{100}$ and $17 \in J$, $100 \in J$;

3 is a rational number because $3 = \dfrac{3}{1}$ and $3 \in J$, $1 \in J$;

$2\dfrac{1}{3}$ is a rational number because $2\dfrac{1}{3} = \dfrac{7}{3}$ and $7 \in J$, $3 \in J$.

Rational numbers are frequently expressed in decimal form by using division. The next three examples illustrate this procedure.

Example 4   Write $\frac{1}{3}$ in decimal form.

Divide the numerator, 1, by the denominator, 3.

```
    0.333
3)1.00
  ↑ 9
    ――
   10
    9
    ――
   10
    9
    ――
    1  ← We started with 1
         The quotient will repeat
```

Notice that the 3 **repeats.** This is indicated by placing a bar over the repeating digit.

$$\frac{1}{3} = 0.33\overline{3} \quad \blacksquare$$

Example 5   Write $\frac{2}{7}$ in decimal form.

Divide 2 by 7.

```
    0.285714
7)2.000000
   1 4
   ――
  ↑ 60
    56
    ――
    40
    35
    ――
    50
    49
    ――
    10
     7
    ――
    30
    28
    ――
     2  ← We started with 2
          The quotient will repeat
```

Here the digits 285714 repeat, so that

$$\frac{2}{7} = 0.\overline{285714}. \quad \blacksquare$$

Example 6   Express $\frac{3}{8}$ in decimal form.

$$
\begin{array}{r}
0.375 \\
8\overline{)3.000} \\
2\ 4 \\
\hline
60 \\
56 \\
\hline
40 \\
40 \\
\hline
\end{array}
$$

0.375 is a terminating decimal. The division process ends.   ∎

In each of the last three examples the decimal form of the rational number either terminated or repeated. A rational number is often defined as one that can be expressed as a repeating or a terminating decimal.

A number that cannot be expressed as a repeating or terminating decimal is said to be **irrational.** The set of irrational numbers is denoted by H. Thus,

0.101001000 . . . is a nonrepeating, nonterminating decimal.
$\sqrt{2}, \sqrt{3}, \sqrt{5}, \sqrt{7}$ can be shown to be nonrepeating, nonterminating decimals.

π is a nonrepeating, nonterminating decimal. It is the ratio of the circumference to the diameter of a circle. Its approximate value is 3.141592 to 6 decimal places.

When the union of the set of rational numbers, Q, and the irrational numbers, H, is formed, the result is the set of real numbers, R. The relationships among the subsets of the real numbers are shown in Figure 1.12 on the facing page. The entries in the diagram are typical examples of the elements of each set.

It is frequently desirable to be able to show the graph of an inequality. To do this, we must first consider the concept of an **interval.** Suppose that x represents a number that lies between two numbers c and d on a number line, as shown in Figure 1.13. This means that $c < x < d$. This would be indicated using **interval notation** as (c, d). The parentheses indicate that the endpoints of the interval are not included in it. The interval is said to be an **open interval.** If the endpoints are included, as in $c \leq x \leq d$, the interval is called a **closed interval** and is written in interval notation as $[c, d]$. The brackets indicate that the endpoints of the interval are to be included. A **half-open interval** is shown by a bracket and parenthesis. The bracket indicates the closed endpoint. Various types of intervals are shown in the table on page 14. The symbols ∞ and $-\infty$ are used to indicate that an interval has no upper or lower boundary. They do not represent real numbers. Unless otherwise indicated, the variable x represents a real number whenever it is used.

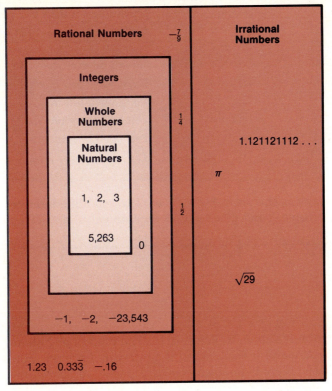

Figure 1.12
$R = \{$real numbers$\}$
$H = \{$irrational numbers$\}$
$Q = \{$rational numbers$\}$
$J = \{$integers$\}$
$W = \{$whole numbers$\}$
$N = \{$natural numbers$\}$

Figure 1.13

Interval Notation	Set-Builder Notation	Interpretation
(c, d)	$\{x \mid c < x < d\}$	Open interval from c to d
$(c, d]$	$\{x \mid c < x \le d\}$	Half-open intervals from c to d
$[c, d)$	$\{x \mid c \le x < d\}$	
$[c, d]$	$\{x \mid c \le x \le d\}$	Closed interval from c to d
$[d, \infty)$	$\{x \mid x \ge d\}$	Right half-open interval
(d, ∞)	$\{x \mid x > d\}$	Right open interval
$(-\infty, c]$	$\{x \mid x \le c\}$	Left half-open interval
$(-\infty, c)$	$\{x \mid x < c\}$	Left open interval
$(-\infty, \infty)$	$\{x \mid x \in R\}$	The real number line

Example 7   Graph $(2, 6) = \{x \mid 2 < x < 6\}$.

The graph is shown in Figure 1.14. The solid line indicates that any real number *between* 2 and 6 lies within the interval.   ■

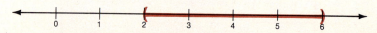

Figure 1.14

Example 8   Graph $[2, 6) = \{x \mid 2 \le x < 6\}$.

The graph is shown in Figure 1.15. The bracket at 2 shows that 2 is a *lower boundary* and is included in the interval. Since 6 is not included (parenthesis) there is no *upper boundary*.   ■

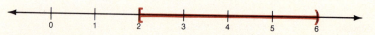

Figure 1.15

Example 9   Graph $[3, \infty) = \{x \mid x \ge 3\}$.

The graph is shown in Figure 1.16. The arrow indicates that all points to the right of 3 are included in the interval.   ■

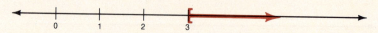

Figure 1.16

Example 10   Convert each of the following to interval notation and graph.

(a) $\{x \mid x < -2\}$

$(-\infty, -2)$ The graph is shown in Figure 1.17(a).

(a)   (b)

Figure 1.17

(b) $\{x \mid -4 \le x \le 2\}$
   $[-4, 2]$ The graph is shown in Figure 1.17(b). ■

Example 11   Graph $(-3, 2] \cap [-1, 5]$.
   The graph will include all points that satisfy *both* intervals. The real numbers from -1 to 2 are in both sets. The graph is shown in Figure 1.18. ■

Figure 1.18

Exercises 1.2

Fill in the blanks with $<$ or $>$ to make each statement correct.

1. 5 _____ 8

2. -6 _____ 3

3. -4 _____ -6

4. $|-3|$ _____ 0

5. $-|-5|$ _____ 0

6. $-|9|$ _____ $|-9|$

7. $\left|-\left(\dfrac{3}{4}\right)\right|$ _____ $\dfrac{1}{4}$

8. $\dfrac{1}{5}$ _____ $-\left|\dfrac{3}{5}\right|$

Fill in the blanks with $\in$ or $\notin$ to make each statement correct.

9. 3 _____ N

10. $\dfrac{1}{5}$ _____ J

11. $0.\overline{333}$ _____ H

12. 0.675 _____ Q

13. $|-5|$ _____ W

14. 0 _____ N

15. $\sqrt{2}$ _____ Q

16. $\sqrt{3}$ _____ H

State the additive inverse of each of the following.

17. 6

18. -9

19. -17

20. a

21. $-|x|$

22. $-|-x|$

23. $-m$

24. $-(-p)$

Fill in the blanks with $\subseteq$ or $\not\subseteq$ to make each statement correct.

25. N _____ W

26. Q _____ R

27. H _____ R

28. Q _____ H

29. J _____ $(J \cap W)$

30. R _____ $(Q \cup H)$

31. W _____ $(J \cap W)$

32. Q _____ $(J \cup W)$

Change each of the following to decimal form. State whether each is a member of the set of rational or irrational numbers.

33. $\dfrac{1}{2}$

34. $\dfrac{4}{5}$

35. $\dfrac{6}{8}$

36. $\dfrac{7}{5}$

37. $\dfrac{3}{7}$

38. $\dfrac{5}{11}$

39. $\dfrac{2}{3}$

40. $\dfrac{1}{9}$

Write each statement using the $>$, $<$, $\geq$ *or* $\leq$ *symbol.*

41. 7 is less than 11.

42. -5 is greater than -8.

43. $x + 5$ is positive.

44. $6t - 3$ is greater than or equal to zero.

45. $3m + 1$ is nonnegative.

46. y is larger than -2 and less than or equal to 4.

47. x is between 4 and 7.

48. $2k + 3$ is nonpositive.

49. $3p$ is greater than or equal to 0 and less than 9.

50. $5y - 7$ is greater than or equal to -6 and less than or equal to 10.

Graph each of the following on a number line and restate each using interval notation.

51. $\{x \mid x < 4\}$

52. $\{y \mid y < 6\}$

53. $\{x \mid x \geq 3\}$

54. $\{p \mid p \leq -1\}$

55. $\{t \mid t < 5 \text{ and } t \geq 1\}$

56. $\{k \mid k > -3 \text{ and } k \leq 5\}$

57. $\{y \mid y < 2 \text{ or } y \geq 5\}$

58. $\{x \mid x < -5 \text{ or } x \geq 0\}$

59. $\{p \mid p \leq 0, p \in R\} \cap \{p \mid p \geq 4, p \in R\}$

60. $\{y \mid y \leq -3, y \in R\} \cap \{y \mid y > 5, y \in R\}$

61. $\{x \mid x \geq -5, x \in R\} \cup \{x \mid x < 3, x \in R\}$

62. $\{t \mid t \geq 4, t \in R\} \cup \{t \mid t > -1, t \in R\}$

Write each of the following using set-builder notation.

63. $(3, 6)$ **64.** $(-7, 0)$ **65.** $[-2, 8)$ **66.** $(-3, 1]$ **67.** $[-1, 0]$

68. $[-4, 2]$ **69.** $[-3, \infty)$ **70.** $[0, \infty)$ **71.** $(-\infty, -2)$ **72.** $(-\infty, 1)$

Graph each of the following.

73. $(-4, 1] \cap [0, 5)$ **74.** $[-2, 0] \cap [-1, 3]$ **75.** $[-4, -1) \cup (1, 3]$

76. $(-5, -3] \cup [0, 2)$ **77.** $\{x \mid x < 5 \text{ and } x \geq 1\}$ **78.** $\{y \mid y \leq -3 \text{ and } y > -5\}$

79. $\{p \mid -2 \leq p < 2 \text{ or } 0 < p \leq 3\}$ **80.** $\{n \mid -5 < n \leq 3 \text{ or } 1 < n \leq 6\}$

1.3   Properties of Real Numbers

In this section several assumptions about real numbers will be made. They will be referred to as the properties of the real numbers as they apply to the operations of addition and multiplication. Throughout the discussion of the properties, a, b, and c will be used to represent real numbers.

Closure Properties	
$a + b$ **is a real number**	Closure property for addition
$a \cdot b$ **is a real number**	Closure property for multiplication

The **closure properties** state that the sum or product of real numbers is a real number.

Example 1   Decide whether each sum or product is a real number.

(a) $9 + 13 = 22$ 22 is a real number.

(b) $6 + 3 + 7 = 16$ 16 is a real number.

(c) $8(1.2) = 9.6$ 9.6 is a real number.

(d) $2 \cdot 5 \cdot 6 = 60$ 60 is a real number.   ■

Commutative Properties

$a + b = b + a$ Commutative property for addition

$a \cdot b = b \cdot a$ Commutative property for multiplication

---

The commutative properties state that the order in which numbers are added or multiplied does not change the sum or product.

Example 2   Complete the following so that the commutative property holds.

(a) $3 + 4 + 2 = 3 + \underline{\hphantom{xxxx}} + 4$

By the commutative property for addition, $3 + 4 + 2 = 3 + 2 + 4$.

(b) $2 \cdot 3 \cdot 5 = 3 \cdot 5 \cdot \underline{\hphantom{xxxx}}$

By the commutative property for multiplication, $2 \cdot 3 \cdot 5 = 3 \cdot 5 \cdot 2$.   ■

Associative Properties

$a + (b + c) = (a + b) + c$ Associative property for addition

$a(bc) = (ab)c$ Associative property for multiplication

The associative properties state that the order in which real numbers are associated (grouped) does not change the sum or product.

Example 3   Use the associative property to check each answer.

(a) $2 + (3 + 5) = 10$,   or $2 + 8 = 10$

$2 + (3 + 5) = (2 + 3) + 5 = 5 + 5 = 10$ True

(b) $2(3 \cdot 4) = 24$,   or $2(12) = 24$

$2(3 \cdot 4) = (2 \cdot 3) \cdot 4 = 6 \cdot 4 = 24$ True   ■

Identity Properties

There are unique numbers 0 and 1 such that

$a + 0 = 0 + a = a$ and $a \cdot 1 = 1 \cdot a = a.$

Zero is called the **additive identity,** and 1 is called the **multiplicative identity.**

To say a number is unique is to say that it is the only one that possesses the property. The identity properties make the following statements true:

$$3 + 0 = 3 \qquad \frac{2}{3} \cdot 1 = \frac{2}{3}$$

$$0 + \sqrt{2} = \sqrt{2} \qquad \pi \cdot 1 = 1 \cdot \pi = \pi.$$

Inverse Properties

There are unique numbers $-a$ and $\frac{1}{a}$ such that

$$a + (-a) = -a + a = 0 \qquad \text{and} \qquad a \cdot \frac{1}{a} = \frac{1}{a} \cdot a = 1 \ (a \neq 0).$$

$-a$ is the **additive inverse** of a, and $\frac{1}{a}$ is the **multiplicative inverse** of a.

Example 4   Simplify the following, using the identity properties.

(a) $3 + (-3) = 0$

(b) $\sqrt{5} \cdot \dfrac{1}{\sqrt{5}} = 1$

(c) $\left(-\dfrac{2}{5}\right) + \dfrac{2}{5} = 0$

(d) $\dfrac{2}{3} \cdot \dfrac{1}{\frac{2}{3}} = \dfrac{2}{3} \cdot \dfrac{3}{2} = 1$ ■

Distributive Property

$$a(b + c) = ab + ac \qquad \text{or} \qquad a(b - c) = ab - ac$$

The distributive property can also be written as

$$a(b + c) = (b + c)a$$

by using the commutative property. Further, b or c could each represent two or more numbers by closure. Thus

$$a(b + c + d + \cdots + z) = ab + ac + ad + \cdots + az.$$

This latter form of the distributive property is called the **extended distributive property**.

Example 5   Use the distributive property to complete the following.

(a) $2(3 + 5) = 2 \cdot 3 + 2 \cdot 5$

$\qquad 2(8) = 6 + 10$

$\qquad\quad 16 = 16$

(b) $4(7 - 2) = 4 \cdot 7 - 4 \cdot 2$

$\qquad 4 \cdot 5 = 28 - 8$

$\qquad\quad 20 = 20$

(c) $5x + 7x = (5 + 7)x$

$\qquad\qquad = 12x$

(d) $9x - 4x = (9 - 4)x$

$\qquad\qquad = 5x$ ■

The given properties of real numbers can be used to prove further properties of real numbers. A property that has been proved will be called a **theorem.** To illustrate how a theorem is proved, consider the following.

Theorem 1.1

The Multiplication Property of Zero

If $a \in R$ then

$$a \cdot 0 = 0 \quad \text{and} \quad 0 \cdot a = 0.$$

To prove this theorem, we will start with a statement that is known to be true and change it through a series of steps, each of which is justified by one of the properties of real numbers.

Proof:

Statement	Justification
$a = a \cdot 1$	Multiplicative identity property
$a = a(1 + 0)$	$1 = 1 + 0$ by the additive identity property
$a = a \cdot 1 + a \cdot 0$	Distributive property
$a = a + a \cdot 0$	Multiplicative identity property

We know that $a = a + 0$, and we have just shown that $a = a + a \cdot 0$. Since zero is unique, we are forced to conclude that

$$a \cdot 0 = 0.$$

Theorem 1.2

Double Negative Property

If $a \in R$, then

$$-(-a) = a.$$

This theorem was made plausible by use of the number line in Section 1.2. We are now ready to prove it formally.

Proof:

Statement	Justification
1. $a + (-a) = 0$	Additive inverse property
2. $-a + [-(-a)] = 0$	Additive inverse property
3. $[-(-a)] + (-a) = 0$	Commutative property

Step 1 shows that a is the additive inverse of $(-a)$; Step 3 shows that $[-(-a)]$ is the additive inverse of $(-a)$. Since the additive inverse of a number is unique, the two forms for the inverse a and $[-(-a)]$ must be equal. Thus,

$$-(-a) = a.$$

Several theorems will be proved in this book, and you will occasionally be asked to prove some theorems in the exercise sets. Although the main thrust of this book is problem solving, it is well worth your time to carefully study the proofs that are given. They provide the foundation for the rules for problem solving.

Example 6   State the property of real numbers that justifies each step in the proof that $x + [y + (-x)] + z = y + z.$

$$\begin{aligned}
x + [y + (-x)] + z &= x + [(-x) + y] + z && \text{Commutative property} \\
&= [x + (-x)] + (y + z) && \text{Associative property} \\
&= 0 + (y + z) && \text{Additive inverse property} \\
&= y + z && \text{Additive identity property} \quad \blacksquare
\end{aligned}$$

To conclude this section we list four additional assumptions that show the meaning of the equality symbol as it applies to the real number system.

Properties of Equality

If a, b, and $c \in R$, the following statements are true.

1. $a = a$	**Reflexive property**
2. If $a = b$, then $b = a$.	**Symmetric property**
3. If $a = b$ and $b = c$, then $a = c$.	**Transitive property**
4. If $a = b$, then a may be replaced by b or b by a without changing the truth or falsity of a statement.	**Substitution property**

State the property illustrated by each of the following.

Example 7   (a) $y + x = y + x$ Reflexive property

(b) If $2 + x = z$, then $z = 2 + x$.   Symmetric property

(c) If $a + b = 2a$ and $2a = 3t$, then $a + b = 3t$.   Transitive property

(d) If $x + y = 4$ and $y = 2$, then $x + 2 = 4$.   Substitution property $\blacksquare$

Exercises 1.3

Use the indicated property of real numbers to complete each equality.

1. $7 + 2 =$ _____ ; closure property

2. $8 \cdot 1 =$ _____ ; identity property

3. $x + (-x) =$ _____ ; additive inverse property

4. $x + y =$ _____ ; commutative property

5. $3(x - t) =$ _____ ; distributive property

6. $9 \cdot 11 =$ _____ ; closure property

7. $7(8y) =$ _____ ; associative property

8. $5m + 5n =$ _____ ; distributive property

9. $a + 0 =$ _____ ; identity property

10. $16 \cdot \dfrac{1}{16} =$ _____ ; inverse property

11. $r \cdot t =$ _____ ; commutative property

12. $5 + (x + 2) =$ _____ ; associative property

Use the indicated property of equality to complete each of the following.

13. If $x = 5 + b$, then _____ $= x$; symmetric property.

14. If $a + b = p$ and $a = 5$, then _____ $+ b = p$; substitution property.

15. $p =$ _____ ; reflexive property.

16. If _____ then $x = 2a + 3b$; symmetric property.

17. If $x = 3$ and $y = 5$, then $x + y =$ _____ ; substitution property.

18. If $a = t$ and $t = b$, then _____ ; transitive property.

19. $x + 4 =$ _____ ; reflexive property.

State whether or not the given set has the closure property with respect to addition, and give an example to illustrate your answer.

20. The set of whole numbers (W)

21. The set of integers (J)

22. The set of odd natural numbers

23. $\{0, 1\}$ (Note: any element may be used more than once.)

24. $\{0\}$ 25. The set of even integers   26. $\{1, 2, 3, 4\}$ 27. $\{-1, 0, 1\}$

Use the distributive property to restate each of the following.

28. $2(a + 3) =$ _____

29. $6(x + 1) =$ _____

30. $3t + 3 \cdot 2 =$ _____

31. $5m + 5 \cdot 3 =$ _____

32. $6x + 6y =$ _____

33. $9a + 9b =$ _____

34. $11(2a + 3b) =$ _____

35. $13(r + 4t) =$ _____

Indicate which property of real numbers is illustrated by each of the following.

36. $15 + 31 = 31 + 15$

37. $(6 + 2) + 9 = 6 + (2 + 9)$

38. $0 + m = m$

39. $x \cdot 3 = 3 \cdot x$

40. $(t + 5) \in R$

41. $y \cdot \dfrac{1}{y} = 1$

42. $(a + b) + c = a + (b + c)$

43. $m \cdot 1 = m$

44. $p + (-p) = 0$

45. $a + b = a + b$

46. $6t + 7t = (6 + 7)t$

47. $5y + 4y = (5 + 4)y$

48. $15r\left(\dfrac{1}{15r}\right) = 1$

49. $(p + 3) \in R$

50. $(6x)y = 6(xy)$

51. $(-x) + x = 0$

Use the distributive property to simplify Exercises 52–57.

Example: $10x + 5x = (10 + 5)x = 15x$

52. $2a + 3a =$ _____

53. $5t + 6t =$ _____

54. $11m - 5m =$ _____

55. $13b - 8b =$ _____

56. $7x + x =$ _____

57. $9y + y =$ _____

58. Give an example to show that subtraction is not commutative.

59. Give an example to show that subtraction is not associative.

60. Give an example to show that division is not commutative.

61. Give an example to show that division is not associative.

Which properties justify each step in the following proofs?

62. Prove that $(x + y) + [(-x) + (-y)] = 0$, $x \in R$, $y \in R$.

$$(x + y) + [(-x) + (-y)] = (y + x) + [(-x) + (-y)]$$
$$= y + [x + (-x)] + (-y)$$
$$= y + [0] + (-y)$$
$$= y + (-y)$$
$$= 0$$

(a) _____

(b) _____

(c) _____

(d) _____

(e) _____

63. Prove that for $a \in R$, $(-1)a = -a$.

$$a + (-1)a = 1 \cdot a + (-1)a$$
$$= a[1 + (-1)]$$
$$= a \cdot 0$$
$$= 0$$
But $a + (-a) = 0$
Therefore $-a = (-1)a$

(a) _____

(b) _____

(c) _____

(d) _____

(e) _____

(f) _____

1.4 A Review of Fractions (Optional)

A fraction is the indicated quotient of two real numbers. Since $\sqrt{2}$ and 5 are real numbers, $\frac{\sqrt{2}}{5}$ is a fraction. Recall from Section 1.2 that if both the numerator and denominator of a fraction are integers, the fraction is also a rational number.

To **multiply two fractions,** find the product of the numerators and divide it by the product of the denominators.

Multiplication of Fractions

If $\dfrac{a}{b}$ and $\dfrac{c}{d}$ are fractions, with b, $d \neq 0$, then

$$\frac{a}{b} \cdot \frac{c}{d} = \frac{a \cdot c}{b \cdot d} = \frac{ac}{bd}.$$

Example 1   Multiply.

$$\frac{3}{4} \cdot \frac{1}{2} = \frac{3 \cdot 1}{4 \cdot 2} = \frac{3}{8} \quad \blacksquare$$

Division of Fractions

If $\dfrac{a}{b}$ and $\dfrac{c}{d}$ are fractions, with b, c, $d \neq 0$, then

$$\frac{a}{b} \div \frac{c}{d} = \frac{a}{b} \cdot \frac{d}{c} = \frac{a \cdot d}{b \cdot c} = \frac{ad}{bc}.$$

Example 2   Divide.

$$\frac{3}{4} \div \frac{5}{7} = \frac{3}{4} \cdot \frac{7}{5} = \frac{3 \cdot 7}{5 \cdot 4} = \frac{21}{20} \quad \blacksquare$$

To **divide fractions,** invert the divisor and follow the rule for multiplication.

To see why the definition is reasonable, we need to make two observations beyond the properties stated for real numbers:

1. Any nonzero number divided by itself is 1.

$$\frac{1}{1} = 1, \ \frac{2}{2} = 1, \ \frac{7}{7} = 1,$$

and $\dfrac{\dfrac{d}{c}}{\dfrac{d}{c}} = 1$ if $c \neq 0, \ d \neq 0$

2. Any number divided by 1 is that same number.

$$\frac{2}{1} = 2, \ \frac{3}{1} = 3,$$

$$\frac{5}{1} = 5, \text{ and } \frac{a}{1} = a$$

Now consider $\dfrac{a}{b} \div \dfrac{c}{d}, \quad b, \ c, \ d \neq 0.$

$$\frac{\dfrac{a}{b}}{\dfrac{c}{d}} = \frac{\dfrac{a}{b}}{\dfrac{c}{d}} \cdot 1 \qquad \text{Multiplicative identity}$$

$$= \frac{\dfrac{a}{b}}{\dfrac{c}{d}} \cdot \frac{\dfrac{d}{c}}{\dfrac{d}{c}} \qquad \text{Substitution property; } 1 = \frac{\dfrac{d}{c}}{\dfrac{d}{c}}$$

$$= \frac{\dfrac{a \cdot d}{b \cdot c}}{\dfrac{c \cdot d}{d \cdot c}} \qquad \text{Definition of multiplication of fractions}$$

$$= \frac{ad}{bc} \qquad \frac{cd}{dc} = 1 \text{ and } \frac{\dfrac{ad}{bc}}{1} = \frac{ad}{bc}$$

Thus $\dfrac{a}{b} \div \dfrac{c}{d} = \dfrac{ad}{bc}.$

Fractions can be written in many ways that are equivalent. To generate **equivalent fractions,** the multiplicative identity is used.

$$\frac{2}{3} = \frac{2}{3} \cdot 1 = \frac{2}{3} \cdot \frac{2}{2} = \frac{2 \cdot 2}{3 \cdot 2} = \frac{4}{6} \qquad \frac{2}{2} = 1$$

$$\frac{2}{3} = \frac{2}{3} \cdot 1 = \frac{2}{3} \cdot \frac{3}{3} = \frac{2 \cdot 3}{3 \cdot 3} = \frac{6}{9} \qquad \frac{3}{3} = 1$$

Thus $\frac{2}{3}$, $\frac{4}{6}$, and $\frac{6}{9}$ are all equivalent fractions.

In general,

$$\frac{a}{b} \quad \text{and} \quad \frac{a \cdot x}{b \cdot x} \qquad (b \neq 0, \ x \neq 0)$$

are equivalent. When the steps outlined above are used in reverse order, the process is called **simplifying** or **reducing fractions to lower terms.**

$$\frac{4}{6} = \frac{2 \cdot 2}{3 \cdot 2} = \frac{2}{3} \cdot \frac{2}{2} = \frac{2}{3} \cdot 1 = \frac{2}{3}$$

$$\frac{6}{9} = \frac{2 \cdot 3}{3 \cdot 3} = \frac{2}{3} \cdot \frac{3}{3} = \frac{2}{3} \cdot 1 = \frac{2}{3}$$

When simplifying fractions, a common factor is removed from both the numerator and denominator. If the numerator and denominator have no common factors other than 1, the fraction is in **simplest terms.**

Fundamental Property of Fractions

$$\frac{a \cdot x}{b \cdot x} = \frac{a}{b} \quad \text{and} \quad \frac{a}{b} = \frac{a \cdot x}{b \cdot x} \qquad b, x \neq 0$$

Example 3 Write each of the following in simplest terms.

(a) $\dfrac{30}{34} = \dfrac{15 \cdot 2}{17 \cdot 2} = \dfrac{15}{17}$ Fundamental property

(b) $\dfrac{35}{49} = \dfrac{5 \cdot 7}{7 \cdot 7} = \dfrac{5}{7}$ Fundamental property

(c) $\dfrac{4.5}{6.0} = \dfrac{3(1.5)}{4(1.5)} = \dfrac{3}{4}$ ∎

When fractions are multiplied or divided, the fundamental property can be used to simplify them before the products or quotients are found.

Example 4 Multiply each of the following.

(a) $\dfrac{15}{4} \cdot \dfrac{2}{9} = \dfrac{3 \cdot 5}{2 \cdot 2} \cdot \dfrac{2}{3 \cdot 3}$ Factor

$= \dfrac{5}{2} \cdot \dfrac{1}{3}$ Fundamental property;
remove common factors

$= \dfrac{5}{6}$ Definition of multiplication

(b) $3 \cdot \dfrac{5}{4} \cdot \dfrac{7}{60} = 3 \cdot \dfrac{5}{4} \cdot \dfrac{7}{3 \cdot 4 \cdot 5}$ Factor

$= 1 \cdot \dfrac{1}{4} \cdot \dfrac{7}{4}$ $\dfrac{5}{5} = 1, \dfrac{3}{3} = 1$

$= \dfrac{7}{16}$ Definition of multiplication

(c) $2\dfrac{1}{2} \cdot 3\dfrac{1}{5} = \dfrac{5}{2} \cdot \dfrac{16}{5}$ Rewrite the mixed numbers
as improper fractions

$= \dfrac{5}{2} \cdot \dfrac{2 \cdot 8}{5}$

$= 8$ ■

Before fractions can be added, they must have a **common** (the same) **denominator.** The sum is found by adding the numerators and writing the result over the common denominator.

$$\dfrac{5}{11} + \dfrac{3}{11} - \dfrac{1}{11} = \dfrac{5 + 3 - 1}{11} = \dfrac{7}{11}$$

When fractions without a common denominator are to be added, use the fundamental property to generate one.

$$\dfrac{1}{2} + \dfrac{1}{3} = \dfrac{1 \cdot 3}{2 \cdot 3} + \dfrac{1 \cdot 2}{3 \cdot 2}$$ Fundamental property

$$= \dfrac{3}{6} + \dfrac{2}{6} = \dfrac{5}{6}$$

The smallest number that can be used as a common denominator is called the **least common denominator (LCD).** In most cases it can be determined by inspection. Other methods to determine it will be developed when needed.

Addition and Subtraction of Fractions

$$\dfrac{a}{c} + \dfrac{b}{c} = \dfrac{a + b}{c} \quad \text{and} \quad \dfrac{a}{c} - \dfrac{b}{c} = \dfrac{a - b}{c} \qquad c \neq 0$$

Example 5 Carry out the indicated operations $\frac{3}{5}x + \frac{5}{6}x + \frac{1}{2}x$ and simplify.
The LCD is 30.

$$\frac{3}{5}x + \frac{5}{6}x + \frac{1}{2}x = \frac{3x \cdot 6}{5 \cdot 6} + \frac{5x \cdot 5}{6 \cdot 5} + \frac{1x \cdot 15}{2 \cdot 15} \qquad \text{Fundamental property}$$

$$= \frac{18x}{30} + \frac{25x}{30} + \frac{15x}{30}$$

$$= \frac{58x}{30} \qquad \text{Definition of addition}$$

$$= \frac{29x}{15} \qquad \blacksquare \qquad \text{Remove a common factor of 2}$$

Example 6 Simplify $\frac{2}{3}\left(\frac{1}{5} + 2\right)$.

$$\frac{2}{3}\left(\frac{1}{5} + 2\right) = \frac{2}{3} \cdot \frac{1}{5} + \frac{2}{3} \cdot 2 \qquad \text{Distributive property}$$

$$= \frac{2}{15} + \frac{4}{3} \qquad \text{Definition of multiplication}$$

$$= \frac{2}{15} + \frac{20}{15} \qquad \text{The LCD is 15}$$

$$= \frac{22}{15} \qquad \blacksquare$$

Example 7 Simplify $\frac{3a}{8} \div \frac{3a}{2} - \frac{1}{3} \cdot \frac{2}{5} + 1$.

▶ **Caution** In arithmetic, multiplication and division are carried out before addition and subtraction.

$$\frac{3a}{8} \div \frac{3a}{2} - \frac{1}{3} \cdot \frac{2}{5} + 1 = \frac{3a}{8} \cdot \frac{2}{3a} - \frac{1}{3} \cdot \frac{2}{5} + 1$$

$$= \frac{1}{4} - \frac{2}{15} + 1 \qquad \frac{a}{a} = 1$$

$$= \frac{15}{60} - \frac{8}{60} + \frac{60}{60} \qquad \text{LCD} = 60$$

$$= \frac{67}{60} \qquad \blacksquare$$

Exercises 1.4

Carry out the indicated operations and simplify.

1. $\dfrac{1}{2} + \dfrac{5}{2}$ 2. $\dfrac{4}{7} + \dfrac{1}{7}$ 3. $\dfrac{11}{8} - \dfrac{3}{8}$

4. $\dfrac{23}{16} - \dfrac{7}{16}$ 5. $\dfrac{1}{3} + \dfrac{4}{3} + 1$ 6. $\dfrac{2}{5} + \dfrac{1}{5} + 2$

7. $\dfrac{7}{11} - \dfrac{3}{11} + 1$

8. $\dfrac{4}{9} - \dfrac{2}{9} + 1$

9. $\left(\dfrac{3}{13} - \dfrac{1}{13}\right) + \dfrac{5}{13}$

10. $\left(\dfrac{6}{17} - \dfrac{2}{17}\right) - \dfrac{1}{17}$

11. $\dfrac{(7-3)}{2} - \dfrac{1}{2}$

12. $\dfrac{(4-1)}{5} + \dfrac{1}{4}$

13. $\dfrac{5}{2}a + \dfrac{1}{4}a$

14. $\dfrac{5}{8}x - \dfrac{1}{2}x$

15. $\dfrac{9}{10}y - \dfrac{2}{5}y$

16. $\dfrac{1}{3}c + \dfrac{1}{2}c + \dfrac{1}{4}c$

17. $\dfrac{2}{5}s + \dfrac{1}{4}s + \dfrac{1}{3}s$

18. $\dfrac{2}{7} + \left(\dfrac{1}{6} + \dfrac{1}{14}\right)$

19. $\dfrac{3}{8} + \left(\dfrac{1}{4} + \dfrac{1}{2}\right)$

20. $\dfrac{5}{7} - \left(\dfrac{1}{7} + \dfrac{1}{4}\right)$

21. $3 + \left(\dfrac{5}{7} - \dfrac{1}{4}\right)$

22. $1 - \left(\dfrac{4}{11} - \dfrac{1}{6}\right)$

23. $3\dfrac{1}{2} + 1\dfrac{1}{4}$

24. $5\dfrac{1}{8} + 2\dfrac{1}{2}$

25. $1\dfrac{3}{8} + 2\dfrac{1}{4} - \dfrac{1}{2}$

26. $5\dfrac{3}{7} + 2\dfrac{1}{4} - 3\dfrac{1}{2}$

27. $3\dfrac{7}{11} + 4\dfrac{1}{3} - 5\dfrac{1}{2}$

28. $\dfrac{(7-2)}{3} + 4 + 3\dfrac{3}{4}$

29. $5 + \dfrac{(8-1)}{8} - 1\dfrac{3}{4}$

30. $14\dfrac{1}{2} - 5 + 7\dfrac{1}{4}$

Multiply or divide as indicated and simplify.

31. $\dfrac{3t}{2} \cdot \dfrac{1}{4t}$

32. $\dfrac{5n}{8} \cdot \dfrac{1}{2n}$

33. $\dfrac{3}{16s} \cdot \dfrac{8s}{9}$

34. $\dfrac{11}{24p} \cdot \dfrac{8p}{22}$

35. $\dfrac{6x}{10} \div \dfrac{2x}{5}$

36. $\dfrac{7y}{13} \div \dfrac{49y}{91}$

37. $\dfrac{51z}{48} \div \dfrac{34z}{32}$

38. $\dfrac{41t}{50} \div \dfrac{17t}{30}$

39. $\left(\dfrac{1}{2} \cdot \dfrac{1}{3}\right) \div \dfrac{1}{6}$

40. $\left(\dfrac{3}{4} \cdot \dfrac{2}{3}\right) \div \dfrac{1}{2}$

41. $\dfrac{5}{8}\left(\dfrac{3}{4} \div \dfrac{7}{4}\right)$

42. $\dfrac{7}{15}\left(\dfrac{5}{21} \div \dfrac{10}{6}\right)$

Carry out the indicated operations and simplify.

43. $\dfrac{1}{2}\left(\dfrac{1}{4} + \dfrac{1}{3}\right)$

44. $\dfrac{5}{8}\left(\dfrac{3}{4} + \dfrac{1}{2}\right)$

45. $\dfrac{7}{16}\left(\dfrac{7}{10} - \dfrac{1}{2}\right)$

46. $\dfrac{4}{11}\left(\dfrac{3}{2} - \dfrac{1}{5}\right)$

47. $2\dfrac{3}{5}\left(1\dfrac{1}{4} + \dfrac{1}{5}\right)$

48. $3\dfrac{1}{8}\left(2\dfrac{1}{3} - 1\dfrac{1}{4}\right)$

49. $\left(\dfrac{1}{2} + \dfrac{1}{3}\right)\left(\dfrac{3}{5} - \dfrac{1}{2}\right)$

50. $\left(\dfrac{3}{7} + \dfrac{2}{5}\right)\left(\dfrac{7}{5} - \dfrac{1}{7}\right)$

51. $\left(1\dfrac{3}{5} - 1\dfrac{1}{4}\right)\left(2\dfrac{7}{10} - \dfrac{3}{20}\right)$

52. $\dfrac{1}{4} + \dfrac{1}{3}\left(\dfrac{3}{5} \div 4\right) + 1$

53. $\dfrac{3}{7} + \dfrac{2}{9}\left(\dfrac{3}{4} \div 2\right) + 1$

54. $\left(\dfrac{1}{8} + \dfrac{1}{5}\right) \div 40 + 1\dfrac{1}{2}$

Write each fraction in simplest form.

55. $\dfrac{150}{225}$

56. $\dfrac{380}{1020}$

57. $\dfrac{282}{564}$

58. $\dfrac{56x}{64x}$

59. $\dfrac{91x}{117x}$

60. $\dfrac{1024}{3072}$

Complete the following to make equivalent fractions.

61. $\dfrac{1}{2} = \dfrac{}{8}$

62. $\dfrac{5}{8} = \dfrac{}{24}$

63. $\dfrac{14}{21} = \dfrac{2}{}$

64. $\dfrac{15}{20} = \dfrac{3}{}$

65. $\dfrac{36}{12} = \dfrac{9}{}$

66. $\dfrac{54}{45} = \dfrac{}{5}$

67. $\dfrac{128}{144} = \dfrac{}{9}$

68. $\dfrac{350}{490} = \dfrac{5}{}$

69. $\dfrac{4}{7} = \dfrac{60}{}$

70. $\dfrac{3}{11} = \dfrac{}{154}$

71. $\dfrac{5}{13} = \dfrac{75}{}$

72. $\dfrac{4}{17} = \dfrac{68}{}$

Solve the following word problems.

73. How long is the average stride (in inches) of a jogger who takes 3168 steps in a mile (5280 ft = 1 mile)? Assume all strides are the same.

74. On a trip Francesca averages 88 feet per second. What is her speed in miles per hour?

1.5   Addition and Subtraction of Real Numbers

In elementary algebra the rules for adding real numbers are generally justified by the use of a number line. Starting at a point representing a real number, an arrow pointing to the right shows that a positive number is added, while an arrow pointing to the left shows that a negative number is added. Consider the sum of two positive numbers such as $2 + 3 = 5$. The total displacement from the starting point gives the sum, as shown in Figure 1.19. The sum of the two negative numbers $-2 + (-3) = -5$ is shown in Figure 1.20.

Figure 1.19

Figure 1.20

In each case both numbers have the same sign (both positive or both negative). Now consider what happens when the numbers have opposite signs, such as $-3 + 5 = 2$ or $-5 + 2 = -3$. See Figures 1.21 and 1.22.

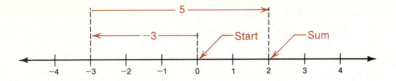

Figure 1.21

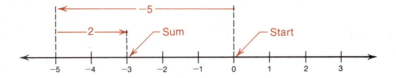

Figure 1.22

When two positive numbers are added, the sum is positive. When two negative numbers are added, the sum is negative. If the two numbers have opposite signs the sum could be positive or negative depending upon which number has the greater absolute value.

> ### Rules for Adding Signed Numbers
> 1. To add two numbers having **like** signs, find the sum of their absolute values and prefix the same sign.
> 2. To add two numbers having **unlike** signs, subtract the smaller absolute value from the larger absolute value and prefix the sign of the number with the greater absolute value.

Example 1 Find the sum $-8 + (-5)$.

First find the sum of their absolute values.

$$|-8| = 8 \quad \text{and} \quad |-5| = 5$$
$$8 + 5 = 13$$

Since both numbers are negative, the sum will be negative.

$$(-8) + (-5) = -13 \quad \blacksquare$$

Example 2 Find the sum $-26 + 8$.

$$|-26| = 26 \quad \text{and} \quad |8| = 8$$
$$26 - 8 = 18$$

Since $|-26| > |8|$, the sum will be negative.

$$-26 + 8 = -18 \quad \blacksquare$$

Example 3 Find the sum $4 + (-30)$.

$$|4| = 4 \quad \text{and} \quad |-30| = 30$$
$$30 - 4 = 26$$

Since $|-30| > |4|$, the sum will be negative.

$$4 + (-30) = -26 \quad \blacksquare$$

Theorem 1.3

If a and $b \in R$, then

$$-a + (-b) = -(a + b).$$

The proof of Theorem 1.3 is left as an exercise in which the steps of the proof are given and the reader is to supply the reasons.

To understand subtraction of signed numbers, first recall from arithmetic that

$$6 - 4 = 2 \quad \text{because} \quad 4 + 2 = 6.$$

In other words, to subtract 4 from 6 we had to find a number which when added to 4 would produce a sum of 6. To subtract (-2) from 6 we must find a number which when added to (-2) will produce a sum of 6. This number is 8.

$$8 + (-2) = 6$$

In each of the above cases the difference could have been found by addition.

$$6 - 4 = 6 + (-4) = 2$$
$$\text{and} \quad 6 - (-2) = 6 + 2 = 8$$

Thus $a - b = c$ implies that $a - b = a + (-b) = c$. This leads to the following definition of subtraction.

If a and $b \in R$, then

$$a - b = a + (-b).$$

The definition of subtraction states that *to subtract one number from another, first change the sign of the number being subtracted and then add.*

Example 4 Find the difference $5 - (-3)$ in the following ways.

(a) On a number line

Since 5 and -3 are 8 units apart on a number line, their difference is 8. See Figure 1.23 on the facing page.

(b) By the definition of subtraction

$$5 - (-3) = 5 + (3) = 8 \quad \text{Change the sign and add} \quad \blacksquare$$

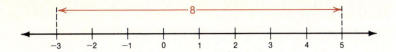

Figure 1.23

Example 5 Find the difference in each of the following.

(a) $8 - (-2) = 8 + 2 = 10$

(b) $-6 - 3 = -6 + (-3) = -9$

(c) $-2 - (-8) = -2 + 8 = 6$

(d) $1.3 - |-1.7| = 1.3 - 1.7 = -0.4$ ■

Both addition and subtraction can appear in the same problem. In this case carry out the operations from left to right.

Example 6 Simplify each of the following.

(a) $6 - (-3) + 9 = 6 + 3 + 9 = 18$

(b) $-6 - 2 - 3 = -6 + (-2) + (-3) = -8 + (-3) = -11$

(c) $3x + 5x + 2y - 4y = (3 + 5)x + (2 - 4)y$

$\qquad\qquad\qquad\qquad = 8x - 2y$ ■

Example 7 Evaluate $x - y + z$ for $x = 3$, $y = -4$ and $z = 0$.
Replace x with 3, y with -4, and z with 0.

$3 - (-4) + 0 = 3 + 4 + 0$

$\qquad\qquad\quad = 7 + 0$

$\qquad\qquad\quad = 7$ ■

Exercises 1.5

Add each of the following.

1. $6 + 4$ 2. $9 + 2$ 3. $10 + (-4)$

4. $18 + (-10)$ 5. $-15 + (-17)$ 6. $-27 + (-31)$

7. $13 + 7 + (-2)$ 8. $26 + (-8) + 4$ 9. $(-11) + 8 + (-5)$

10. $(-1) + 17 + (-16)$ 11. $(-2) + (-10) + (-3)$ 12. $(-5) + (-1) + (-4)$

Subtract each of the following.

13. $8 - 2$ 14. $17 - 4$ 15. $16 - 34$ 16. $8 - 52$

17. $-8 - 15$ 18. $-1 - 35$ 19. $-3 - (-5) - 8$ 20. $-2 - (-9) - 6$

21. $-18 - (-2) - (-13)$ 22. $-41 - (-9) - (-6)$

Add or subtract as indicated.

23. $16 + (-5) - (-7)$

24. $17 + (-8) - (-10)$

25. $-15 + 6 - 9 + 8$

26. $-13 + 11 - 12 + 5$

27. $-6 - (-5 + 9) + 16 - (-3)$

28. $-1 - [16 - (-3)] + 27 - (-6)$

29. $-|-3| + |6|$

30. $|23| - |-5|$

31. $|-30| - [6 + 11] - (-15)$

32. $-(32 - 18) - (-18 + 2) - 9$

33. $-5 - (-3) - (2) - (-1)$

34. $-7 - (-1) + (-5) - (2)$

35. $6x + 5x$

36. $7a + 2a$

37. $8p - 3p$

38. $16t - 11t$

39. $-13m - 11m$

40. $-28k - 7k$

41. $-16p - (-7p)$

42. $6y + 3 + 2y$

43. $2r + 5 + 17r$

44. $-x + 7y + 3x - 4y$

45. $-16a + 18b - (-11a) + 2b$

46. $-60r - (-6t) + 75r - 15t$

47. $-43m - (-13p) - 81p - (-30m)$

Evaluate each of the following.

48. $a + b + c$ for $a = 2$, $b = -5$, $c = 7$

49. $x - (y - z)$ for $x = -9$, $y = 8$, $z = 5$

50. $-m + (-n) - (-p)$ for $m = 2$, $n = -11$, $p = -8$

51. $-r - [(-s) - (-t)]$ for $r = -12$, $s = 6$, $t = 0$

Find the distance between the indicated points.

52. -13 and 7

53. -56 and -20

54. Mt. McKinley (20,270 feet above sea level) and Death Valley (282 feet below sea level).

55. Mt. Whitney (14,494 feet above sea level) and the Salton Sea (235 feet below sea level).

Use a calculator to evaluate each of the following.

56. Teri had a share draft balance of $6.82 on Monday. She wrote checks of $18.41, $37.89, $7.37, and $121.87 before making a deposit of $148.64 on Friday. What was her balance?

57. Patti is keeping a record of the temperature changes in Lafayette. The first day it was $-7°$F. Then it fell $5°$, rose $18°$, fell $2°$, rose $13°$, fell $22°$, rose $3°$ on the next six consecutive days. What was the temperature after the last change?

58. $6.125 + (-1.823) - (-16.791)$

59. $-|-3.642| + (-607.211) + |16.184|$

60. $-(54.027) - (-0.888) + |-481.005|$

61. $-0.028 - (-5.832) - (101.700)$

State the property of real numbers that justifies each step in the proofs below.

62. If $a, b \in R$, then $(-a) + (-b) = -(a + b)$.

$(a + b) + [-(a + b)] = 0$ **(a)** _____

$a + (-a) = 0;\ b + (-b) = 0$ **(b)** _____

$a + (-a) + b + (-b) = 0$ **(c)** _____

$a + b + (-a) + (-b) = 0$ **(d)** _____

$(a + b) + [(-a) + (-b)] = 0$ **(e)** _____

$(-a) + (-b) = -(a + b)$ **(f)** The additive inverse is unique. Step (a) and step (e)

63. If $x, y \in R$, then $(x - y) = -(y - x)$.

$x + (-x) = 0; y + (-y) = 0$ **(a)** _____

$x + (-x) + y + (-y) = 0$ **(b)** _____

$x + (-y) + y + (-x) = 0$ **(c)** _____

$x - y + y - x = 0$ **(d)** _____

$(x - y) + (y - x) = 0$ **(e)** _____

$(y - x) + [-(y - x)] = 0$ **(f)** _____

$(x - y) = -(y - x)$ **(g)** The additive inverse is unique. Step (e) and step (f)

1.6 Multiplication and Division of Real Numbers and the Order of Operations

In arithmetic **the product of two positive numbers is positive.** For example,

$$5(3) = 15$$
$$(1.2)(3.6) = 4.32$$
$$\frac{1}{3} \cdot 8 = \frac{8}{3}.$$

In elementary algebra you learned that **the product of a positive number and a negative number is negative,** as stated below.

Theorem 1.4

> If a and $b \in R$, then
>
> $$a(-b) = -(ab).$$

To prove this theorem, we need to use another theorem that will be proven in Section 2.1.

Theorem

> **The Multiplicative Property of Equality**
>
> If a, b, and $c \in R$ and $a = b$, then
>
> $$ac = bc.$$

We are now ready to proceed to the proof of Theorem 1.4.

Statement	Justification
1. $b + (-b) = 0$	Additive inverse property
2. $a[b + (-b)] = a \cdot 0$	Multiplicative property of equality
3. $ab + a(-b) = 0$	Distributive property and Theorem 1.1
4. $ab + [-(ab)] = 0$	Additive inverse property

From Steps 3 and 4, $a(-b)$ and $-(ab)$ are both additive inverses of ab. Since the additive inverse of a number is unique, we have

$$a(-b) = -(ab).$$

Example 1 Multiply.

(a) $(-1)(3) = -(1 \cdot 3) = -3$

(b) $4(-5) = -(4 \cdot 5) = -20$

(c) $(-4)(4)(2) = -(4 \cdot 4 \cdot 2) = -32$ ■

Although we choose not to do so, it is possible to prove that the product of two negative numbers is positive. This fact is stated below.

Theorem 1.5

If a and $b \in R$, then

$$(-a)(-b) = a \cdot b.$$

Example 2 Multiply.

(a) $(-8)(-9) = 72$

(b) $(-4)(-5)(-2) = -40$

(c) $(-2)\left(\dfrac{3}{5}\right)(-7) = \dfrac{42}{5}$ ■

The results of the theorems on multiplication can be summarized as follows.

To find the product of two signed numbers, multiply their absolute values. If the numbers have **like signs,** the product is positive. If they have **unlike signs,** the product is negative.

Recall that subtraction was defined in terms of addition. In a similar manner we will define division in terms of multiplication.

If a, b and q are real numbers, then

$$\frac{a}{b} = q \quad \text{only if} \quad b \cdot q = a; \ b \neq 0.$$

Example 3   Simplify. Note the signs in each of the following.

(a) $\dfrac{8}{4} = 2$ because $4 \cdot 2 = 8$.

The two numbers have the same sign,
and the quotient is positive

(b) $\dfrac{-8}{-4} = 2$ because $(-4)(2) = -8$.

(c) $\dfrac{-8}{4} = -2$ because $4(-2) = -8$.

The two numbers have unlike signs,
and the quotient is negative

(d) $\dfrac{8}{-4} = -2$ because $(-4)(-2) = 8$. ■

The results of the division problems lead to the following rules for dividing signed numbers.

To find the **quotient** of two signed numbers, find the quotient of their absolute values. If the numbers have **like signs**, the quotient is positive. If they have **unlike signs**, the quotient is negative.

Using the definition of division, it can be seen that

$$\frac{-a}{b} = \frac{a}{-b} = -\frac{a}{b}; \ b \neq 0.$$

In this book all negative answers will be written in the form $\frac{-a}{b}$ or $-\frac{a}{b}$.

Example 4   Write $\frac{-5}{6}$ in two other forms.

$$\frac{-5}{6} = \frac{5}{-6} = -\frac{5}{6} \quad ■$$

When two real numbers are divided, the divisor **can never** *be zero.* To see this consider the quotient

$$\frac{a}{0} = q; \ a \neq 0.$$

By the definition of division,

$$\frac{a}{0} = q \quad \text{only if} \quad a = 0 \cdot q.$$

But $0 \cdot q = 0$ and $a \neq 0$. Therefore the division is not possible.

$$\text{If } a = 0, \quad \text{then} \quad \frac{0}{0} = q \quad \text{only if} \quad 0 = 0 \cdot q.$$

This is true for any real number q, so no unique result can be determined. We will use the term **undefined** when division by zero is indicated. (Try dividing by zero on your calculator.)

Many times an arithmetic or an algebra problem involves more than one operation. When this is the case, a definite order must be followed if a unique result is to be obtained. For example, $2 \cdot 8 + 3 = 16 + 3 = 19$ if the multiplication is completed first and $2 \cdot 11 = 22$ if the addition is completed first. The first result is the accepted one. The following outline specifies the accepted **order of operations**.

Order of Operations

1. Carry out all operations within symbols of grouping, working from the innermost symbols outward.
2. Carry out all multiplications and divisions, working from left to right.
3. Carry out all additions and subtractions, working from left to right.

The most common symbols of grouping are the **( )** parentheses, **[ ]** brackets, **{ }** braces, and the fraction bar ——— .

Example 5   Simplify $4(8 + 3) + 9$.

$4(\mathbf{8 + 3}) + 9$

$4(\mathbf{11}) + 9$ Symbols of grouping first

$44 + 9$ Multiplication

53 Addition ■

Example 6   Simplify $\dfrac{5(-3) + (-5)}{2(-8) + 12}$.

The fraction bar is a symbol of grouping. The operations above and below the bar will be carried out first.

$$\frac{5(-3) + (-5)}{2(-8) + 12} = \frac{-15 + (-5)}{-16 + 12} \quad \text{Multiplication first}$$

$$= \frac{-20}{-4} \quad \text{Addition}$$

$$= 5 \quad ■$$

Example 7   Simplify $2[8 + 6(4 - 2) - 5] - 9$.

Working within the innermost grouping symbols first yields

$$2[8 + 6(4 - 2) - 5] - 9 = 2[8 + 6(2) - 5] - 9$$
$$= 2[8 + 12 - 5] - 9$$
$$= 2[15] - 9$$
$$= 21. \quad \blacksquare$$

Example 8   Simplify $\dfrac{\dfrac{1}{2}[15 - (-7)] + \dfrac{1}{2} \cdot 10}{\dfrac{5}{3}[-7 + 4]}$.

$$\frac{\dfrac{1}{2}[15 - (-7)] + \dfrac{1}{2} \cdot 10}{\dfrac{5}{3}[-7 + 4]} = \frac{\dfrac{1}{2}[22] + \dfrac{1}{2} \cdot 10}{\dfrac{5}{3}(-3)}$$

$$= \frac{11 + 5}{-5}$$

$$= \frac{16}{-5} = \frac{-16}{5} \quad \blacksquare$$

Example 9   Simplify $3(2x + 7) + 5(x + 2)$.

Begin by using the distributive property.

$$3(2x + 7) + 5(x + 2) = 6x + 21 + 5x + 10$$
$$= 11x + 31 \quad \blacksquare$$

Example 10   Simplify $3(x + y) - (x - 2y)$.

The expression $-(x - 2y)$ can be thought of as $-1(x - 2y)$. When the distributive property is used, this becomes $-x + 2y$.

$$3(x + y) - (x - 2y) = 3x + 3y - x + 2y$$
$$= 2x + 5y \quad \blacksquare$$

Exercises 1.6

Find each of the indicated products.

1. $11(3)$

2. $12(5)$

3. $6(-13)$

4. $-8(15)$

5. $-9(-8)$

6. $-10(-7)$

7. $-3(-2)(8)$

8. $-8(-10)(2)$

9. $-5(-8)(-11)$

10. $-7(-13)(-6)$

11. $(-1)(-4)(-7)(-3)$

12. $(-9)(-7)(-3)(-5)$

Simplify where possible. If not possible, write undefined.

13. $\dfrac{8}{4}$

14. $\dfrac{25}{5}$

15. $\dfrac{-36}{12}$

16. $\dfrac{-81}{-27}$

17. $\dfrac{-256}{-8}$

18. $\dfrac{-250}{-30}$

19. $\dfrac{-8 - 4(-2)}{15}$

20. $\dfrac{28 - 7(4)}{13(-2)}$

21. $\dfrac{15 + 6(-5)}{14 + 2(-7)}$

22. $\dfrac{27 - 6(-3)}{-56 - 8(-7)}$

Simplify using the correct order of operations. If division is not possible, write undefined.

23. $-6(8 - 4) \div 8$

24. $-5(13 - 8) \div -25$

25. $-2[-6 + (-3)(-2)] \div 15$

26. $-[-5 + 2(4 - 1) - 1] \div -17$

27. $-33(18 - 15)(0)$

28. $-54(17 - 14)(0)$

29. $[-3 + (-2)][-6 - (-4)]$

30. $[6 + (-4)][-8 - (-7)]$

31. $-13 + (-5)(6)$

32. $-18 + (-2)(-5)$

33. $-11 - 4 \cdot 8 + 30$

34. $-16 - 3 \cdot 9 + 26$

35. $\dfrac{8 + (-2)(5 - 6)}{-5}$

36. $\dfrac{11 + (-3)(1 - 3)}{-17}$

37. $\dfrac{(-9 - 6)(-4 + 9)}{-5 + (-4)(-5)}$

38. $\dfrac{(-11 + 5)(-18 + 15)}{-2 + (-1)(4)}$

39. $(-3 - 2)(1 - 5)(2 - 7)$

40. $(4 - 1)(2 - 5)(7 - 10)$

41. $-2[6 + (-1)(5 - 4)] \div 0$

42. $-3[8 - (5)(4 - 6)] \div 0$

43. $15 \div 3 \div 4 \div (-10)$

44. $28 \div (-7) \div (-3) \div (-6)$

45. $56(18 - 48 \div 6)$

46. $-72(4 - 96 \div 48)$

47. $\dfrac{-2[-15 - 3(8 - 3) + 5]}{-5[-31 - 15(-2) + 1]}$

48. $\dfrac{-3[-24 - 2(4 - 8) + 8]}{-8[-18 + 5(10 - 13) + 30]}$

49. $\dfrac{-\dfrac{1}{5}(-11 + 21)}{-\dfrac{1}{7}(4)(9 - 2)}$

50. $\dfrac{-\dfrac{3}{8}[-12 - (11 - 7)]}{-\dfrac{5}{9}[-16 + (-4 + 15)]}$

Simplify by first using the distributive property.

51. $-3(x - 4y) + 5(3x + y)$

52. $-7(3a - b) + 4(a - 5b)$

53. $6(-r - 7t) - (3r - 4t)$

54. $-(m - 7p) - (-3m + 4p)$

55. $-5(-7a - 4b) + (-6a + b)$

56. $-9(-5x - 4y) - (7x - 9y)$

Use a calculator to evaluate each of the following. Round all answers to one decimal place.

57. $\dfrac{1.8(-15.32 + 6.14)}{-5.4(-11.1 - 0.812)}$

58. $\dfrac{-9.51(0.018 - 8.935)}{-0.849(81.02 - 0.914)}$

59. $-5.3x + 9.52y$ if $x = -0.814$ and $y = 6.145$

60. $\dfrac{3.142a}{5.415b} + \dfrac{0.915a}{0.8746b}$ if $a = -1.951$ and $b = 3.142$

61. $\dfrac{5.832n - 6.852p}{2.164n + 0.875p}$ where $n = 0.987$ and $p = -2.918$

62. $$\dfrac{1.7r + 8 - t\left(5.3 + \dfrac{4.7}{r}\right)}{6.915t - 5 + r\left(1.827 - \dfrac{6.93}{t}\right)} \quad \text{if } r = 3 \text{ and } t = 0.4$$

Answer each of the following.

63. If $a > b$, then the sign of $b - a$ is _____ .

64. Is $|a - b| = |b - a|$ for all $a, b \in R$?

65. If x and $y > 0$, does $\dfrac{|x|}{|y|} = \dfrac{x}{y}$?

66. If $m < n$, then the sign of $n - m$ is _____ .

67. If a and $b < 0$, does $\dfrac{|a|}{|b|} = \dfrac{a}{b}$?

68. If $r = 0$, $t \neq 0$ then $\dfrac{|r|}{|t|} =$ _____ .

69. If $a \in N$, $b \in J$, $c = 0$ then $|-a(-b) \cdot c| =$ _____ .

70. If $a \in (N \cap W)$, $b < 0$ then the sign of $\dfrac{a}{b}$ is _____ .

1.7 Translating Verbal Statements from Words to Symbols

To use algebra as a problem-solving tool, you must learn how to translate a verbal statement into one involving variables and mathematical symbols. The first step is to look for key words and phrases. They may be as simple as **sum, difference, product,** and **quotient,** or they may be phrases indicating that more than one operation is involved. The table below indicates many common phrases and their interpretations. Notice that several phrases have exactly the same interpretation.

Common Algebraic Expressions

Operation	Expression in Words	Expression in Symbols
Addition	a plus b a increased by b b more than a The sum of a and b	$a + b$
Subtraction	The difference of a and b b less than a a decreased by b a minus b b fewer than a a diminished by b a less b b subtracted from a	$a - b$
Multiplication	a multiplied by b The product of a and b a times b	$a \cdot b$, ab, and others
Division	The quotient of a and b a divided by b	$\dfrac{a}{b}$, $a \div b$, and others

Now consider a variety of expressions that illustrate the same four operations.

Verbal Statement	Algebraic Expression
5 more than a number	$x + 5$
A number decreased by 2	$n - 2$
The quotient of 7 and a number	$\dfrac{7}{a}$
The quotient of two numbers	$x \div y$ or $\dfrac{x}{y}$
The product of two numbers	xy or $x \cdot y$
Twice a number increased by 5	$2x + 5$
Twice the sum of a number and 5	$2(x + 5)$
The quotient of a number and 6, less 8	$\dfrac{n}{6} - 8$
The sum of a number and 8, divided by 12	$\dfrac{y + 8}{12}$
6 more than the quotient of a number and 15	$\dfrac{n}{15} + 6$

Punctuation is often the key to proper interpretation of verbal statements. The product of 4 and x, plus 9 means

$4x + 9$.   Comma after x

The product of 4, and x plus 9 means

$4(x + 9)$.   Comma after 4

In the table of expressions you saw that many different letters were used to represent the unknown number. When translating from a verbal statement to one using symbols, choose letters that remind you of what they represent:

d = number of dollars

m = Mary's age

t = number of tickets sold

w = width of a rectangle.

More examples of translation follow.

Example 1   Change each verbal statement to one using symbols.

(a) Seven times Cynthia's age
     Let c = Cynthia's age.

     $7c$

(b) Three years less than Lee's age

Let L = Lee's age.

$L - 3$ ■

Example 2 Find an expression for the number of theater tickets that remain unsold if the theater seats 600 and t tickets have been sold.

$600 - t$ ■

Translation of verbal statements is usually needed to solve applied problems. Many translations involve the equals sign or a symbol of inequality such as "$<$". Such expressions are called **mathematical sentences.** Notice in the following that the verb "**is**" signals the point where either $=$ or an inequality symbol is to be used.

Example 3 Translate each of the following into a mathematical sentence.

Verbal statement	Translation
(a) The sum of a number and 6 is 5.	$n + 6 = 5$
(b) Twice a number added to 3 is greater than 9.	$3 + 2n > 9$
(c) The quotient of a number and 10 more than the number is 40.	$\dfrac{x}{x + 10} = 40$
(d) A student's test average based on test scores of x and y is between 70 and 80.	$70 < \dfrac{x + y}{2} < 80$ ■

Example 4 Find a mathematical sentence that indicates that the sum of two consecutive even integers is 82. Consecutive even integers such as 6 and 8 or 14 and 16 differ by 2.

Let n = first even integer

$n + 2$ = next consecutive even integer

$n + (n + 2) = 82$ ■

Example 5 When a number is increased by 85, the sum is 4 less than twice the number. Write a mathematical sentence describing this relationship.

Let n = the number

$n + 85 = 2n - 4$ ■

The number is increased by 85

Four less than twice the number

Exercises 1.7

Name the variable to be used, then translate into an algebraic expression.

1. Twice a number
2. Three times a number
3. One-half of a number
4. One-fourth of a number
5. Seven more than a number
6. Six more than a number
7. Eight plus a number
8. A number plus 16
9. A number increased by 1
10. The sum of 11 and a number
11. Thirty less than a number
12. A number decreased by 4
13. Six times a number
14. Eight times a number
15. A number divided by 6
16. Forty decreased by a number
17. Eighty decreased by a number
18. Sixteen divided by a number
19. The product of 19 and a number
20. A number multiplied by 20

Translate each of the following into an algebraic expression.

21. x multiplied by 19
22. x divided by 25
23. 6 divided by twice n
24. 64 multiplied by 5 times p
25. 26 less than $3h$
26. 7 more than $5m$
27. The quotient of y and 11
28. The quotient of 15 and x
29. Eight more than one-fourth of a
30. Thirty-seven less than one-fifth of b
31. Fifty-one subtracted from six times r
32. m subtract 29
33. 36 subtract x

*Name the **variable** to be used, then translate the following into algebraic expressions.*

34. The cost of five hamburgers
35. Wages for eight hours work
36. The value of six identical paintings
37. The cost of 18 gallons of gas
38. The value of some nickels
39. The cost of six 30-pound bags of fertilizer
40. The sum of two consecutive integers
41. The sum of two consecutive even natural numbers
42. Five less than the sum of two consecutive odd whole numbers
43. Six more than the sum of two consecutive real numbers
44. The sum of three consecutive integers
45. The sum of three consecutive even integers
46. Seventeen more than the sum of three consecutive odd whole numbers
47. Thirty less than the sum of three consecutive even natural numbers
48. Two natural numbers whose sum is 17
49. Two whole numbers whose sum is 83
50. Two integers whose sum is -28
51. The sum of two integers, if one is 1 more than twice the other
52. The sum of two integers, if one is 6 less than one-third the other

Name the variable to be used and translate the following into mathematical sentences.

53. A number added to 8 is 10.

54. A number less 5 is 6.

55. Eleven more than three times a number is 24.

56. Fifteen less than eight times a number is 41.

57. The area of a square is 64 square centimeters.

58. A rectangle with length 3 more than the width has a perimeter of 48.

59. The perimeter of an equilateral triangle (with all sides of equal length) is 81 inches.

60. The value of 15 coins, where x are dimes and the rest are nickels, is $1.05.

61. The interest earned on $1000 invested for one year at 8% is I.

62. The distance traveled at an average rate r for seven hours is 385 miles.

63. The value of y coins, six of which are nickels and the rest dimes, is $2.75.

64. The distance walked in t hours, at an average rate of 4 miles per hour, is 24 miles.

65. The number of dresses purchased is found by taking the quotient of the total cost of the dresses ($134.85) and the average cost per dress of $44.95.

66. The cost per shirt is found by taking the quotient of the total cost of the shirts ($51.00) and the number of shirts, 3.

67. Jadene earned a grade of x% on her first test, 10% higher on her second test, so her average was less than 90.

68. Polly's test average on scores of y and $y - 6$ was between 85 and 93.

69. When a number is increased by 10, the sum is six less than three times the number.

70. When twice a number is decreased by 11, the difference is five less than four times the number.

Chapter 1 Summary

[1.1]   A **set** is a well-defined collection of objects. Each object is an **element** or **member** of the set. Sets are generally designated by capital letters, and the members are enclosed within **braces.** There are two ways to indicate the members of a set: the **listing method,** which lists each element of a set individually, and **set-builder notation** (also known as the rule method), which gives a verbal description of the elements.

A set with a limited number of elements is a **finite set.** A set with an unlimited number of elements is an **infinite set.** A set with no elements is an **empty set,** denoted by $\emptyset$.

The **union** of sets A and B, written $A \cup B$, includes all the elements in A, B, or both A and B. The **intersection** of sets A and B, written $A \cap B$, includes all the elements that are common to both A and B.

Set A is a **subset** of set B, written $A \subseteq B$, if every element of A is an element of B.

[1.2]   The **absolute value** of a number is its distance from zero on the number line. A more formal definition is

$$|x| = x \text{ if } x \geq 0$$
$$|x| = -x \text{ if } x < 0.$$

The set of **real numbers** consists of several subsets.

Natural numbers	$N = \{1, 2, 3, \ldots\}$
Whole numbers	$W = \{0, 1, 2, 3, \ldots\}$
Integers	$J = \{\ldots, -3, -2, -1, 0, 1, 2, 3, \ldots\}$
Rational numbers	$Q = \left\{\dfrac{a}{b} \middle| a \in J, b \in J, b \neq 0\right\}$
Irrational numbers	$H =$ nonrepeating, nonterminating decimals

Interval notation is used to indicate the portion of a number line that satisfies an inequality. A parenthesis indicates that a point does not satisfy an inequality, while a bracket indicates that it does.

[1.3] **Properties of the Real Numbers**

If $a, b, c \in R$, the following properties are assumed to be true.

Closure	$a + b \in R$	$a \cdot b \in R$
Commutative	$a + b = b + a$	$a \cdot b = b \cdot a$
Associative	$a + (b + c) = (a + b) + c$	$a(b \cdot c) = (a \cdot b)c$
Identity	$a + 0 = 0$	$a \cdot 1 = 1 \cdot a = a$
Inverse	$a + (-a) = 0$	$a \cdot \dfrac{1}{a} = 1, a \neq 0$
Distributive	$a(b + c) = ab + ac$	

A **theorem** is a statement that can be proven. The following theorems were either stated or proven in this chapter.

Multiplicative Property of Zero   If $a \in R$, then $a \cdot 0 = 0 \cdot a = 0$.

Double Negative Property   If $a \in R$, then $-(-a) = a$.

[1.6] **Multiplicative Property of Equality**   If $a, b, c \in R$ and $a = b$, then $ac = bc$.

If $a, b \in R$, then $(-a) + (-b) = -(a + b)$.

If $a, b \in R$, then $(a)(-b) = -(ab)$.

If $a, b \in R$, then $(-a)(-b) = ab$.

[1.3] **Properties of Equality:**

	Reflexive	$a = a$
	Symmetric	If $a = b$, then $b = a$.
	Transitive	If $a = b$ and $b = c$, then $a = c$.
	Substitution	If $a = b$, then a may be replaced by b or b by a without changing the truth or falsity of the statement.

[1.4]   Fractions are multiplied, divided, added, and subtracted as follows.

$$\frac{a}{b} \cdot \frac{c}{d} = \frac{a \cdot c}{b \cdot d} \qquad b, d \neq 0$$

$$\frac{a}{b} \div \frac{c}{d} = \frac{a}{b} \cdot \frac{d}{c} \qquad b, c, d \neq 0$$

$$\frac{a}{c} + \frac{b}{c} = \frac{a + b}{c} \qquad c \neq 0$$

$$\frac{a}{c} - \frac{b}{c} = \frac{a - b}{c} \qquad c \neq 0$$

Fractions are simplified by using the **fundamental property,** which states that

$$\frac{a \cdot x}{b \cdot x} = \frac{a}{b} \quad \text{or} \quad \frac{a}{b} = \frac{a \cdot x}{b \cdot x} \qquad b, x \neq 0.$$

To add fractions that do not have a common denominator, use the second part of the fundamental property to generate equivalent fractions having a common denominator.

Signed numbers are added, subtracted, multiplied, and divided according to the following rules.

[1.5]   **Addition**   To add two numbers of **like** signs, find the sum of their absolute values and prefix the same sign.

To add two numbers of **unlike** signs, find the difference of their absolute values and prefix the sign of the number with the greater absolute value.

Subtraction   To subtract two numbers, change the sign of the number being subtracted and follow the rules for addition.

[1.6]   **Multiplication**   To find the product of two signed numbers, multiply their absolute values. If the numbers have **like** signs, the product is positive. If they have **unlike** signs, the product is negative.

Division   To find the quotient of two signed numbers, find the quotient of their absolute values. If the numbers have **like** signs, the quotient is positive. If they have **unlike** signs, the quotient is negative.

Order of Operations

1. Carry out all operations within symbols of grouping, working from the innermost symbols outward.
2. Carry out all multiplications and divisions, working from left to right.
3. Carry out all additions and subtractions, working from left to right.

Chapter 1   Review

Complete the following statements.

1. A well-defined collection of objects is called a _____ .

2. The set N is an example of an _____ set.

3. The set of letters in the word *irrational* is _____ .

4. The subsets of $\{1, 3\}$ are _____ .

5. List $B = \{x \mid x \in J,\ x > -4,\ x < 10,\ x \text{ an even number}\}$.

6. $\{x \mid x \in A \text{ and } x \in B\}$ represents the _____ of sets A and B.

7. If $A \cap B = \emptyset$ then A has _____ elements that are also in B.

8. A is said to be a _____ of B if whenever $x \in A$ then $x \in B$.

9. $a \not\equiv b$ means a _____ b.

10. If $a > b$ then b is _____ of a on the number line.

11. x and $-x$ are called _____ of each other.

12. The $\{\ldots -3, -2, -1, 0, 1, 2, 3, \ldots\}$ is called the set of _____ .

13. The $\left\{ \dfrac{a}{b} \,\middle|\, a \in J,\ b \in J,\ b \neq 0 \right\}$ is called the set of _____ numbers.

14. $\pi \in$ _____ numbers.

15. $\dfrac{3}{11}$ in decimal form is _____ .

16. The statement, "If $x, y \in R$, then $x \cdot y$ is a real number" is true because of the _____ property of multiplication.

17. $6 + y = y + 6$ is true because of the _____ property of addition.

18. $6x - 6y = 6(x - y)$ illustrates the _____ property.

19. The statement, "If $a, b \in R$ and if $a = b$, then $b = a$" is true because of the _____ property of equality.

Simplify each of the following.

20. $-\dfrac{3}{5} + \dfrac{2}{7} - \dfrac{1}{2}$

21. $-\dfrac{5}{8} \cdot \dfrac{1}{4} \div \dfrac{1}{16}$

22. $-5\dfrac{7}{9} \div 2\dfrac{1}{8}$

23. $-6 + \dfrac{2}{3}\left(2 - \dfrac{7}{8}\right)$

24. $\dfrac{49}{5} - 2 + \dfrac{1}{3} \div \dfrac{3}{2}$

25. $-4\dfrac{1}{9} + \dfrac{1}{2}\left(6\dfrac{1}{2} + \dfrac{1}{3}\right) - 1$

26. $\dfrac{156}{72} \div \left(\dfrac{-90}{48}\right)$

27. $-1\dfrac{1}{5}\left(-3\dfrac{1}{8}\right) \div \left(-\dfrac{1}{4}\right)$

28. The sum of (-11) and (7) is _____ .

29. The product of (-5) and (9) is _____ .

30. $-16 - (-3)$ equals _____ .

31. The value of $x + y - z$ for $x = -5$, $y = 7$ and $z = -2$ is _____ .

32. The sign of the product of two numbers with unlike signs is _____ .

33. $\dfrac{-8(6)}{-12}$

34. Division by zero is _____ .

35. $\dfrac{6(-8 + 4) + (-3)}{15(-5 + 4) + 6}$

36. $\dfrac{\dfrac{1}{3}[16 - (-2)] + \dfrac{1}{2}(-8)}{\dfrac{1}{5}[18 + (-3)] - \dfrac{1}{4}(-12)}$

37. $-4(x - 2y) + 5(-3x + y)$

Write a mathematical sentence for each of the following.

38. The sum of two consecutive integers is -33.

39. When a number is decreased by 5, the result is 4.

40. Three more than twice a number is 18.

41. Distance is the rate (55 miles per hour) multiplied by the time (3 hours).

42. The value of 18 coins (some dimes and the rest nickels) is $1.40.

43. The perimeter of a rectangle with width 6 less than the length is 48 inches.

44. The area of a rectangle with length one more than the width is 132 square feet.

45. Julie's test average on scores of x and $x + y$ was between 88 and 93.

2

Linear Equations and Inequalities

One of the main reasons for studying mathematics is its usefulness in everyday life. The fact that you are reading this paragraph indicates that you have chosen a field of study or career that requires more than a knowledge of arithmetic and elementary algebra.

At some point you will need to translate real-life problems into mathematical expressions that can be used to solve them. Such expressions usually take the form of equations or inequalities. This chapter is devoted to helping you meet this need.

2.1 Solving Linear Equations in One Variable

A **linear equation** in one variable is one that can be written in the form

$$ax + b = c$$

where a, b, $c \in R$, with $a \neq 0$. A **solution** to an equation involving a single variable is a number that makes the equation true when the variable is replaced by it.

In the equation $2x + 7 = 13$ the solution is 3, since

$$2 \cdot 3 + 7 = 13$$
$$6 + 7 = 13$$
$$13 = 13.$$

The set of all solutions to an equation is called its **solution set.** The solution set of the equation $2x + 7 = 13$ is {3}.

Linear equations are solved by rewriting them in simpler forms called **equivalent equations** until the solution set can be determined by inspection.

Equivalent equations are equations with exactly the same solution set.

To generate equivalent linear equations, four theorems are used. They are the **addition, subtraction, multiplication** and **division properties of equality.**

Theorem 2.1

Addition Property of Equality

If a, b, and c represent real numbers and

$$a = b,$$

then

$$a + c = b + c.$$

In simplest terms, Theorem 2.1 states that *if the same real number is added to each side of an equation, the result will be an equivalent equation.*

The proof of Theorem 2.1 is based on the properties of equality. We start by using the reflexive property.

$$a + c = a + c \qquad \text{Reflexive property}$$

Since $a = b$ is given, we can use the substitution property to replace a with b on the right side of the equation, giving

$$a + c = b + c. \qquad \text{Substitute } b \text{ for } a \text{ on the right side}$$

Recall from Chapter 1 that $a - c = a + (-c)$. Adding the negative of a number to both sides of an equation will yield the same result as subtracting the number. This leads to the subtraction property.

Theorem 2.2

Subtraction Property of Equality

If a, b, and c represent real numbers and

$$a = b,$$

then

$$a - c = b - c.$$

In simplest terms, Theorem 2.2 states that *if the same real number is subtracted from each side of an equation, the result is an equivalent equation.*

Before using Theorems 2.1 and 2.2 to solve equations, the concept of combining like terms must be considered.

Two terms that are exactly alike except for their numerical coefficients are called **like terms.** For example, $3x$ and $5x$ are considered to be like terms because the variable, x, is exactly the same. Like terms are combined by using the distributive property.

Example 1   Combine like terms.

(a) $3x + 5x = (3 + 5)x = 8x$

(b) $7x - 2x = (7 - 2)x = 5x$ ⟵ Distributive property

(c) $17y - 5y + 2x = (17 - 5)y + 2x$
$$= 12y + 2x \quad ■$$

Example 2   Solve the equation $x + 5 = 7$.

    To solve this equation we need to generate an equivalent equation in which the variable x stands alone on one side. Use the subtraction property of equality.

$$x + 5 = 7$$
$$x + 5 - 5 = 7 - 5 \qquad \text{Subtract 5 from each side}$$
$$x = 2$$

    To **check** the solution, substitute the result for the variable in the original equation.

$$x + 5 = 7$$
$$2 + 5 \overset{?}{=} 7 \qquad x = 2$$
$$7 \overset{\checkmark}{=} 7$$

Since the solution satisfies the original equation, the solution set is {2}.   ■

Example 3   Solve $3x - 3 - x = x + 8$.

    First combine like terms on each side of the equation, as shown below. Each equation is equivalent to the original equation.

$$3x - 3 - x = x + 8$$
$$(3 - 1)x - 3 = x + 8$$
$$2x - 3 = x + 8 \qquad \text{Combine like terms}$$
$$2x - 3 - x = x + 8 - x \qquad \text{Subtract } x \text{ from each side}$$
$$x - 3 = 8 \qquad \text{Combine like terms}$$
$$x - 3 + 3 = 8 + 3 \qquad \text{Add 3 to each side}$$
$$x = 11$$

Check: $\qquad 3x - 3 - x = x + 8$
$$3 \cdot 11 - 3 - 11 \overset{?}{=} 11 + 8 \qquad x = 11$$
$$33 - 3 - 11 \overset{?}{=} 19$$
$$19 \overset{\checkmark}{=} 19.$$

The solution set is {11}.   ■

    The equations in Examples 2 and 3 were solved by using the addition or subtraction properties of equality to generate simpler equivalent equations. Some equations cannot be solved by using only these properties. Two such equations are $6x + 5 = 8$ and $\frac{t}{2} - \frac{3t}{7} = 9$. Their solutions require the multiplication and division properties of equality.

Theorem 2.3

Multiplication Property of Equality

If a, b, and c represent real numbers and

$$a = b,$$

then

$$a \cdot c = b \cdot c.$$

In simplest terms, Theorem 2.3 states that *if each side of an equation is multiplied by the same nonzero real number, the result is an equivalent equation.*

Theorem 2.4

Division Property of Equality

If a, b, and c represent real numbers and

$$a = b,$$

then

$$\frac{a}{c} = \frac{b}{c}, \quad c \neq 0.$$

In simplest terms, Theorem 2.4 states that *if each side of an equation is divided by the same nonzero real number, the result is an equivalent equation.*

Example 4 Solve $6x + 5 = 8$.

First write an equivalent equation in which the term involving the variable stands alone on one side.

$$6x + 5 = 8$$
$$6x + 5 - 5 = 8 - 5 \qquad \text{Subtract 5 from each side}$$
$$6x = 3$$

To solve for x, use the division property to write another equivalent equation.

$$\frac{6x}{6} = \frac{3}{6} \qquad \text{Divide each side by 6}$$
$$x = \frac{1}{2}$$

The solution set is $\left\{\frac{1}{2}\right\}$. ∎

Example 5   Solve $2(m + 1) - 4(m - 6) = 6$.

First use the distributive property to remove symbols of grouping.

$$2(m + 1) - 4(m - 6) = 6$$

$$2m + 2 - 4m + 24 = 6 \qquad \text{Distributive property}$$

$$-2m + 26 = 6 \qquad \text{Combine like terms}$$

$$-2m + 26 - 26 = 6 - 26 \qquad \text{Subtract 26 from each side}$$

$$-2m = -20$$

$$\frac{-2m}{-2} = \frac{-20}{-2} \qquad \text{Divide each side by } -2$$

$$m = 10$$

The solution set is {10}. ■

Example 6   Solve $\frac{t}{2} - \frac{3t}{7} = 9$.

To clear of fractions multiply each side by the least common multiple of 2 and 7, which is 14. The number 14 is the smallest number that can be divided by both 2 and 7.

$$14\left(\frac{t}{2} - \frac{3t}{7}\right) = 14(9) \qquad \text{Multiply each side by 14}$$

$$14 \cdot \frac{t}{2} - 14 \cdot \frac{3t}{7} = 14 \cdot 9 \qquad \text{Distributive property}$$

$$7t - 6t = 126 \qquad \text{Simplify}$$

$$t = 126 \qquad \text{Combine like terms}$$

Check: $\dfrac{126}{2} - 3 \cdot \dfrac{126}{7} = 9 \qquad t = 126$

$$63 - 54 \overset{?}{=} 9$$

$$9 \overset{\checkmark}{=} 9.$$

The solution set is {126}. ■

Example 7   Solve $\frac{x}{x-3} + 1 = \frac{3}{x-3}$.

Begin by noting that $x \neq 3$. If $x = 3$, then $x - 3 = 0$. **Division by zero is not defined.** If each side of the equation is multiplied by the LCD, $x - 3$, the equation can be cleared of fractions.

$$(x - 3)\left(\frac{x}{x-3} + 1\right) = (x - 3)\left(\frac{3}{x-3}\right) \qquad \text{Multiply by } x - 3$$

$$(x - 3) \cdot \frac{x}{x-3} + (x - 3) \cdot 1 = (x - 3) \cdot \left(\frac{3}{x-3}\right) \qquad \text{Distributive property}$$

$$x + x - 3 = 3 \qquad \text{Simplify}$$

$$2x - 3 = 3 \qquad \text{Combine like terms}$$

$$2x = 6 \qquad \text{Add 3 to each side}$$

$$x = 3 \qquad \text{Divide each side by 2}$$

Remember $x \neq 3$. The equation has no solution. The solution set is empty. ■

As you gain confidence in your ability to solve equations you will find that you check your solutions less frequently. It is essential that you note the following caution.

▶ *Caution*   When an equation involves variables in the denominator, always note any restrictions on the variable before solving the equation. No replacement for a variable can result in a denominator being zero.

Example 8   State the restrictions on the variable.

	Equation	Restriction
(a)	$\dfrac{6}{x+6} + 1 = \dfrac{-x}{x+6}$	$x \neq -6$
(b)	$\dfrac{3}{2y-5} = \dfrac{5}{4y+7}$	$y \neq \dfrac{5}{2}$ or $\dfrac{-7}{4}$
(c)	$\dfrac{5}{p} + \dfrac{2}{p-2} = \dfrac{4}{p-2}$	$p \neq 0$ or 2 ■

In contrast to equations that have no solution are those that have infinitely many solutions.

Example 9   Solve $x + (3x + 2) - 4 = 2x + 2(x - 1)$.

$$x + 3x + 2 - 4 = 2x + 2x - 2 \qquad \text{Associative and distributive properties}$$

$$4x - 2 = 4x - 2 \qquad \text{Combine like terms}$$

Since $4x - 2 = 4x - 2$ is true for any real number replacement for x, the solution set is written as

$\{x \mid x \in R\}$.   ■

An equation that is true for any permissible replacement for the variable is called an **identity**. Example 9 is an identity. If an equation is true only for specified values of the variable, it is called a **conditional equation**. Examples 2–7 are conditional equations.

As the examples in this section show, a series of steps are followed to solve linear equations in one variable. The steps are summarized on the next page. Not all steps are necessary to solve every linear equation.

To Solve Linear Equations:
1. Note any restrictions on the variable.
2. Clear the equation of fractions by using the multiplication property.
3. Remove symbols of grouping by using the distributive property.
4. Combine like terms on each side.
5. Isolate the variable by using the addition or subtraction property.
6. Use the multiplication or division property to solve for the variable.

The next two examples will show how mathematical sentences can be used to solve applied problems.

Example 10   The sum of two consecutive integers is 167. Find the integers. (Consecutive integers are integers that differ by 1. Examples are 13 and 14, and -5 and -6.)

Let x represent the smaller integer, and let $x + 1$ represent the larger integer.

The sum of two consecutive integers is 167.

$$x + x + 1 = 167$$

$2x + 1 = 167$	Combine like terms
$2x = 166$	Subtract 1 from each side
$x = 83$	Divide each side by 2

Since $x = 83$, $x + 1 = 84$.

The integers are 83 and 84. ∎

The check of an applied problem should always be done using the words of the original problem. In Example 10 the two integers 83 and 84 are consecutive, and their sum is 167, as required. The solution checks.

Example 11   The length L of a rectangular field is 15 meters more than the width W. If the perimeter P is 310 meters, find the dimensions.

The formula for the perimeter of a rectangle is $2W + 2L = P$.

Let W = width of the field; then $W + 15$ = length of the field.

The width is W The length is $W + 15$ The perimeter is 310

$$2W + 2(W + 15) = 310$$
$$2W + 2W + 30 = 310$$
$$4W + 30 = 310$$
$$4W = 280$$

Dividing both sides by 4 gives

$$W = 70,$$

so $W + 15 = 85.$

The width is 70 meters, and the length is 85 meters.

Check: The length is 85 meters, 15 meters longer than the width, which is 70 meters. The perimeter, $2(70 + 85)$, is 310 meters. The solution checks. ∎

Exercises 2.1

Solve each equation. Check each solution.

1. $x - 3 = 5$
2. $y + 4 = 7$
3. $m + 8 = 3$
4. $p - 7 = 2$
5. $3a - 2 = 7$
6. $4s - 1 = 3$
7. $4y - 9 = -3$
8. $5x + 7 = -4$
9. $3p + 5 = -6$
10. $6a - 4 = -1$
11. $-2s + 1 = 9$
12. $-3m - 2 = -5$
13. $2r + 3 = r - 1$
14. $5q - 2 = 2q + 1$
15. $5K - 2 + K = 3K - 1$
16. $6y - 5 = 3y + 1 - 2y$
17. $11x - 5 - 4x = 7 - x$
18. $2c - 11 - 8c = 5c - 11$
19. $16a - 5 + 7a = 6a - 5$
20. $-s + 3 - 2s = -3s + 3 - s$
21. $7r - 5 + 2r = -6r + 5 + 14r$
22. $-8m + 8 + 9m = 5 + 2m + 3$

State whether each equation is a conditional equation *or* identity *and solve.*

23. $x + 4 = 3$
24. $3x + 4 = 2x - 1$
25. $a + 2 = a + 2$
26. $m - 3 = -3 + m$
27. $8K - 3 = 6K + 3 + K$
28. $7r - 5 = 7r + 5 - r$
29. $3(y - 1) = 3y - 3$
30. $-2(3 - s) = 2s - 6$

Use the distributive property to remove the grouping symbols, combine like terms and solve.

31. $2(a + 3) - 3(a - 4) = 8$
32. $3t + 4(1 - t) = 5$
33. $-2(x - 5) + 4(2x - 1) = x + 3$
34. $-5(y + 1) + 2(3 - 2y) = y - 9$
35. $2(3x - 4) + 3(2 - x) = -(x - 1)$
36. $3(s - 5) + 4(-2s + 1) = 2(s - 4)$
37. $6 + 3(K - 1) - 2(2K + 4) = 5 - 3(2K + 5)$
38. $6m - (3m - 4) - 2 = 11 - (4m + 1) - 3m$
39. $5r + (6 - 7r) - 5 = 8 - (2 - 5r) + 8r$
40. $t + (t + 2) + 3(t + 4) = 50$
41. $-3\{2 - [p - (3p + 2)]\} = 4[-5 - (-3p + 7)] + 2$
42. $-2\{-5 + [-4 - 3(4 - 2r)]\} = 7[-2 - (-4r + 3)] - 5$
43. $-\{-6 - 2[3 - (2y - 4)] + 3\} = -[-5 - 3(4 - y) + 2] + 1$
44. $-\{1 - [2 - 3(4 - x) + 3x] - 5\} = -[2 + 7(6 - x) - 4x] - 8$

Solve each equation. Check each solution.

45. $\dfrac{x}{2} = 4$
46. $\dfrac{a}{3} = 1$
47. $\dfrac{2m}{3} = -5$
48. $\dfrac{4s}{7} = -2$
49. $\dfrac{2}{5}K = \dfrac{3}{7}$
50. $\dfrac{1}{3}r = \dfrac{1}{2}$
51. $-\dfrac{2}{9}y = \dfrac{5}{7}$
52. $-\dfrac{1}{4}b = \dfrac{2}{3}$

53. $4a = \dfrac{3}{2}$ **54.** $7x = -\dfrac{2}{3}$ **55.** $-4r = -\dfrac{7}{4}$ **56.** $-9b = -\dfrac{4}{3}$

57. $-\dfrac{1}{2}r - \dfrac{1}{3} = \dfrac{2}{3}$ **58.** $\dfrac{5m}{7} + \dfrac{2}{5} = -\dfrac{3}{5}$ **59.** $\dfrac{7m}{3} - \dfrac{4m}{2} = 5$

60. $\dfrac{1}{3}y + \dfrac{1}{4} = \dfrac{1}{2}y - \dfrac{2}{3}$ **61.** $\dfrac{3}{5}K - \dfrac{2}{3} = \dfrac{1}{4}K + \dfrac{1}{2}$ **62.** $\dfrac{1}{7}a - \dfrac{5}{3} = \dfrac{3}{8}a + \dfrac{2}{5}$

Solve each equation. Note any restrictions on the variable. Check each solution. (Some may not have solutions.)

63. $\dfrac{1}{x} = 4$ **64.** $\dfrac{1}{y} = 2$ **65.** $\dfrac{3}{K} = -1$ **66.** $\dfrac{2}{m} = -5$

67. $\dfrac{3}{r} = \dfrac{2}{5}$ **68.** $-\dfrac{4}{b} = -\dfrac{1}{5}$ **69.** $\dfrac{1}{K+1} = 2$ **70.** $\dfrac{2}{t-3} = -3$

71. $\dfrac{2y+3}{5} - \dfrac{1}{2} = \dfrac{y-4}{4}$ **72.** $\dfrac{a-3}{2} - \dfrac{1}{3} = \dfrac{a+4}{4}$

73. $\dfrac{3}{m+5} + 2 = \dfrac{1}{m+5}$ **74.** $\dfrac{6}{t-1} = \dfrac{3}{t-1} + 1$

75. $\dfrac{8}{p+2} + \dfrac{12}{p+2} = 5$ **76.** $\dfrac{1}{x-4} + \dfrac{-3}{x-4} = \dfrac{1}{2}$

77. $\dfrac{1}{a+1} = 2 + \dfrac{1}{a+1}$ **78.** $\dfrac{2}{y+3} + \dfrac{-2}{y+3} = -3$

Use a calculator to solve each equation. Round all solutions to one decimal place.

79. $0.3(x - 2) = 0.85$ **80.** $-2.1(1.3y - 1.4) = 3.85$

81. $0.1K - 0.1(0.5 - 3K) = 1.2$ **82.** $1.1 - 0.2(-0.7 + 3t) = -0.4t$

Translate into an equation and solve.

83. Find two consecutive integers whose sum is 107.

84. Find two consecutive odd natural numbers whose sum is 136.

85. One whole number is 6 more than the other. Their sum is 108. Find the numbers.

86. One integer is 5 less than the other. Their sum is −13. Find the integers.

87. The length of a rectangular garden is twice the width. Find the dimensions if the perimeter is 120 meters.

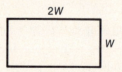

88. The perimeter of a rectangular pasture is 80 kilometers. If the width is 10 kilometers less than the length, find the dimensions.

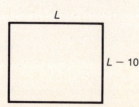

89. The cost of mailing a package by first-class mail is given by $C = 20 + 19(x - 1)$ where C is the cost in cents and x is the number of ounces. If a package costs 96 cents to mail, what does it weigh?

90. The cost of mailing a package by second-class mail is given by $C = 0.17 + 0.15(W - 1)$, where C is the cost in dollars and W is the weight in ounces. If mailing a package second-class costs \$2.72, what does it weigh?

2.2 Solving Linear Equations Involving Absolute Value

Recall from Chapter 1 that the *absolute value* of a number is defined as its distance from the origin on a number line. To solve an equation such as

$$|x| = 9$$

locate the points on a number line that are 9 units from the origin. (See Figure 2.1.) Since both 9 and -9 are nine units from the origin, the solution set is $\{-9, 9\}$.

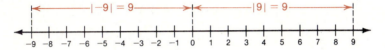

Figure 2.1

The equation can also be solved by using the more formal definition

$$|x| = x \text{ if } x \geq 0 \quad \text{or} \quad |x| = -x \text{ if } x < 0.$$

When the latter definition is used, $|x| = 9$ translates into

$x = 9$ if $x \geq 0$ or $-x = 9$ if $x < 0$

$x = 9$ or $x = -9$ Multiply each side by -1

The solution set, $\{9, -9\}$, agrees with the one found by the distance definition.

To solve the general absolute value equation $|x| = p$ where p is a positive real number, we find that $x = p$ if $x \geq 0$ or $x = -p$ if $x < 0$. The solution set is $\{p, -p\}$.

Theorem 2.5

If p is a positive real number, then

$$|ax + b| = p$$

is equivalent to

$$ax + b = p \quad \text{or} \quad ax + b = -p.$$

Example 1   Solve $|a + 2| = 6$.

$a + 2 = 6$ or $a + 2 = -6$ Theorem 2.5

      $a = 4$ $a = -8$ Subtract 2 from each side

Check the solutions.

$$|4 + 2| \overset{?}{=} 6 \qquad a = 4 \qquad\qquad |-8 + 2| \overset{?}{=} 6 \qquad a = -8$$
$$|6| \overset{?}{=} 6 \qquad\qquad\qquad\qquad |-6| \overset{?}{=} 6$$
$$6 \overset{\checkmark}{=} 6 \qquad\qquad\qquad\qquad\quad 6 \overset{\checkmark}{=} 6$$

The solution set is $\{4, -8\}$. ■

Example 2 Solve $|2x + 4| = 9$.

$2x + 4 = 9$	or	$2x + 4 = -9$	Theorem 2.5
$2x = 5$		$2x = -13$	Subtract 4 from each side
$x = \dfrac{5}{2}$		$x = -\dfrac{13}{2}$	Divide each side by 2

Check the solutions.

$$\left|2 \cdot \frac{5}{2} + 4\right| \overset{?}{=} 9 \qquad x = \frac{5}{2} \qquad\qquad \left|2 \cdot \left(-\frac{13}{2}\right) + 4\right| \overset{?}{=} 9 \qquad x = \frac{-13}{2}$$
$$|5 + 4| \overset{?}{=} 9 \qquad\qquad\qquad\qquad |-13 + 4| \overset{?}{=} 9$$
$$|9| \overset{?}{=} 9 \qquad\qquad\qquad\qquad\qquad |-9| \overset{?}{=} 9$$
$$9 \overset{\checkmark}{=} 9 \qquad\qquad\qquad\qquad\qquad\quad 9 \overset{\checkmark}{=} 9$$

The solution set is $\{\frac{5}{2}, \frac{-13}{2}\}$. ■

Example 3 Solve $|x + 3| + 2 = 8$.

First isolate $|x + 3|$ by subtracting 2 from each side.

$$|x + 3| + 2 - 2 = 8 - 2$$
$$|x + 3| = 6$$

Now $x + 3 = 6$	or	$x + 3 = -6.$	Theorem 2.5
$x = 3$		$x = -9$	Subtract 3 from each side

The solution set is $\{3, -9\}$.
The check is left to the reader. ■

Example 4 Solve $|x + 6| = -2$.

Since the absolute value of a number can never be negative, this equation has no solution. ■

When both sides of an equation involve absolute values such as $|a| = |b|$, Theorem 2.5 can be extended so that it applies. Since the absolute value of a number is always positive or zero,

$$|a| = |b| \quad \text{means that} \quad a = b \quad \text{or} \quad a = -b.$$

Example 5   Solve $|2x + 3| = |x + 5|$.

We know that $|x + 5|$ is positive or zero.

$$2x + 3 = x + 5 \qquad \text{or} \qquad 2x + 3 = -(x + 5)$$
$$2x - x = 5 - 3 \qquad\qquad\qquad 2x + 3 = -x - 5$$
$$x = 2 \qquad\qquad\qquad\qquad 3x = -8$$
$$x = \frac{-8}{3}$$

Check the solutions.

$$|2 \cdot 2 + 3| \overset{?}{=} |2 + 5| \qquad x = 2 \qquad\qquad \left|2\left(\frac{-8}{3}\right) + 3\right| \overset{?}{=} \left|\frac{-8}{3} + 5\right| \qquad x = \frac{-8}{3}$$
$$|4 + 3| \overset{?}{=} |2 + 5| \qquad\qquad\qquad\qquad \left|\frac{-16}{3} + \frac{9}{3}\right| \overset{?}{=} \left|\frac{-8}{3} + \frac{15}{3}\right|$$
$$|7| \overset{?}{=} |7| \qquad\qquad\qquad\qquad\qquad\quad \left|\frac{-7}{3}\right| \overset{?}{=} \left|\frac{7}{3}\right|$$
$$7 \overset{\checkmark}{=} 7 \qquad\qquad\qquad\qquad\qquad\qquad \frac{7}{3} \overset{\checkmark}{=} \frac{7}{3}$$

The solution set is $\{\frac{-8}{3}, 2\}$.   ■

Example 6   Determine whether $|-6x| = 6|x|$ is true for $x \in \{-5, 5\}$.

For $x = -5$ For $x = 5$

$$|-6(-5)| \overset{?}{=} 6|-5| \qquad\qquad |-6 \cdot 5| \overset{?}{=} 6|5|$$
$$|30| \overset{?}{=} 6 \cdot 5 \qquad\qquad\qquad |-30| \overset{?}{=} 6 \cdot 5$$
$$30 \overset{\checkmark}{=} 30 \quad \text{True} \qquad\qquad 30 \overset{\checkmark}{=} 30 \quad \text{True}   ■$$

Example 7   Describe the set of numbers that are 6 units from the origin as an absolute value equation.

The numbers that are 6 units from the origin are 6 and -6. Since $|x| = 6$ is equivalent to $x = 6$ or $x = -6$, the equation is $|x| = 6$.   ■

Exercises 2.2

Find the solution set for each of the following. Indicate those for which no solution exists.
Check your answers.

1. $|x| = 4$ **2.** $|y| = 25$ **3.** $|p| = -1$ **4.** $|k| = -9$

5. $|t| = 169$ **6.** $|m| = 121$ **7.** $|r| = \frac{1}{4}$ **8.** $|n| = \frac{3}{25}$

9. $|x| = \frac{7}{36}$ **10.** $|y| = \frac{5}{9}$ **11.** $-|2x| = 16$ **12.** $-|3p| = 81$

13. $|t + 2| = 9$ **14.** $|m - 1| = 3$ **15.** $|k + 5| = -5$ **16.** $|2r + 3| = -11$

17. $-|3 - y| = -5$ **18.** $-|2x - 4| = -3$ **19.** $|m - 3| + 2 = 5$ **20.** $|4 + p| - 3 = 7$

21. $|3 + 6t| = 0$ **22.** $|k - 3| = 0$ **23.** $-|5 - 3x| = 1$ **24.** $-|2 - 5y| = 2$

25. $\left|\dfrac{r - 2}{3}\right| = 4$ **26.** $\left|\dfrac{p + 5}{7}\right| = 1$ **27.** $\left|\dfrac{2x - 3}{5}\right| = 6$ **28.** $\left|\dfrac{6m - 7}{8}\right| = \dfrac{3}{8}$

29. $\left|\dfrac{5k - 3}{7}\right| = \dfrac{5}{7}$ **30.** $\left|\dfrac{11p - 9}{26}\right| = \dfrac{3}{13}$ **31.** $\left|\dfrac{2y + 9}{18}\right| = \dfrac{5}{6}$

Find each solution set. Indicate those for which no solution exists.

32. $|x + 1| = |2x - 3|$ **33.** $|p - 3| = |3p|$

34. $|2t + 3| = 2|t - 1|$ **35.** $|3m + 4| = 3|m + 1|$

36. $|r + 3| = |1 - r|$ **37.** $|2y - 3| = |2y - 3|$

38. $|x - 1| = x; \ x > 0$ **39.** $|k + 3| = 3k; \ k > 0$

40. $|m| = -m; \ m > 0$ **41.** $|6a - 1| = |9 - 3a|$

42. $|2b + 3| = 8 - b; \ b < 0$ **43.** $|p| = -p; \ p > 0$

44. $|3a - 2| = 4 + a; \ a > 0$ **45.** $|2 + 3m| = -(7 - 2m)$

46. $|2k + 7| = |2k + 7|$ **47.** $|p - 4| = p - 4; \ p > 4$

48. $|5r - 1| = |1 - 5r|$ **49.** $|3y - 5| = |5 - 3y|$

50. $|s - 3| = |2s + 3|$ **51.** $|2x - 1| = |3x + 2|$

52. $|5x - 8| = -(8 - 5x); \ x < \dfrac{-8}{5}$ **53.** $|4a - 3| = -|3 - 4a|$

54. $|x + 3| = |x + 2|$ **55.** $|y - 5| = |6 - y|$

56. $|2k + 5| = |6 - 2k|$ **57.** $|2m - 1| = |5 - 3m|$

Determine whether each equation is true for the specified values of the variables.

58. $|5x| = 5|x|; \ x = -4$ **59.** $|ab| = ab; \ a = -3, \ b = 2$

60. $\left|\dfrac{x}{y}\right| = \dfrac{x}{y}; \ x = -8, \ y = 3$ **61.** $\left|\dfrac{k}{10}\right| = \dfrac{|k|}{10}; \ k = 0$

62. $|-6p| = 6|p|; \ p = 7$ **63.** $|(-a)(-b)| = |ab|; \ a = 1, \ b = -9$

64. $\left|\dfrac{x}{y}\right| = \dfrac{|x|}{|y|}; \ x = 0, \ y = -6$ **65.** $|xy| = |x||y|; \ x = 0, \ y = -11$

Use a calculator to solve each of the following. Round all answers to one decimal place.

66. $|-6.85x| = 3.27$ **67.** $|3.01k| = 9.03$

68. $|2.7y - 5.8| = 11.64$ **69.** $|3.91p + 2.01| = 1.95$

70. $|0.08t - 1.89| = |6.01t + 3.25|$ **71.** $|11.32 - 9.63m| = |2.79m - 5.26|$

Rewrite each of the following as an absolute value equation and solve. Indicate those for which no solution exists.

72. The absolute value of a number is 1.

73. The absolute value of a number is 8.

74. If 6 is subtracted from a number, the absolute value of the result is 10.

75. Three is subtracted from twice a number and this quantity is divided by 4. The absolute value of the result is 0.

76. Five is added to one-third of a number. The absolute value of the result is -8.

77. One is added to three-fourths of a number. The absolute value of the result is -10.

Write the following as absolute value equations.

78. The numbers that are 2 units from the origin

79. The numbers that are 4 units from the origin

80. The numbers that are 6 units from zero

81. The numbers that are 3 units from zero

2.3 Solving Linear Inequalities

A **linear inequality** is one that can be written in the form

$$ax + b < c$$

where a, b, and $c \in R$ and $a \neq 0$. The expressions $ax + b > c$, $ax + b \leq c$, and $ax + b \geq c$ are also examples of linear inequalities. Throughout this section all rules and theorems will be stated in terms of $<$, but they are equally true for any other inequality symbol, such as $>$, $\leq$, or $\geq$.

Linear inequalities are solved in a manner similar to that used to solve linear equations. That is, a series of simpler *equivalent inequalities* are generated until the solution set can be determined by inspection. **Equivalent inequalities** are inequalities that have the same solution set.

To solve linear equations we used the addition, subtraction, multiplication, and division properties of equality. Similar properties are used to solve linear inequalities.

Theorem 2.6

Addition and Subtraction Properties of Inequality

For real numbers a, b, and c,

if $a < b$, then $a + c < b + c$ by the **addition property**;

if $a < b$, then $a - c < b - c$ by the **subtraction property**.

It is not difficult to illustrate the truth of these properties. Consider the following.

(a)
$$-2 < 3$$
$$-2 + 2 < 3 + 2 \quad \text{Addition property}$$
$$0 < 5$$

(b)
$$-2 < 3$$
$$-2 - 2 < 3 - 2 \quad \text{Subtraction property}$$
$$-4 < 1$$

Example 1 Solve and graph the inequality $x + 7 < 9$.

$$x + 7 < 9$$
$$x + 7 - 7 < 9 - 7 \quad \text{Subtraction property}$$
$$x < 2$$

The solution set is $(-\infty, 2)$. The graph is shown in Figure 2.2.

The parenthesis at 2 indicates that 2 is not in the solution set. The arrow pointing to the left indicates that all real numbers less than 2 are in the solution set. ■

Figure 2.2

Example 2   Solve and graph the inequality $x - 3 \geq 13$.

$$x - 3 \geq 13$$
$$x - 3 + 3 \geq 13 + 3 \qquad \text{Addition property}$$
$$x \geq 16$$

The solution set is $[16, \infty)$. The graph is shown in Figure 2.3.

The bracket at 16 indicates that 16 is a part of the solution set. The arrow indicates that all real numbers greater than 16 are in the solution set. ■

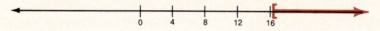

Figure 2.3

Example 3   Solve $2t + 2(t + 3) \leq 3(t - 1)$, $t \in J$ and graph. Recall that J represents the set of integers.

To solve this inequality, follow the same steps that were appropriate for solving equations.

$$2t + 2(t + 3) \leq 3(t - 1)$$
$$2t + 2t + 6 \leq 3t - 3 \qquad \text{Remove symbols of grouping}$$
$$4t + 6 \leq 3t - 3 \qquad \text{Combine like terms}$$
$$4t - 3t \leq -3 - 6 \qquad \text{Subtraction property}$$
$$t \leq -9 \qquad \text{Combine like terms}$$

Since t is an integer, the solution set is $\{ \ldots \ -12, -11, -10, -9\}$ or $\{t \mid t \leq -9, t \in J\}$. (See Figure 2.4.) ■

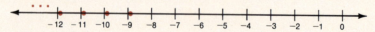

Figure 2.4

Inequalities such as $\frac{x}{2} \geq 5$ and $-3x < 12$ cannot be solved by use of the addition or subtraction properties. Their solution requires the multiplication or division property of inequality.

Theorem 2.7

Multiplication Property of Inequality

For real numbers a, b, and c,

if $a < b$, then $a \cdot c < b \cdot c$ if c is positive;

if $a < b$, then $a \cdot c > b \cdot c$ if c is negative.

It is easy to see the truth of this theorem by considering numerical examples.

(a) $-1 < 3$

$2(-1)$? $2 \cdot 3$ Multiply each side by 2

$-2 < 6$

(b) $-1 < 3$

$-2(-1)$? $-2 \cdot 3$ Multiply each side by -2

$2 > -6$

Part (a) illustrates that when each side of an inequality is multiplied by a *positive* real number the order (the direction the inequality symbol points) *remains the same*.

Part (b) illustrates that when each side of an inequality is multiplied by a *negative* real number the order is *reversed*.

Example 4 Solve $\frac{1}{3}x + 2 > 3$ and graph.

Clear the inequality of fractions by multiplying each side of the inequality by 3. Since 3 is positive, the order will remain the same.

$$3\left(\frac{1}{3}x + 2\right) > 3 \cdot 3$$

$x + 6 > 9$ Distributive property

$x > 3$ Subtraction property

The solution set is $(3, \infty)$. The graph is shown in Figure 2.5. ■

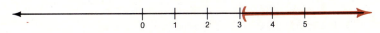

Figure 2.5

Example 5 Solve $2(x + 5) - 6 < 3(x + 4)$.

$2(x + 5) - 6 < 3(x + 4)$

$2x + 10 - 6 < 3x + 12$ Distributive property

$2x + 4 < 3x + 12$ Combine like terms

$-x < 8$ Subtract 3x and 4 from each side

To solve for x, multiply each side by -1.

$-1(-x) < (-1)8$

$x > -8$ Multiplying by -1 reverses the order

The solution set is $(-8, \infty)$. The graph is shown in Figure 2.6. ■

Figure 2.6

Theorem 2.8

Division Property of Inequality

For real numbers a, b, and c, with $c \neq 0$,

$$\text{if } a < b, \quad \text{then} \quad \frac{a}{c} < \frac{b}{c} \quad \text{if } c \text{ is positive;}$$

$$\text{if } a < b, \quad \text{then} \quad \frac{a}{c} > \frac{b}{c} \quad \text{if } c \text{ is negative.}$$

As numerical examples of this theorem, consider the following.

(a) $6 > -3$

$\dfrac{6}{3} \; ? \; -\dfrac{3}{3}$ Divide each side by 3

$2 > -1$

(b) $6 > -3$

$\dfrac{6}{-3} \; ? \; \dfrac{-3}{-3}$ Divide each side by -3

$-2 < 1$

Part (a) illustrates that when each side of an inequality is divided by a *positive* real number the order *remains the same*.

Part (b) illustrates that when each side of an inequality is divided by a *negative* real number the order is *reversed*.

Example 6   Solve $3(x + 2) \geq 5(x - 9) + 4$.

$$3(x + 2) \geq 5(x - 9) + 4$$
$$3x + 6 \geq 5x - 45 + 4 \qquad \text{Distributive property}$$
$$3x + 6 \geq 5x - 41 \qquad \text{Combine like terms}$$
$$3x - 5x \geq -41 - 6 \qquad \text{Subtract } 5x \text{ and } 6 \text{ from each side}$$
$$-2x \geq -47$$
$$\frac{-2x}{-2} \leq \frac{-47}{-2} \qquad \text{Dividing by } -2 \text{ reverses the order}$$
$$x \leq \frac{47}{2}$$

The solution set is $(-\infty, \frac{47}{2}]$. ■

One frequently encounters inequalities in which the value of the variable lies between two numbers, such as $-2 \leq x \leq 3$. Such an inequality is called a **compound inequality.** In general, a compound inequality is solved by using the properties of inequality to isolate the variable between two numbers.

Example 7   Solve and graph $8 < 2x + 4 \leq 16$.

$$8 < 2x + 4 \leq 16$$

$$8 - 4 < 2x + 4 - 4 \leq 16 - 4 \qquad \text{Subtract 4 from each part}$$

$$4 < 2x \leq 12$$

$$\frac{4}{2} < \frac{2x}{2} \leq \frac{12}{2} \qquad \text{Divide each part by 2}$$

$$2 < x \leq 6$$

The solution set is (2, 6]. The graph is shown in Figure 2.7.   ■

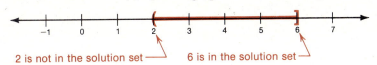

2 is not in the solution set ⟶   6 is in the solution set ⟶

Figure 2.7

Example 8   Solve and graph $\frac{5}{2} > \frac{-x}{3} + 2 > 1$, $x \in J$.

Begin by multiplying each part by 6 to clear the inequality of fractions.

$$6 \cdot \frac{5}{2} > 6\left(\frac{-x}{3} + 2\right) > 6 \cdot 1$$

$$15 > -2x + 12 > 6 \qquad \text{Distributive property}$$

$$15 - 12 > -2x + 12 - 12 > 6 - 12 \qquad \text{Subtract 12 from each part}$$

$$3 > -2x > -6 \qquad \text{Combine like terms}$$

$$\frac{3}{-2} < \frac{-2x}{-2} < \frac{-6}{-2} \qquad \begin{array}{l}\text{Divide each part by } -2 \\ \text{and reverse the order}\end{array}$$

$$-\frac{3}{2} < \quad x \quad < 3$$

Recall that *J* represents the set of integers. The solution set is $\{x | -\frac{3}{2} < x < 3,$ $x \in J\}$ or $\{-1, 0, 1, 2\}$. The graph is shown in Figure 2.8.   ■

Figure 2.8

Example 9   The number of male students in an algebra class is 6 more than the number of female students. If there are no more than 44 students in the class, find the maximum number of male and female students.

Let m = number of male students

$m - 6$ = number of female students.

The sum of the number of male and female students must be less than or equal to 44.

$$m + m - 6 \leq 44$$

$$2m - 6 \leq 44 \qquad \text{Combine like terms}$$

$$2m \leq 50 \qquad \text{Add 6 to each side}$$

$$m \leq 25 \qquad \text{Divide each side by 2}$$

The maximum number of females is 6 less than the maximum number of males. Thus,

number of males ≤ 25

number of females ≤ 19. ■

Example 10   Solve and graph $\{K \mid K > 3\} \cap \{K \mid K \leq 6\}$.

The solution set for $K > 3$ is $(3, \infty)$. The graph is shown in Figure 2.9. The solution set for $K \leq 6$ is $(-\infty, 6]$. The graph is shown in Figure 2.10. The **intersection** ($\cap$) of the two sets is $(3, 6]$. The graph is shown in Figure 2.11. ■

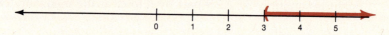

Figure 2.9

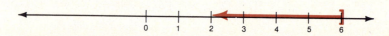

Figure 2.10

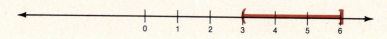

Figure 2.11

Example 11   Solve and graph $\{x \mid x > 3\} \cup \{x \mid x \leq -2\}$.

Since we are considering the **union** ($\cup$) of two sets, any x that satisfies $x > 3$ or $x \leq -2$ is in the solution set. The graph is shown in Figure 2.12. ■

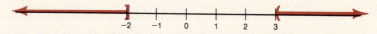

Figure 2.12

Exercises 2.3

Solve and graph each of the following. Write your answers in interval or set-builder notation.
(Recall that J is the set of integers, W the set of whole numbers, N the set of natural numbers,
and R the set of real numbers.)

1. $x - 3 < 1$

2. $x + 4 > -1$, $x \in J$

3. $x - 5 \leq -1$, $x \in N$

4. $x - 7 \geq -5$, $x \in W$

5. $3x \geq 8$

6. $5x \leq -15$, $x \in N$

7. $2x - 3 > -4$

8. $6x + 2 < 6$, $x \in N$

9. $7x - 5 \geq 2$, $x \in J$

10. $2x - 7 \leq 3$, $x \in W$

11. $\dfrac{3x}{5} \geq -1$

12. $\dfrac{2}{7}x \leq \dfrac{3}{7}$, $x \in J$

13. $-6 < 2x \leq 8$

14. $9 > x + 1 \geq 5$, $x \in N$

15. $8 \geq x + 5 \geq 2$, $x \in J$

16. $-2 < 5 - 6x \leq 3$

17. $4 > 2 - x > -1$, $x \in W$

18. $\dfrac{3(x + 3)}{5} \geq 5 - x$

19. $\dfrac{2(x - 4)}{7} + 3 \leq x$

20. $\dfrac{6(x + 5)}{7} - 1 \geq \dfrac{-2(3 - x)}{7}$

Solve each of the following.

21. $-3(2y - 5) + 5y < -4y + 2(7 - y)$, $y \in R$

22. $-(5 - 7k) - 3k \geq -8k + 3(1 - k)$, $k \in W$

23. $\dfrac{x}{4} - \dfrac{x}{3} \geq 4 - \dfrac{x - 2}{2}$, $x \in J$

24. $\dfrac{3a}{7} - \dfrac{2a}{7} > 4 + \dfrac{a - 4}{3}$, $a \in R$

25. $\dfrac{2p}{5} - \dfrac{1}{3}(p - 4) \leq \dfrac{4}{3}p - \dfrac{1}{5}(p + 2)$, $p \in W$

26. $\dfrac{t}{6} + \dfrac{1}{2}(3t - 1) \geq \dfrac{1}{3}t + \dfrac{3}{4}(2t - 5)$, $t \in N$

27. $2(6x - 5) - 3(3x + 2) < -(3x - 5) + 2x$, $x \in J$

28. $\dfrac{3p - 2}{4} - \dfrac{2p - 1}{3} < \dfrac{p}{12} - \dfrac{1}{4}$, $p \in R$

29. $\dfrac{3y + 7}{5} - \dfrac{4y + 1}{3} > \dfrac{y}{6} - \dfrac{1}{5}$, $y \in R$

30. $\dfrac{3(2x - 5)}{4} - \dfrac{2(5x - 1)}{3} \leq \dfrac{2x}{3} + \dfrac{1}{12}$, $x \in R$

Solve and graph each of the following. Assume all variables represent real numbers.

31. $\{x \mid x > 2\} \cap \{x \mid x < 5\}$

32. $\{y \mid y > -3\} \cap \{y \mid y \leq 5\}$

33. $\{p \mid p \leq 3\} \cup \{p \mid p > 1\}$

34. $\{m \mid m > 3\} \cup \{m \mid m \leq 2\}$

35. $\{k \mid -3 \leq k + 2 < 6\}$

36. $\{r \mid 4 < r + 4 \leq 5\}$

37. $\left\{ y \left| \dfrac{1 - 2y}{3} + \dfrac{3y - 2}{4} > \dfrac{4y - 1}{12} \right. \right\}$

38. $\left\{ q \left| \dfrac{7 - 2q}{4} - 3 \leq \dfrac{-7q}{18} - \dfrac{q + 5}{9} \right. \right\}$

39. $\{t \mid 3t - 7 > 3\} \cup \{t \mid -(t - 4) > 5\}$

40. $\{y \mid -(4y - 1) \leq -4\} \cup \{y \mid 2y + 6 \leq 3\}$

Use a calculator to solve each of the following. Round answers to one decimal place.

41. $\{x \mid 0.089x < 1.932, x \in R\}$

42. $\{y \mid -6.871y \geq 1.888, y \in R\}$

43. $\{p \mid 168.3 - 1.008p \leq 3.095, p \in R\}$

44. $\left\{q \;\middle|\; \dfrac{3.82q - 0.006}{1.284} \geq 6.81q + 5.000, q \in R\right\}$

45. $\left\{t \;\middle|\; \dfrac{3.8t - 6.9}{5.2} + 4.6 \geq \dfrac{1.84t - 0.01}{4.8}, t \in R\right\}$

46. $\left\{k \;\middle|\; \dfrac{-8.18k + 6.92}{0.034} - 1.1 \geq \dfrac{3.7k + 4.92}{1.82}, k \in R\right\}$

Translate each of the following into an inequality and solve. Interpret your answer in words.

47. The sum of a number and 5 more than the number is to be no more than 59.

48. Eight times a number is between -24 and 24.

49. One third of a number is between -5 and 7.

50. If Ruth earned 86 on her first test, what must she earn on her second test for her average to be at least 90?

51. The perimeter of a square must be less than 160 feet. What is the maximum length of a side in feet?

52. Marlene has no more than $3.45 in her purse for lunch. If her soft drink costs 75¢, what is the most she can pay for a sandwich?

53. Judy's boutique makes a profit provided her revenue, $R = 25n$, exceeds her cost, $C = 60 + 10n$. What is the smallest number n of units that must be sold in order to make a profit?

54. Jeff tries to keep his caloric intake at less than 3000 calories per day. If his food intake per day contains 2250 calories, what is the maximum number of full glasses of juice he can have (at 120 calories per glass) per day and still stay under the desired caloric intake?

55. Nett expects to have an average score on two tests of less than 90 and more than or equal to 80. If his score on the first test was 88, what score would he need to earn on the second test?

56. Tim would like to keep his average golf score between 65 and 75. If his scores on the first two rounds were 72 and 68, what should his score be on the third round?

2.4 Solving Linear Inequalities Involving Absolute Value

Inequalities such as $|x| < 1$ and $|x| > 1$ are encountered with reasonable frequency. They are nothing more than simple ways to write statements involving compound inequalities.

If we consider $|x| < 1$ with respect to the definition of absolute value, two cases result.

Case 1 If $x \geq 0$ then $|x| = x$. Thus $|x| < 1$ becomes $x < 1$. (See Figure 2.13.)

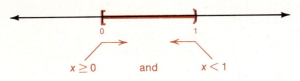

Figure 2.13

Case 2 If $x < 0$ then $|x| = -x$. Thus $|x| < 1$ becomes $-x < 1$ or $x > -1$. (See Figure 2.14.)

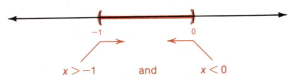

Figure 2.14

When these two cases are shown on the same number line, Figure 2.15, a simple interpretation results. $|x| < 1$ means that x is any real number whose distance from zero is less than 1.

Figure 2.15

Thus

$|x| < 1$ is equivalent to $-1 < x < 1$.

Now consider $|x| > 1$. Other than the points where $x = \pm 1$, the inequality $|x| > 1$ represents all the real numbers that are not solutions to $|x| < 1$. In other words, $|x| > 1$ is any real number whose distance from zero is greater than 1. (See Figure 2.16.)

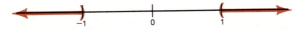

Figure 2.16

Thus,

$|x| > 1$ is equivalent to $x < -1$ *or* $x > 1$.

The results of these two examples can be generalized as follows.

Theorem 2.9

For any positive number p,

$|x| < p$ is equivalent to $-p < x < p$;

$|x| > p$ is equivalent to $x < -p$ or $x > p$.

▶ *Caution* $x > p$ cannot be written as $p < x < -p$ or $-p > x > p$. Such notation would imply that a positive number is less than a negative number (as in the first case) or that a negative number is greater than a positive number (as in the second case).

Theorem 2.9 is equally true when $<$ is replaced by $\leq$ and $>$ by $\geq$. The solution set would contain p and $-p$ when $\leq$ or $\geq$ are used.

Example 1 Solve and graph $|x| \leq 5$.
The inequality

$|x| \leq 5$

is equivalent to

$-5 \leq x \leq 5$.

The solution set is $[-5, 5]$. The graph is shown in Figure 2.17. ■

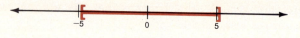

Since the symbol is $\leq$, the points 5 and -5 are in brackets

Figure 2.17

Example 2 Solve and graph $|x| \geq 4$.
The inequality

$|x| \geq 4$

is equivalent to

$x \leq -4$ or $x \geq 4$.

The solution set is $(-\infty, -4] \cup [4, \infty)$. The graph is shown in Figure 2.18. ■

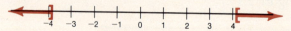

Figure 2.18

To solve an inequality such as $|a + 2| < 3$, think of $a + 2$ as the x in Theorem 2.9. Since $|x| < 3$ is equivalent to $-3 < x < 3$, the inequality $|a + 2| < 3$ is equivalent to $-3 < a + 2 < 3$.

Example 3   Solve and graph $|2a + 3| < 4$.
The inequality $|2a + 3| < 4$ is equivalent to $-4 < 2a + 3 < 4$.

$$-4 < 2a + 3 < 4$$
$$-4 - 3 < 2a + 3 - 3 < 4 - 3 \quad \text{Subtract 3 from each part}$$
$$-7 < 2a < 1 \quad \text{Combine like terms}$$
$$-\frac{7}{2} < a < \frac{1}{2} \quad \text{Divide each part by 2}$$

The solution set is $(\frac{-7}{2}, \frac{1}{2})$. The graph is shown in Figure 2.19. ■

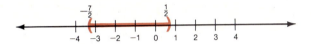

Figure 2.19

Example 4   Solve $|2m - 4| > 1$.
Again, think of $2m - 4$ as the x in Theorem 2.9. The inequality $|2m - 4| > 1$ is equivalent to

$$2m - 4 < -1 \quad \text{or} \quad 2m - 4 > 1$$
$$2m < 3 \quad \text{or} \quad 2m > 5 \quad \text{Add 4 to each side}$$
$$m < \frac{3}{2} \quad \text{or} \quad m > \frac{5}{2}. \quad \text{Divide each side by 2}$$

The solution set is $(-\infty, \frac{3}{2}) \cup (\frac{5}{2}, \infty)$. ■

Example 5   Solve $|x + 5| \leq 2$, $x \in J$.
The inequality $|x + 5| \leq 2$ is equivalent to $-2 \leq x + 5 \leq 2$.

$$-2 \leq x + 5 \leq 2$$
$$-7 \leq x \leq -3$$

Since x must be an integer, the solution set is

$$\{-7, -6, -5, -4, -3\} \quad \text{or} \quad \{x | -7 \leq x \leq -3, x \in J\}. \quad ■$$

Example 6   Solve $|2x + 3| < -5$.
The absolute value of any real number is always positive or zero. In order for $|2x + 3| < -5$ to be true, $|2x + 3|$ would have to be a negative number. This is not possible. The solution set is ∅. ■

Example 7   Solve $|2x + 3| > -5$.
The absolute value of any real number is always positive or zero. Every positive real number and zero are greater than -5; thus $|2x + 3|$ is positive or zero for any real number replacement of x. The solution set is $(-\infty, \infty)$. ■

Example 8   Write $-8 < x < 8$ using absolute value symbols.
              The inequality $-8 < x < 8$ is equivalent to $|x| < 8$ by Theorem 2.9. ∎

Example 9   Write $-7 \le a + 3 \le 7$ using absolute value symbols.
              Theorem 2.9 states that $|x| \le p$ is equivalent to $-p \le x \le p$. If we let
              $x = a + 3$, we have $|a + 3| \le 7$. ∎

Exercises 2.4

Rewrite each of the following as a compound inequality without absolute value symbols.

1. $|x| < 3$
2. $|y| < 5$
3. $|p| \le 4$
4. $|t| \le 0$
5. $|m| > 3$
6. $|q| > 0$
7. $|x - 5| \le 8$
8. $|3y + 4| \le 7$
9. $|2m - 1| \ge 4$
10. $|3r - 7| > 3$

Write each of the following using absolute value symbols.

11. $-2 < p < 2$
12. $-5 \le t \le 5$
13. $m < -3$ or $m > 3$
14. $y < -1$ or $y > 1$
15. $-6 \le x + 1 \le 6$
16. $-2 < 3r - 2 < 2$
17. $5q - 1 < -3$ or $5q - 1 > 3$
18. $1 - x \le -6$ or $1 - x \ge 6$
19. $-3 < p < 3$
20. $-6 \le p \le 6$
21. $-5 \le y - 3 \le 5$
22. $-8 < 2x + 3 < 8$
23. $m + 3 < -5$ or $m + 3 > 5$
24. $2r - 5 < -8$ or $2r - 5 > 8$
25. $6q - 2 < -4$ or $6q - 2 > 4$
26. $3p + 4 < -11$ or $3p + 4 > 11$

Solve and graph each of the following, if possible.

27. $|x| < 2$
28. $|y| < 3$
29. $|x| > 5,\ x \in J$
30. $|p| \ge 1,\ p \in J$
31. $|m| \le 0,\ m \in N$
32. $|r| \le -5,\ r \in N$
33. $|q| \le -3$
34. $|t| \ge -2$
35. $|p| \ge -3,\ p \in W$
36. $|m + 2| < 4,\ m \in W$
37. $|3y - 2| > 1,\ y \in J$
38. $|2r - 3| \ge 0,\ r \in J$
39. $|2t - 5| \le 3$
40. $|4m - 1| < -5$
41. $|5x + 1| \ge -3$
42. $\left|\dfrac{1}{2}p + 1\right| \le \dfrac{3}{2}$
43. $\left|\dfrac{1}{3}q - 2\right| < \dfrac{1}{4}$
44. $|3 - y| \ge -2,\ y \in J$
45. $-1 > |6 - t|,\ t \in J$
46. $1 + |3 - 4m| \le 3$
47. $6 + |2 - 7x| \ge 7$
48. $0 > |3 - 4h|,\ h \in W$
49. $-2 \ge |3 - 5k|,\ k \in W$
50. $2 > \left|5 - \dfrac{4}{3}y\right|,\ y \in J$
51. $|3u - 8| - 9 \ge -6,\ u \in J$
52. $|7t - 5| + 2 \le 4$
53. $|-3x + 1| + 3 \le 8,\ x \in W$
54. $|-7m - 5| - 1 \ge 3,\ m \in W$
55. $|6q + 5| - 4 < 0$
56. $|3h + 9| \ge -1$

Use a calculator to solve each of the following. Round answers to two decimal places.

57. $|0.385 + 1.84p| \le 3.298$
58. $|-2.151 - 6.82k| < 4.912$
59. $18.973 \le |3.041 + 6.009m|$
60. $75.01 \ge |2.11y - 6.878|$
61. $|-0.83x + 4.05| \le -8.018$
62. $|-11.11t + 0.99| \ge -3.05$

Translate each of the following into a mathematical sentence and solve.

63. The absolute value of a natural number is less than 8.

64. The absolute value of a negative integer is at least 1.

65. If $5.00 is added to twice the price of a pen, the absolute value of the result is no more than $12.00.

66. If 25¢ is added to the price of a dozen eggs, the absolute value of the sum is at least 95¢.

67. If $6.00 is subtracted from the price of a shirt, the absolute value of the difference is more than $8.00.

68. If 10% is deducted from a student's grade, the absolute value of the result is no less than 78%.

69. Show that $|a + b| = |a| + |b|$ for the negative real numbers $a = -3$ and $b = -7$.

70. Show that $|a + b| \neq |a| + |b|$ for the real numbers $a = -10$ and $b = 41$.

2.5   Literal Equations and Formulas

In areas such as statistics, physics, and business it is often useful to be able to solve a given equation or formula for a particular variable. This is done by using the addition, subtraction, multiplication, and division properties of equality that were developed in Section 2.1. For example, consider the common formula

distance = rate · time

$$d = r \cdot t.$$

As it stands, the formula allows you to consider the distance traveled if the rate (speed) and the time are known. However, it is often desirable to solve for one of the variables such as t in terms of r and d. To solve $d = rt$ for t, divide each side by r.

$$d = rt$$

$$\frac{d}{r} = \frac{rt}{r} \qquad \text{Division property of equality: divide each side by } r, r \neq 0$$

$$\frac{d}{r} = t$$

Thus

$$\text{time} = \frac{\text{distance}}{\text{rate}}.$$

Equations such as $d = rt$ are called **literal equations.** In addition to the areas mentioned at the beginning of this section, they are also found in many other fields such as chemistry, auto mechanics, electricity, and economics. The examples in this section show how to work with literal equations.

Example 1   The volume of a rectangular solid is $V = lwh$. Solve for h. (See Figure 2.20.)

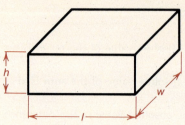

Figure 2.20

$$V = lwh$$

$$\frac{V}{lw} = \frac{lwh}{lw} \qquad \text{Division property: divide each side by } lw$$

$$\frac{V}{lw} = h$$

The height (h) of a rectangular solid is equal to the volume (V) divided by the product of the length and width (lw).   ■

Example 2   The volume of a cone is $V = \frac{1}{3}\pi r^2 h$. (See Figure 2.21.) Find a formula for the height of the cone in terms of the other dimensions.

Figure 2.21

First clear the equation of fractions by using the multiplication property of equality.

$$V = \frac{1}{3}\pi r^2 h$$

$$3V = 3 \cdot \frac{1}{3}\pi r^2 h \qquad \text{Multiply each side by 3}$$

$$3V = \pi r^2 h$$

Now use the division property to isolate h.

$$\frac{3V}{\pi r^2} = \frac{\pi r^2 h}{\pi r^2} \qquad \text{Divide each side by } \pi r^2$$

$$\frac{3V}{\pi r^2} = h \quad ■$$

Example 3   The formula $C = \frac{5}{9}(F - 32)$ is used to convert degrees Fahrenheit (°F) to degrees Celsius (°C). Use the properties of equality to find a formula for converting degrees Celsius to degrees Fahrenheit. Use the new formula to convert 100°C to °F.

$$C = \frac{5}{9}(F - 32)$$

$$\frac{9}{5}C = \frac{9}{5} \cdot \frac{5}{9}(F - 32) \qquad \text{Multiplication property}$$

$$\frac{9}{5}C = F - 32$$

$$\frac{9}{5}C + 32 = F - 32 + 32 \qquad \text{Addition property}$$

$$\frac{9}{5}C + 32 = F$$

$$\text{or} \qquad F = \frac{9}{5}C + 32$$

To convert 100°C to °F, substitute 100 for C in the formula.

$$F = \frac{9}{5}(100) + 32$$

$$F = 180 + 32$$

$$F = 212°$$

The temperature 100°C or 212°F is the boiling point of water at sea level. ∎

Example 4   A person's intelligence quotient is 100 times the person's mental age (M) divided by his or her chronological age (C). Thus

$$IQ = \frac{100M}{C}.$$

Find a formula that would predict the chronological age of a person whose mental age and intelligence quotient are known.

$$IQ = \frac{100M}{C}$$

$$C \cdot IQ = C \cdot \frac{100M}{C} \qquad \text{Multiply each side by } C$$

$$C \cdot IQ = 100M$$

$$\frac{C \cdot IQ}{IQ} = \frac{100M}{IQ} \qquad \text{Divide each side by IQ}$$

$$C = \frac{100M}{IQ}$$

A girl with an IQ of 140 and mental age of 14 years would be 10 years old.

$$C = \frac{100 \cdot M}{IQ} = \frac{100 \cdot 14}{140} = 10 \text{ years old} \quad \blacksquare$$

Example 5   The formula to determine the area of a trapezoid is $A = \frac{h}{2}(B + b)$. The h, B, and b are the dimensions shown in Figure 2.22. Solve the formula for b. Begin by multiplying each side by 2.

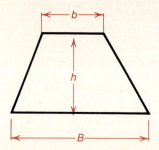

Figure 2.22

$$2 \cdot A = 2 \cdot \frac{h}{2}(B + b) \qquad \text{Multiply each side by 2}$$

$$2A = h(B + b)$$

$$\frac{2A}{h} = \frac{h(B + b)}{h} \qquad \text{Divide each side by } h$$

$$\frac{2A}{h} = B + b$$

$$\frac{2A}{h} - B = B + b - B \qquad \text{Subtract } B \text{ from each side}$$

$$\frac{2A}{h} - B = b \quad \blacksquare$$

An equivalent form is

$$\frac{2A - Bh}{h} = b,$$

which is obtained by collecting the left side over the LCD, h.

Example 6   Solve for x if $y = \frac{3x + 4}{2x - 2}$.

Begin by clearing the equation of fractions by multiplying each side by the LCD, $2x - 2$.

$$y = \frac{3x + 4}{2x - 2}$$

$$(2x - 2)y = 3x + 4$$

$$2xy - 2y = 3x + 4 \qquad \text{Distributive property}$$

Now isolate all terms involving x on one side of the equation.

$$2xy - 3x = 4 + 2y \qquad \text{Subtract 3x and add 2y to each side}$$
$$(2y - 3)x = 4 + 2y \qquad \text{Distributive property}$$
$$x = \frac{4 + 2y}{2y - 3} \qquad \text{Divide each side by 2y - 3} \quad \blacksquare$$

Exercises 2.5

Solve for the specified variable in each literal equation and simplify.

1. $I = prt$ for r (interest = principal · rate · time)
2. $A = \frac{sr}{2}$ for s $\left(\text{area of a circular sector} = \frac{\text{arc length} \cdot \text{radius}}{2}\right)$
3. $ax + b = c$ for a (linear equation in one variable)
4. $C = 2\pi r$ for r (circumference of a circle = 2π · radius)
5. $P = 2(L + W)$ for L (perimeter of a rectangle = 2(length + width))
6. $ax + by = c$ for y (linear equation in two variables)
7. $S = r\theta$ for θ (arc length = radius · measure of angle in radians)
8. $S = 2(lw + lh + wh)$ for w (surface area of a rectangular box)
9. $y = mx + b$ for m (slope-intercept form of the equation of a line)
10. $S = 2\pi rh$ for h (lateral surface area of a right circular cylinder)
11. $A = \frac{h}{2}(B + b)$ for B (area of a trapezoid)
12. $d = \frac{c}{\pi}$ for π (diameter of a circle)
13. $E = mC^2$ for m (energy = mass · speed of light squared)
14. $A = \frac{1}{2}bh$ for b (area of a triangle)
15. $D = \frac{1}{2}gt^2$ for g (falling body)
16. $A = \frac{1}{2}bh$ for h (area of a triangle)
17. $A = P(1 + rt)$ for t (amount of money accumulated if P dollars are invested at r rate for t years)
18. $h = vt - 16t^2$ for v (height of a projectile at time t)
19. $\frac{1}{f} = \frac{1}{a} + \frac{1}{b}$ for f (optics, focal length)
20. $V = \pi r^2 h$ for h (volume of a right circular cylinder of radius r and height h)
21. $\frac{P_1 V_1}{T_1} = \frac{P_2 V_2}{T_2}$ for T_1 (gas law—pressure, volume, temperature)
22. $E = IR$ for I (electrical circuits, Ohm's law)
23. $C = \frac{100B}{L}$ for L (anthropology—cephalic index)
24. $F = \frac{GM_1 M_2}{d^2}$ for G (force of gravitational pull between two masses)

25. $\dfrac{1}{R} = \dfrac{1}{R_1} + \dfrac{1}{R_2}$ for R (electrical circuits)

26. $\dfrac{12dS}{w} = CD$ for w (cutting speed of a saw)

27. $D = \dfrac{11}{5}(P - 15)$ for P (pressure is related to the depth in feet below the surface of the ocean)

28. $H = \dfrac{62.4NS}{33,000}$ for N (horsepower H generated by N cubic feet of water flowing over a dam S feet high)

29. $M = \dfrac{f_o}{f_e}$ for f_e (magnifying power of a telescope)

30. $K = \dfrac{1}{2}MV^2$ for M (kinetic energy)

31. $F = \dfrac{9}{5}C + 32$ for C (temperature conversion)

32. $A = 2\pi r^2 + 2\pi rh$ for h (surface area of a right circular cylinder)

33. $A = S(1 - DN)$ for N (proceeds of a discounted loan)

34. $y - y_1 = m(x - x_1)$ for x (point-slope form of the equation of a line)

35. $A = \dfrac{1}{2}nal$ for n (lateral surface area of a regular pyramid with slant height l, length of one side of the base a and n sides)

36. $V = \dfrac{1}{3}\pi h^2(3r - h)$ for r (volume of a spherical segment)

37. $V = \dfrac{1}{3}\pi r^2 h$ for r^2 (volume of a right circular cone)

38. $V = \dfrac{1}{6}H(S_0 + 4S_1 + S_2)$ for S_2 (prismoidal formula)

39. $A = \dfrac{1}{3}h(y_0 + 4y_1 + y_2)$ for h (Simpson's rule)

40. $y^2 = 2px$ for p (equation of a parabola)

Solve for x in each equation. (Hint: In Exercises 46–48 use the distributive property.)

41. $3x + y = 4$

42. $\dfrac{1}{2}x - 2y = 7$

43. $y = mx + b$

44. $y - y_1 = m(x - x_1)$

45. $m = \dfrac{y - y_1}{x - x_1}$

46. $y = \dfrac{2x - 3}{x + 4}$

47. $y = \dfrac{5x - 7}{2x + 4}$

48. $y = \dfrac{a - xL}{1 - x}$

Evaluate each equation for the remaining variable.

49. $I = prt$; $I = 240$, $p = 1000$, $t = 2$

50. $V = lwh$; $V = 800$, $l = 20$, $h = 10$

51. $A = \dfrac{h}{2}(B + b)$; $A = 100$, $h = 2$, $B = 5$

52. $I = \dfrac{PN}{(RN + A)}$; $P = 400$, $N = 10$, $R = 20$, $A = 5$

53. $\dfrac{1}{R} = \dfrac{1}{R_1} + \dfrac{1}{R_2}$; $R_1 = 40$, $R_2 = 20$

54. $a_n = a_1 + (n - 1)d$; $d = 2$, $n = 15$, $a_n = 60$

55. $\dfrac{1}{f} = \dfrac{1}{a} + \dfrac{1}{b}$; $a = 20$ cm, $b = 60$ cm

56. $s = \dfrac{a}{1 - r}$; $a = 10$, $s = 20$

Use a calculator to evaluate each of the following. Find an appropriate formula from Exercises 1–40. Round answers to one decimal place, if necessary.

57. Find the surface area of a rectangular box with length of 1.3 meters, width of 0.7 meter, and height of 1.1 meters.

58. Find the circumference of a circle with radius of 1.68. Use 3.14 for π.

59. Find the a_n term of an arithmetic sequence if the first term a_1 is 16, the number of terms n is 5 and the difference d is 2. (See Exercise 54.)

60. Find the height of a right circular cylinder with volume V of 300 cubic inches and radius r of 1.5 inches. Use 3.1416 for π.

61. Find the temperature in Fahrenheit (°F) if the temperature in Celsius (°C) is 48.6°.

62. Find the force of the gravitational pull F between two masses of 28.3 and 51.9 if the distance d between them is 6.75, and $G = 62.4$.

63. Find the pressure P on an object at a depth D of 111.5 feet below the surface of the ocean.

64. Find the amount A of money accumulated if $1000 is invested at 13% for two and one half years.

65. Find the height h of a projectile if its velocity V is 60 ft/second when $t = 2$ seconds.

66. Find the number of sides of a regular pyramid with lateral surface area = 31.11 cm^2, length of one side of the base = 6.1 cm and slant height = 1.7 cm.

67. Find the volume of a right circular cone with $h = 4$ ft, $r = 3$ ft. Use $\pi = 3.14$.

68. Find the surface area of a right circular cylinder with $r = 5$ inches, $h = 2$ inches and use $\pi = 3.14$.

2.6 More Applications

Applied problems are an essential part of a course in intermediate algebra. Most students find them easier to handle once they have developed a formal plan of attack to use in their solutions, such as the plan outlined below.

To Solve Applied Problems

Step 1   Read the problem through completely as many times as necessary, until you understand what the problem is about.

Step 2   Re-read the problem and determine what you are asked to find.

Step 3   Represent what you are asked to find by a variable or variables.

Step 4   Establish the relationship between the variable and other parts of the problem. The relationship may be an equation or an inequality.

Step 5   Solve the equation or inequality and check your result in the words of the original problem.

Example 1 **Step 1** Read completely.

Patty is one year older than Ron, and the sum of their ages is 99 years. How old is each?

Step 2 What are you asked to find?

Patty is 1 year older than Ron and the sum of their ages is 99 years.   **How old is each?**

Step 3 Represent the unknown quantities.

Let x = Ron's age

Then $x + 1$ = Patty's age.   One year older than Ron

Step 4 Establish the relationship.

Patty's age + Ron's age = 99
$(x + 1)$ + x = 99

Step 5 Solve the equation and check.

$$(x + 1) + x = 99$$
$$2x + 1 = 99$$
$$2x = 98$$
$$x = 49$$

Patty is 50. Ron is 49. The sum of their ages is 99. Since $50 + 49 = 99$ and 50 is 1 greater than 49, the solution checks. ■

Example 2 **Step 1** Read completely.

A student has test scores of 70 and 83. To obtain an average of at least 80, what score must she make on her next test?

Step 2 What are you asked to find?

A student has test scores of 70 and 83. To obtain an average of at least 80, **What score must she make on her next test?**

Step 3 Represent the unknown quantity.

Let s represent her next test score.

Step 4 Establish the relationship.

An average score is the sum of the scores divided by the number of scores. We want this average to be at least 80.

$$\frac{70 + 83 + s}{3} \geq 80$$

Step 5 Solve the equation and check.

$$\frac{70 + 83 + s}{3} \geq 80$$

$$70 + 83 + s \geq 80 \cdot 3 \qquad \text{Multiply each side by 3}$$

$$153 + s \geq 240$$

$$x \geq 87$$

We will check using only one value, 87. If 87 will produce an average of at least 80, then any score above 87 also will produce an average of at least 80.

$$\frac{70 + 83 + 87}{3} \overset{?}{\geq} 80$$

$$\frac{240}{3} \overset{?}{\geq} 80$$

$$80 \overset{\checkmark}{\geq} 80 \quad \blacksquare$$

Example 3 The sum of 3 consecutive odd integers is 111. Find the integers.

Consecutive odd or even integers differ by 2.

Let x = the smallest odd integer
$x + 2$ = the middle odd integer
$x + 4$ = the largest odd integer.

Since their sum is 111, we have

$$x + (x + 2) + (x + 4) = 111$$

$$3x + 6 = 111$$

$$3x = 105$$

$$x = \ \ 35$$

$$x + 2 = \ \ 37$$

$$x + 4 = \ \ 39.$$

The consecutive odd integers are 35, 37, and 39. The check is left to the reader. ◼

The next two examples are **mixture problems.** In a mixture problem two or more quantities are combined to form a final mixture. The key to solving a mixture problem is the determination of what remains constant throughout the problem.

Example 4 The owner of the Nut Shop received a new shipment of peanuts and cashews. Rather than sell them separately, she decides to create a mixture to sell for $2.30 a pound. If the peanuts would normally sell for $1.50 a pound and the cashews for $3.50 a pound, how many pounds of each should she use if she wants to have 30 pounds of the mixture?

In this example the money received by selling the mixture must be the same as the amount that would be received by selling the varieties separately. Begin by letting x represent the number of pounds of peanuts in the mixture. Since the mixture contains both peanuts and cashews, the number of pounds of cashews will be the number of pounds in the mixture minus the number of pounds of peanuts in the mixture. This would be represented as $30 - x$.

Type	Number of Pounds	Price Per Pound	Cash Value
Peanuts	x	$1.50	$1.50x$
Cashews	$30 - x$	$3.50	$3.50(30 - x)$
Mixture	30	$2.30	$2.30(30)$

Cash value of peanuts	+	Cash value of cashews	=	Cash value of mixture
$1.50x$	+	$3.50(30 - x)$	=	$2.30(30)$

$$1.50x + 3.50(30 - x) = 2.30(30)$$
$$15x + 35(30 - x) = 23(30) \quad \text{Multiply each side by}$$
$$15x + 1050 - 35x = 690 \quad \text{10 to eliminate decimals}$$
$$-20x = -360$$
$$x = 18$$
$$30 - x = 12$$

The mixture should contain 18 pounds of peanuts and 12 pounds of cashews. The check is left to the reader. ■

Example 5   Ron McGurk had a very successful year as a salesman and received a $20,000 bonus. He invested part of it in a second mortgage paying 20% and the rest in municipal bonds paying 12%. Both of the investments pay simple interest and the total yearly income is $3,260. How much did he invest at each rate?

Recall that simple interest is calculated by use of the formula $I = prt$. This is a mixture problem involving investments.

$I = prt$

Amount invested   Interest rate   Time for investment

Let x represent the amount invested at 20% then the remainder of the bonus, $20,000 - x$, will be invested at 12%.

Investment Medium	p = Amount Invested	r = Interest Rate	t = Time in Years	I = Interest
Mortgage	x	20% = 0.20	1	$0.20 \times (1)$
Bonds	$20,000 - x$	12% = 0.12	1	$0.12(20,000 - x)(1)$

The total interest received is $3260. Thus

Mortgage interest   +   Bond interest   =   Total interest

$$0.20x \quad + \quad 0.12(20{,}000 - x) \quad = \quad 3260.$$

$$100[0.20x + 0.12(20{,}000 - x)] = 100 \cdot 3260 \qquad \text{Multiply by 100 to eliminate decimals}$$

$$20x + 12(20{,}000 - x) = 326{,}000$$

$$20x + 240{,}000 - 12x = 326{,}000$$

$$8x = 86{,}000$$

$$x = 10{,}750$$

$$20{,}000 - x = 9{,}250$$

Ron invested $10,750 in the second mortgage and $9,250 in municipal bonds. ■

In Section 2.5 the formula $d = r \cdot t$, distance = rate · time, was introduced. Example 6 shows one way it can be used to solve applied problems.

Example 6   Two joggers are headed toward each other on the same road. One jogs at 5 miles per hour, while the other jogs at 8 miles per hour. If they start 6.5 miles apart, how long will it take them to meet? After passing, how long will it be before they are 9 miles apart?

Let t = time it takes for them to meet. Since $d = r \cdot t$, the first jogger travels $5t$ miles and the second jogger $8t$ miles. When they meet the sum of these distances will be 6.5 miles.

$$5t + 8t = 6.5$$

$$13t = 6.5$$

$$t = \frac{6.5}{13}$$

$$t = \frac{1}{2}$$

They will meet in one-half hour. After they pass and the sum of the distances traveled is 9 miles, solve for time t as follows.

$$5t + 8t = 9$$

$$13t = 9$$

$$t = \frac{9}{13}$$

They will be 9 miles apart $\frac{9}{13}$ hours after they have met. ■

Exercises 2.6

Write a mathematical sentence for each of the following, solve, and check your answer.

1. Find three consecutive integers whose sum is 168.

2. Find two consecutive even integers such that 8 times the smaller equals 6 times the larger.

3. One-half of a natural number plus one-third of the next consecutive natural number is 12. Find the number.

4. Four-thirds of an even integer less one-half the next consecutive even integer is 9. Find the integers.

5. The average size of the three angles of a triangle is 60°. If one angle is 40° and the second is 20° larger, what is the measure of the third angle?

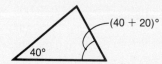

6. Lori earned 79, 83, and 80 on her first three tests. What is the least she can earn on the fourth test for her average to be at least 81?

7. Steve earned $272, $310, $364 and $284 in 4 consecutive weeks. What is the most he can earn in the fifth week if his average earning is to be no more than $276?

8. The average length of the four sides of a trapezoid is 13.5 inches. If the second side is 3 inches longer than the first, the third side is 6 inches longer than the first, and the fourth side is twice as long as the first, find the length of each of the sides.

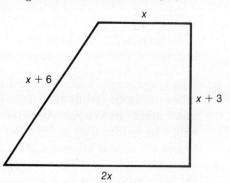

9. Bill is now four times as old as his son. Six years ago he was five times as old as his son was then. Determine his son's age now.

10. Sam is two years older than three times his daughter's age now. Twice the sum of their ages is 68. Find their ages.

11. The length of a rectangular backyard is three meters less than twice the width. If the perimeter is 66 meters, find the dimensions.

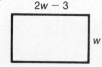

12. Find the dimensions of a rectangle with a perimeter of 112 centimeters if its width is 10 centimeters less than three-eighths its length.

13. If one side of a triangle is $\frac{1}{3}$ the perimeter, P, the second side is 5 feet long, and the third side is $\frac{1}{4}$ the perimeter, what is the perimeter?

14. A telephone pole is located in a pond. If one-eighth of the height of the pole, H, is in cement, 3 meters in water, and one-half of the height above the water, what is the total height of the pole?

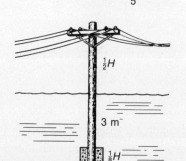

15. One hundred tickets were sold for the school play for a total of $333.00. If the tickets sold for $3.00 and $4.50, how many of each kind were sold?

16. Cindy wishes to realize a return of 11% on her total investments. If she has $10,000 invested at 9%, how much additional money should she invest at 15%?

17. Les invested $25,000 in two lots last year. He made a profit of 15% on one lot but lost 5% on the second lot. If his profit was equivalent to a return of 8% on the total investment, how much was invested in each lot?

18. Tom deals in antique cars. He bought two cars for a total of $6,000. He sold one of the cars for a profit of 25%, but lost 10% on the second car. If his gross profit was $940, how much was invested in each car?

19. It takes you $\frac{3}{5}$ hour to ride your bicycle to school when you ride at 15 miles per hour for part of the way and 10 miles per hour the rest of the way. If the school is seven miles from your house, how many minutes did you travel at each rate?

20. Two bicyclists leave Cuesta College at the same time, traveling in opposite directions. One travels at 17 miles per hour and the other at 13 miles per hour. How long does it take for them to become 60 miles apart?

21. A coffee shop blends a $5.00 per pound coffee with a $6.50 per pound coffee to produce a blend to sell for $6.00 per pound. How many pounds of each should be used to obtain 75 pounds of the new blend?

22. Dee's Meat Market mixes 600 pounds of sausage to sell for 93¢ a pound. How many pounds of $1.29 a pound lean hamburger meat should be mixed with other meat at 69¢ a pound to make up the sausage?

23. A 30-pound weight (W_1) is 3 feet from the fulcrum on the left side of a balance beam. What weight (W_2) at a distance of 6 feet on the right side of the fulcrum will balance it? For the weights to balance $W_1 \cdot d_1 = W_2 \cdot d_2$.

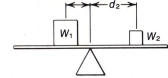

24. A 24-gram weight is placed 3 centimeters farther from the fulcrum on one side of a balance beam than a 36-gram weight on the other side. How far is each weight from the fulcrum?

25. Where should the fulcrum of a 9-foot beam be placed so that a 180-pound man will balance a 540-pound weight?

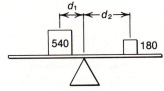

26. Where should the fulcrum of a 3-meter beam be placed so that a 18-kilogram weight would balance a 51-kilogram weight?

27. How many liters of a 30% salt solution must be added to 40 liters of a 12% salt solution to obtain a 20% solution?

28. How many centiliters of pure alcohol must be added to 35 centiliters of 20% solution to obtain a 40% solution?

29. How many quarts of 40% antifreeze solution must be drained from a 12-quart radiator and replaced with pure antifreeze to obtain a 50% solution?

30. How many liters of a 30% chemical solution should be drained from a 60-liter tank and replaced with a 80% chemical solution to obtain a 50% chemical solution?

Chapter 2   Summary

[2.1]   A **linear equation** in one variable is one that can be written in the form $ax + b = c$, a, b, $c \in R$ with $a \neq 0$. A **solution** to a linear equation is the value that makes it true, and the set of all values that make it true is called the **solution set. Equivalent equations** are ones with the same solution set.

Linear equations are solved by applying the properties of equality.

Addition Property of Equality	If $a = b$, then $a + c = b + c$.
Subtraction Property of Equality	If $a = b$, then $a - c = b - c$.
Multiplication Property of Equality	If $a = b$, then $a \cdot c = b \cdot c$.
Division Property of Equality	If $a = b$, then $\dfrac{a}{c} = \dfrac{b}{c}$, $c \neq 0$.

Linear equations are solved by applying a series of six steps.

1. Note any restrictions on the variable.
2. Clear the equation of fractions by using the multiplication property.
3. Remove symbols of grouping by using the distributive property.
4. Combine like terms on each side.
5. Isolate the variable by using the addition or subtraction property.
6. Use the multiplication or division property to solve for the variable.

An equation that is true for any permissible replacement for the variable is called an **identity.** One that is true only for specified values is called a **conditional equation.**

[2.2]   Equations involving **absolute value** are solved by writing them in equivalent forms free of absolute value.

$$|x| = p, \ p > 0$$

is equivalent to

$$x = p \quad \text{or} \quad x = -p.$$

[2.4]   In a similar manner, inequalities involving absolute value are solved by writing them in equivalent forms.

$$|x| < p, \ p > 0 \quad \text{is equivalent to} \quad -p < x < p$$

$$|x| > p, \ p > 0 \quad \text{is equivalent to} \quad x < -p \quad \text{or} \quad x > p$$

If $|x| > -p$, $p > 0$, the solution set is the set of all real numbers, and if $|x| < -p$, $p \geq 0$, the solution set is empty.

[2.3]   **Linear inequalities** exhibit properties similar to the properties of equality.

1. If $a < b$, then $a + c < b + c$.
2. If $a < b$, then $a - c < b - c$.

3. If $a < b$, then $a \cdot c < b \cdot c$ if c is positive.

4. If $a < b$, then $a \cdot c > b \cdot c$ if c is negative.

5. If $a < b$, then $\dfrac{a}{c} < \dfrac{b}{c}$ if c is positive.

6. If $a < b$, then $\dfrac{a}{c} > \dfrac{b}{c}$ if c is negative.

[2.5]   Equations such as $d = rt$, $v = \frac{1}{2}bh$ and $A = LW$ are known as **literal equations.** Literal equations often occur as well-known formulas that can be solved for one of the variables by using the techniques of this chapter.

Applied problems are stated in a verbal format. To solve such a problem the verbal statement must be translated into a mathematical equation or inequality.

Chapters 1 and 2   Review

Complete each of the following.

1. A _____ is a well-defined collection of objects.

2. $\in$ is the symbol for _____ .

3. $\subseteq$ is the symbol for _____ .

4. { } are the symbols for _____ .

5. $\{x \,|\, x < 3,\ x \in J\}$ is an example of _____ notation.

6. $\cup$ is the symbol for _____ of sets.

7. $\varnothing$ is the symbol for an _____ set.

8. If $A \cap B = B$, A and B are said to be _____ sets.

9. If whenever $x \in A$ then $x \in B$, A _____ B.

10. The set $\{x, y, z\}$ has _____ subsets.

11. W is used to represent the set of _____ numbers.

12. For all $a \in R$, $-(-a) = $ _____ .

13. $|-6| = $ _____ .

14. $\{ \ldots\ -3, -2, -1, 0, 1, 2, 3,\ \ldots \}$ is called the set of _____ .

15. $\left\{ \dfrac{a}{b} \,\middle|\, a, b \in J,\ b \neq 0 \right\}$ is called the set of _____ numbers.

16. $\dfrac{2}{7}$ in decimal form is _____ .

17. is the graph of $x \in$ _____ .

18. is the graph of _____ .

19. $x + (3 + y) = (x + 3) + y$ is an example of the _____ property.

20. x and $-x$ are additive _____ of each other.

21. $\dfrac{3}{a}$ and $\dfrac{a}{3}$ are multiplicative _____ of each other.

22. The multiplicative identity is _____ .

23. The multiplication property states that if a, b, and $c \in R$ and $a = b$, then _____ .

24. By the definition of subtraction $a - b =$ _____ .

25. The solution for $3x - 5 = 8$ is _____ .

26. The solution set for $|p| = 4$ is _____ .

27. The solution set for $|t - 3| \leq 3$ is _____ .

28. The solution for $\dfrac{1}{y - 5} + 3 = \dfrac{2}{y - 5}$ is _____ .

29. The sign of the product of three negative numbers is _____ .

30. Simplified, $\dfrac{3(x - 5) - 2(x + 3)}{15 + (-8)}$ is _____ .

31. The translation of "twice the difference of a number and 6" is _____ .

32. The restrictions on the variable in $\dfrac{1}{x - 2} + \dfrac{3}{x} = \dfrac{5}{x + 4}$ are _____ .

33. The solution set for $|x + 4| \leq -5$ is _____ .

34. $-4 < x + 2 < 4$ written in absolute value form is _____ .

35. The number of women out for sports is 8 more than the number of men. If there are 62 out for sports, how many are men?

36. If $A = 2(lw + lh + hw)$ and $l = 13$ in, $w = 4$ in, and $h = 6$ in, find A.

37. In $A = \dfrac{h}{2}(B + b)$, solve for B.

38. If $a < 0$, $b < 0$, the sign of $|a| + |b|$ is _____ .

39. The sum of Gary's and Mary's ages is 72. If Gary is four years older than Mary, determine each of their ages.

40. The sum of three consecutive numbers is 126. Find the numbers.

Solve and check each of the following. Note any restrictions on the variables. Graph all inequalities.

41. $x - 5 = 7$

42. $6p - 3 = 11$

43. $\dfrac{2}{t - 5} = -6$

44. $\dfrac{1}{r - 4} + 1 = \dfrac{1}{r - 4}$

45. $|m| = 6$

46. $|6a + 1| = 5$

47. $|2y - 1| = |y + 4|$

48. $|k| < 3$

49. $p + 3 \leq 4$, $p \in W$

50. $\dfrac{3}{4}x \geq -2$

51. $\dfrac{2x + 7}{3} - \dfrac{3x - 1}{2} \geq 1$, $x \in J$

52. $|3r - 1| \geq -2$, $r \in N$

53. $|3y - 5| \leq 1$

54. $\dfrac{3y + 2}{3} + \dfrac{y - 1}{2} \geq 2$

Solve for the specified variable in each exercise.

55. $y = mx + b$ for x

56. $I = prt$ for t

57. $A = p(1 + rt)$ for r

58. $\dfrac{1}{f} = \dfrac{1}{a}$ for a

59. $y = \dfrac{4x - 5}{x + 7}$ for x

60. $s = \dfrac{a}{1 - r}$ for r

61. Rewrite $\{x \mid -5 \le x \le 5\}$ using absolute value.

62. Rewrite $\{y \mid y \le -5 \text{ or } y \ge 5\}$ using absolute value.

Translate each of the following into a mathematical sentence and solve.

63. The sum of three consecutive even numbers is 114. Find the numbers.

64. Thirty coins, some nickels and the rest quarters, are worth $4.50. How many nickels and quarters are there?

65. A 40-pound mixture of candy has a value of $153.00. Some of the candy sells for $3.75 a pound and the rest for $4.00 a pound. How many pounds of each are in the mixture?

3

Linear Equations and Inequalities in Two Variables

In Chapter 2 methods were developed for solving linear equations and inequalities in one variable. These methods were then used to solve applied problems. This chapter introduces linear equations and inequalities in two variables and shows how to use them to solve additional types of applied problems.

3.1 The Rectangular Coordinate System: Graphing Linear Equations in Two Variables

Consider the equation

$$y = 2x + 4.$$

The value of y is related to the value of x. When the value of one variable is known, the value of the other can be found by substitution.

Example 1 Given $x = 5$, find y for the equation $y = 2x + 4$.

$y = 2x + 4$
$y = 2 \cdot 5 + 4$ Substitute 5 for x
$y = 10 + 4$
$y = 14$ ∎

A **solution** to an equation in two variables consists of values for each variable that together make the equation true. One solution to Example 1 is $x = 5$, $y = 14$. This solution is written as the *ordered pair* (5, 14). An **ordered pair** consists of two numbers written in a specified order within parentheses. In an ordered pair the x-value is always given first and the y-value second.

A linear equation in two variables has infinitely many solutions. For each new value of x, a new ordered pair results.

Example 2   Find solutions for $y = 3x + 5$ for $x = 2$, $x = 4$, and $x = -2$.

$$y = 3x + 5 \qquad\qquad y = 3x + 5 \qquad\qquad y = 3x + 5$$
$$y = 3 \cdot 2 + 5 \qquad\qquad y = 3 \cdot 4 + 5 \qquad\qquad y = 3(-2) + 5$$
$$y = 11 \qquad\qquad y = 17 \qquad\qquad y = -1$$

The solutions are $(2, 11)$, $(4, 17)$, and $(-2, -1)$. ∎

The relationship between the x- and y-values in an equation can be visualized by plotting the ordered pairs that satisfy the equation, using a **rectangular (Cartesian) coordinate system.** This system was invented by French mathematician and philosopher René Descartes (1596–1650).

Descartes' system involves the construction of two perpendicular number lines, one horizontal and one vertical, whose point of intersection is called the **origin.** The area to the right and above the origin is considered the positive direction, while the area to the left and below the origin is considered the negative direction. The horizontal line is called the x-**axis** and the vertical line is called the y-**axis.** The two axes divide a plane into four **quadrants** that are numbered in a counterclockwise direction, as shown in Figure 3.1. The axes themselves are not a part of any quadrant.

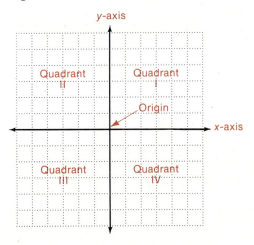

Figure 3.1

Every point in the plane can be described in terms of an ordered pair (x, y). The value for x indicates the distance to the right or left of the origin, while y indicates the distance above or below the origin. The x and y distances are known as the **coordinates** of the point. The coordinates of the origin are $(0, 0)$. In the point $(3, 4)$, 3 is the x-coordinate and 4 is the y-coordinate.

$$\underset{\text{x-coordinate} \qquad\qquad \text{y-coordinate}}{(3,\ 4)}$$

To locate the point (3, 4), move 3 units to the right of the origin and then move up 4 units. This point has x-coordinate 3 and y-coordinate 4. The point (3, 4) and several other points are shown in Figure 3.2.

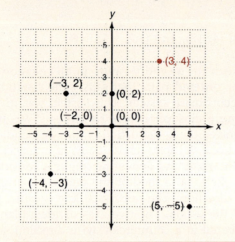

Figure 3.2

Various names are used to describe the x- and y-axes. The x-axis is often called the horizontal axis or **axis of abscissas.** The y-axis is called the vertical axis or **axis of ordinates.** Using the latter terminology would require that the coordinates of the point (x, y) be referred to as the **abscissa** and the **ordinate.**

$$(x, y)$$

Abscissa   Ordinate

We opened this section by observing that solutions to equations in two variables are ordered pairs. We now show how these ordered pairs relate to each other on a rectangular coordinate system.

Example 3   Find three ordered pairs that are solutions of the equation $2x + 3y = 6$.

Any real number can be chosen for x or y. Two of the easier choices are $x = 0$ and $y = 0$.

$$2x + 3y = 6$$
$$2 \cdot \mathbf{0} + 3y = 6 \qquad x = 0$$
$$3y = 6$$
$$y = 2$$

The ordered pair is (0, 2).

$$2x + 3y = 6$$
$$2x + 3 \cdot \mathbf{0} = 6 \qquad y = 0$$
$$2x = 6$$
$$x = 3$$

The ordered pair is (3, 0).

As a third choice, let $x = -3$.

$$2(-3) + 3y = 6 \qquad x = -3$$
$$-6 + 3y = 6$$
$$3y = 12$$
$$y = 4$$

The ordered pair is $(-3, 4)$. ■

When the three points found in Example 2 are graphed (plotted) on a rectangular coordinate system, all of them lie on the same straight line, as shown in Figure 3.3. The points where the graph crosses the axes are called the **x-** and **y-intercepts**. To find the x-intercept, let $y = 0$. To find the y-intercept, let $x = 0$. Note: All y-values are 0 on the x-axis. All x-values are 0 on the y-axis.

It will be shown in Section 3.3 that any equation that can be put in the form

$$ax + by = c$$

will have a graph that is a straight line. Equations whose graphs are straight lines are called *linear equations*.

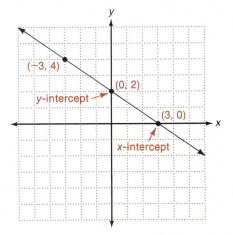

Figure 3.3

A **linear equation in two variables** is one that can be put in the form

$$ax + by = c$$

where a, b, and c are real numbers, a and b not both 0.

Only two points are needed to graph a straight line. Obtaining a third point provides a check on the accuracy of the other two.

Example 4   Graph $-2x + y = 4$ using the x- and y-intercepts together with a check point.

$$-2x + y = 4$$
$$-2x + 0 = 4 \qquad y = 0$$
$$x = -2 \qquad \text{x-intercept}$$

The ordered pair is $(-2, 0)$.

$$-2x + y = 4$$
$$-2 \cdot 0 + y = 4 \qquad x = 0$$
$$y = 4 \qquad \text{y-intercept}$$

The ordered pair is $(0, 4)$.

As a check point let $x = 1$.

$$-2x + y = 4$$
$$-2 \cdot 1 + y = 4 \qquad x = 1$$
$$-2 + y = 4$$
$$y = 6$$

The ordered pair is $(1, 6)$. The graph is shown in Figure 3.4.   ■

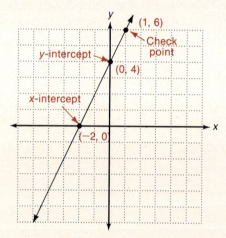

Figure 3.4

Example 5   Graph the lines $x = 2$ and $y = 1$ on the same coordinate system.

The equation $x = 2$ can be thought of as $x + 0y = 2$. The value of x must be 2 for any value of y. The three points $(2, -2)$, $(2, 0)$, and $(2, 4)$ meet the requirement.

Similarly, $y = 1$ can be thought of as $0x + y = 1$. The points $(-2, 1)$, $(0, 1)$, and $(2, 1)$ all lie on the line $y = 1$. The graphs of the lines $x = 2$ and $y = 1$ are shown in Figure 3.5.   ■

The graphs in Figure 3.5 illustrate that the graph of a line where x is a constant is a **vertical line.** The graph of a line where y is a constant is a **horizontal line.**

There are numerous variations involving the graphs of linear equations. One such variation occurs when absolute value becomes a part of the equation.

Example 6   Graph $y = |x + 2|$.

To graph this equation, consider the definition of absolute value together with three ordered pairs that are solutions to each part of the definition. If $x + 2 \geq 0$ then $x \geq -2$ and $|x + 2| = x + 2$. If $x + 2 < 0$, then $x < -2$ and $|x + 2| = -(x + 2)$. This leads to the following equations.

$$y = x + 2 \quad \text{if} \quad x \geq -2 \qquad \text{and} \qquad y = -(x + 2) \quad \text{if} \quad x < -2$$

Three ordered pairs satisfying each condition are

$$(-2, 0),\ (0, 2),\ (4, 6) \qquad \text{and} \qquad (-8, 6),\ (-6, 4),\ (-4, 2).$$

The graph is shown in Figure 3.6. Since $y = |x + 2|$ and $|x + 2|$ is positive or zero for any value of x, the graph will not extend below the x-axis. ∎

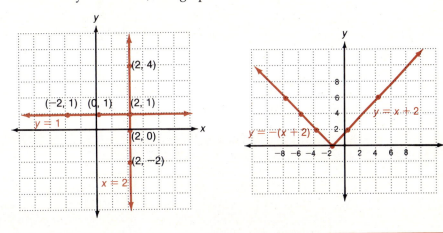

Figure 3.5   **Figure 3.6**

Example 7   A company manufactures n articles each day that are sold for $3.00 each. Costs associated with the manufacturing process are $25.00 daily. If p represents the daily profit and n represents the number of units manufactured, find the equation relating p to n and sketch its graph.

Profit = Income − Cost

Since n items are sold for $3.00 each, the income is $3n$. The cost is $25.00.

$$p = 3n - 25$$

The graph can be found by letting p be the vertical axis and n be the horizontal axis. Such a choice is arbitrary. We could have let n be the vertical axis and p be the horizontal axis. Three points are $(0, -25)$, $(10, 5)$, $(15, 20)$. The graph is shown in Figure 3.7. No negative values of n were used to construct the graph, since the company cannot manufacture a negative number of units. ■

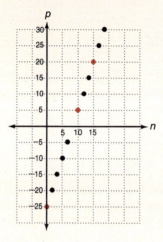

Figure 3.7

In Example 7 note that although a straight line would pass through the points satisfying the equation, none was drawn. Only the n-values that represent a completed item, such as $n = 1, 2, 3, \ldots, 25, \ldots$ would apply here.

The relationship between the solutions of a linear equation and the graph of the line can be described as follows.

> If a point **lies on the graph** of a line, its coordinates **satisfy** (are solutions to) the equation of the line. If the coordinates of a point **satisfy** the equation of a line, the point **lies on the line**.

In other words, a graph is a pictorial description of the solutions to the equation.

Example 8 Which of the following points lie on the line $2x - 3y = 6$?

(a) $(3, 0)$

$$2 \cdot 3 - 3 \cdot 0 \stackrel{?}{=} 6$$

$$6 \stackrel{\checkmark}{=} 6$$

The point $(3, 0)$ lies on the line.

(b) (5, 1)

$$2 \cdot 5 - 3 \cdot 1 \overset{?}{=} 6$$

$$7 \neq 6$$

The point (5, 1) does not lie on the line.

(c) $\left(\dfrac{1}{2}, -\dfrac{5}{3} \right)$

$$2\left(\dfrac{1}{2} \right) - 3\left(-\dfrac{5}{3} \right) \overset{?}{=} 6$$

$$6 \overset{\checkmark}{=} 6$$

The point $\left(\frac{1}{2}, -\frac{5}{3} \right)$ lies on the line.   ■

Exercises 3.1

State the quadrant in which each point is located.

1. (2, 3)

2. (4, 1)

3. (−2, −5)

4. (3, −7)

5. (−8, 3)

6. (−1, −1)

7. $(x, y); x < 0, y < 0$

8. $(x, y); x > 0, y < 0$

9. $(-x, y); x < 0, y > 0$

10. $(-x, -y); x < 0, y < 0$

Graph each equation by finding three ordered pairs.

11. $y = x + 1$

12. $y = x + 2$

13. $x - y = 5$

14. $2x - 3y = 6$

15. $-3x + 4y = 12$

16. $-x + 2y = 2$

17. $2x - 5y = -10$

18. $5x - 3y = -15$

19. $y = 6$

20. $y = -3$

21. $x = -4$

22. $x = 7$

23. $y = x$

24. $y = -x$

25. $x - 2y = 0$

26. $3x + y = 0$

Which of the following points lie on the given line?

27. $7x + y = 8$;   **(a)** (1, 1),   **(b)** (2, −6),   **(c)** (0, 8),   **(d)** (−1, −1)

28. $m - 3n = 6$;   **(a)** (0, 2),   **(b)** (0, −2),   **(c)** (6, 0),   **(d)** $\left(\dfrac{3}{2}, \dfrac{-9}{2} \right)$

29. $3x - 4y = 6$;   **(a)** (2, 0),   **(b)** $\left(0, \dfrac{3}{2} \right)$,   **(c)** $\left(1, \dfrac{-3}{4} \right)$,   **(d)** $\left(4, \dfrac{3}{2} \right)$

30. $-a + 3b = 4$;   **(a)** (1, 1),   **(b)** (2, 2),   **(c)** $\left(0, \dfrac{4}{3} \right)$,   **(d)** $\left(-5, \dfrac{-1}{3} \right)$

Graph each equation by using the x- and y- intercepts and one check point.

31. $x + 2y = 6$

32. $5x - y = 10$

33. $3x + y = 9$

34. $4x - y = 2$

35. $-x + 6y = 3$

36. $y = 2x$

37. $x - 4y = 0$

38. $x + 5y = 0$

39. $6x - y = 0$

40. $3x = -y$

41. $4x + 3y = 0$

42. $2x + 7y = 14$

43. $8x + y = 8$

44. $x - 9y = 3$

45. $-x + y = -4$

Find three ordered pairs for each of the following. Answers may vary.

46. $\{(x, y)\,|\,y = 2x\}$

47. $\{(a, b)\,|\,b = 3a\}$

48. $\{(m, n)\,|\,m - 5 = 2n\}$

49. $\{(r, t)\,|\,2r + 3 = 4t\}$

50. $\{(h, k)\,|\,3h - 4k = 7\}$

51. $\{(c, d)\,|\,4c - 5d = -2\}$

Translate each statement into an equation. Find three ordered pairs satisfying each equation and graph each line.

52. The y-value equals the x-value.

53. The x-value is three times the y-value.

54. Twice the y-value subtracted from three times the x-value is -6.

55. If five is subtracted from five times the y-value, the result is twice the x-value.

Graph each equation, using the given values to calculate the ordered pairs.

56. $y = |x|$ for $x = -3, -2, -1, 0, 1, 2, 3$

57. $y = -|x|$ for $x = -3, -2, -1, 0, 1, 2, 3$

58. $y = |x| - 2$ for $x = -3, -2, -1, 0, 1, 2, 3$

59. $y = |x| + 3$ for $x = -3, -2, -1, 0, 1, 2, 3$

60. $y = |x + 1|$ for $x = -4, -3, -2, -1, 0, 1, 2, 3$

61. $y = |x + 1|$ for $x = -4, -3, -2, -1, 0, 1, 2, 3$

Use a calculator to carry out the indicated operations. Round each answer to two decimal places.

62. Find the x- and y-intercepts of $6.92x - 3.76y = 2.39$.

63. Find the x- and y-intercepts of $0.697x - 3.125y + 1.61 = 0$.

64. Determine whether the ordered pair $(1.25, 0.80)$ satisfies the equation $3.84x + 5.21y = 8.97$.

65. Determine whether the ordered pair $(2.10, -1.13)$ satisfies the equation $-0.81x + 8.12y = 10.88$.

Translate each statement into an equation and answer parts (a)–(d).

66. Carl's transfer company bases the total charge to the customer on the following: $80 initial fee, $0.50 per mile of traveling distance and $45 per hour for loading and unloading time.

  (a) Find the cost C of a trip involving n miles of traveling and 6 hours of loading and unloading.

  (b) If $C = \$400$ and the loading and unloading time is 6 hours, find the number of miles n.

  (c) Find the total cost C of a 100-mile trip with a loading and unloading time of 3 hours.

  (d) Construct a graph to illustrate the cost of traveling 10 miles, 20 miles, 40 miles, 60 miles, and 80 miles if the loading and unloading time remains constant at 2 hours. Use the graph to estimate the cost of traveling 30 miles.

67. Berna's Flowers has a constant cost of $270 per week for rent and materials. Berna can make and sell n bouquets of flowers each week at $9.00 per bouquet. Assume her profit P is equal to her income less her cost.

  (a) What would her profit be if she sold 5 bouquets in one week?

  (b) How many bouquets would she have to sell to break even ($P = 0$)?

  (c) How many bouquets would she have to sell to earn a profit of $90.00 in one week?

  (d) How much money would she lose if she sold only 20 bouquets in one week?

3.2 The Slope of a Line

In the last section the graphs of many lines were sketched. In some cases the graph of the line rose when moving from left to right. In other cases the graph fell or was horizontal or vertical. When a number is assigned that measures the change between any two points on a line, that number is called the **slope** of the line.

Suppose that (x_1, y_1) and (x_2, y_2) are any two distinct points on a line, as shown in Figure 3.8. When we move from the first point (x_1, y_1) to the second point (x_2, y_2) the x-value changes from x_1 to x_2 and the y-value changes from y_1 to y_2. The amounts of change are $x_2 - x_1$ and $y_2 - y_1$.

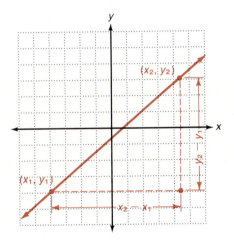

Figure 3.8

The ratio of the change in y to the change in x is called the slope of a line. The letter m is generally used to designate slope.

$$m = \text{slope} = \frac{\text{change in } y}{\text{change in } x} = \frac{y_2 - y_1}{x_2 - x_1}; \quad x_2 \neq x_1$$

Example 1 Find the slope of the line through $(2, 5)$ and $(7, 9)$.

To find the slope, let $(x_1, y_1) = (2, 5)$ and $(x_2, y_2) = (7, 9)$.

$$m = \frac{y_2 - y_1}{x_2 - x_1} = \frac{9 - 5}{7 - 2} \qquad \text{Substitute } y_2 = 9, \ y_1 = 5$$
$$\qquad\qquad\qquad\qquad\qquad x_2 = 7, \ x_1 = 2$$

$$m = \frac{4}{5} \quad \blacksquare$$

The choice of (x_1, y_1) and (x_2, y_2) in Example 1 was arbitrary. If we let $(x_1, y_1) = (7, 9)$ and $(x_2, y_2) = (2, 5)$, then

$$m = \frac{5 - 9}{2 - 7} = \frac{-4}{-5} = \frac{4}{5}.$$

The two results are the same.

Example 2   Find the slope of the line $x + y = 5$ and show the slope on its graph.

To find the slope, first determine two points that lie on the line. If we let $y = 2$, then $x + 2 = 5$ and $x = 3$. Similarly, when $x = 0$, $y = 5$. Use these two points to obtain the slope.

$$m = \frac{y_2 - y_1}{x_2 - x_1} = \frac{2 - 5}{3 - 0} \qquad (x_1, y_1) = (0, 5)$$
$$\qquad\qquad\qquad\qquad (x_2, y_2) = (3, 2)$$
$$m = \frac{-3}{3} = -1$$

The graph is shown in Figure 3.9.

Using two other points on the line, $(1, 4)$ and $(4, 1)$, gives

$$m = \frac{y_2 - y_1}{x_2 - x_1} = \frac{4 - 1}{1 - 4} = \frac{3}{-3} = -1. \quad \blacksquare$$

Any two points on a line can be used to determine its slope. In Figure 3.10, triangle *ABC* is similar to triangle *DEF*. Geometry tells us that the ratios of their corresponding sides are equal.

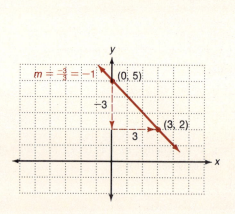

Figure 3.9

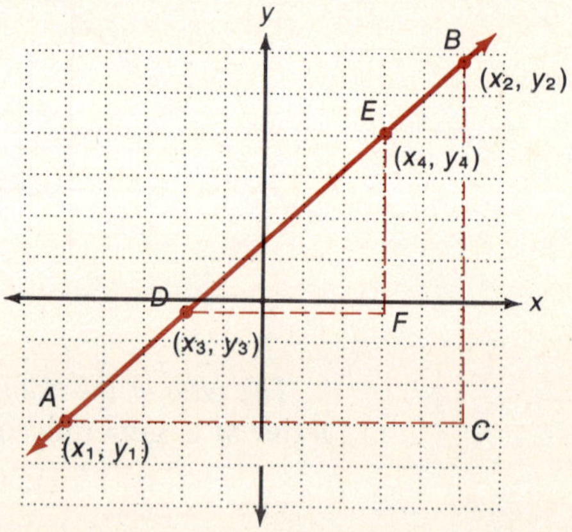

Figure 3.10
$$m = \frac{y_2 - y_1}{x_2 - x_1} = \frac{y_4 - y_3}{x_4 - x_3}$$

Example 3   Graph the line through $(-2, -3)$ with slope $\frac{2}{5}$.

To do this locate a second point by letting y change by 2 and x change by 5. Since 2 and 5 are positive, both changes will be in a positive direction from $(-2, -3)$. See Figure 3.11.

The slope $\frac{2}{5}$ can also be interpreted as $\frac{-2}{-5}$. To sketch the line using the slope $\frac{-2}{-5}$, start at the point $(-2, -3)$ and move 2 units down and 5 units to the left. Doing this gives a third point on the line, as shown in Figure 3.11. ■

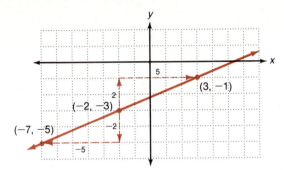

Figure 3.11
The second point has
x-coordinate $-2 + 5 = 3$ and
y-coordinate $-3 + 2 = -1$

Example 4   Graph the line through $(-2, 3)$ with slope $\frac{-4}{3}$.
      Find a second point on the line by letting y change by 4 units in the negative direction and x change by 3 units in the positive direction from $(-2, 3)$. The graph is shown in Figure 3.12. ■

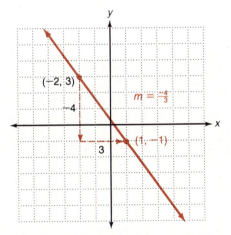

Figure 3.12
The second point has
x-coordinate $-2 + 3 = 1$ and
y-coordinate $3 + (-4) = -1$

Figure 3.11 illustrates that a line with positive slope **rises** from left to right, and Figure 3.12 illustrates that a line with negative slope **falls** from left to right.

If a line is horizontal, the y-coordinates of any two points on the line are equal. Thus

$$m = \frac{y_2 - y_1}{x_2 - x_1} = \frac{y_2 - y_2}{x_2 - x_1} = \frac{0}{x_2 - x_1} = 0. \qquad y_2 = y_1$$

A horizontal line has **zero slope.** On a vertical line the x coordinates of any two points are equal.

$$m = \frac{y_2 - y_1}{x_2 - x_2} = \frac{y_2 - y_1}{0} \qquad x_2 = x_1$$

Since division by zero is not defined, a vertical line is said to have **no slope** or **undefined slope.**

Example 5 Find the slope of each line.

(a) $x = 2$

Since the value of x is always 2, it is a vertical line. The line has **no slope.**

(b) $y = 1$

Since the value of y is always 1, it is a horizontal line. The slope of the line is **zero.**

The graphs of the lines $x = 2$ and $y = 1$ are shown in Figure 3.5. ■

In general, if k represents a constant, the line $x = k$ has no slope (undefined slope) while the line $y = k$ has zero slope.

*If two lines L_1 and L_2 are **parallel,** written as $L_1 \| L_2$, their slopes m_1 and m_2 are equal.* One way to establish that two lines are parallel is to compute their slopes by finding two points that satisfy the equation for each line.

Example 6 Show that L_1: $x + 2y = 6$ and L_2: $2x + 4y = 8$ are parallel.
Two points satisfying L_1 are (0, 3) and (6, 0). Thus

$$m_1 = \frac{0 - 3}{6 - 0} = \frac{-1}{2}.$$

Two points satisfying L_2 are (0, 2) and (4, 0). Thus

$$m_2 = \frac{0 - 2}{4 - 0} = \frac{-1}{2}.$$

Since $m_1 = m_2$, $L_1 \| L_2$. The graphs of L_1, and L_2 are shown in Figure 3.13. ■

*If two nonvertical lines L_1 and L_2 are **perpendicular** (meet at right angles), their slopes are negative reciprocals. If m_1 is the slope of L_1 and m_2 is the slope of L_2, then*

$$m_1 = \frac{-1}{m_2} \qquad or \qquad m_1 \cdot m_2 = -1.$$

Example 7   Show that the line L_1 through $(5, -2)$ and $(-2, 3)$ is perpendicular to line L_2 through $(-2, -1)$ and $(3, 6)$.

$$m_1 = \frac{y_2 - y_1}{x_2 - x_1} = \frac{3 - (-2)}{-2 - 5} = \frac{5}{-7} = \frac{-5}{7}$$

$$m_2 = \frac{y_2 - y_1}{x_2 - x_1} = \frac{6 - (-1)}{3 - (-2)} = \frac{7}{5}$$

$$m_1 \cdot m_2 = \left(\frac{-5}{7}\right)\left(\frac{7}{5}\right) = -1$$

Thus L_1 is perpendicular to L_2. Their graphs are shown in Figure 3.14. ■

Example 8   Use slopes to determine whether the points $A(-1, 2)$, $B(0, 4)$, and $C(2, 8)$ are **collinear,** that is, whether the points lie on the same line. If the points lie on the same line the slopes between any two ordered pairs must be the same.

$$m_{AB} = \frac{4 - 2}{0 - (-1)} = \frac{2}{1} = 2 \qquad m_{AB} = \text{slope from } A \text{ to } B$$

$$m_{BC} = \frac{8 - 4}{2 - 0} = \frac{4}{2} = 2 \qquad m_{BC} = \text{slope from } B \text{ to } C$$

The three points A, B, and C are collinear.   ■

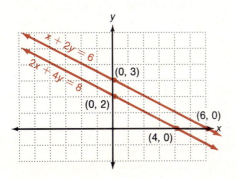

Figure 3.13

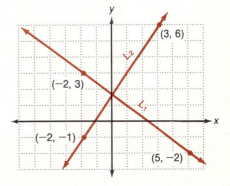

Figure 3.14

Exercises 3.2

Find the slope of the line joining each pair of points. Some may have no slope or undefined slope.

1. $(3, 4)$, $(6, 5)$ **2.** $(1, 5)$, $(3, 8)$ **3.** $(4, 1)$, $(3, 2)$

4. $(8, 5)$, $(6, 7)$ **5.** $(0, 0)$, $(4, -1)$ **6.** $(3, -2)$, $(0, 0)$

7. $(3, 9)$, $(4, 9)$ **8.** $(-5, 2)$, $(3, 2)$ **9.** $(4, -7)$, $(4, 8)$

10. $(-6, -1)$, $(-6, -2)$ **11.** $(-5, -3)$, $(-1, 0)$ **12.** $(-7, -7)$, $(-4, -1)$

13. $\left(\dfrac{1}{2}, \dfrac{4}{3}\right)$, $\left(\dfrac{3}{2}, \dfrac{1}{3}\right)$ **14.** $\left(\dfrac{-1}{5}, \dfrac{3}{8}\right)$ $\left(\dfrac{1}{4}, \dfrac{-1}{2}\right)$ **15.** (a, b), (c, d)

16. (r, s), (t, v) **17.** (a_1, b_1), (a_2, b_2) **18.** (r_2, s_2), (r_1, s_1)

Determine the slope of each linear equation. State whether the graph of the equation is a horizontal line, a vertical line, or whether it rises or falls.

19. $y = 6$ **20.** $y = -8$ **21.** $x = -4$ **22.** $x = 7$

23. $y = x + 2$ **24.** $y = x - 3$ **25.** $x + y = 1$ **26.** $2x + y = 5$

27. $y = x$ **28.** $x + y = 0$

Use slopes to determine whether the three points in each group are collinear.

29. $(3, -5)$, $(10, 7)$, $(5, 1)$ **30.** $(2, 1)$, $(0, 7)$, $(1, 3)$

31. $(3, 6)$, $(-2, -1)$, $(7, 8)$ **32.** $(0, 0)$, $(4, -2)$, $(10, 1)$

33. $\left(\dfrac{1}{2}, 1\right)$, $\left(\dfrac{1}{4}, 2\right)$, $(1, -1)$ **34.** $\left(\dfrac{5}{8}, \dfrac{1}{4}\right)$, $\left(\dfrac{3}{8}, \dfrac{1}{2}\right)$, $\left(\dfrac{7}{8}, -1\right)$

Determine whether the lines L_1 and L_2 are parallel, perpendicular, *or* neither *in Exercises 35–42 and graph each pair of lines.*

35. L_1 through $(1, 4)$, $(3, 6)$ **36.** L_1 through $(-7, 3)$, $(1, 11)$
      L_2 through $(-3, 1)$, $(2, 6)$ L_2 through $(-2, -5)$, $(3, 0)$

37. L_1: $x + 2y = 7$ **38.** L_1: $5x - 2y = 0$
      L_2: $2x - y = 5$ L_2: $2x - 3 = -5y$

39. L_1 through $(3, -5)$, $(-2, 4)$ **40.** L_1 through $\left(\dfrac{1}{2}, \dfrac{1}{4}\right)$, $(-2, -3)$
      L_2 through $(0, 6)$, $(-3, 1)$

                                       L_2 through $\left(\dfrac{-1}{3}, 2\right)$, $\left(4, \dfrac{-1}{5}\right)$

41. L_1: $3x - y = -9$ **42.** L_1: $8x - 4y = -1$
      L_2: $3y = -x + 2$ L_2: $2x + 4y = 5$

43. Prove that the four points $A(8, 4)$, $B(6, 1)$, $C(3, 4)$, and $D(1, 1)$ are vertices of a parallelogram by showing that the pairs of opposite sides are parallel.

44. Prove that the four points $A(-3, 5)$, $B(0, 8)$, $C(3, 6)$, and $D(-3, 0)$ are vertices of a trapezoid by showing that one pair of opposite sides is parallel.

45. Prove that the three points $A(-1, 2)$, $B(-3, 5)$, and $C(0, 7)$ are vertices of a right triangle by showing that two sides are perpendicular.

46. Prove that the four points $A(5, 8)$, $B(-3, 2)$, $C(0, -2)$, and $D(8, 4)$ are vertices of a rectangle by showing that the pairs of adjacent sides are perpendicular.

47. The ordinate of a point is 3; the line through it and the point $(-2, 5)$ is parallel to the line through the points $(8, -4)$ and $(-1, 2)$. Find the abscissa of the point.

48. The abscissa of a point is 2; the line through it and the point (4, 9) is perpendicular to the line through the points (6, 7) and (8, 0). Find the ordinate of the point.

49. The abscissa of a point is −3; the line through it and the point (2, −3) is perpendicular to the line through the points (0, 7) and (−3, 2). Find the ordinate of the point.

50. The ordinate of a point is $\frac{-1}{2}$; the line through it and the point $(-2, \frac{5}{2})$ is parallel to the line through the points $(-\frac{1}{3}, -4)$ and $(\frac{2}{3}, -1)$. Find the abscissa of the point.

Use a calculator to find the slope of the line joining each pair of points. Round each answer to three decimal places.

51. (3.823, 1.791), (0.121, −5.632)

52. (−8.174, 0.931), (3.147, −5.693)

53. (0.001, 8.543), (5.182, 3.077)

54. (−6.152, 3.514), (1.017, 9.001)

55. (−11.018, −2.019), (3.567, 2.981)

56. (−1.998, −4.113), (2.079, −0.025)

3.3 Solving Linear Systems of Equations by Graphing

A **system of linear equations** involves two or more linear equations. In this section we will examine several types of linear systems and concentrate our efforts on finding solutions to the systems. A **solution** to a system of equations is that set of values for the variables that are solutions to every equation in the system.

Suppose that you were asked to find two numbers whose sum is 10 and whose difference is 2. This problem can be solved by trial and error. A more systematic approach is to let x represent the larger number and y represent the smaller number.

$x + y = 10$ The sum of the numbers is 10

$x - y = 2$ The difference of the numbers is 2

When both equations are graphed on the same coordinate system, it appears that the lines intersect at (6, 4), as shown in Figure 3.15. To establish that (6, 4) is the solution to both equations we need to show that it satisfies each equation.

$$x + y = 10 \qquad\qquad x - y = 2$$
$$6 + 4 \overset{?}{=} 10 \qquad x = 6, y = 4 \qquad 6 - 4 \overset{?}{=} 2$$
$$10 \overset{\checkmark}{=} 10 \qquad\qquad 2 \overset{\checkmark}{=} 2$$

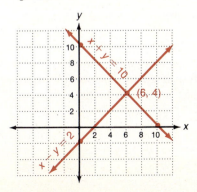

Figure 3.15

There are many other points that satisfy one equation but not the other. The ordered pair (6, 4) is the only solution of the system.

Example 1   Solve the system $2x + y = 6$, $3x - 2y = 2$ by graphing.

Recall that two of the easiest points to find are the intercepts. The intercepts and a check point for each equation are listed below.

$2x + y = 6$		$3x - 2y = 2$
(0, 6)	*y*-intercepts	(0, −1)
(3, 0)	*x*-intercepts	$\left(\dfrac{2}{3}, 0\right)$
(2, 2)	Check points	(4, 5)

When the two lines are graphed on the same coordinate system, the solution appears to be the ordered pair (2, 2). See Figure 3.16.

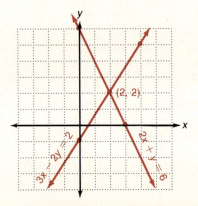

Figure 3.16

Check: $2x + y = 6$ $3x - 2y = 2$

$2 \cdot 2 + 2 \overset{?}{=} 6$ $x = 2, y = 2$ $3 \cdot 2 - 2 \cdot 2 \overset{?}{=} 2$ $x = 2, y = 2$

$6 \overset{\checkmark}{=} 6$ $2 \overset{\checkmark}{=} 2.$

Since the ordered pair (2, 2) satisfies both equations, it is the solution to the system. ■

The solution to a system of linear equations is the point of intersection of their graphs. Set notation is often used to describe the intersection. In the next example the notation is read as shown below.

$$\{(x, y) \quad | \quad 2x + 3y = 8\}$$

| The set of ordered pairs | such that | $2x + 3y = 8$ |

Recall that the symbol ∩ means intersection.

Example 2   Find the solution for

$$\{(x, y)\,|\,2x + 3y = 8\} \cap \{(x, y)\,|\,2x - 3y = -4\}$$

by graphing.

First, determine three points that lie on each line.

$$2x + 3y = 8 \qquad\qquad\qquad 2x - 3y = -4$$

$$\left(0, \frac{8}{3}\right),\ (4, 0),\ \left(2, \frac{4}{3}\right) \qquad \left(0, \frac{4}{3}\right),\ (-2, 0),\ \left(2, \frac{8}{3}\right)$$

The graphs are shown in Figure 3.17.

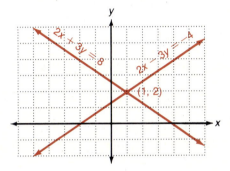

Figure 3.17

The solution appears to be the ordered pair (1, 2). Verify this observation by substitution.

$$2x + 3y = 8$$
$$2 \cdot 1 + 3 \cdot 2 \overset{?}{=} 8$$
$$8 \overset{\checkmark}{=} 8$$

$$2x - 3y = -4$$
$$2 \cdot 1 - 3 \cdot 2 \overset{?}{=} -4$$
$$-4 \overset{\checkmark}{=} -4$$

The solution set is {(1, 2)}. ■

A linear system consisting of two equations in two variables may not always have a unique solution. If the graphs of the two equations are parallel, the lines will not intersect. When this happens there is no point that is a solution to each equation. At the other extreme, the graphs of the two lines may coincide (one line traces over the other). When this happens any point on either line, of which there are an infinite number, is a solution to the system. A system of equations with a unique solution is said to be

independent [Figure 3.18(a)]. A system involving parallel lines is said to be **inconsistent** [Figure 3.18(b)] and one that involves lines that coincide is said to be **dependent** [Figure 3.18(c)].

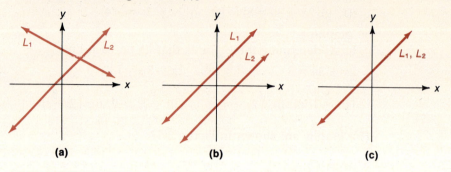

(a)   (b)   (c)

Figure 3.18
Unique solution   No solution   Infinitely many solutions
(independent)   (inconsistent)   (dependent)

Example 3   Solve the system L_1: $x + y = -2$, L_2: $x + y = 3$ by graphing.

Three ordered pairs that are solutions to $x + y = -2$ are $(0, -2)$, $(-2, 0)$, and $(1, -3)$. Similarly, three solutions to $x + y = 3$ are $(0, 3)$, $(3, 0)$, and $(2, 1)$. When the lines are graphed they appear to be parallel. (See Figure 3.19.)

To confirm that the lines are parallel their slopes can be found from two of the ordered pairs used to graph them.

$$m_1 = \frac{0 - (-2)}{-2 - 0} = \frac{2}{-2} = -1$$

$$m_2 = \frac{0 - 3}{3 - 0} = \frac{3}{-3} = -1$$

$$m_1 = m_2$$

The lines are parallel, and the system is inconsistent. ■

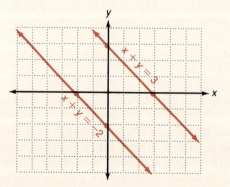

Figure 3.19

Example 4 Solve the system $2x - y = 2$, $4x - 2y = 4$ by graphing.

Three ordered pairs that are solutions to each equation are $(0, -2)$, $(1, 0)$, and $(2, 2)$. Since the same three ordered pairs satisfy each system, the lines will coincide. The system is dependent. (See Figure 3.20.) ■

Notice that when two lines coincide, one equation is a multiple of the other. Two times $2x - y = 2$ is $2(2x - y) = 2 \cdot 2$ or $4x - 2y = 4$. Any point that lies on the first line is a solution to both equations. An infinite number of solutions exist.

In general, a dependent system will result whenever one equation is a multiple of the other. This fact provides an easy way to recognize dependent systems.

Example 5 Lois and Larry both worked to clear the snow from their yard in Laramie. Lois worked twice as long as Larry and together they worked six hours. Use a graph to find how long each worked.

Let x = the number of hours Larry worked

y = number of hours Lois worked.

Since Lois worked twice as long as Larry, her time y is twice Larry's time x, or

$y = 2x.$

The sum of their working time is 6 hours. A second equation is

$x + y = 6.$

Three points satisfying $y = 2x$ are $(0, 0)$, $(1, 2)$, and $(3, 6)$. Three points satisfying $x + y = 6$ are $(0, 6)$, $(6, 0)$, and $(2, 4)$. The solution set is $\{(2, 4)\}$. See Figure 3.21.

Lois worked 4 hours and Larry worked 2 hours. Neither graph includes negative value of x or y, because it is impossible to work for a negative number of hours. ■

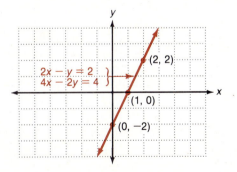

Figure 3.20

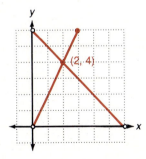

Figure 3.21

Exercises 3.3

Solve each system of equations by graphing. Indicate whether the system is independent, dependent *or* inconsistent. *Estimate answers if necessary.*

1. $x + y = 3$
    $2x + 2y = 6$

2. $x - y = 4$
    $3x - 3y = 12$

3. $x + y = 2$
    $x - y = 4$

4. $2x - y = 4$
    $3x - 2y = 6$

5. $3x - 2y = 6$
    $2x + 3y = -6$

6. $2x + 3y = 6$
    $x - 2y - -4$

7. $x + 3y = 3$
    $2x \quad 6y = 6$

8. $x + 7y = 8$
    $2x - y = 1$

9. $2x - y = 8$
    $x - 2y = -2$

10. $x = 3$
    $x = -4$

11. $3x = 3$
    $4y = 8$

12. $2x - 4y = 4$
    $5x - 10y = 10$

13. $2x - y = -3$
    $8x - 4y = -12$

14. $2x + 3y = 6$
    $x - y = 3$

15. $4x - y = 9$
    $x + y = 1$

16. $5x - 3y = 2$
    $-5x + 3y = -2$

17. $3x - 4y = 12$
    $-6x + 8y = 24$

18. $2x + y = 4$
    $x - y = 2$

19. $8x + 3y = 24$
    $4x - 8 = -y$

20. $5x = 1 - y$
    $3y = 3 - x$

21. $y = -2x + 2$
    $4x = 2y + 12$

22. $7x = 3y + 7$
    $x = 1 - 8y$

Solve each system by graphing.

23. $\{(x, y) \mid x = 2y + 1\} \cap \{(x, y) \mid 3y = x - 3\}$

24. $\{(x, y) \mid 1 + 3y = x\} \cap \{(x, y) \mid 2 - y = x\}$

25. $\{(x, y) \mid 1 + 2y = x\} \cap \{(x, y) \mid x - 8 = 2y\}$

26. $\{(x, y) \mid y + 5 = x\} \cap \{(x, y) \mid 5y + 5 = x\}$

27. $\{(x, y) \mid y + 7 = x\} \cap \{(x, y) \mid y + 1 = -x\}$

28. $\{(x, y) \mid 3y - x = -6\} \cap \{(x, y) \mid 3y + 6 = -2x\}$

Use a calculator to determine whether the given ordered pair satisfies the system of equations.

29. (4, 3); $3.1x - 1.1y = 9.1$
    $1.1x + 2.2y = 11$

30. (1.5, 0.5); $x - 2.5y = 34.5$
    $0.05x - 1.05y = 0.8$

31. (1.1, 0.9); $0.6x - 0.3y = -0.18$
    $0.8x + 2.0y = 2.16$

32. (1.1, 0.3); $0.6x - 1.5y = 0.21$
    $3.2x - 1.2y = 3.16$

Solve each system of equations by graphing. Label your axes by using the variables in the system.

33. Kyle's Appliances buys a total of 25 blenders from two different companies. The store buys 3 more of brand *A* than brand *B*. The number of each brand purchased can be found by solving the following system of equations.

    $A + B = 25$ Total of both brands is 25
    $A - B = 3$ *A* is 3 more than *B*

34. The velocity-time equation of Linda's car is $v = 5t - 20$ where *t* is the time in seconds and *v* is the velocity of the car in feet per second. The velocity-time equation of Barbara's car is $v = 2t + 12$ ft/sec. At what time do both cars have the same velocity? What is the velocity at that time?

35. The perimeter of a rectangle is 48 inches ($2l + 2w = 48$). If the width of the rectangle is doubled, the perimeter is 60 inches ($2l + 4w = 60$). Find the dimensions by solving the system of equations.

36. Three hamburgers and 2 orders of french fries cost $3.70 ($3h + 2f = 3.70$). Two hamburgers and 4 orders of french fries cost $3.80 ($2h + 4f = 3.80$). Find the cost of each by solving the system of equations.

3.4 Solving Linear Systems of Equations by Elimination or Substitution

In Section 3.3 graphs were used to solve linear systems of equations. Although the method allows one to visualize the solution, it has the drawback that the exact point of intersection may be impossible to determine. In this section two alternate means for solving such systems will be developed. They are known as the **elimination method** (or addition method) and the **substitution method.**

To use the *elimination method,* the *coefficients* of one of the variables must be additive inverses. In the term $3x$, 3 is the **numerical coefficient** or simply the **coefficient** of x.

Example 1   Solve the system $x + 2y = 7$, $x - 2y = 3$.

The coefficients of y are 2 and -2, which are additive inverses. If we add the left and right hand members of the two equations, y will be eliminated.

$$\begin{array}{rl}
\text{I} & x + 2y = 7 \\
\text{II} & \underline{x - 2y = 3} \\
& 2x \quad\;\; = 10 \qquad \text{Add I to II} \\
& x \quad\;\; = 5
\end{array}$$

Substituting the value of x into either of the original equations will determine y.

$$\begin{array}{ll}
x + 2y = 7 & x - 2y = 3 \\
5 + 2y = 7 \quad x = 5 & 5 - 2y = 3 \\
\quad\; 2y = 2 & \quad -2y = -2 \\
\quad\;\; y = 1 & \quad\quad y = 1
\end{array}$$

The solution set is $\{(5, 1)\}$. The ordered pair $(5, 1)$ represents the point of intersection of the graphs of the system. ■

If the coefficients of one variable are not additive inverses, then the multiplicative property of equality can be used to generate equivalent equations in which the coefficients are additive inverses. For example, $(3, 2)$ is a solution to $2x + 3y = 12$ since $2 \cdot 3 + 3 \cdot 2 = 12$. When both sides of the equation are multiplied by any constant, say k, an equivalent equation results. Recall that equivalent equations have the same solution set.

$$\begin{array}{ll}
2x + 3y = 12 & (3, 2) \text{ is a solution} \\
(2x + 3y)k = 12k & \text{Multiply each side by } k \\
(2 \cdot 3 + 3 \cdot 2)k = 12k & \\
12k = 12k & (3, 2) \text{ is still a solution}
\end{array}$$

Example 2   Solve the system $2x + 3y = 11$, $x - 5y = -14$.

Neither the coefficients of x nor y are additive inverses. If both sides of the second equation are multiplied by -2, however, the coefficients of x will be additive inverses.

$$
\begin{array}{ll}
\textbf{I} & 2x + 3y = 11 \\
\textbf{II} & \underline{-2x + 10y = 28} \qquad \text{Multiply each side of } x - 5y = -14 \text{ by } -2 \\
& \qquad 13y = 39 \qquad \text{Add I to II} \\
& \qquad \ \ y = 3
\end{array}
$$

Substitute $y = 3$ into either of the original equations to find x. We use Equation I.

$$2x + 3y = 11$$
$$2x + 3 \cdot 3 = 11$$
$$x = 1$$

The solution set is $\{(1, 3)\}$.

Check:
$$
\begin{array}{ll}
2x + 3y = 11 & \qquad\qquad x - 5y = -14 \\
2 \cdot 1 + 3 \cdot 3 \overset{?}{=} 11 \qquad x = 1, y = 3 & \quad 1 - 5 \cdot 3 \overset{?}{=} -14 \\
11 \overset{\checkmark}{=} 11 & \qquad\qquad -14 \overset{\checkmark}{=} -14 \quad \blacksquare
\end{array}
$$

▶ **Caution**   Always check the solution in the original system. An error could have been made when multiplying to establish coefficients that are additive inverses. If the new incorrect system is solved correctly from this point on, the solution would check in this system. However, it would not be a solution to the original system.

The solution to a linear system of equations is also a solution to the equation formed by their sum. When the equations in Example 2 are added directly, we have

$$
\begin{array}{l}
2x + 3y = 11 \\
\underline{x - 5y = -14} \\
3x - 2y = -3.
\end{array}
$$

Their solution, $(1, 3)$, satisfies this new equation.

$$3 \cdot 1 - 2 \cdot 3 \overset{?}{=} -3$$
$$-3 \overset{\checkmark}{=} -3$$

When the coefficients are additive inverses, however, the equation that represents their sum will involve a single variable. As you have seen in Examples 1 and 2, the solution to this equation is then used to find the value of the other variable.

Example 3   Solve the system $2x + 3y = 5$, $3x + 8y = 5$.

Multiply the first equation by 3 and the second equation by -2, so that the coefficients of x are additive inverses.

$$
\begin{array}{lll}
\text{I} & 6x + 9y = 15 & \text{Multiply } 2x + 3y = 5 \text{ by } 3 \\
\text{II} & \underline{-6x - 16y = -10} & \text{Multiply } 3x + 8y = 5 \text{ by } -2 \\
& -7y = 5 & \text{Add I to II} \\
& y = -\dfrac{5}{7} &
\end{array}
$$

Solve for x by substituting $y = -\frac{5}{7}$ into $2x + 3y = 5$.

$$2x + 3\left(-\frac{5}{7}\right) = 5$$

$$2x - \frac{15}{7} = 5$$

$$14x - 15 = 35$$

$$14x = 35 + 15$$

$$x = \frac{50}{14} = \frac{25}{7}$$

The solution set is $\{(\frac{25}{7}, -\frac{5}{7})\}$.

$$
\begin{array}{ccc}
\text{Check:} \quad 2x + 3y = 5 & & 3x + 8y = 5 \\
2 \cdot \dfrac{25}{7} + 3\left(-\dfrac{5}{7}\right) \overset{?}{=} 5 \qquad x = \dfrac{25}{7}; y = -\dfrac{5}{7} & & 3 \cdot \dfrac{25}{7} + 8\left(-\dfrac{5}{7}\right) \overset{?}{=} 5 \\
5 \overset{\checkmark}{=} 5 & & 5 \overset{\checkmark}{=} 5 \quad \blacksquare
\end{array}
$$

Recall that when a system of equations is inconsistent, the lines are parallel and no solution exists. Trying to solve such a system results in an outcome like the following.

$$
\begin{array}{ll}
\text{I} & x - y = 5 \\
 & \qquad\qquad \text{Both lines have a slope of 1} \\
\text{II} & \underline{-x + y = 3} \\
 & 0 = 8 \qquad \text{Add I to II}
\end{array}
$$

A false statement such as $0 = 8$ will always result if the elimination method is used on an inconsistent system.

If a system of equations is dependent, the two lines coincide. They have the same solution set. This condition occurs when one equation is a multiple of the other equation. The system $x + 2y = 7$, $2x + 4y = 14$ is dependent since the second equation can be obtained from the first by multiplying both sides by 2.

A second method for solving systems of equations is the *substitution method*, as shown in the next two examples.

Example 4   Solve the system $2x + 3y = 11$, $x - 5y = -14$.

If the system has a unique solution, the ordered pair that represents the solution satisfies both equations. Thus an alternative to the elimination method is to first solve one of the equations for either variable and then substitute that result into the remaining equation.

Solving $x - 5y = -14$ for x yields

$$x = 5y - 14.$$

When this is substituted in $2x + 3y = 11$, we have

$$2(5y - 14) + 3y = 11$$
$$10y - 28 + 3y = 11$$
$$13y = 39$$
$$y = 3.$$

Substituting $y = 3$ into $x - 5y = -14$ yields

$$x - 5 \cdot 3 = -14$$
$$x = 1.$$

The solution set is $\{(1, 3)\}$.   ∎

In Example 4 we first solved for x because its coefficient was 1. This allowed us to avoid fractions in the solution process.

Example 5   Solve the system $2x + 3y = 5$, $3x + 6y = 5$.

Since no coefficients are 1, we will solve the first equation for y.

$$2x + 3y = 5$$
$$3y = 5 - 2x \qquad \text{Subtract } 2x \text{ from both sides}$$
$$y = \frac{5 - 2x}{3} \qquad \text{Divide both sides by 3}$$

This value is now substituted for y in the second equation.

$$3x + 6y = 5$$
$$3x + 6\left(\frac{5 - 2x}{3}\right) = 5$$
$$3x + 10 - 4x = 5$$
$$-x + 10 = 5$$
$$x = 5$$

The value for x is now substituted into $2x + 3y = 5$.

$$2 \cdot 5 + 3y = 5$$
$$3y = -5$$
$$y = \frac{-5}{3}$$

The solution set is $\{(5, \frac{-5}{3})\}$.

If we had solved the first equation for x, we would have had

$2x + 3y = 5$

$$x = \frac{5 - 3y}{2}.$$

Substituting into the second equation yields

$$3\left(\frac{5 - 3y}{2}\right) + 6y = 5$$

$3(5 - 3y) + 12y = 10$ Multiply both sides by 2

$15 - 9y + 12y = 10$

$3y = -5$

$$y = \frac{-5}{3}. \quad \blacksquare$$

When an attempt is made to solve an inconsistent system by substitution, a false statement will result. When the first equation of the inconsistent system $x - y = 5$, $-x + y = 3$ is solved for x and substituted in the second equation, the false statement is $-5 = 3$. The reader is encouraged to discover what results when an attempt is made to solve a dependent system by substitution.

Example 6   Determine the values of a and b so that $ax + by = 5$ passes through $(1, 1)$ and $(-2, 3)$.

If a line passes through a point, the coordinates of the point must satisfy the equation of the line.

$a(1) + b(1) = 5$ $(1, 1)$ satisfies the equation $ax + by = 5$

$a + b = 5$

$a(-2) + b(3) = 5$ $(-2, 3)$ satisfies the equation $ax + by = 5$

$-2a + 3b = 5$

Thus we have a system of equations

$a + b = 5$

$-2a + 3b = 5.$

Solving by elimination yields $a = 2$ and $b = 3$. These are the values of a and b so that $ax + by = 5$ passes through the points $(1, 1)$ and $(-2, 3)$. The equation is $2x + 3y = 5$. $\blacksquare$

Example 7   Solve the system $\frac{3}{x} - \frac{2}{y} = 5$, $\frac{2}{x} + \frac{3}{y} = -1$ by elimination.

$$\frac{9}{x} - \frac{6}{y} = 15 \qquad \text{Multiply the first equation by 3}$$

$$\frac{4}{x} + \frac{6}{y} = -2 \qquad \text{Multiply the second equation by 2}$$

$$\frac{13}{x} = 13$$

$$x = 1$$

Substitute $x = 1$ into the first equation.

$$\frac{3}{1} - \frac{2}{y} = 5$$

$$-\frac{2}{y} = 2$$

$$y = -1$$

The solution set is $\{(1, -1)\}$. ■

Example 8   Solve the system $2(x - 3) = 5(y + 4)$, $3x + 2(y - 3) = 5(x - 1)$.

Simplify the first equation: Simplify the second equation:

$$2x - 6 = 5y + 20 \qquad\qquad 3x + 2y - 6 = 5x - 5$$
$$2x - 5y = 26. \qquad\qquad\quad -2x + 2y = 1.$$

Solve by elimination.

$$2x - 5y = 26$$
$$-2x + 2y = 1$$
$$-3y = 27$$
$$y = -9$$

Substitute $y = -9$ into $2x - 5y = 26$.

$$2x - 5(-9) = 26$$
$$2x + 45 = 26$$
$$2x = -19$$
$$x = -\frac{19}{2}$$

The solution set is $\{(\frac{-19}{2}, -9)\}$. ■

Example 9   Two different types of cattle feed contain crude protein in the amount of 45% and 20% respectively. How many pounds of each type must a feed lot manager mix together to obtain 10,000 pounds of a mixture that is 25% protein?

Let x = pounds of feed containing 45% protein

y = pounds of feed containing 20% protein.

Since 10,000 pounds of the mixture are desired,

$x + y = 10,000.$

A second equation can be found by considering that the number of pounds of protein in the mixture is equal to the sum of the number of pounds that is in the two feeds from which it is made.

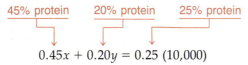

$$0.45x + 0.20y = 0.25\,(10,000)$$

Multiplying each side of the second equation by 100 produces the following system of equations.

$$x + \quad y = \quad 10,000$$
$$45x + 20y = 250,000.$$

When the first equation is multiplied by -20 and added to the second equation, we have

$$25x = 50,000$$
$$x = 2000.$$

Since the mixture contains 10,000 pounds, $y = 8000$. The mixture contains 2000 pounds of the feed with 45% protein and 8000 pounds of the feed with 20% protein. ■

To Solve a Linear System of Equations in Two Variables

By Elimination:
1. If the coefficients of one variable are not additive inverses, use the multiplicative property of equality to generate equivalent equations that satisfy this condition.
2. Add the respective members of the equations to eliminate one variable, then solve for the remaining variable.
3. Substitute the result of Step 2 into either of the original equations and solve for the remaining variable.

By Substitution:
1. Solve for either variable, say x, in either equation. If possible solve for a variable whose coefficient is 1 or -1.
2. Substitute the value of x from Step 1 in the remaining equation and solve for y.
3. Substitute the value for y in Step 2 into either of the original equations or their equivalent and solve for x.

Exercises 3.4

Solve each system of equations by the elimination method. If the system is not independent, indicate whether it is dependent *or* inconsistent. *Write answers in alphabetical order.*

1. $x + y = 1$
   $x - y = 3$

2. $-x + y = 5$
   $x + y = 3$

3. $s - 2t = 3$
   $s + 2t = 3$

4. $3m - 4n = 6$
   $m + 4n = 2$

5. $-4a + 3b = 9$
   $4a + 2b = 1$

6. $3r + q = -6$
   $-3r + q = 6$

7. $-2c + 5d = 4$
   $3c + 2d = 13$

8. $2x - y = 3$
   $-6x + 3y = -9$

9. $a - b = -2$
   $-4a + 4b = 2$

10. $3s + t = 15$
    $2s - 5t = 10$

11. $4r - 3q = 12$
    $3r - 4q = 2$

12. $5m - 3n = 1$
    $-3m + 7n = 2$

13. $-x - 2y = 0$
    $5x - y = 27$

14. $2a - 3b = 4$
    $-4a + 6b = 2$

15. $3m + 7n = 1$
    $4m - 2n = 3$

16. $10c + 4d = 10$
    $-4c + 6d = -42$

17. $5a + 2b = 1$
    $3a + 5b = 6$

18. $15x - 7y = 2$
    $4x - 2y = 5$

Solve each system of equations by the substitution method. If the system is not independent, indicate whether it is dependent *or* inconsistent.

19. $x - y = 1$
    $x + y = 5$

20. $a + 2b = 8$
    $3a - 4b = 4$

21. $r + 3q = 7$
    $2r + 6q = 14$

22. $2s - t = 5$
    $-4s + 2t = 5$

23. $3m - 6n = 0$
    $2m + 5n = 8$

24. $6x - 5y = 11$
    $5x - 10y = 0$

25. $a = 3 - 2b$
    $5a + 7b = 4$

26. $3c + d = 7$
    $2c - 4d = 14$

27. $12s + 4t = 28$
    $-3s = t + 9$

28. $3r + 2q = 10$
    $-6q = 9r + 8$

29. $3x - y = 18$
    $\dfrac{3}{2}x - \dfrac{5}{2}y = 1$

30. $2m - n = 6$
    $\dfrac{1}{2}m - \dfrac{3}{2}n = 2$

Solve each system of equations.

31. $\dfrac{1}{x} + \dfrac{1}{y} = 2$

    $\dfrac{1}{x} - \dfrac{1}{y} = 3$

32. $\dfrac{1}{x} + \dfrac{1}{y} = \dfrac{5}{6}$

    $\dfrac{1}{x} + \dfrac{2}{y} = \dfrac{7}{6}$

33. $\dfrac{5}{a} - \dfrac{2}{b} = 4$

    $\dfrac{3}{a} + \dfrac{3}{b} = 1$

34. $\dfrac{7}{a} - \dfrac{1}{b} = -2$

    $\dfrac{3}{a} + \dfrac{5}{b} = 0$

35. $\dfrac{7}{p} - \dfrac{3}{q} = \dfrac{1}{3}$

    $\dfrac{5}{p} + \dfrac{4}{q} = -\dfrac{1}{4}$

36. $\dfrac{6}{m} - \dfrac{5}{n} = \dfrac{3}{2}$

    $\dfrac{11}{m} + \dfrac{2}{n} = -\dfrac{1}{5}$

Determine a and b so that each line passes through the given points.

37. $ax + by - 8 = 0$; $(2, 2)$, $(6, 2)$

38. $ax + by = 0$; $(2, 2)$, $(-4, 1)$

39. $ax + by - 6 = 0$; $(2, 3)$, $(1, -3)$

40. $ax + by - 15 = 0$; $(5, 1)$, $(10, -1)$

41. $ax + by = 6$; *x-intercept of 2, y-intercept of 3*

42. $ax + by = 12$; *x-intercept of 3, y-intercept of -4*

After simplifying, solve each system of equations by any method.

43. $3(y - 1) - 2(x + 2) = 10 + y + 2x$
$2(y + 3) - 6(x + 1) = 9 - 2y$

44. $3(a - 2) + 2(b - 1) = 11 + 3b$
$4(a - 5) + 2(b + 3) = -1 + 2a - b$

45. $2m - 3n - 20 = 0$
$5m - \dfrac{3}{4}n - 5 = 0$

46. $t - 4 = 8(s - 3)$
$2t = 3s + 12$

47. $x = 12(2y - 4) - 19(y - 1)$
$2x + 3y = 11$

48. $m = 3(2n - 7) - 5(n - 4)$
$2m - 5n = 10$

49. $a = 3b$
$8(a - 1) + 2(b - 4) = 6(a + 2) - 3(4 - b)$

50. $13x - 17y = 51$
$18x - 7y = 5$

51. $\{(x, y) \mid x = 3y\} \cap \{(x, y) \mid 5x + y = 16\}$

52. $\{(a, b) \mid 2a = 5b\} \cap \{(a, b) \mid 4a + b = 9\}$

53. $\{(s, t) \mid s + 5t = 6\} \cap \{(s, t) \mid -s + t = 0\}$

54. $\{(x, y) \mid x + 3y = 4\} \cap \{(x, y) \mid -x + y = 0\}$

55. The sum of two numbers is 74, and their difference is 10. Find the numbers.

56. Two types of silver ore contain varying amounts of silver. When 10 pounds of Ore 1 are processed with 20 pounds of Ore 2, 6 pounds of silver are obtained. When 20 pounds of Ore 1 are processed with 10 pounds of Ore 2, 9 pounds of silver are obtained. Find to the nearest percent the amount of silver in each of the two ores.

3.5   Systems of Linear Equations in Three or More Variables

An equation such as $3x + 4y + z = 19$ is said to be a **linear equation in three variables.** A solution to such an equation consists of values for x, y, and z that satisfy the equation. One solution to $3x + 4y + z = 19$ is **$x = 1$, $y = 2$, $z = 8$.**

$$3 \cdot \mathbf{1} + 4 \cdot \mathbf{2} + \mathbf{8} = 19$$
$$19 = 19$$

Solutions to equations in three variables are called **ordered triples.** The numbers (1, 2, 8) in the order (x, y, z) is a solution to the given equation.

Linear equations in three variables have an infinite number of solutions. The ordered triples (3, −1, 14) and (2, −3, 25) also are solutions, since

$$3 \cdot 3 + 4(-1) + 14 = 19$$

and $3 \cdot 2 + 4(-3) + 25 = 19.$

The ordered triple (1, 4, 6) is not a solution, since

$$3 \cdot 1 + 4 \cdot 4 + 6 \neq 19.$$

A system of equations such as

$$
\begin{array}{rl}
\mathbf{I} & x + y + z = 6 \\
\mathbf{II} & 2x + y - z = 1 \\
\mathbf{III} & 3x + 2y + z = 11
\end{array}
$$

is called a **system of three linear equations in three variables.** An ordered triple (x, y, z) that is a solution to each equation in the system is a solution to the system itself. The **solution set** to the system above is $\{(3, -1, 4)\}$, since

$$3 + (-1) + 4 = 6$$
$$2 \cdot 3 + (-1) - 4 = 1$$
$$3 \cdot 3 + 2(-1) + 4 = 11.$$

To solve a system of three linear equations in three variables, either the elimination method or the substitution method can be used. The elimination method is generally considered to be easier, so our efforts will be directed toward its use. Example 1 shows how elimination is used to solve the above system of equations.

Example 1   Use elimination to solve the system below.

$$\begin{array}{ll} \textbf{I} & x + y + z = 6 \\ \textbf{II} & 2x + y - z = 1 \\ \textbf{III} & 3x + 2y + z = 11 \end{array}$$

To solve the system, we will eliminate one of the variables, z, by adding the left and right members of Equations I and II.

$$\begin{array}{lll} \textbf{I} & x + y + z = 6 & \\ \textbf{II} & \underline{2x + y - z = 1} & \\ \textbf{IV} & 3x + 2y = 7 & \text{IV is the sum of I and II} \end{array}$$

The result, IV, is an equation in two variables. A second equation in the same two variables must be found using a combination of the original equations other than I and II. Notice that z can also be eliminated if the members of equations II and III are added, since the coefficients of z are additive inverses.

$$\begin{array}{lll} \textbf{II} & 2x + y - z = 1 & \\ \textbf{III} & \underline{3x + 2y + z = 11} & \\ \textbf{V} & 5x + 3y = 12 & \text{V is the sum of II and III} \end{array}$$

The new equations IV and V form a linear system in two variables. The methods of Section 3.4 can be used to find its solution. To eliminate y, multiply both sides of equation IV by -3 and both sides of V by 2. Add the results.

$$\begin{array}{ll} -9x - 6y = -21 & \text{Multiply IV by } -3 \\ \underline{10x + 6y = 24} & \text{Multiply V by 2} \\ x = 3 & \end{array}$$

Substitute 3 for x in either IV or V to find y. Choosing IV yields

$$\begin{array}{ll} \textbf{IV} & 3x + 2y = 7 \\ & 3 \cdot 3 + 2y = 7 \qquad x = 3 \\ & 9 + 2y = 7 \\ & y = -1. \end{array}$$

Finally, substitute $x = 3$ and $y = -1$ into any of the original equations and solve for z. Choosing I yields

I $\qquad x + y + z = 6$

$\qquad\quad 3 + (-1) + z = 6 \qquad x = 3, y = -1$

$\qquad\qquad\quad 2 + z = 6$

$\qquad\qquad\qquad z = 4.$

The solution set is $\{(3, -1, 4)\}$.

To check, the solution $(3, -1, 4)$ would have to be shown to satisfy each of the original equations. This was done at the beginning of this section. ∎

A linear equation in two variables is a line. A solution to a linear system in two variables was shown to lie on the intersection of the graphs of the two lines. The graph of a linear equation in three variables is a **plane.** The solutions to a linear system in three variables lie on the intersection of the graphs of the planes. We will not attempt to graph a specific system but will use the graphs in Figure 3.22 to illustrate the possibilities.

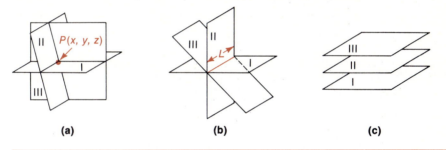

(a)   **(b)**   **(c)**

Figure 3.22

In Figure 3.22(a) the planes intersect in a single point P. There is one solution. In Figure 3.22(b) the planes intersect along the line L. There will be infinitely many solutions. In Figure 3.22(c) the planes are parallel. No solutions exist. If the three equations are equivalent the planes they represent will coincide. In this case any point that lies on the plane will be a solution to the system. Infinitely many solutions exist.

The problems in this section (if consistent) will be restricted to those with a single solution.

Example 2 Solve the following system.

I $\qquad x + y \qquad\quad = \quad 4$

II $\qquad\qquad y + z = \quad 6$

III $\qquad x \qquad - 2z = -2$

If y is eliminated from Equations I and II, the resulting equation will have the same variables as Equation III.

$$
\begin{array}{lll}
\text{I} & x + y \quad = 4 & \\
& -y - z = -6 & \text{Multiply both sides of II by } -1 \\ \hline
\text{IV} & x \quad - z = -2 & \text{Add}
\end{array}
$$

Now eliminate z from III and IV. (We could choose to eliminate x instead of z.)

$$
\begin{array}{lll}
\text{III} & x - 2z = -2 & \\
& -2x + 2z = 4 & \text{Multiply both sides of IV by } -2 \\ \hline
& -x \quad = 2 & \text{Add} \\
& x \quad = -2 &
\end{array}
$$

To find the values of y and z, substitute the value of x into Equations I and III.

$$
\begin{array}{ll}
\qquad\text{I} & \qquad\quad\text{III} \\
x + y = 4 & x - 2z = -2 \\
-2 + y = 4 \quad x = -2 & -2 - 2z = -2 \\
y = 6 & z = 0
\end{array}
$$

The solution set is $\{(-2, 6, 0)\}$.

Check:

$$
\begin{array}{lll}
x + y = 4 & x = -2 & \rightarrow -2 + 6 = 4 \\
y + z = 6 & y = 6 & \rightarrow 6 + 0 = 6 \\
x - 2z = -2 & z = 0 & \rightarrow -2 - 2\cdot 0 = -2 \quad \blacksquare
\end{array}
$$

Example 3   Solve the system below.

$$
\begin{array}{lll}
\text{I} & 2x + y + z = 7 \\
\text{II} & 4x + 2y - z = 8 \\
\text{III} & 6x - 3y + 4z = -1
\end{array}
$$

First eliminate z since two of the z-coefficients are additive inverses.

$$
\begin{array}{lll}
\text{I} & 2x + y + z = 7 & \\
\text{II} & 4x + 2y - z = 8 & \\ \hline
\text{IV} & 6x + 3y \quad = 15 & \text{IV} = \text{I} + \text{II}
\end{array}
$$

Now eliminate z from Equations II and III by multiplying both sides of II by 4 and adding the result to III.

$$
\begin{array}{lll}
& 16x + 8y - 4z = 32 & \text{Multiply II by 4} \\
\text{III} & 6x - 3y + 4z = -1 & \\ \hline
\text{V} & 22x + 5y \quad = 31 & \text{V} = 4(\text{II}) + \text{III}
\end{array}
$$

To eliminate y from IV and V multiply both sides of IV by 5 and both sides of V by -3.

$$
\begin{array}{ll}
30x + 15y = 75 & \text{Multiply IV by 5} \\
-66x - 15y = -93 & \text{Multiply V by } -3 \\
\hline
-36x \qquad\quad = -18 & 5(\text{IV}) + (-3)(\text{V}) \\
\end{array}
$$

$$x = \frac{1}{2}$$

Substituting $\frac{1}{2}$ for x in IV yields

IV $6x + 3y = 15$

$$6\left(\frac{1}{2}\right) + 3y = 15 \qquad x = \frac{1}{2}$$

$$3 + 3y = 15$$
$$3y = 12$$
$$y = 4$$

Substituting $\frac{1}{2}$ for x and 4 for y in I yields

$$2x + y + z = 7$$
$$2\left(\frac{1}{2}\right) + 4 + z = 7$$
$$5 + z = 7$$
$$z = 2.$$

The solution set is $\{(\frac{1}{2}, 4, 2)\}$. The check is left to the reader. ■

Example 4 A coin collection of nickels, dimes, and quarters has a face value of $2.00. The value of the nickels is $0.90 less than that of the dimes and quarters together. How many of each are there if the collection contains 21 coins?

Let n = the number of nickels
d = the number of dimes
q = the number of quarters.

The number of coins is 21.

$n + d + q = 21$

The value of the collection is 200 cents.

$5n + 10d + 25q = 200$

The value of the nickels	is	90 less than the value of the dimes and quarters together
$5n$	$=$	$10d + 25q - 90$

First write each equation in **standard form** (all variables stand alone on the left side of the equals sign.)

$$\begin{array}{rl} \text{I} & n + d + q = 21 \\ \text{II} & 5n + 10d + 25q = 200 \\ \text{III} & 5n - 10d - 25q = -90 \end{array}$$

We begin by eliminating d.

$$\begin{array}{rll} & 10n + 10d + 10q = 210 & 10(\text{I}) \\ \text{III} & \underline{5n - 10d - 25q = -90} & \\ \text{IV} & 15n \quad\quad - 15q = 120 & \text{IV} = 10(\text{I}) + \text{III} \end{array}$$

$$\begin{array}{rll} \text{II} & 5n + 10d + 25q = 200 & \\ \text{III} & \underline{5n - 10d - 25q = -90} & \\ \text{V} & 10n \quad\quad\quad\quad = 110 & \text{V} = \text{II} + \text{III} \\ & n \quad\quad\quad\quad = 11 & \end{array}$$

Substituting 11 for n in IV yields

$$\begin{aligned} 15n - 15q &= 120 \\ 15(11) - 15q &= 120 \\ 165 - 15q &= 120 \\ -15q &= -45 \\ q &= 3. \end{aligned}$$

Substituting $n = 11$ and $q = 3$ in I yields

$$\begin{array}{rl} \text{I} & n + d + q = 21 \\ & 11 + d + 3 = 21 \\ & d \quad\quad = 7. \end{array}$$

The solution is the ordered triple (11, 7, 3) in the order (n, d, q). There are 11 nickels, 7 dimes, and 3 quarters. ■

Example 5   When Ruth, Nell, and Dee work together they can complete a job in 5 hours. When Ruth works alone for 2 hours and Nell works alone for 5 hours they can complete half of the job. When Dee works alone for 8 hours and Nell works alone for 3 hours they can complete three-fifths of the job. How long would it take each to do the job alone?

Work problems are typically solved by determining what fractional part of a job a person can do in one hour. For example, if a woman could do a job in 3 hours, she should do $\frac{1}{3}$ of the job in 1 hour. If she could do a job in n hours she could do $\frac{1}{n}$ of the job in 1 hour. In 2 hours she could do $2(\frac{1}{n}) = \frac{2}{n}$ of the job. Let

x = the number of hours for Ruth to do the job alone

y = the number of hours for Nell to do the job alone

z = the number of hours for Dee to do the job alone.

Based on the discussion above, Ruth can do $\frac{1}{x}$ of the job in 1 hour, Nell can do $\frac{1}{y}$ of the job in 1 hour, and Dee can do $\frac{1}{z}$ of the job in 1 hour.

$$\underbrace{\text{Ruth + Nell + Dee}}\ \ \underbrace{\text{can do}}\ \ \underbrace{\frac{1}{5}\ \text{of the job in 1 hour}}$$

$$\frac{1}{x} + \frac{1}{y} + \frac{1}{z}\ =\ \frac{1}{5}$$

$$\underbrace{\text{Ruth (2 hrs) + Nell (5 hrs)}}\ \ \underbrace{\text{can do}}\ \ \underbrace{\frac{1}{2}\ \text{of the job}}$$

$$\frac{2}{x} + \frac{5}{y}\ =\ \frac{1}{2}$$

$$\underbrace{\text{Nell (3 hrs) + Dee (8 hrs)}}\ \ \underbrace{\text{can do}}\ \ \underbrace{\frac{3}{5}\ \text{of the job}}$$

$$\frac{3}{y} + \frac{8}{z}\ =\ \frac{3}{5}$$

I $\quad \dfrac{1}{x} + \dfrac{1}{y} + \dfrac{1}{z} = \dfrac{1}{5}$

II $\quad \dfrac{2}{x} + \dfrac{5}{y} \phantom{+ \dfrac{1}{z}} = \dfrac{1}{2}$

III $\quad \phantom{\dfrac{2}{x} +} \dfrac{3}{y} + \dfrac{8}{z} = \dfrac{3}{5}$

The elimination method can be used to solve for $\frac{1}{x}$, $\frac{1}{y}$, and $\frac{1}{z}$.

IV $\quad \dfrac{8}{x} + \dfrac{5}{y} = 1 \qquad 8(\text{I}) + (-1)\text{III} = \text{IV}$

$\qquad\quad \dfrac{6}{x} \phantom{+ \dfrac{5}{y}} = \dfrac{1}{2} \qquad \text{IV} + (-1)\text{II}$

$\qquad\quad x \phantom{+ \dfrac{5}{y}} = 12$

Substitute $x = 12$ into II.

$$\frac{2}{12} + \frac{5}{y} = \frac{1}{2}$$

$$\frac{5}{y} = \frac{1}{3} \qquad \frac{1}{2} - \frac{2}{12} = \frac{1}{3}$$

$$y = 15$$

Substitute $y = 15$ into III.

$$\frac{3}{15} + \frac{8}{z} = \frac{3}{5}$$

$$\frac{8}{z} = \frac{2}{5}$$

$$z = 20$$

Ruth can do the job alone in 12 hours, Nell can do the job alone in 15 hours, and Dee can do the job alone in 20 hours. ■

To Solve a Linear System of Equations in Three Variables

1. If necessary, rewrite the equations in standard form.
2. Use the elimination method to eliminate any variable from two equations.
3. Eliminate the same variable from any other combination of equations different from those in Step 2.
4. Solve the system of equations in two variables from Steps 2 and 3 by elimination or substitution.
5. Substitute the result of Step 4 into any of the three original equations to find the value of the third variable.

Exercises 3.5

Solve each system of equations and check. If inconsistent, write Ø.

1. $x - 4y - 4z = -13$
$2x + 9y + z = 2$
$-x + 3y + 2z = 4$

2. $x + y + 3z = 10$
$2x + z = 7$
$x + y - z = -2$

3. $2a - b + c = -1$
$a - 4b - 3c = -4$
$-a - b + 2c = -5$

4. $a + b - c = 6$
$a - 3b - c = 10$
$a + b + c = 12$

5. $5r + 4s - 6t = -5$
$-2r + 3s - 5t = -11$
$4r - 7s + 8t = 14$

6. $2r + 3s + t = 1$
$-3r + s - 2t = -5$
$5r + s + 4t = 8$

7. $y + z = 7$
$x + z = 6$
$x + y = 5$

8. $y - z = 1$
$x - z = 5$
$x - y = 3$

9. $12a + 3b - 8c = -12$
$3a - 3b - 2c = -8$
$-9a + 6b - 2c = 8$

10. $3a + b + 6c = 1$
$-4a - 3b - 5c = -5$
$2a + 3b + 3c = 9$

11. $9r - 5t = 18$
$3r + 7s + 4t = -45$
$5r - 9s - 6t = 39$

12. $r + 5s - 3t = 4$
$2r + 10s + 6t = 8$
$4r + 3s - 7t = 10$

13. $-2x - 9y = -8$
$x + 8z = 3$
$- 5y + 17z = 7$

14. $-5b - c = -1$
$2a + 3c = 4$
$-3a + 20b = 1$

15. $-\dfrac{2}{3}x + 2y - \dfrac{2}{3}z = -\dfrac{1}{3}$
$\dfrac{3}{4}x - \dfrac{5}{4}y + z = \dfrac{1}{2}$
$\dfrac{1}{2}x - \dfrac{3}{2}y + \dfrac{1}{2}z = \dfrac{5}{2}$

16. $\dfrac{3}{2}x - 3y + 2z = -2$
$-\dfrac{5}{8}x + \dfrac{12}{9}y + \dfrac{14}{6}z = -\dfrac{7}{3}$
$3x - \dfrac{8}{3}y - z = 1$

17. $\dfrac{1}{2}a + b + \dfrac{6}{4}c = \dfrac{5}{2}$
$a + 2b + 3c = 4$
$\dfrac{1}{4}a + \dfrac{3}{12}b - \dfrac{2}{8}c = 2$

18. $r - \dfrac{7}{4}s - 2t = \dfrac{11}{2}$
$r + \dfrac{6}{5}s + \dfrac{2}{10}t = \dfrac{-132}{35}$
$2r - \dfrac{5}{2}s + t = \dfrac{55}{7}$

19. $0.3r + 0.4s - 0.2t = 0.5$
     $0.60r + 0.80s - 0.4t = 1.0$
     $r + 0.5\ s - 0.5t = 1$

20. $0.05a - 0.01b + 0.04c = 0.16$
     $-0.5a + 0.2\ b - 0.6\ c = -1.8$
     $0.2a - 0.5\ b - 0.3\ c = -1.2$

21. $\dfrac{1}{x} + \dfrac{4}{y} + \dfrac{3}{z} = 0$
     $\dfrac{3}{x} - \dfrac{1}{y} + \dfrac{1}{z} = -2$
     $\dfrac{2}{x} - \dfrac{3}{y} + \dfrac{1}{z} = -7$

22. $\dfrac{2}{x} - \dfrac{1}{y} + \dfrac{4}{z} = -2$
     $\dfrac{3}{x} - \dfrac{2}{y} - \dfrac{1}{z} = 2$
     $\dfrac{1}{x} + \dfrac{1}{y} + \dfrac{3}{z} = 4$

23. $\dfrac{2}{x} + \dfrac{3}{y} - \dfrac{6}{z} = 11$
     $\dfrac{3}{x} - \dfrac{4}{y} + \dfrac{18}{z} = -19$
     $\dfrac{6}{x} - \dfrac{9}{y} - \dfrac{2}{z} = -1$

24. $\dfrac{1}{x} + \dfrac{1}{3y} + \dfrac{2}{3z} = \dfrac{5}{3}$
     $\dfrac{2}{5x} - \dfrac{3}{5y} + \dfrac{2}{5z} = -1$
     $\dfrac{1}{3x} + \dfrac{1}{y} + \dfrac{1}{3z} = \dfrac{11}{3}$

Write a system of three equations with three variables for each exercise and solve.

25. A collection of nickels, dimes, and quarters is worth $5.55. There are nine more dimes in the collection than the number of nickels and quarters combined. The number of nickels is one less than three times the number of quarters. Find the number of nickels, dimes, and quarters in the collection.

26. The sum of the measures of the angles of a triangle is 180°. The measure of the first angle is three times the measure of the second angle. The third angle measures 20° more than the sum of the measures of the first two angles. What is the measure of each angle?

27. Barbara invests $25,000 for one year. Part of it is invested at 5 percent, another part is invested at 6 percent, and the rest is invested at 8 percent. The total income from the investments is $1600. The income from the 8 percent investment is the same as the sum of the incomes from the other two investments. Find the amount invested at each rate.

28. Adam invests $21,900 for one year. Part is invested at 9 percent, part is invested at 10 percent, and the rest is invested at 14 percent. The total income from the investments is $2409. The income from the investment at 14 percent is $164 more than the income from the investment at 10 percent. Find the amount invested at each rate.

29. Kim is considering the purchase of a duplicating machine. The salesman supplies the following information on three available models. If all three machines are operating, a particular job can be completed in 50 minutes. If machine A operates for 20 minutes and machine B for 50 minutes, one-half of the job is finished. If machine B operates for 30 minutes and machine C for 80 minutes, three-fifths of the job is completed. Which is the fastest machine, and how long does it take for this machine to finish the whole job?

30. Carlos, Sam, and Bennie are doing a certain job. The job could be completed if Carlos and Sam work together for 2 hours and Bennie works alone for 1 hour or if Carlos and Sam work together for 1 hour and Bennie works alone for $2\frac{1}{2}$ hours or if Carlos works alone for 3 hours, Sam works alone for 1 hour and Bennie works alone for $\frac{1}{2}$ hour. Determine the time it takes for each man working alone to complete the job.

3.6 Determinants (Optional)

The solution to a system of linear equations can be found by several methods other than the ones you have already studied. One such method uses *determinants** and another uses the *matrix*. Although we shall develop only the first method, we will begin this section by defining a matrix, since the definition of a determinant will be given in terms of the definition of a matrix.

A **matrix** is a rectangular array of numbers enclosed within parentheses. For example,

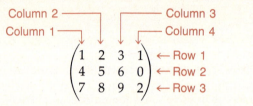

is a matrix with 3 rows and 4 columns, a 3×4 matrix. In an **m $\times$ n matrix** the number of rows, m, is always listed first and the number of columns, n, is listed second. If $m = n$, that is, if the matrix has the same number of rows as columns, it is said to be a **square matrix.** Associated with each square matrix is a number called its **determinant.** The 2×2 matrix

$$\begin{pmatrix} a & b \\ c & d \end{pmatrix} \quad \text{has as its determinant} \quad \begin{vmatrix} a & b \\ c & d \end{vmatrix}.$$

The 3×3 matrix

$$\begin{pmatrix} a_1 & b_1 & c_1 \\ a_2 & b_2 & c_2 \\ a_3 & b_3 & c_3 \end{pmatrix} \quad \text{has as its determinant} \quad \begin{vmatrix} a_1 & b_1 & c_1 \\ a_2 & b_2 & c_2 \\ a_3 & b_3 & c_3 \end{vmatrix}.$$

We are now ready to define the number that represents the value of a determinant. In Section 3.7 we will show why the definition is useful when solving systems of linear equations.

The Value of a 2 × 2 Determinant

If a, b, c, and d represent real numbers, then

$$\begin{vmatrix} a & b \\ c & d \end{vmatrix} = ad - bc.$$

---

*Historically, determinants were known as early as 1683 in Japan. Leibnitz, the co-inventor of calculus, is known to have used them to solve systems of equations as early as 1693.

Example 1   Find the value of each 2×2 determinant.

(a) $\begin{vmatrix} 1 & 3 \\ 2 & 5 \end{vmatrix}$

$\begin{vmatrix} 1 & 3 \\ 2 & 5 \end{vmatrix} = 1 \cdot 5 - 3 \cdot 2 = 5 - 6 = -1 \qquad (a = 1, b = 3, c = 2, d = 5)$

(b) $\begin{vmatrix} 3 & 1 \\ 3 & -2 \end{vmatrix}$

$\begin{vmatrix} 3 & 1 \\ 3 & -2 \end{vmatrix} = 3(-2) - 1 \cdot 3 = -6 - 3 = -9 \qquad (a = 3, b = 1, c = 3, d = -2)$

(c) $\begin{vmatrix} -4 & 5 \\ -2 & \dfrac{1}{2} \end{vmatrix}$

$\begin{vmatrix} -4 & 5 \\ -2 & \dfrac{1}{2} \end{vmatrix} = -4 \cdot \dfrac{1}{2} - 5(-2) = -2 + 10 = 8$ ■

Example 2   Find x in each 2×2 determinant.

(a) $\begin{vmatrix} 2 & x \\ 3 & 4 \end{vmatrix} = 20$

(b) $\begin{vmatrix} x & 3 \\ -2 & 4 \end{vmatrix} = 15$

$\begin{vmatrix} 2 & x \\ 3 & 4 \end{vmatrix} = 20$

$\begin{vmatrix} x & 3 \\ -2 & 4 \end{vmatrix} = 15$

$2 \cdot 4 - 3 \cdot x = 20 \qquad\qquad 4 \cdot x - 3(-2) = 15$

$8 - 3x = 20 \qquad\qquad\qquad 4x + 6 = 15$

$x = -4 \qquad\qquad\qquad\qquad x = \dfrac{9}{4}$ ■

The Value of a 3×3 Determinant

If a_1, a_2, a_3, b_1, b_2, b_3, c_1, c_2, and c_3 represent real numbers, then

$$\begin{vmatrix} a_1 & b_1 & c_1 \\ a_2 & b_2 & c_2 \\ a_3 & b_3 & c_3 \end{vmatrix} = a_1 b_2 c_3 - a_1 b_3 c_2 - a_2 b_1 c_3 + a_3 b_1 c_2 + a_2 b_3 c_1 - a_3 b_2 c_1.$$

The definition above would be difficult to memorize. In practice, a 3×3 determinant is **expanded** and evaluated in terms of three 2×2 determinants.

$$\begin{vmatrix} a_1 & b_1 & c_1 \\ a_2 & b_2 & c_2 \\ a_3 & b_3 & c_3 \end{vmatrix} = a_1 \begin{vmatrix} b_2 & c_2 \\ b_3 & c_3 \end{vmatrix} - b_1 \begin{vmatrix} a_2 & c_2 \\ a_3 & c_3 \end{vmatrix} + c_1 \begin{vmatrix} a_2 & b_2 \\ a_3 & b_3 \end{vmatrix}$$

When the 2×2 determinants are evaluated, we have

$$\begin{vmatrix} a_1 & b_1 & c_1 \\ a_2 & b_2 & c_2 \\ a_3 & b_3 & c_3 \end{vmatrix} = a_1 \begin{vmatrix} b_2 & c_2 \\ b_3 & c_3 \end{vmatrix} - b_1 \begin{vmatrix} a_2 & c_2 \\ a_3 & c_3 \end{vmatrix} + c_1 \begin{vmatrix} a_2 & b_2 \\ a_3 & b_3 \end{vmatrix}$$

$$= a_1(b_2c_3 - b_3c_2) - b_1(a_2c_3 - a_3c_2) + c_1(a_2b_3 - a_3b_2)$$

or $\qquad a_1b_2c_3 - a_1b_3c_2 - a_2b_1c_3 + a_3b_1c_2 + a_2b_3c_1 - a_3b_2c_1,$

which agrees with the definition.

Each 2×2 determinant was obtained from the 3×3 by dropping the row and column in which the multipliers a_1, b_1, and c_1 appear. The sign before a_1, b_1, and c_1 alternates. The 2×2 determinants that result from these operations are called the **minors** of a_1, b_1, and c_1 respectively. In fact, the minor of any element in a 3×3 determinant is found by crossing out all elements in the same row and column as the multiplier.

Example 3   Find the minor of each element in the second column of

$$\begin{vmatrix} 1 & 2 & 3 \\ 4 & 5 & 6 \\ 7 & 8 & 9 \end{vmatrix}$$

$$\begin{vmatrix} 1 & 2 & 3 \\ 4 & 5 & 6 \\ 7 & 8 & 9 \end{vmatrix} \qquad \text{The minor of } 2 \text{ is } \begin{vmatrix} 4 & 6 \\ 7 & 9 \end{vmatrix}.$$

$$\begin{vmatrix} 1 & 2 & 3 \\ 4 & 5 & 6 \\ 7 & 8 & 9 \end{vmatrix} \qquad \text{The minor of } 5 \text{ is } \begin{vmatrix} 1 & 3 \\ 7 & 9 \end{vmatrix}.$$

$$\begin{vmatrix} 1 & 2 & 3 \\ 4 & 5 & 6 \\ 7 & 8 & 9 \end{vmatrix} \qquad \text{The minor of } 8 \text{ is } \begin{vmatrix} 1 & 3 \\ 4 & 6 \end{vmatrix}. \qquad ■$$

The value of any 3×3 determinants can be found by using the minors of any row or column. The correct signs to be used before each minor are shown below.

$$\begin{matrix} + & - & + \\ - & + & - \\ + & - & + \end{matrix}$$

For example, if the second row were used in an expansion the signs would be $-$ $+$ $-$. If the third column were used that would be $+$ $-$ $+$.

The value of a determinant remains the same when it is expanded about any row or column.

Example 4 Evaluate

$$\begin{vmatrix} 1 & 2 & -3 \\ 2 & 1 & -1 \\ 3 & 4 & 3 \end{vmatrix}$$

about (a) Row 1 (b) Column 3 (c) Row 3.

(a) $\begin{vmatrix} 1 & 2 & -3 \\ 2 & 1 & -1 \\ 3 & 4 & 3 \end{vmatrix} = 1\begin{vmatrix} 1 & -1 \\ 4 & 3 \end{vmatrix} - 2\begin{vmatrix} 2 & -1 \\ 3 & 3 \end{vmatrix} + (-3)\begin{vmatrix} 2 & 1 \\ 3 & 4 \end{vmatrix}$

$= 1(3 + 4) - 2(6 + 3) - 3(8 - 3)$

$= 7 - 18 - 15$

$= -26$

(b) $\begin{vmatrix} 1 & 2 & -3 \\ 2 & 1 & -1 \\ 3 & 4 & 3 \end{vmatrix} = -3\begin{vmatrix} 2 & 1 \\ 3 & 4 \end{vmatrix} - (-1)\begin{vmatrix} 1 & 2 \\ 3 & 4 \end{vmatrix} + 3\begin{vmatrix} 1 & 2 \\ 2 & 1 \end{vmatrix}$

$= -3(8 - 3) + 1(4 - 6) + 3(1 - 4)$

$= -15 - 2 - 9$

$= -26$

(c) $\begin{vmatrix} 1 & 2 & -3 \\ 2 & 1 & -1 \\ 3 & 4 & 3 \end{vmatrix} = 3\begin{vmatrix} 2 & -3 \\ 1 & -1 \end{vmatrix} - 4\begin{vmatrix} 1 & -3 \\ 2 & -1 \end{vmatrix} + 3\begin{vmatrix} 1 & 2 \\ 2 & 1 \end{vmatrix}$

$= 3(-2 + 3) - 4(-1 + 6) + 3(1 - 4)$

$= 3 - 20 - 9$

$= -26$ ■

Example 5 Find x.

$$\begin{vmatrix} 1 & 0 & 4 \\ 2 & 4 & x \\ 5 & 6 & 7 \end{vmatrix} = -16$$

If we choose the row or column containing the zero there will be one less minor to evaluate. Choose Row 1.

$$1\begin{vmatrix} 4 & x \\ 6 & 7 \end{vmatrix} - 0\begin{vmatrix} 2 & x \\ 5 & 7 \end{vmatrix} + 4\begin{vmatrix} 2 & 4 \\ 5 & 6 \end{vmatrix} = -16$$

$$1(28 - 6x) - 0 + 4(12 - 20) = -16$$

$$28 - 6x - 32 = -16$$

$$-6x = -12$$

$$x = 2 \quad ■$$

Exercises 3.6

Evaluate each determinant.

1. $\begin{vmatrix} 1 & 2 \\ 1 & 3 \end{vmatrix}$

2. $\begin{vmatrix} 2 & 0 \\ 3 & 1 \end{vmatrix}$

3. $\begin{vmatrix} 5 & \dfrac{1}{2} \\ 4 & \dfrac{2}{5} \end{vmatrix}$

4. $\begin{vmatrix} \dfrac{3}{4} & -1 \\ \dfrac{2}{3} & -4 \end{vmatrix}$

5. $\begin{vmatrix} -6 & 0 \\ 8 & 0 \end{vmatrix}$

6. $\begin{vmatrix} -7 & 8 \\ 0 & 0 \end{vmatrix}$

7. $\begin{vmatrix} -11 & \dfrac{5}{8} \\ -8 & 0 \end{vmatrix}$

8. $\begin{vmatrix} 0 & -\dfrac{11}{4} \\ -4 & -\dfrac{5}{9} \end{vmatrix}$

9. $\begin{vmatrix} x & y \\ x & y \end{vmatrix}$

10. $\begin{vmatrix} r & s \\ s & -r \end{vmatrix}$

Solve for the variable in each equation.

11. $\begin{vmatrix} 10 & x \\ 3 & -2 \end{vmatrix} = 1$

12. $\begin{vmatrix} 10 & 12 \\ y & 5 \end{vmatrix} = 14$

13. $\begin{vmatrix} 7 & 5 \\ 3x & -1 \end{vmatrix} = 1$

14. $\begin{vmatrix} r & 3 \\ r & 5 \end{vmatrix} = 8$

15. $\begin{vmatrix} 2 & m \\ -5 & -2 \end{vmatrix} = 6$

16. $\begin{vmatrix} -7 & p \\ 5 & 0 \end{vmatrix} = -10$

Evaluate each determinant.

17. $\begin{vmatrix} 0 & 7 & -5 \\ -2 & 6 & 8 \\ 3 & 4 & 1 \end{vmatrix}$

18. $\begin{vmatrix} 1 & 0 & 0 \\ 0 & 1 & 3 \\ 0 & 4 & 6 \end{vmatrix}$

19. $\begin{vmatrix} 2 & 4 & -1 \\ 3 & -1 & 2 \\ 5 & -1 & 0 \end{vmatrix}$

20. $\begin{vmatrix} 8 & -1 & 0 \\ 2 & 3 & 0 \\ 0 & 0 & -2 \end{vmatrix}$

21. $\begin{vmatrix} 4 & 2 & 5 \\ -1 & 3 & -2 \\ 1 & -3 & -1 \end{vmatrix}$

22. $\begin{vmatrix} -2 & -3 & -4 \\ 6 & 1 & 5 \\ 2 & 1 & -1 \end{vmatrix}$

Solve for the variable in each equation.

23. $\begin{vmatrix} r & 1 & 0 \\ 0 & 2 & -8 \\ 3 & 0 & 1 \end{vmatrix} = -16$

24. $\begin{vmatrix} a & 0 & 1 \\ 3 & 2 & 0 \\ 2 & -1 & 3 \end{vmatrix} = -1$

25. $\begin{vmatrix} 2 & -1 & 2 \\ 1 & 2 & -3 \\ 3 & 1 & p \end{vmatrix} = -25$

26. $\begin{vmatrix} 5 & 1 & -6 \\ 1 & 2 & m \\ 2 & -3 & 1 \end{vmatrix} = 119$

27. $\begin{vmatrix} x & -2 & 3 \\ x & 1 & 2 \\ 4 & -8 & 12 \end{vmatrix} = 0$

28. $\begin{vmatrix} y & y & 2 \\ 1 & 3 & 3 \\ 3 & 1 & -2 \end{vmatrix} = 0$

29. $\begin{vmatrix} k & -1 & k \\ 3 & 2 & 3 \\ 1 & -1 & -2 \end{vmatrix} = -36$

30. $\begin{vmatrix} t & -1 & 2 \\ 3 & 1 & -2 \\ t & 3 & -1 \end{vmatrix} = 35$

Show that each statement is true.

31. $\begin{vmatrix} x & 0 \\ y & 0 \end{vmatrix} = 0$

32. $\begin{vmatrix} x & 3 \\ x & 1 \end{vmatrix} = -2x$

33. $\begin{vmatrix} a & b \\ kc & kd \end{vmatrix} = k \begin{vmatrix} a & b \\ c & d \end{vmatrix}$

34. $\begin{vmatrix} a & b \\ c & d \end{vmatrix} = - \begin{vmatrix} c & d \\ a & b \end{vmatrix}$

35. $\begin{vmatrix} a & b & c \\ d & e & f \\ g & h & i \end{vmatrix} = \begin{vmatrix} a+kd & b+ke & c+kf \\ d & e & f \\ g & h & i \end{vmatrix}$

36. $\begin{vmatrix} a & b & c \\ 3a & 3b & 3c \\ d & e & f \end{vmatrix} = 0$

37. Show that

$$\begin{vmatrix} x & y & 1 \\ 1 & 2 & -1 \\ 3 & 4 & 1 \end{vmatrix} = 0$$

is the equation of a line containing the points $(1, 1)$ and $(-3, -5)$.

38. Show that

$$\begin{vmatrix} x & y & 0 \\ 3 & 5 & -2 \\ 2 & 0 & 1 \end{vmatrix} = 35$$

is the equation of a line containing the points $(0, -5)$ and $(7, 0)$.

Review Exercises

Solve by any method.

39. $5x + 2y = -2$
$3x + 4y = -1$

40. $\dfrac{1}{2}a - \dfrac{2}{3}b = \dfrac{3}{4}$
$\dfrac{1}{3}a + 2b = \dfrac{5}{6}$

41. $r + \quad 2t = 6$
$-r - 2s + 3t = 8$
$-3r + 4s + 6t = 30$

42. $-3x - 2y + 7z = 4$
$2x + 5y - z = 3$
$- x + 3y + 6z = 0$

3.7 Cramer's Rule (Optional)

As stated in Section 3.6, determinants can be used to solve linear systems of equations. The technique that is used is known as **Cramer's rule.** Before stating the rule as a formal theorem we will solve the system of equations

$$a_1 x + b_1 y = c_1$$
$$a_2 x + b_2 y = c_2$$

by the elimination method. Cramer's rule summarizes the results of the work.

$\quad$ **I** $\quad a_1 x + b_1 y = c_1$
$\quad$ **II** $\quad a_2 x + b_2 y = c_2$

To eliminate x, multiply I by $-a_2$ and II by a_1 so that the coefficients of x are additive inverses.

$$-a_2 a_1 x - a_2 b_1 y = -a_2 c_1 \qquad \text{Multiply I by } -a_2$$
$$\underline{a_1 a_2 x + a_1 b_2 y = a_1 c_2} \qquad \text{Multiply II by } a_1$$

III $\qquad a_1 b_2 y - a_2 b_1 y = a_1 c_2 - a_2 c_1 \qquad \text{Add}$

Applying the distributive property to the left side of III gives

$$(a_1 b_2 - a_2 b_1) y = a_1 c_2 - a_2 c_1.$$

If $a_1 b_2 - a_2 b_1 \neq 0$, then we can divide both sides by it.

$$y = \frac{a_1 c_2 - a_2 c_1}{a_1 b_2 - a_2 b_1}$$

To eliminate y, multiply equation I by b_2 and II by $-b_1$.

$$a_1 b_2 x + b_1 b_2 y = b_2 c_1 \qquad \text{Multiply I by } b_2$$
$$\underline{-b_1 a_2 x - b_1 b_2 y = -b_1 c_2} \qquad \text{Multiply II by } -b_1$$
$$a_1 b_2 x - a_2 b_1 x = b_2 c_1 - b_1 c_2 \qquad \text{Add}$$
$$(a_1 b_2 - a_2 b_1) x = b_2 c_1 - b_1 c_2 \qquad \text{Distributive property}$$

$$x = \frac{b_2 c_1 - b_1 c_2}{a_1 b_2 - a_2 b_1}$$

Both x and y can be expressed in terms of determinants. The solutions will be written in the form

$$x = \frac{D_x}{D}, \qquad y = \frac{D_y}{D}.$$

In this form D is the determinant formed with the coefficients of the variables of the two equations. The elements of the determinant D_x are found by replacing the coefficients of x, in D, by the constants. The elements in D_y are obtained by replacing the coefficients of y, in D, by the constants.

$$x = \frac{b_2 c_1 - b_1 c_2}{a_1 b_2 - a_2 b_1} = \frac{\begin{vmatrix} c_1 & b_1 \\ c_2 & b_2 \end{vmatrix}}{\begin{vmatrix} a_1 & b_1 \\ a_2 & b_2 \end{vmatrix}} = \frac{D_x}{D}$$

$$y = \frac{a_1 c_2 - a_2 c_1}{a_1 b_2 - a_2 b_1} = \frac{\begin{vmatrix} a_1 & c_1 \\ a_2 & c_2 \end{vmatrix}}{\begin{vmatrix} a_1 & b_1 \\ a_2 & b_2 \end{vmatrix}} = \frac{D_y}{D}$$

We now state Cramer's rule for a system of linear equations in two variables.

Theorem
3.1(a)

Cramer's Rule

The solution to the system of equations

$$a_1 x + b_1 y = c_1$$
$$a_2 x + b_2 y = c_2$$

is

$$x = \frac{D_x}{D}, \qquad y = \frac{D_y}{D}, \qquad D \neq 0,$$

where

$$D_x = \begin{vmatrix} c_1 & b_1 \\ c_2 & b_2 \end{vmatrix}, \qquad D_y = \begin{vmatrix} a_1 & c_1 \\ a_2 & c_2 \end{vmatrix}, \qquad D = \begin{vmatrix} a_1 & b_1 \\ a_2 & b_2 \end{vmatrix}.$$

Example 1   Use Cramer's rule to solve the system of equations

$$3x + 4y = 11$$
$$2x + 7y = 16.$$

$$D = \begin{vmatrix} 3 & 4 \\ 2 & 7 \end{vmatrix} = 21 - 8 = 13$$

$$D_x = \begin{vmatrix} 11 & 4 \\ 16 & 7 \end{vmatrix} = 77 - 64 = 13$$

$$D_y = \begin{vmatrix} 3 & 11 \\ 2 & 16 \end{vmatrix} = 48 - 22 = 26$$

$$x = \frac{D_x}{D} = \frac{13}{13} = 1, \qquad y = \frac{D_y}{D} = \frac{26}{13} = 2$$

The solution set is $\{(1, 2)\}$.   ■

Notice in Example 1 that we solved for D first. If D had been equal to zero there would be no need to proceed further. Division by zero is undefined. A system in which $D = 0$ is either dependent or inconsistent. If D_x and D_y are also zero, the system is dependent.

Example 2   Use Cramer's rule to solve the system

$$x + 3y = -2$$
$$2x + y = 7.$$

$$D = \begin{vmatrix} 1 & 3 \\ 2 & 1 \end{vmatrix} = 1 - 6 = -5$$

$$D_x = \begin{vmatrix} -2 & 3 \\ 7 & 1 \end{vmatrix} = -2 - 21 = -23$$

$$D_y = \begin{vmatrix} 1 & -2 \\ 2 & 7 \end{vmatrix} = 7 - (-4) = 11$$

$$x = \frac{D_x}{D} = \frac{-23}{-5} = \frac{23}{5}, \qquad y = \frac{D_y}{D} = \frac{11}{-5} = \frac{-11}{5}$$

The solution set is $\{(\frac{23}{5}, \frac{-11}{5})\}$. ∎

Cramer's rule can be shown to hold for a linear system of equations in three or more variables. The determinants are computed in a manner parallel to Theorem 3.1(a).

Theorem 3.1(b)

The solution to the system of equations

$$a_1x + b_1y + c_1z = k_1$$
$$a_2x + b_2y + c_2z = k_2$$
$$a_3x + b_3y + c_3z = k_3$$

is

$$x = \frac{D_x}{D}, \qquad y = \frac{D_y}{D}, \qquad z = \frac{D_z}{D}, \qquad D \neq 0,$$

where

$$D_x = \begin{vmatrix} k_1 & b_1 & c_1 \\ k_2 & b_2 & c_2 \\ k_3 & b_3 & c_3 \end{vmatrix}, \qquad D_y = \begin{vmatrix} a_1 & k_1 & c_1 \\ a_2 & k_2 & c_2 \\ a_3 & k_3 & c_3 \end{vmatrix}, \qquad D_z = \begin{vmatrix} a_1 & b_1 & k_1 \\ a_2 & b_2 & k_2 \\ a_3 & b_3 & k_3 \end{vmatrix},$$

$$D = \begin{vmatrix} a_1 & b_1 & c_1 \\ a_2 & b_2 & c_2 \\ a_3 & b_3 & c_3 \end{vmatrix}.$$

Example 3   Use Cramer's rule to solve the system

$$x + y + z = 2$$
$$3x - y + z = 6$$
$$2x \quad\;\; + 3z = 8.$$

Notice that the coefficient of y in the third equation is 0 because y is missing.

$$D = \begin{vmatrix} 1 & 1 & 1 \\ 3 & -1 & 1 \\ 2 & 0 & 3 \end{vmatrix} \qquad D_x = \begin{vmatrix} 2 & 1 & 1 \\ 6 & -1 & 1 \\ 8 & 0 & 3 \end{vmatrix}$$

$$D_y = \begin{vmatrix} 1 & 2 & 1 \\ 3 & 6 & 1 \\ 2 & 8 & 3 \end{vmatrix} \qquad D_z = \begin{vmatrix} 1 & 1 & 2 \\ 3 & -1 & 6 \\ 2 & 0 & 8 \end{vmatrix}$$

Use row 3 to find D.

$$D = 2\begin{vmatrix} 1 & 1 \\ -1 & 1 \end{vmatrix} - 0\begin{vmatrix} 1 & 1 \\ 3 & 1 \end{vmatrix} + 3\begin{vmatrix} 1 & 1 \\ 3 & -1 \end{vmatrix}$$
$$= 2[1 - (-1)] + 0 + 3(-1 - 3)$$
$$= 4 - 12$$
$$= -8$$

When minors are used further, $D_x = -8$, $D_y = 8$, and $D_z = -16$. Thus

$$x = \frac{-8}{-8} = 1, \qquad y = \frac{8}{-8} = -1, \qquad z = \frac{-16}{-8} = 2.$$

The solution set is $\{(1, -1, 2)\}$. ∎

Example 4 Use Cramer's rule to determine whether a unique solution exists for the system

$$r - s + 3t = 3$$
$$3r - 5s + 2t = -1$$
$$2r - s + 6t = 8.$$

Treat r as x, s as y, and t as z. Expand about row 1.

$$D = \begin{vmatrix} 1 & -1 & 3 \\ 3 & -5 & 2 \\ 2 & -1 & 6 \end{vmatrix} = 1\begin{vmatrix} -5 & 2 \\ -1 & 6 \end{vmatrix} - (-1)\begin{vmatrix} 3 & 2 \\ 2 & 6 \end{vmatrix} + 3\begin{vmatrix} 3 & -5 \\ 2 & -1 \end{vmatrix}$$
$$= 1(-30 + 2) + 1(18 - 4) + 3(-3 + 10)$$
$$= -28 + 14 + 21$$
$$= 7$$

Since $D \neq 0$, a unique solution exists. ∎

Example 5   Solve the following system of equations by Cramer's rule.

$$\frac{1}{x} + \frac{1}{y} - \frac{1}{z} = 2$$

$$\frac{1}{x} - \frac{1}{y} + \frac{1}{z} = -1$$

$$\frac{-2}{x} + \frac{2}{y} \quad\;\; = 0$$

To use Cramer's rule we solve the system for $\frac{1}{x}$, $\frac{1}{y}$, and $\frac{1}{z}$.

$$D = \begin{vmatrix} 1 & 1 & -1 \\ 1 & -1 & 1 \\ -2 & 2 & 0 \end{vmatrix} = -4, \qquad D_{1/x} = \begin{vmatrix} 2 & 1 & -1 \\ -1 & -1 & 1 \\ 0 & 2 & 0 \end{vmatrix} = -2$$

Thus $\frac{1}{x} = \frac{-2}{-4}$ so that $x = 2$. When $D_{1/y}$ and $D_{1/z}$ are found, we have $y = 2$ and $z = -1$. The solution set is $\{(2, 2, -1)\}$. ■

Exercises 3.7

Solve each system of equations by Cramer's rule.

1. $3x - 4y = 0$
$-6x + 7y = -3$

2. $2x + y = 0$
$x + 3y = 4$

3. $3a - b = 2$
$a + 2b = 5$

4. $4a + b = 3$
$8a - 3b = -6$

5. $r + 3s = -14$
$r - 2s = 8$

6. $2r - s = 12$
$3r - 6s = -6$

7. $\frac{1}{3}x - \frac{1}{2}y = 0$
$\frac{1}{2}x + \frac{1}{4}y = 4$

8. $\frac{2}{3}x + 3y = -5$
$x + \frac{1}{3}y = 4$

Consider each of the following as a system in $\frac{1}{x}$ and $\frac{1}{y}$; solve by Cramer's rule.

9. $\dfrac{3}{x} + \dfrac{6}{y} = \dfrac{3}{4}$
$\dfrac{4}{3x} + \dfrac{3}{2y} = \dfrac{5}{8}$

10. $\dfrac{5}{x} - \dfrac{2}{y} = \dfrac{1}{3}$
$\dfrac{8}{x} - \dfrac{4}{y} = \dfrac{1}{5}$

11. $\dfrac{2}{x} + \dfrac{5}{y} = \dfrac{3}{4}$
$\dfrac{1}{3x} - \dfrac{1}{2y} = \dfrac{1}{5}$

12. $\dfrac{3}{4x} - \dfrac{1}{2y} = \dfrac{1}{2}$
$\dfrac{1}{5x} + \dfrac{1}{6y} = \dfrac{3}{10}$

Solve each system by Cramer's rule. Some may be inconsistent.

13. $-a + 3b - 2c = 5$
$2a + b + 3c = 3$
$-a - b - c = -2$

14. $4x + 8y - z = -6$
$x + 7y - 3z = -8$
$3x - 3y + 2z = 0$

15. $a + b = 2$
$2b - 3c = -1$
$2a - c = 1$

16. $r + 4t = 3$
$2r + 5s - 5t = -5$
$s + 3t = 9$

17. $2x - 4y + 4z = 1$
$3x - 3y - 3z = 4$
$x - 2y + 2z = 3$

18. $a - b + 6c = 21$
$2a + 2b + c = 1$
$3a + 2b - c = -4$

19. $5r + s - 3t = 1$
$3r + s + t = 5$
$-2r - 2s + t = 4$

20. $5x - y + z = 6$
$y + z = 5$
$2x + 4y - 3z = 1$

Use Cramer's rule to show whether or not a unique solution exists. Find the solution if it exists.

21. $\begin{aligned} 5a + 2b - c &= 0 \\ 7a + b + 2c &= 4 \\ 2a - b + 3c &= 1 \end{aligned}$

22. $\begin{aligned} 5a + 6b \phantom{{}- c} &= 1 \\ 3a - b - c &= 0 \\ 2a + 7b + c &= 0 \end{aligned}$

23. $\begin{aligned} 2x - y - z &= 0 \\ x - y \phantom{{}- 2z} &= 0 \\ x - 3y - 2z &= 4 \end{aligned}$

24. $\begin{aligned} \frac{x}{2} + \frac{y}{4} + z &= 0 \\ y - z &= -2 \\ x + y \phantom{{}- z} &= 1 \end{aligned}$

25. $\begin{aligned} \frac{1}{2}r - \frac{1}{3}s - \frac{1}{4}t &= 0 \\ \frac{12}{3}r + 5s - \frac{6}{2}t &= -1 \\ 2r - \frac{2}{3}s + t &= 2 \end{aligned}$

26. $\begin{aligned} \frac{1}{8}r \phantom{{}+ 3s} - \frac{1}{2}t &= -\frac{1}{8} \\ \frac{1}{4}r + \frac{3}{12}s \phantom{{}- t} &= \frac{1}{6} \\ r \phantom{{}+ 3s} + \frac{4}{3}t &= 5 \end{aligned}$

Consider each of the following as a system of equations in terms of $\frac{1}{x}$, $\frac{1}{y}$, and $\frac{1}{z}$. Solve by Cramer's rule.

27. $\begin{aligned} \frac{1}{x} + \frac{1}{y} + \frac{1}{z} &= 1 \\ \frac{1}{x} - \frac{1}{y} - \frac{1}{z} &= 4 \\ -\frac{1}{x} - \frac{1}{y} \phantom{{}- \frac{1}{z}} &= -2 \end{aligned}$

28. $\begin{aligned} \frac{2}{x} + \frac{3}{y} \phantom{{}+ \frac{2}{z}} &= 0 \\ \frac{2}{x} - \frac{4}{y} + \frac{2}{z} &= 3 \\ -\frac{1}{x} + \frac{2}{y} + \frac{3}{z} &= 5 \end{aligned}$

29. $\begin{aligned} \frac{5}{2v} - \frac{1}{w} &= 2 \\ \frac{3}{4u} - \frac{1}{v} \phantom{{}+ \frac{3}{2w}} &= 5 \\ \frac{2}{7u} \phantom{{}- \frac{1}{v}} + \frac{3}{2w} &= 1 \end{aligned}$

30. $\begin{aligned} \frac{1}{3a} + \frac{1}{b} \phantom{{}- \frac{2}{3c}} &= 4 \\ \frac{1}{a} \phantom{{}+ \frac{1}{b}} - \frac{2}{3c} &= -1 \\ \frac{4}{5b} - \frac{1}{c} &= 2 \end{aligned}$

Write each exercise as a system of two or three equations and solve using Cramer's rule.

31. Starr swims to help lose weight. Doing the crawl stroke for 1 minute burns up 10 calories, while doing the breaststroke for 1 minute burns up 8 calories. If she burns up 140 calories by swimming 15 minutes, how much time was spent on each type of stroke?

32. Kim jogs and walks to help keep physically fit. If walking burns 7 calories per minute while jogging burns 10 calories per minute, and he burns 164 calories in 20 minutes, how many minutes were spent jogging?

33. Sherrie, Jason, and Lisa go out to lunch together. By pooling all their money they come up with $12. Twice Sherrie's contribution less twice Jason's is equal in amount to Lisa's. Lisa's lunch cost $\frac{1}{2}$ her contribution, Jason's lunch cost $1\frac{1}{2}$ times his, and Sherrie's lunch cost $\frac{3}{5}$ of hers. The total cost for their lunch was $7.95. How much money did each one originally have?

34. Helen often goes to the post office to buy three different types of stamps (*A*, *B*, and *C*) for mailing parcels. One time she bought 30 of the *A* stamps, 3 of the *B* stamps, and 2 of the *C* stamps at a cost of $5.41. Another time she bought 40 of *A*, 12 of *B*, and 3 of *C* at a cost of $8.64. Today she bought 15 of *A*, 5 of *B*, and 4 of *C* at a cost of $3.90. What is the cost of each type of stamp?

3.8   Graphing Linear Inequalities in Two Variables

In Section 3.4 it was pointed out that graphing is not the best method for solving a system of linear equations. *Linear inequalities* are quite a different matter. Graphing is the best method for displaying the solution set. In this section we consider graphs of linear inequalities such as

$$y > x + 1, \quad x \le 2, \quad \text{or} \quad 2x + y \le 4$$

as well as systems of linear inequalities such as

$$2x + y > 4 \quad \text{and} \quad y < x + 1.$$

A **linear inequality,** like a linear equation, has solutions that are ordered pairs. For example, $(3, 7)$ is a solution to $y > x + 2$ because a true statement results when x is replaced by 3 and y by 7.

$y > x + 2$

$7 > 3 + 2$ $x = 3, y = 7$

$7 > 5$ True

There are an infinite number of points other than $(3, 7)$ whose coordinates satisfy $y > x + 2$. The following discussion shows how they can be found and **displayed graphically.**

Consider the equation $y = x + 2$. Let $x = x_1$ be any vertical line whose graph intersects the graph of $y = x + 2$ at $P(x_1, y_1)$, as in Figure 3.23. Since $P(x_1, y_1)$ lies on $y = x + 2$ its coordinates must satisfy the equation. Thus

$$y_1 = x_1 + 2.$$

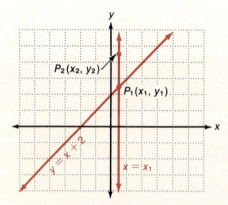

Figure 3.23

If $P_2(x_1, y_2)$ is any other point on $x = x_1$ such that $y_2 > y_1$, then $y_2 > x_1 + 2$. This means that any point on $x = x_1$ that lies above (x_1, y_1) is a solution to the inequality $y > x + 2$. Thus, that region of the plane above the line $y = x + 2$ contains the points whose coordinates form the solution set. An argument parallel to the one above would show that points below the line have coordinates that satisfy the inequality $y < x + 2$.

Solution sets to linear inequalities in two variables lie in a region on a graph called a **half-plane,** since any line passing through a plane divides it into two half-planes. The half-plane may or may not include the line as a boundary. When the boundary is not included it is shown as a broken line. The region satisfying the inequality is indicated by shading. The region containing the solution set to $y > x + 2$ is shown in Figure 3.24.

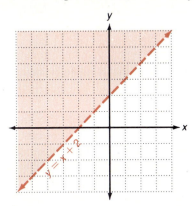

Figure 3.24

An inequality need not be in the form $y < ax + b$ or $y > ax + b$ to determine the half-plane that contains its solution set. Example 1 shows an alternate method.

Example 1   Use a graph to illustrate the solution set of $2x + 3y \leq 6$.
   Begin by drawing the graph of $2x + 3y = 6$, as in Figure 3.25. Three ordered pairs satisfying the equation are $(0, 2)$, $(3, 0)$ and $(6, -2)$. A solid line is drawn since the solution to $2x + 3y \leq 6$ includes the solution to $2x + 3y = 6$.

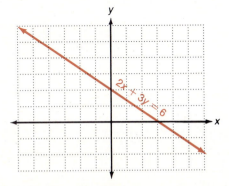

Figure 3.25

To decide which half-plane contains the solution set, a **test point** is selected. A test point is any point in the plane that clearly lies on one side of the line. If the test point satisfies the inequality (makes it true), it is a part of the half-plane that contains the solution set. If the test point does not satisfy the inequality (makes it false) the solution set lies in the half-plane on the other side of the line. Choose the origin (0, 0) as the test point.

$$2x + 3y \le 6$$
$$2 \cdot 0 + 3 \cdot 0 \overset{?}{\le} 6 \qquad \text{Test point } x = 0, y = 0$$
$$0 \le 6 \qquad \text{True}$$

The desired half-plane contains the test point since $0 \le 6$. (See Figure 3.26.) ■

The easiest test point to use is the origin, unless the line passes through it.

It is common practice with respect to linear inequalities to ask one to "graph" an inequality as a method of indicating its solution set. We will use this practice throughout the remainder of the text.

Example 2   Graph $y > 2x$.

Since $y > 2x$ does not include $y = 2x$, the boundary will be shown as a broken line. Three ordered pairs that are solutions to $y = 2x$ are (0, 0), (2, 4) and (−2, −4). The line passes through the origin, so a different test point must be used. Try (2, −2).

$$y > 2x$$
$$-2 \overset{?}{>} 2 \cdot 2 \qquad \text{Test } x = 2, y = -2$$
$$-2 > 4 \qquad \text{False}$$

The test resulted in a false statement. The desired half-plane is on the other side of the line, as shown in Figure 3.27. ■

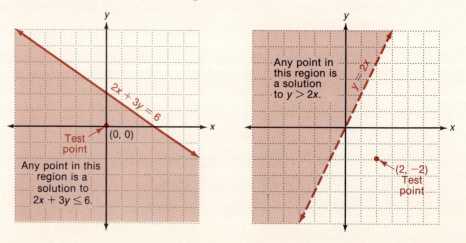

Figure 3.26   **Figure 3.27**

Example 3   Graph $x > -1$.

The line $x = -1$ is not a part of the graph. Any point that lies to the right of $x = -1$ has an x-coordinate greater than -1. The graph is shown in Figure 3.28. ∎

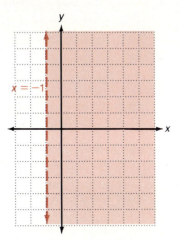

Figure 3.28

In general, inequalities of the form $x > c$ or $x < c$ have graphs that are half-planes to the right or left of the vertical line $x = c$. Graphs of $y > c$ or $y < c$ are half-planes above or below the horizontal line $y = c$.

Systems of linear inequalities are solved graphically by using the same procedure that is used to solve a single linear inequality.

Example 4   Graph the system of inequalities $y \le x + 1$ and $2y + x \ge 4$.

First graph each boundary line on the same set of axes. (See Figure 3.29.) Neither line passes through the origin, so (0, 0) can be used as a test point.

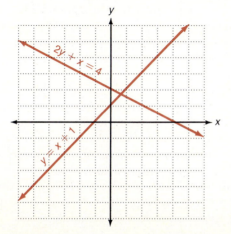

Figure 3.29

$$y \leq x + 1 \qquad\qquad 2y + x \geq 4$$
$$0 \overset{?}{\leq} 0 + 1 \qquad\qquad 2 \cdot 0 + 0 \overset{?}{\geq} 4$$
$$0 \leq 1 \qquad \text{True} \qquad\qquad 2 \geq 4 \qquad \text{False}$$

Now shade the origin side of $y \leq x + 1$ and the side opposite the origin for $2y + x \geq 4$. The region with *double shading* is the desired graph. (See Figure 3.30.) ■

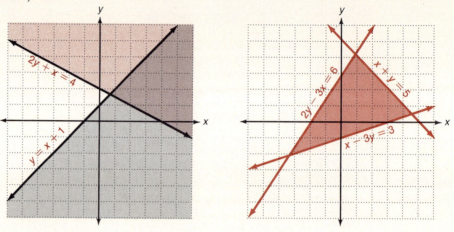

Figure 3.30   Figure 3.31

Example 5   Graph the system of inequalities $2y - 3x \leq 6$, $x - 3y \leq 3$ and $x + y \leq 5$.
     Sketch each line associated with the inequality and shade the region that forms its solution set. The region where the shading from each of the inequalities intersects forms the solution set, as shown in Figure 3.31. ■

     Inequalities are often used to solve problems related to business. The goal of business is to provide a quality product for the consumer that returns a fair profit to the manufacturer. There are often limitations on a plant's capacity to produce certain items, and these limitations affect manufacturing decisions. If the limitations can be expressed in terms of a system of inequalities and the objective is to maximize profits, the problem falls into a class of problems solved by **linear programming**. The next example shows how the conditions arise.

Example 6   Chris Cornelius opened a plant that manufactures personalized hockey skates. Two models are available: a custom model that requires 2 hours for measuring and shaping and 4 hours for detail work, and a standard model that requires 2 hours for measuring and shaping and 2 hours for detail work. His company has 2 persons assigned to measuring and shaping and 3 persons assigned to detail work, each of whom works a maximum of 40 hours per week. Write a system of inequalities describing his possible output and graph the region of possible production schedules.

     Let x = the number of custom models made
         y = the number of standard models made.

	Custom Model	Standard Model
Measuring and shaping time	2	2
Detailing time	4	2

Two persons are assigned to measuring and shaping. Each works a maximum of 40 hours per week, so a maximum of 80 work hours are available for this task. How is the time used? The x number of custom models and y standard models require 2 hours each. Thus $2x$ represents shaping time for the custom models and $2y$ the time for the standard models. The sum of the times is less than or equal to 80 hours.

$$2x + 2y \leq 80$$

Three persons are assigned to detailing at 40 hours each for a maximum of $3 \cdot 40 = 120$ hours. The x number of custom models require 4 hours each for a total of $4x$ hours. At 2 hours each the y standard models will require $2y$ hours. Thus

$$4x + 2y \leq 120.$$

It also is evident that $x \geq 0$ and $y \geq 0$, or Chris would not be in business.

The region that indicates Chris' options for production must satisfy each of the following inequalities.

$$2x + 2y \leq 80$$
$$4x + 2y \leq 120$$
$$x \geq 0$$
$$y \geq 0$$

Since $x \geq 0$ lies to the right of, or on the y-axis, and $y \geq 0$ lies above, or on the x-axis, the graph will lie entirely in the first quadrant, as shown in Figure 3.32. Any point (x, y) in the shaded region represents a possible production schedule for x (custom models) and y (standard models). ■

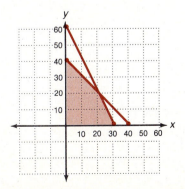

Figure 3.32

Exercises 3.8

Graph the following systems of inequalities.

1. $y \geq 2$
$x \leq 4$

2. $y \geq -3$
$x \geq 2$

3. $x + y > 3$
$x - y > 2$

4. $y - x > 0$
$y - x \leq 3$

5. $x + y \leq 3$
$2x + 2y \geq 0$

6. $3x - y > 0$
$x \quad < y$

7. $x > -y$
$x - 2y < 0$

8. $x \geq 0$
$y \leq 0$

9. $x < 0$
$y > 0$

10. $x + y < 5$
$x - y \geq 5$

11. $3x + y \geq 4$
$2x - y < 1$

12. $4x - y > 3$
$4x - 2y \leq 6$

13. $3x - 5y > 2$
$x + 2y < 0$

14. $5x + 2y < 10$
$-10x - 4y < 4$

15. $4x - 3y < 8$
$8x - 6y \leq -9$

16. $x + y \leq 3$
$x - y \leq 1$
$x \geq 0$
$y \geq 0$

17. $x - y < 0$
$x + y \geq 1$
$x < 4$

18. $x + y < 2$
$y - 3x \leq 3$
$x - y < 0$

19. $x + 3y \leq 4$
$-2x - 6y \leq 3$
$0 \leq x \leq 3$

20. $x + y \leq 4$
$-3x - 3y \leq 3$
$0 \leq y \leq 3$

21. $x - y \leq 0$
$x - y \leq 3$
$-2 \leq y \leq 0$

22. $x - y \leq 2$
$3x - 3y \geq 0$
$-2 \leq x \leq 2$

Graph the following systems of inequalities. Label the axes as indicated by the ordered pairs.

23. $\{(x, y) \mid x - y > 3\} \cap \{(x, y) \mid 2x + y \leq 8\}$

24. $\{(x, y) \mid y - x < 6\} \cap \{(x, y) \mid x + y < -2\}$

25. $\{(s, t) \mid 3s - t > 3\} \cap \{(s, t) \mid 6s - 2t > 9\}$

26. $\{(m, n) \mid 2m + n < 3\} \cap \{(m, n) \mid 4m - 2n > 11\}$

Determine whether or not the given ordered pair is in the solution set of the system of inequalities.

27. $(1, 1)$; $x - y < 2$
$x + y > 1$

28. $(3, 1)$; $y - 3x < 3$
$6x - 2y < 12$

29. $(2, 6)$; $y > x + 3$
$y > -x + 3$
$x > 1$
$y > 5$

30. $(1, 1)$; $x < y + 2$
$x > y - 1$
$x > 0$
$y \geq 0$

31. $(1, 1)$; $y - x < 0$
$y - x \geq 3$
$-2 \leq x \leq 2$
$-2 \leq y \leq 2$

32. $(3, 0)$; $x + y \leq 4$
$x - y > 0$
$y \geq 0$
$-6 \leq x \leq 6$

Write a system of inequalities and graph each solution. Use a calculator (if needed).

33. Tony is preparing for a wilderness backpacking trip on which he plans to eat a combination of cereals and nuts. The recommended daily requirement is 3000 calories and 2.36 ounces of protein. Each ounce of cereal supplies 120 calories and 0.04 ounces of protein. Each ounce of nuts supplies 60 calories and 0.12 ounces of protein. Show by a graph the possible combinations of cereal and nuts that meet or exceed the daily requirements of calories and protein.
(Hint: Let x = the ounces of cereal in his daily diet.
y = the ounces of nuts in his daily diet.
Then $120x + 60y \geq 3000$
$0.04x + 0.12y \geq 2.36$
$x \geq 0$
$y \geq 0$.)

34. Leslie, a dietician, is planning a dinner that is to consist of a cheese dish and a green vegetable. She wants the dinner to supply 420 milligrams of calcium and 4 milligrams of iron. Each ounce of the cheese dish supplies 140 milligrams of calcium and 0.25 milligrams of iron, while each ounce of the green vegetable supplies 20 milligrams of calcium and 0.5 milligrams of iron. Show by a graph the possible combinations of each that would meet or exceed the desired amounts of calcium and iron.

3.9 More Applications

Many applied problems are easier to solve when two or more variables are used instead of one. All that is necessary is to be able to write two or more equations or inequalities. You should continue to use the five steps for solving word problems that were listed in Section 2.6.

Example 1   The sum of two integers is 6. Twice the smaller integer is 24 less than the larger one. Find the integers.

Let x represent the smaller integer and y represent the larger one. Since their sum is 6, one equation is

I $x + y = 6.$

Twice the smaller integer is $2x$. When this product is subtracted from the larger integer, the difference is 24. A second equation is

II $y - 2x = 24.$

Solve the system by substitution. Solve Equation I for y and substitute the result in Equation II.

$$y = 6 - x$$
$$6 - x - 2x = 24$$
$$-3x = 18$$
$$x = -6$$

Since $y = 6 - x$, we have $y = 6 - (-6) = 12$. The smaller integer is -6 and the larger is 12.   ∎

Example 2   Rachel, a lab technician, has two different concentrations of hydrochloric acid in stock. One is a 15% solution, and the other is a 40% solution. Mr. Brundage's Chem 1A class is conducting an experiment that requires 200 milliliters of a 25% solution. If the 25% solution is to be made by mixing the two solutions, how many milliliters of each must be used? If two variables are to be used, two equations must be found.

Let x = the number of milliliters of the 15% solution

y = the number of milliliters of the 40% solution.

Since 200 milliliters of the solution are needed, and this quantity is obtained by mixing the 15% solution and the 40% solution, the following equation must be satisfied.

I $x + y = 200$

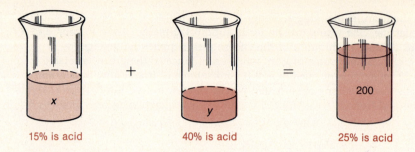

x	*y*	200
15% is acid	40% is acid	25% is acid

Figure 3.33

A second equation is obtained by observing that the amount of acid present before mixing remains the same after mixing. Its concentration does change. Figure 3.33 suggests the following equation.

II $0.15x$ $+$ $0.40y$ $=$ $0.25(200)$

Acid in x ml   plus   Acid in y ml $=$ Acid in mixture

If we multiply both sides of II by 100, an equivalent equation free of decimals is obtained.

III $15x + 40y = 5000$

I $x + y = 200$

Solving the system by elimination yields

III $15x + 40y = 5000$

IV $\underline{-15x - 15y = -3000}$ Multiply both sides of I by -15

$25y = 2000$ III + IV

$y = 80.$

Since the mixture contains 200 milliliters,

$x = 120.$

Rachel will use 120 milliliters of the 15% solution and 80 milliliters of the 40% solution. ■

Example 3 A manufacturer produces items that require casting and finishing. The maximum number of hours available for casting each week is 48 and the maximum number available for finishing is 32. Two items, A and B, are to be produced that require casting and finishing. The times required on each machine are outlined below.

	Item A	Item B	Maximum Hours Available
Casting	6 hours	4 hours	48
Finishing	5 hours	2 hours	32

If the machines work at capacity, how many items of each type can be produced?

Let a = the number of item A that will be produced weekly

b = the number of item B that will be produced weekly.

The total time available for casting is 48 hours. Item A requires 6 hours for each of the a units, a total of $6a$ hours. Item B requires 4 hours for each of the b units, a total of $4b$ hours. One equation is

I $6a + 4b = 48.$

Similarly, the 32 hours available for finishing is consumed at a rate of 5 hours for each unit of item A, a total of $5a$ hours. Two hours of finishing time are required for each unit of item B, a total of $2b$ hours. A second equation is

II $5a + 2b = 32.$

The equations are solved by elimination.

I	$6a + 4b = 48$	The sum of the casting times is 48 hours
II	$5a + 2b = 32$	The sum of the finishing times is 32 hours

$$6a + 4b = 48$$
$$-10a - 4b = -64$$
Multiply Equation II by -2, and add the result to Equation I
$$-4a = -16$$
$$a = 4$$
$$6 \cdot 4 + 4b = 48$$
$$b = 6$$

The number of units of item A produced is 4 and the number of units of item B is 6. ■

Example 4   Suppose that the demand for the items in Example 3 does not always require maximum use of the machines. Graph the area that shows all possible combinations of output for items A and B.

The equations now become inequalities.

$$6a + 4b \le 48$$
$$5a + 2b \le 32$$
$$a \ge 0 \quad \text{A negative number of items cannot be produced}$$
$$b \ge 0$$

The graph is shown in Figure 3.34. Any ordered pair in the shaded region represents a possible production schedule. ■

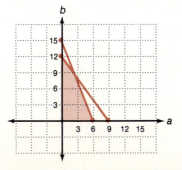

Figure 3.34

Example 5   An airplane can fly 360 miles into the wind in 3 hours. If the plane reverses direction and the wind conditions remain the same, the return trip to the airport takes 2 hours. Find the speed of the wind and the speed of the plane in still air.

One of the best ways to organize the information in motion problems of this type is to summarize the information in a table. We will first show the table and then explain how the information was found.

Let w = the speed of the wind

p = the speed of the plane in still air.

	Distance d	Rate r	Time t
Against the wind	360	$p - w$	3
With the wind	360	$p + w$	2

Distance $d = 360$ for both trips, because the plane flew 360 miles into the wind, and it returned covering the same distance.

When an airplane flies against the wind its speed in still air will be *decreased* by an amount equal to the speed of the wind. Its speed (rate r) will be $p - w$. When an airplane flies with the wind, its speed in still air will be *increased* by the speed of the wind. Its speed (rate r) will be $p + w$.

The flying time (t) against the wind is 3 hours. When the plane is flying with the wind only 2 hours are needed.

Two equations can be found using the relationship

Rate · Time = Distance

The equations are

I	$3(p - w) = 360$	
II	$2(p + w) = 360$	
III	$p - w = 120$	Divide both sides of I by 3
IV	$p + w = 180$	Divide both sides of II by 2

Solving equations III and IV by elimination yields

$2p = 300$	III + IV
$p = 150$	
$w = 30$	III − IV

The speed of the plane in still air is 150 miles per hour and the speed of the wind is 30 miles per hour. ∎

Example 6   Last year Mrs. Smyth had her life savings of $100,000 invested in three different ways: mutual funds, government bonds, and the money market. The mutual funds returned 7% on her investment, the bonds returned 8%, and the money market 10%. She had as much in the money market as in the other two investments together. Her total income from the investments was $8800. How much was invested at each rate?

Let x = the investment in mutual funds

y = the investment in bonds

z = the investment in the money market.

Medium	Amount invested	Rate of Return	Income
Mutual funds	x	7%	$0.07x$
Government bonds	y	8%	$0.08y$
Money market	z	10%	$0.10z$

The total amount invested was $100,000

I $\qquad x + y + z = 100,000$

Money market is equal to the sum of the others

II $\qquad z = x + y$

The total income was $8800

III $\qquad 0.07x + 0.08y + 0.10z = 8800$

Rewriting the equations in a convenient form for solving gives

$$
\begin{aligned}
x + y + z &= 100,000 \\
-x - y + z &= 0 \\
7x + 8y + 10z &= 880,000. \qquad \text{100(III)}
\end{aligned}
$$

Verify by elimination or determinants that $x = \$20,000$, $y = \$30,000$ and $z = \$50,000$. ■

Exercises 3.9

Write a system of equations or inequalities for each of the following and solve.

1. The sum of two numbers is −7. If the second number is subtracted from the first, the difference is 10. Find the numbers.

2. The difference of two integers is 36. If twice the smaller integer is added to three times the larger, the resulting sum is −117. Find the integers.

3. The sum of two consecutive odd integers is less than 37. Their difference is less than 3. Find the two largest consecutive odd integers satisfying these conditions.

4. The sum of two consecutive even integers is at least 46. If the larger is subtracted from twice the smaller, the resulting difference is at least 20. Find the two smallest consecutive even integers satisfying these conditions.

5. A boat can travel 60 miles downstream in 3 hours and 30 miles upstream in 2 hours. What is the speed of the current and the speed of the boat in still water?

6. In 2 hours an airplane can travel 600 miles with the wind and 550 miles against the wind. Find the speed of the wind and the speed of the plane in still air.

7. Two cars leave a gas station and travel in opposite directions. The difference in the speeds of the two cars is 10 kilometers per hour. At what speed is each car traveling if the cars are 200 kilometers apart after two hours?

8. When driving his truck, Ron averaged 10 kilometers per hour faster in level country than he did in the mountains. After traveling for 4 hours in level country and 2 hours through the mountains, Ron had traveled 295 kilometers. What was his speed in level country? What was his speed in the mountains?

9. Marie bought 3 small colas and 2 large colas for $3.60. If she had bought 2 small colas and 3 large colas, the cost would have been $3.90. What was the cost of each?

10. Cuesta Theater sold $410 worth of tickets. Adult tickets cost $2.00 each and children's tickets cost $1.00 each. If 310 tickets were sold, how many of each did the theater sell?

11. Jan invested all of her $3000 savings, with part at an annual rate of 6% and the rest at an annual rate of 9%. Her annual income from both of these investments is equivalent to 8% of her total investment. How much is invested at 9%?

12. Iva invests her money at annual rates of 8% and 9% and receives $260 per year from these investments. If she reversed the amount invested at each rate, her yearly income would be $250. How much did she invest at each rate?

13. Mr. Asire needs 2 liters of a 65% nitric acid solution. He has a 40% and an 80% solution on hand to use to make the mixture. How much of each should be used?

14. How many gallons of gasoline and gasohol (a mixture that is 90% gasoline and 10% alcohol) must be used to make a 20-gallon mixture that is 7% alcohol and 93% gasoline?

15. Peanuts worth $1.29 per pound are to be mixed with pecans worth $1.99 per pound in order to make a 10-pound mixture worth $1.71 per pound. How many pounds of each are needed?

16. The main floor of the Topliff Office Building contains 20,000 square feet and is divided into two office complexes. The rent on one office complex is $1.00 per square foot, and the other is $1.20 per square foot. If the total rent for both is $21,000, how many square feet are in each office complex?

17. A 24-inch bookshelf holds 12 dictionaries and 6 encyclopedias. A 15-inch bookshelf holds 4 dictionaries and 15 encyclopedias. How thick is each dictionary? How thick is each encyclopedia?

18. Four cartons of note pads and 6 cartons of paper together weigh 98 pounds. Three cartons of paper and 6 cartons of note pads weigh 93 pounds. How much does a carton of note pads weigh?

19. Anne sells insurance for two different companies. She receives a 15% commission of all the premiums from her sales paid to the first company. She receives a salary of $800 per month plus a 5% commission of all the premiums from her sales paid to the second company. The total premiums from her sales amount to $7000 for one month. Her income was $1400. What were the premiums paid to each company?

Write a system of three equations in three variables for each of the following and solve.

20. Lester has a total of $15,000 invested. One investment pays 6% annual interest, a second pays 5% annual interest, and the third is in a business. Two years ago he was assessed 3% of his investment in business to cover business losses and his total income was $400. Last year the business paid 9% dividends on his investment and his income from the three investments was $1000. How much does he have invested in each?

21. Nina has $25,000 invested. Part of it is invested at 5%, another part at 6% and the rest at 8%. The total yearly income from the three investments is $1600. The income from the 8% investment is the same as the sum of the incomes from the other two investments. How much is invested at each rate?

Write a system of inequalities for each of the following and graph.

22. Wood's Publishing Company publishes a single title in both paperback and hardback. Producing the paperback yields a profit of $400 per day while the hardback yields $700 per day. The company uses only one printing crew, which works a five-day week. Equipment limitations require that the paperback be produced no more than 3 days a week, whereas hardback production is limited to no more than 4 days a week.
    (a) Determine by graphing the possible production schedules.
    (b) Locate the ordered pairs of the corner points (where the boundary lines meet).
    (c) Calculate the profit at each of these corner points. These points produce the maximum or minimum profits.
    (d) If you owned the company, how would you assign your printing crew?

23. Trett's Machine Shop makes both drum and disc brakes. Both types require a lathe and grinder in production. Drum brakes require 2 hours on the lathe and 4 hours on the grinder; disc brakes require 4 hours on the lathe and 2 hours on the grinder. The shop makes a profit of $14 on each set of drum brakes and $18 on each set of disc brakes. It owns one lathe and one grinder, and both machines are used a maximum of 16 hours each day.
    (a) Determine the possible outcomes by graphing.
    (b) Locate the ordered pairs of the corner points (where the boundary lines meet).
    (c) Calculate the profit at each corner point. These points produce the maximum or minimum profits.
    (d) If you owned the company, how many of each would you produce?

Chapter 3   Summary

[3.1]   The **solution** to an equation in two or more variables consists of the values for each variable that together make the equation true. The solution to an equation in two variables is called an ordered pair, since the order in which the entries appear is important. The first element in the ordered pair is known as the **abscissa,** and the second as the **ordinate.**

Equations of the form $ax + by = c$, with a, b not both zero, are known as **linear equations.** Their graphs are straight lines.

[3.2]   Parallel lines have the same **slope.** The slope of a line between any two points (x_1, y_1) and (x_2, y_2) on it is given by

$$m = \frac{y_2 - y_1}{x_2 - x_1}, \qquad x_2 \neq x_1.$$

A vertical line has **no slope,** and a horizontal line has **zero slope.**

[3.3]   A **system** of linear equations involves two or more linear equations. A **solution** to a system of equations is that set of values for the variables that are solutions to every equation in the system. If the graphs of a linear system intersect in a single point, the system is **independent.** If they are parallel, the system is **inconsistent,** and if they coincide, the system is **dependent.**

[3.5]   The graph of a linear equation in three variables is a **plane.** A system of linear equations in three variables may have one or an infinite number of solutions.

[3.3] [3.4]   Systems of linear equations can be solved graphing, elimination,
[3.6] substitution, or by using determinants. A **determinant** is a square array of numbers enclosed by vertical bars, such as

$$\begin{vmatrix} a & b \\ c & d \end{vmatrix} = ad - bc.$$

The value of a 3×3 determinant is defined in terms of its **minors** as they apply to any row or column.

[3.7]   When determinants are used to solve linear systems, the process is known as **Cramer's rule.**

[3.8]   The graph of a **linear inequality** in two variables is a half-plane on one side of the line obtained by replacing the inequality symbol with an equals sign. The line is a part of the solution if the original symbol was $\leq$ or $\geq$. A system of linear inequalities has as its solution the area of the plane that is the intersection of the **half-planes** satisfying either inequality.

Chapter 3   Review

Complete the following.

1. The slope of a horizontal line is _____.
2. The formula for the slope $m = $ _____.
3. Two lines with the same slope are said to be _____.
4. The slope of a vertical line is _____.
5. If two lines have slopes that are negative reciprocals, the lines are _____.
6. The slope of the line through (2, 4) and (−3, 5) is _____.
7. The slope of $x - 2y = 3$ is _____.
8. The line through (−2, 3) and (3, 2) is _____ to the line through (0, 6) and (−10, 8).
9. Find a point $A(\ \ ,\ \ )$ so that A, B, and C are collinear for $B(0, 0)$ and $C(2, 2)$. Answers will vary.
10. The ordinate of a point is −2; the line through it and the point (2, 4) is parallel to the line $x - y = 4$.
The abscissa of the point is _____.
11. The solution to the system of equations $3x - y = 5$, $2x + 3y = 6$ is _____.

12. When the graphs of a system of equations coincide, the system is said to be

_____.

13. When the graphs of a system of equations do not intersect, the system is said to be _____ .

14. When the graphs of a system of equations intersect in a unique point, the system is said to be _____ .

Use the x-intercept, the y-intercept, and a check point to graph the equations in Exercises 15–18.

15. $y = 3x + 5$

16. $2x - 5y = 10$

17. $2(x - 5) + 3 = 3(y + 2) - 1$

18. $\dfrac{2x}{3} + \dfrac{7x}{3} = 4$

19. Solve $4x + 2y = 6$, $3x + y = 2$. Graph the solution.

20. Solve $2x + y = 4$, $4x - y = 1$ by substitution.

21. Solve $4a + 3b = 11$, $a - 2b = 0$ by elimination.

22. Solve $4m + 2n = 3$, $m - n = 5$ by substitution.

23. Graph the system of inequalities $3x + 2y \le 6$, $6x + 4y \ge 8$.

24. Solve $6s + t = 10$, $10s - 2t = 8$ by elimination.

25. Graph the system of inequalities $2x + y \le 1$, $x - y > 1$.

26. Graph the system of inequalities $y < x$, $y > 0$, $x \le 2$.

27. Solve $2a - 4b - c = 10$
$a - 2b + c = 2$
$2a - b + c = 3$.

28. Solve $x + 3y = 8$
$2x - 3y = 4$.

29. Graph $\{(x, y) \mid y \le x + 3\} \cap \{(x, y) \mid y \ge -x - 2\} \cap \{(x, y) \mid x \le 3\} \cap \{(x, y) \mid y \ge -1\}$

30. Solve $\dfrac{5}{x} - \dfrac{2}{y} = \dfrac{1}{2}$

$\dfrac{3}{x} + \dfrac{3}{y} = \dfrac{1}{4}$.

31. Given $5x - y = 10$ as the equation of a line, complete the following.
(a) $(0, \quad)$ is a point on the line.
(b) _____ is the slope of the line.
(c) _____ is the x-intercept.
(d) _____ is the y-intercept.
(e) $(\quad , 12)$ is a point on the line.

Write as a system of equations and solve.

32. Lonnie's Farms, Inc., sold 25 bushels of oats and 30 bushels of rye to the local co-op for $257.50. It also sold 50 bushels of oats and 10 bushels of rye (at the same price) to a local feed store for $315.00. What was the price per bushel of each kind of grain?

4

Polynomials and the Laws of Exponents

Recall from elementary algebra that the symbol a^2 means $a \cdot a$. What does the symbol a^n mean? Must n be a natural number? What meaning is assigned to the symbols

$$a^0, \quad 3^{-2}, \quad 4 \times 10^{-3}?$$

(In science the average distance to the sun is written as 9.3×10^7 miles; the mass of a hydrogen atom is about 1.67×10^{-24} grams.)

This chapter will explain each of the symbols above by extending the concept of an exponent beyond the natural numbers. It will also list some of the rules governing exponents and their use in algebraic expressions.

4.1 Properties of Exponents

The expression b^4 means $b \cdot b \cdot b \cdot b$. The letter b is the **base,** and 4 is the **exponent.**

$$\overset{\text{exponent}}{\underset{\text{base}}{b^4}}$$

When $b \cdot b \cdot b \cdot b$ is written as b^4, it is said to be in **exponential form.**

Recall from arithmetic that if a number is written as the product of two or more numbers, it is *factored.* Since $6 = 2 \cdot 3$, 2 and 3 are factors of 6. Six is also equal to $6 \cdot 1$, $(-2)(-3)$ and $(-6)(-1)$. Thus 6, 3, 2, 1, -1, -2, -3, and -6 are all factors of 6.

A natural number exponent indicates the number of times the base is used as a factor. As such it provides a convenient way to indicate a product of repeated factors.

Exponent is 3 $b^3 = b \cdot b \cdot b$ b is used as a factor 3 times

Exponent is n $b^n = b \cdot b \cdot b \cdots b$ b is used as a factor n times

When the exponent is 1 it usually is not written, since $b^1 = b$.

The expression b^4 is read, "b to the fourth power." The expression b^n is read, "b to the nth power." The expressions b^2 and b^3 can be read as "b to the second power" and "b to the third power" respectively, but they are also known by shorter names:

b^2 is called "b **squared**", and

b^3 is called "b **cubed.**"

The expressions b^2 and b^3 can be illustrated geometrically, as shown in Figures 4.1 and 4.2. Figure 4.1 illustrates that the area of a square of side b is $b \cdot b$, or b^2. Figure 4.2 illustrates that the volume of a cube of side b is $b \cdot b \cdot b$, or b^3.

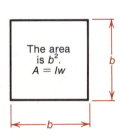

The area is b^2.
$A = lw$

The volume is b^3.
$V = lwh$

Figure 4.1   **Figure 4.2**

Example 1   Evaluate each of the following.

(a) $5^2 = 5 \cdot 5 = 25$

(b) $3^3 = 3 \cdot 3 \cdot 3 = 27$

(c) $\left(\dfrac{1}{2}\right)^4 = \dfrac{1}{2} \cdot \dfrac{1}{2} \cdot \dfrac{1}{2} \cdot \dfrac{1}{2} = \dfrac{1}{16}$

(d) $5^1 = 5$

(e) $2 \cdot 3^2 = 2 \cdot \mathbf{3} \cdot \mathbf{3} = 18$ The exponent applies to 3 only

(f) $2^3 \cdot 3^2 = (2 \cdot 2 \cdot 2) \cdot (3 \cdot 3) = 8 \cdot 9 = 72$ ∎

Example 2   Identify the base and exponent in each of the following.

(a) 7^2 The base is 7 and the exponent is 2.

(b) p The base is p and the exponent is **1**, since $p^1 = p$.

(c) $(3x)^2$ The base is $3x$ and the exponent is 2.

(d) -2^3 Think of -2^3 as $-1 \cdot 2^3$. The base is 2 and the exponent is 3. The exponent does not apply to the factor -1.

(e) $(-2)^4$ The parentheses indicate that the exponent applies to the negative sign as well as the 2. The base is -2 and the exponent is 4.

(f) $(x + 7)^6$ The base is $x + 7$ and the exponent is 6.   ∎

▶ *Caution* Unless symbols of grouping indicate otherwise, an exponent applies only to the quantity immediately to its left.

Right	*Wrong*
$2x^3 = 2 \cdot x \cdot x \cdot x$	$2x^3 = 2x \cdot 2x \cdot 2x$
$-5^2 = -1(5)^2 = -1 \cdot 5 \cdot 5 = -25$	$-5^2 = (-5)(-5) = 25$
$-3 \cdot 4^2 = -3 \cdot 4 \cdot 4 = -48$	$-3 \cdot 4^2 = (-3 \cdot 4)(-3 \cdot 4) = 144$

Example 3 Write each of the following in exponential form.

(a) $x \cdot x \cdot x \cdot x = x^4$

(b) $x \cdot x + 3 \cdot y \cdot y = x^2 + 3y^2$

(c) $3t \cdot 3t \cdot 3t = (3t)^3$

(d) $-2x \cdot 2x \cdot 2x \cdot 2x = -1 \cdot 2x \cdot 2x \cdot 2x \cdot 2x = -1(2x)^4 = -(2x)^4$

(e) $(a + b + c)(a + b + c) = (a + b + c)^2$

(f) $(-3x)(-3x)(-3x)(-3x) = (-3x)^4$ ■

Several definitions and theorems govern the way that exponents can be used to simplify expressions. For example, consider the product of b^2 and b^3.

$$b^2 \cdot b^3 = (b \cdot b)(b \cdot b \cdot b) \qquad \text{Definition of an exponent}$$
$$= b \cdot b \cdot b \cdot b \cdot b \qquad \text{Associative property}$$
$$= b^5 \qquad \text{Definition of an exponent}$$

This observation leads to Theorem 4.1.

Theorem 4.1

Product Rule for Exponents
If b is any real number, then
$$b^m \cdot b^n = b^{m+n} \qquad m, n \in J.$$

We will prove the theorem for the case where $m, n \in N$.

Proof:

By the definition of a natural number exponent,

$$b^m = \underbrace{(b \cdot b \cdot b \cdots b)}_{m \text{ factors}} \qquad b^n = \underbrace{(b \cdot b \cdot b \cdots b)}_{n \text{ factors}}$$

$$b^m \cdot b^n = \underbrace{(b \cdot b \cdot b \cdots b)}_{m \text{ factors}}\underbrace{(b \cdot b \cdot b \cdots b)}_{n \text{ factors}}$$

$$= \underbrace{(b \cdot b \cdot b \cdots b)}_{m+n \text{ factors}} \qquad\qquad \text{Associative property}$$

$$= b^{m+n}. \qquad\qquad\qquad \text{Definition of an exponent}$$

Example 4 Use the product rule to simplify each of the following:

(a) $x^7 \cdot x^9 = x^{7+9} = x^{16}$

(b) $a^2 \cdot a^3 \cdot a^5 = a^{2+3+5} = a^{10}$

(c) $(x + y)^3(x + y)^4 = (x + y)^{3+4} = (x + y)^7$

(d) $-2y^2 \cdot 3y^3 = (-2 \cdot 3)(y^2 \cdot y^3)$ Associative and commutative properties

$$= -6y^{2+3}$$

$$= -6y^5$$

(e) $10^3 \cdot 10^5 = 10^{3+5} = 10^8$ ∎

▶ *Caution* The product rule can be used only when the bases are the same.

Right	Wrong
$a^2 \cdot b^3 = a^2 b^3$	$a^2 b^3 = ab^5$
$a^3 \cdot a^3 \cdot b^5 = a^6 b^5$	$a^3 \cdot a^3 \cdot b^5 = ab^{11}$

Example 4e illustrates that even when a base is numerical the product is found by adding the exponents and retaining the same base.

Exponential expressions involving quotients can be simplified in much the same way as those involving products. For example, if $b \neq 0$, then

$$\frac{b^5}{b^2} = \frac{b \cdot b \cdot b \cdot b \cdot b}{b \cdot b}$$ Definition of an exponent

$$= \frac{\overset{1}{\cancel{b}} \cdot \overset{1}{\cancel{b}} \cdot b \cdot b \cdot b}{\cancel{b} \cdot \cancel{b}}$$ Reduce common factors

$$= 1 \cdot 1 \cdot b \cdot b \cdot b = b^3.$$

Thus $\dfrac{b^5}{b^2} = b^3$.

This result could easily have been achieved by subtracting the exponent in the denominator from the exponent in the numerator.

$$\frac{b^5}{b^2} = b^{5-2} = b^3$$

Notice that the exponent in the numerator is greater than the exponent in the denominator. What would result if this were not the case? Can we still divide exponential numbers with like bases by subtracting exponents? Consider

$$\frac{b^2}{b^4} = \frac{b \cdot b}{b \cdot b \cdot b \cdot b}$$ Definition of an exponent

$$= \frac{\overset{1}{\cancel{b}} \cdot \overset{1}{\cancel{b}}}{\cancel{b} \cdot \cancel{b} \cdot b \cdot b}$$ Reduce common factors

$$= \frac{1}{b^2}.$$

If we carry out the division by subtraction of exponents, then

$$\frac{b^2}{b^4} = b^{2-4} = b^{-2}.$$

Thus we see that negative integers are meaningful as exponents and that b^{-2} must be interpreted as $\frac{1}{b^2}$. The generalized form of this interpretation leads to the following definition.

> If $b \neq 0$ and $n \in J$,
>
> then $b^{-n} = \frac{1}{b^n}$.

Now consider the case where the numerator and denominator have the same exponent. Remember that any nonzero quantity divided by itself is 1. Thus

$$\frac{b^n}{b^n} = 1, \qquad b \neq 0.$$

If this division is carried out by subtracting exponents, we have

$$\frac{b^n}{b^n} = b^{n-n} = b^0.$$

The last two results will be meaningful if $b^0 = 1$. Since b can represent any nonzero algebraic expression, we have the following definition.

> If $b \neq 0$, then $b^0 = 1$.
> Note: 0^0 is not defined.

With the last two definitions in mind, we now state Theorem 4.2.

Theorem 4.2

> **Quotient Rule for Exponents**
> If b is any real number except zero, then
>
> $$\frac{b^m}{b^n} = b^{m-n} \qquad m, n \in J.$$

Example 5   Evaluate each of the following.

(a) $10^3 = 10 \cdot 10 \cdot 10 = 1000$

(b) $10^2 = 10 \cdot 10 = 100$

(c) $10^1 = 10$

(d) $10^0 = 1$

(e) $10^{-1} = \dfrac{1}{10^1} = \dfrac{1}{10}$

(f) $10^{-2} = \dfrac{1}{10^2} = \dfrac{1}{100}$

(g) $10^{-3} = \dfrac{1}{10^3} = \dfrac{1}{1000}$ ■

Example 6   Express 5642.324 using integral powers of 10.

$$5642.324 = 5000 + 600 + 40 + 2 + \frac{3}{10} + \frac{2}{100} + \frac{4}{1000}$$
$$= 5 \cdot 10^3 + 6 \cdot 10^2 + 4 \cdot 10^1 + 2 \cdot 10^0 + 3 \cdot 10^{-1}$$
$$+ 2 \cdot 10^{-2} + 4 \cdot 10^{-3} ■$$

Example 7   Simplify.

(a) $\dfrac{x^7}{x^2} = x^{7-2} = x^5$

(b) $\dfrac{14x^8}{-7x^6} = \dfrac{14}{-7} \cdot \dfrac{x^8}{x^6} = -2 \cdot x^{8-6} = -2x^2$

(c) $\dfrac{(x + y)^{12}}{(x + y)^8} = (x + y)^{12-8} = (x + y)^4$

(d) $(x + 9)^0 = 1$ ■

> If an algebraic expression involves negative exponents, it is **not** considered to be in simplest form.

Example 8   Simplify.

(a) $-2x^{-2} = -2 \cdot \dfrac{1}{x^2} = \dfrac{-2}{x^2}$

(b) $(2a)^{-1} = \dfrac{1}{(2a)^1} = \dfrac{1}{2a}$

(c) $\dfrac{(x + y)^4}{(x + y)^6} = (x + y)^{4-6}$

$$= (x + y)^{-2} = \dfrac{1}{(x + y)^2} ■$$

Work with negative exponents can frequently be made easier with the following observation.

$$\frac{a^{-n}}{b^{-m}} = \frac{\dfrac{1}{a^n}}{\dfrac{1}{b^m}}$$ Definition of a negative exponent

$$= \frac{1}{a^n} \cdot \frac{b^m}{1}$$ Definition of division of fractions

$$= \frac{b^m}{a^n}$$ Definition of multiplication of fractions

Thus $\dfrac{a^{-n}}{b^{-m}} = \dfrac{b^m}{a^n}.$

Three additional rules for exponents should be considered at this point.

Theorem 4.3

Power Rule for Exponents

If a and b represent real numbers and $m, n \in J$, then

(a) $(b^m)^n = b^{m \cdot n}$

(b) $(a \cdot b)^m = a^m \cdot b^m$

(c) $\left(\dfrac{a}{b}\right)^m = \dfrac{a^m}{b^m}, \qquad b \neq 0.$

We will prove parts (a) and (b) for $n \in N$.

(a) $(b^m)^n = \underbrace{b^m \cdot b^m \cdots b^m}_{n \text{ factors of } b^m}$ Definition of the exponent n

$\qquad = b^{\underbrace{m+m+\cdots+m}_{n \text{ times}}}$ Theorem 4.1
Add the exponents

$\qquad = b^{m \cdot n}$ There are n exponents of m

(b) $(a \cdot b)^m = \underbrace{(a \cdot b)(a \cdot b) \cdots (a \cdot b)}_{m \text{ factors of } (a \cdot b)}$ Definition of the exponent m

$\qquad = \underbrace{(a \cdot a \cdots a)}_{m \text{ factors}}\underbrace{(b \cdot b \cdots b)}_{m \text{ factors}}$ Associative and commutative
properties

$\qquad = a^m \cdot b^m$ Definition of an exponent

Theorem 4.3 is often interpreted verbally as follows.

(a) To raise a power to a power, multiply the exponents.
$$(b^m)^n = b^{m \cdot n}$$

(b) The mth power of a product is the product of the mth powers.
$$(a \cdot b)^m = a^m \cdot b^m$$

(c) The mth power of a quotient is the quotient of the mth powers.
$$\left(\frac{a}{b}\right)^m = \frac{a^m}{b^m}$$

Example 9   Simplify.

(a) $(a^2)^4 = a^{2 \cdot 4} = a^8$

(b) $(3^2)^2 = 3^{2 \cdot 2} = 3^4 = 3 \cdot 3 \cdot 3 \cdot 3 = 81$

(c) $(3^{-2})^2 = 3^{-2 \cdot 2} = 3^{-4} = \dfrac{1}{3^4} = \dfrac{1}{81}$

(d) $\left(\dfrac{a}{b}\right)^2 = \dfrac{a^2}{b^2}, \quad b \neq 0$

(e) $(x^2 y^3)^2 = (x^2)^2 (y^3)^2 = x^{2 \cdot 2} y^{3 \cdot 2} = x^4 y^6$

(f) $(-3x^2 y^3)^2 = (-3)^2 (x^2)^2 (y^3)^2$
$\qquad\qquad\quad = (-3)(-3) x^{2 \cdot 2} y^{3 \cdot 2}$
$\qquad\qquad\quad = 9x^4 y^6$

(g) $\left(\dfrac{-a^2}{b^3}\right)^2 = \left(\dfrac{-1 \cdot a^2}{b^3}\right)^2 = \dfrac{(-1)^2 (a^2)^2}{(b^3)^2}$

$\qquad\qquad\qquad = \dfrac{(-1)(-1)(a^{2 \cdot 2})}{b^{3 \cdot 2}}$

$\qquad\qquad\qquad = \dfrac{a^4}{b^6} \qquad\qquad b \neq 0 \quad \blacksquare$

Any or all of the rules for simplifying exponential expressions can occur in any problem. The order in which they are applied makes little difference as long as they are applied correctly and the order of algebraic operations is followed.

Rules of Exponents

If a and b are real numbers and m, n are integers, the following rules hold true.

1. $b^m \cdot b^n = b^{m+n}$ **Product Rule**

2. $\dfrac{b^m}{b^n} = b^{m-n}, \quad b \neq 0$ **Quotient Rule**

3. $(b^m)^n = b^{m \cdot n}$ **Power Rules**

   $(a \cdot b)^m = a^m \cdot b^m$

   $\left(\dfrac{a}{b}\right)^m = \dfrac{a^m}{b^m}, \quad b \neq 0$

4. $b^{-n} = \dfrac{1}{b^n}, \quad b \neq 0$ **Definition of a negative exponent**

5. $b^0 = 1, \quad b \neq 0$ **Definition of a zero exponent**

Example 10   Simplify. Write all answers with positive exponents only. Assume x, $y \neq 0$.

(a) $\dfrac{x^{-3}y^5}{x^{-2}y^{-3}} = x^{-3-(-2)}y^{5-(-3)}$ $\dfrac{b^m}{b^n} = b^{m-n}$

            $= x^{-3+2}y^{5+3}$

            $= x^{-1}y^8$

            $= \dfrac{y^8}{x}$ $b^{-n} = \dfrac{1}{b^n}$

(b) $\left(\dfrac{x^2y^2}{x^{-2}y^3}\right)^{-2} = (x^{2-(-2)}y^{2-3})^{-2}$ $\dfrac{b^m}{b^n} = b^{m-n}$

                $= (x^{2+2}y^{-1})^{-2}$

                $= (x^4y^{-1})^{-2}$

                $= x^{4 \cdot (-2)}y^{(-1) \cdot (-2)}$ $(b^m)^n = b^{m \cdot n}$

                $= x^{-8}y^2$

                $= \dfrac{y^2}{x^8}$ $b^{-n} = \dfrac{1}{b^n}$

(c) $\dfrac{-2^5y^4(y^{-4})^{-2}}{2^2y^{-2}} = \dfrac{-2^5y^4 \cdot y^{(-4) \cdot (-2)}}{2^2y^{-2}}$ $(b^m)^n = b^{m \cdot n}$

                 $= \dfrac{(-1) \cdot 2^5y^4y^8}{2^2y^{-2}}$

                 $= (-1) \cdot 2^{5-2}y^{4+8-(-2)}$ $b^m \cdot b^n = b^{m+n}; \quad \dfrac{b^m}{b^n} = b^{m-n}$

$$= (-1)2^3 y^{4+8+2}$$

$$= -8y^{14}$$

(d) $\dfrac{x^{2n}y^{n+1}}{x^n y^{n+2}} = x^{2n-n}y^{(n+1)-(n+2)}$

$$= x^n y^{-1} = \dfrac{x^n}{y} \quad \blacksquare$$

Exercises 4.1

Evaluate each of the following.

1. 2^3

2. 3^2

3. 10^{-1}

4. 10^{-3}

5. 10^0

6. $(6 \cdot 3)^0$

7. -2^5

8. -10^2

9. $(-10)^2$

10. $(-2)^5$

11. -5^{-1}

12. -6^{-2}

13. $(-3)^{-2}$

14. $(-2)^{-4}$

15. $(0.2)^2$

16. $(0.3)^2$

17. $\left(\dfrac{2}{3}\right)^2$

18. $\left(\dfrac{3^2}{4}\right)^2$

19. $\left(\dfrac{3^2}{5}\right)^{-2}$

20. $\left(-\dfrac{2}{3}\right)^{-3}$

21. $\dfrac{4^5}{4^3}$

22. $\dfrac{10^{-2}}{10^4}$

23. $\dfrac{10^{-3}}{10^{-4}}$

24. $\dfrac{2^{-5}}{2^{-5}}$

25. $(5^2 - 4^2)^{-2} \cdot 3^4$

26. $(6^2 - 5^2)^{-1} \cdot 11^2$

27. $\left(\dfrac{1}{3}\right)^{-2} \cdot 5^{-3} \cdot 10^0$

28. $\left(\dfrac{1}{3}\right)^{-3} \cdot 3^{-2} \cdot 3$

29. $\left(\dfrac{6^{-8} \cdot 5^{-3} \cdot 2^{-4}}{6^{-8} \cdot 5^{-4} \cdot 2^{-5}}\right)^{-1}$

30. $\left(\dfrac{10^{-7} \cdot 10^5 \cdot 10^{-3}}{10^{-5} \cdot 10^4 \cdot 10^{-4}}\right)^0$

31. $\left(\dfrac{8^2 \cdot 9^2 \cdot 25^2}{2^5 \cdot 3^3 \cdot 5^3}\right)^0$

Write each of the following in exponential form.

32. $x \cdot x \cdot x$

33. $5 \cdot 5 \cdot 5 \cdot 5 \cdot 5$

34. $(-6x)(-6x)(-6x)(-6x)$

35. $-8 \cdot 8 \cdot 8 \cdot 8 \cdot 8$

36. $3 \cdot 3 \cdot 3 \cdot (-x)(-x)(-x)(-x)$

37. $(x - y)(x - y)(-7)(-7)(-7)$

Write each of the following in powers of 10 form.

38. 956.3

39. 129.02

40. 1035.218

41. 8701.906

Simplify each of the following. Final results should contain positive exponents only. Assume no variable in the denominator is equal to zero.

42. $x^2 \cdot x^3$

43. $a^4 \cdot a^5$

44. $m^7 \cdot m^{-5}$

45. $t^3 \cdot t^{-2}$

46. $(d^2)^3$

47. $(p^{-2})^{-4}$

48. $\dfrac{y^5}{y^2}$

49. $\dfrac{s^9}{s^5}$

50. $(b^{-4})^2$

51. $(c^{-5})^3$

52. $\left(\dfrac{x^2}{x^7}\right)^{-1}$

53. $(ab)^4$

54. $\left(\dfrac{c^5}{c^9}\right)^3$

55. $s^6 \cdot t^{-5} \cdot s^{-4} \cdot t^6$

56. $(2p^4)(3r^{-5})^2(2p^{-3})(r^{11})$

57. $\dfrac{(3x^{-5})^2(2y^{-7})(2x^{10})^2(y^{13})}{9x^6y^7}$

58. $\dfrac{(9a^3)^2(-3a)^{-2}(6a)^{-3}}{(-3a)^3(12a)^{-1}}$

59. $\dfrac{(-2x^3)^3(-3x^2)^2}{-9x^{15}}$

60. $\dfrac{-x^2}{(-x)^2}$

61. $\dfrac{-y^3}{(-y)^3}$

62. $\dfrac{(-2a^2)^3}{(2^2a)^4}$

63. $\dfrac{(-p^2)^2}{(-p^3)^2}$

64. $\dfrac{(18m^3)^0}{(3m^2)^{-2}}$

65. $\dfrac{8a^{-3}b^{-1}}{6a^2b^{-4}}$

66. $\dfrac{9s^{-4}t^3}{12s^{-1}t^{-1}}$

67. $\dfrac{2m^6n^{-2}}{(-2)^4m^{-3}n^{-4}}$

68. $\left(\dfrac{6ab^{-2}}{3a^{-1}b^2}\right)^{-3}$

69. $\left(\dfrac{2p^{-3}q^{-2}}{(-2)^2pq^{-1}}\right)^{-2}$

70. $\left[\left(\dfrac{a^{-2}b^3c}{a^{-3}b^{-2}c^2}\right)^2\right]^{-1}$

Simplify each of the following. Assume no denominator is equal to zero.

71. $x^n \cdot x^m$

72. $y^n \cdot y^{2n}$

73. $a^{2n+1} \cdot a^{3n-2}$

74. $\dfrac{t^m}{t^n}$

75. $\dfrac{p^{m+1}}{p^{m-1}}$

76. $\dfrac{u^{n+3}}{u^{2n-1}}$

77. $(c^{n+1})^n$

78. $(r^{3n+2})^n$

79. $\dfrac{(a^n \cdot b^{n+1})^2}{a^{2n+1} \cdot b^{2n-2}}$

80. $\left(\dfrac{u^{3n-4} \cdot v^{2n-1}}{u^{2n+3} \cdot v^{3n-1}}\right)^{-2}$

81. $\dfrac{(x^{-n}y^{3n})^{-1}(x^{3n}y^{-2n})^{-1}}{x^{-5n}y^{-3n}}$

82. $\dfrac{(s^{-3n}t^{2n})^{-2}(s^nt^{-2n})^{-1}}{s^{-4n+1}t^{n+2}}$

Simplify each of the following. Assume no denominator is equal to zero.

83. $\dfrac{(a+b)^3}{(a+b)^2}$

84. $\dfrac{(x-y)^5}{(x-y)^3}$

85. $\dfrac{(m-n-p)^{-2}}{(m-n-p)^{-2}}$

86. $\dfrac{(r+s-t)^{-1}}{(r+s-t)^{-2}}$

87. $(x+y)^5 \div (x+y)^{-2}$

88. $(a-b+c)^{-4} \div (a-b+c)^3$

89. $\dfrac{(2x-y)^3(a-3b)^4}{(a-3b)^3(2x-y)^3}$

90. $\dfrac{(5s+2t)^{-2}(m-4n)^{-3}}{(m-4n)^{-5}(5s+2t)^{-3}}$

91. $\left(\dfrac{a^2}{a+b}\right)^4 \cdot \left(\dfrac{ab}{a+b}\right)^{-2}$

92. $\left(\dfrac{3m}{m-2}\right)^{-3} \cdot \left(\dfrac{m}{m-2}\right)^2$

93. Show that $\left(\dfrac{x}{y}\right)^{-m} = \left(\dfrac{y}{x}\right)^m$.

94. Show that $\dfrac{1}{a^{-m}} = a^m$.

95. Show by numerical example that $(a+b)^2 \neq a^2 + b^2$.

96. Show by numerical example that $(a+b)^3 \neq a^3 + b^3$.

Use a calculator to evaluate each of the following. Round all answers to four decimal places.

97. $A = \pi r^2$ for $\pi = 3.1416$ and $r = 0.3071$ inches

98. $A = 4\pi r^2$ for $\pi = 3.1416$ and $r = 0.1582$ meters

99. $V = \dfrac{4}{3}\pi r^3$ for $\pi = 3.1416$ and $r = 1.896$ centimeters

100. $V = \pi r^2 h$ for $\pi = 3.1416$, $r = 3.8503$ feet and $h = 2.1950$ feet

101. $Q = (x^{3n-2})^n$ for $x = 0.9212$ and $n = 2$

102. $u = \dfrac{p^{3a-4}}{p^{a+2}}$ for $p = 0.9991$ and $a = -1$

Review Exercises

103. Solve
$$3x - 2y = 6$$
$$x + 3y = 2$$
by graphing.

104. Solve
$$5x + 4y = 1$$
$$x - 6y = 3$$
by elimination.

Graph each of the following.

105. $-x + y > -3$
$\quad\quad x \geq 0$

106. $x + 3y \leq 4$
$\quad\quad x \geq -2$

4.2   Adding and Subtracting Polynomials

The earlier chapters often referred to algebraic expressions. A *polynomial* is a special type of algebraic expression. Before defining a polynomial, we first need to consider the concept of an algebraic term. A **term** is the product of a number and one or more variables raised to positive integer or zero powers. Some examples of terms are

$$5, \quad 3x^2, \quad \frac{1}{4}x^5y, \quad 9x^5yz, \quad \text{and} \quad -2x.$$

In the term $5xy$ the 5 is called the **numerical coefficient** of xy. If a term does not have a written numerical coefficient, such as in x^2y, it is understood to be 1. Thus $x^2y = 1x^2y$. When one or more terms are added, the result is a polynomial.

> A **polynomial** is the sum of one or more algebraic terms satisfying the following conditions.
>
> 1. **All exponents must be whole numbers.**
> 2. **No term may contain a variable in the denominator.**
> 3. **When simplified, no variable may appear under a radical sign.**

Thus $13x$, $3x + 5$, $x^3 - 7x + \dfrac{1}{2}$,   and $x^6 + 3x^2 - \dfrac{x}{2} + 7$ are all

polynomials since they satisfy the conditions of the definition. The following algebraic expressions are not polynomials.

$x^2 + 2x^{-1}$ The exponent, (-1), is not a whole number

$\sqrt{x} + 3xy + x^2$ A variable appears under a radical sign

$6xy^2 + 4y - \dfrac{6}{x^2}$ A variable appears in the denominator

Polynomials can contain a single term or any number of terms. Specific names are given to polynomials with one, two, or three terms.

A **monomial** is a polynomial with one term.

A **binomial** is a polynomial with two terms.

A **trinomial** is a polynomial with three terms.

Example 1   Classify each polynomial.

(a) $6x^2$, $3y$ and 7 are monomials.

(b) $-2x + \frac{1}{2}$ and $\frac{1}{5}x^2y + y^3$ are binomials.

(c) $-6x^2 + 9x - 8$ is a trinomial.

(d) $4x^2 + 9xy + 3x^2$ at first appears to be a trinomial. However, when simplified it is $7x^2 + 9xy$, which is a binomial.   ■

▶ *Caution*   Before a polynomial is classified as a monomial, binomial, or trinomial, it must first be simplified.

Polynomials are also classified by their degree. If only one variable is present, the **degree of a polynomial** is the largest exponent that appears on the variable. When a polynomial is written in descending powers of a variable, it is in **standard form.** This means that the term with the largest exponent appears first, the term with the second largest exponent appears next, and so on.

Example 2   Write each of the following in standard form and determine its degree.

(a) $x^5 + 2x^3 + 5$ is a fifth-degree trinomial in standard form. The first term, x^5, is a fifth-degree term; the second term, $2x^3$, is a third-degree term, and the third term, 5, is a zero-degree term.

(b) $\frac{2}{7}x^7 - \frac{1}{8}x^6 + 5x^2 - 8$ is a seventh-degree polynomial in standard form.

(c) 3 is a zero-degree monomial since $3 = 3x^0$.

(d) $-x^2 + 7x + x^9 + 5$, when written in standard form, is $x^9 - x^2 + 7x + 5$ and is a ninth-degree polynomial.

(e) $-x + 8$ is a first-degree binomial in standard form.   ■

When the terms of a polynomial contain more than one variable, the degree of any one term is the sum of the exponents on the variables. Thus $3x^2y^5z$ is an eighth-degree term, since the sum of the exponents 2, 5, and 1 is 8.

Polynomials are added or subtracted by combining **like** or **similar** terms. Like or similar terms differ only by their numerical coefficients. Except for order, their variable parts must be identical.

Example 3   Determine whether the terms are like or unlike terms.

(a) $3x^2$, $\frac{1}{2}x^2$, and $9x^2$ are like terms.

(b) $-5xy$, $4yx$, and $-\frac{3}{7}xy$ are like terms, since $xy = yx$ by the commutative property.

(c) $4x^2y$ and $7xy^5$ are unlike terms, since x^2y and xy^5 are not identical.

(d) $13x$ and $13y$ are unlike terms, since x is not identical to y.   ■

If we assume that the coefficients of the variables in a polynomial represent real numbers, the properties of real numbers stated in Chapter 1 can be used to add or subtract them. For example,

$$2x^2 + 5x^2 = (2 + 5)x^2 = 7x^2.\qquad \text{Distributive property}$$

Example 4   Add $2x^2 + 7x + 9$ to $4x^2 + 5x + 7$.

$$2x^2 + 7x + 9 + 4x^2 + 5x + 7$$
$$= 2x^2 + 4x^2 + 7x + 5x + 9 + 7 \qquad \text{Commutative property}$$
$$= (2x^2 + 4x^2) + (7x + 5x) + (9 + 7) \qquad \text{Associative property}$$
$$= (2 + 4)x^2 + (7 + 5)x + (9 + 7) \qquad \text{Distributive property}$$
$$= 6x^2 + 12x + 16 \quad ■$$

Example 5   Add $-x^3 + 3x^2$ to $5x^3 + x^2 - x - 7$.

$$-x^3 + 3x^2 + 5x^3 + x^2 - x - 7$$
$$= -x^3 + 5x^3 + 3x^2 + x^2 - x - 7 \qquad \text{Commutative property}$$
$$= (-x^3 + 5x^3) + (3x^2 + x^2) - x - 7 \qquad \text{Associative property}$$
$$= (-1 + 5)x^3 + (3 + 1)x^2 - x - 7 \qquad \text{Distributive property}$$
$$= 4x^3 + 4x^2 - x - 7 \quad ■$$

Example 6   Add $(x^2 + xy) + (3x^2 - 2xy + 5y^2) + y^3$.

$$(x^2 + xy) + (3x^2 - 2xy + 5y^2) + y^3$$
$$= (x^2 + 3x^2) + (xy - 2xy) + 5y^2 + y^3$$
$$= 4x^2 + (-xy) + 5y^2 + y^3$$
$$= 4x^2 - xy + 5y^2 + y^3 \quad ■$$

▶ **Caution**   Avoid the temptation to add terms such as $2x^3$ and $3x^2$ to get $5x^5$. Notice that while x^2 and x^3 do have the same base, they have different exponents. They are not like terms.

Polynomials can be subtracted as well as added. To discover one method for doing this, recall that

$$-(a + b) = (-a) + (-b)$$
$$= -a - b.$$

This shows that to remove a symbol of grouping that is preceded by a negative sign, all signs within the grouping symbol are changed. Recall also that in terms of multiplication

$$a - (b + c) = a + (-1)(b + c)$$
$$= a + (-b) + (-c)$$
$$= a - b - c.$$

Example 7   Remove the grouping symbols and change the signs accordingly.

(a) $-(7x^2 + 5x - 3) = -7x^2 - 5x + 3$

(b) $-(-4y^2 + 6xy - 9y) = 4y^2 - 6xy + 9y$ ■

Extending this concept leads to a rule for subtracting polynomials. Subtraction is carried out in a manner similar to that used to subtract real numbers.

> To **subtract one polynomial from another,** change the sign of every coefficient in the polynomial being subtracted and add.

Example 8   Subtract $4x^3 - 3x^2 + 2x$ from $6x^3 + 4x^2 + x - 9$.

$$6x^3 + 4x^2 + x - 9 - [4x^3 - 3x^2 + 2x]$$

Change the sign of every term in the polynomial being subtracted, then add.

$$= 6x^3 + 4x^2 + x - 9 + [-4x^3 + 3x^2 - 2x]$$
$$= [6x^3 + (-4x^3)] + (4x^2 + 3x^2) + [x + (-2x)] - 9 \qquad \text{Commutative and associative properties}$$

$$= 2x^3 + 7x^2 - x - 9 \quad ■$$

Example 9   Subtract $(3x^3 + 2x - 4) - (-5x^3 + x^2 - 6x + 9)$.

$$(3x^3 + 2x - 4) - (-5x^3 + x^2 - 6x + 9)$$
$$= 3x^3 + 2x - 4 + [+5x^3 + (-x^2) + 6x + (-9)] \qquad \text{Change signs}$$
$$= (3x^3 + 5x^3) + (-x^2) + (2x + 6x) + [-4 + (-9)]$$
$$= 8x^3 + (-x^2) + (8x) + (-13) \qquad \text{Combine like terms}$$
$$= 8x^3 - x^2 + 8x - 13 \quad ■$$

In practice, many of the steps shown in Example 9 are carried out mentally. Compare the following example with Example 9 to see which steps can usually be eliminated.

Example 10   Subtract $3x^3 + 2x - 4 - (-5x^3 + x^2 - 6x + 9)$.

$$3x^3 + 2x - 4 - (-5x^3 + x^2 - 6x + 9)$$
$$= 3x^3 + 2x - 4 + 5x^3 - x^2 + 6x - 9 \qquad \text{Change signs}$$
$$= 8x^3 - x^2 + 8x - 13 \qquad \text{Combine like terms} \quad ■$$

Example 11 Simplify $-\{x^2 + x - [2x^2 + 2x - (3x^2 + 3x)]\}$.

Begin by removing the *innermost* grouping symbol first. Be careful to change all signs when the symbol is preceded by a negative sign.

$-\{x^2 + x - [2x^2 + 2x - (3x^2 + 3x)]\}$

$= -\{x^2 + x - [2x^2 + 2x - 3x^2 - 3x]\}$ Change signs

$= -\{x^2 + x - [-x^2 - x]\}$ Combine like terms

$= -\{x^2 + x + x^2 + x\}$ Change signs

$= -\{2x^2 + 2x\}$ Combine like terms

$= -2x^2 - 2x$ Change signs ∎

Polynomials are frequently represented by symbols such as $P(x)$, $Q(x)$, or $M(y)$, to name just a few. The expression $P(x)$ does not mean P times x, but rather a polynomial in the variable x. $M(y)$ represents a polynomial in the variable y. Some examples of polynomials are

$P(x) = x^2 + 2x + 5$

$M(y) = -y^2 + 9$

$R(t) = t^3 - t^2 - 2t + 5.$

Polynomial notation is often used in problems asking for the value of the polynomial for a specific value of the variable.

If $P(x) = x^2 + 2x + 5$, then $P(2)$ stands for the value of the polynomial when $x = 2$.

$P(x) = x^2 + 2x + 5$

$P(2) = (2)^2 + 2 \cdot 2 + 5$ Replace each x by 2

$P(2) = 4 + 4 + 5$

$P(2) = 13$

Thus, for $x = 2$ the value of the polynomial $P(x)$ is 13.

Example 12 $Q(y) = 2y^3 + 6y - 1$. Find $Q(3)$.

$Q(3) = 2 \cdot 3^3 + 6 \cdot 3 - 1$ Replace each y by 3

When a term involves an exponent, the operation indicated by the exponent is carried out first.

$Q(3) = 2 \cdot 27 + 6 \cdot 3 - 1$

$= 54 + 18 - 1$

$= 71$

For $y = 3$ the value of the polynomial $Q(y)$ is 71. ∎

Example 13   A potter makes and sells x vases at a cost, $C(x)$, expressed in dollars as $C(x) = 7x + 50$. Her profit, $P(x)$, in dollars is $P(x) = x^2 - 11x - 5$.

(a) Find a polynomial to express her revenue, $R(x)$, if the revenue equals the cost plus the profit.

$$R(x) = C(x) + P(x)$$
$$= (7x + 50) + (x^2 - 11x - 5) \quad \text{Substitute for } C(x) \text{ and } P(x)$$
$$= x^2 - 4x + 45$$

Her revenue $R(x)$ is $x^2 - 4x + 45$.

(b) Find the revenue in dollars if $x = 25$.
When $x = 25$,

$$R(25) = (25)^2 - 4(25) + 45$$
$$= 625 - 100 + 45$$
$$= 570.$$

Her revenue, $R(x)$, is $570. ∎

Exercises 4.2

Determine whether or not each of the following is a polynomial.

1. $x + 2$

2. $-3y + 4$

3. $\dfrac{1}{p}$

4. $\dfrac{3}{m} + 2$

5. $t^6 - 5t + 1$

6. $m^7 + 3m^2 - 5$

7. $a^{-3} + 5$

8. $c^{-4} + c^2 + 1$

9. $\dfrac{2s^3}{3} + \dfrac{4s}{3} - 2$

10. $\dfrac{-5r^2}{4} + \dfrac{7r}{3} - \dfrac{1}{5}$

11. $3^x + 5$

12. $8\sqrt{x} - 11$

Write each of the following in standard form, simplify, and (a) determine its degree; (b) determine the coefficient of the highest-degree term; (c) state whether the polynomial is a monomial, binomial, or trinomial.

13. $5p^2 + p$

14. $t^2 - 7t$

15. $19r + 4r$

16. $-16m^3 + 7m^3$

17. $5q^0 - 9q^3 - q^5$

18. $b^0 - 8b^4 - 11b^5$

19. $x^2 - 5x + 2x^3 - 4x + 4x^2$

20. $3a - a^2 + 4a - a^3 + 2a^2 + 4a^3$

Perform the indicated operations for each of the following.

21. $6a + a$

22. $8x + 3x$

23. $9y - 6y$

24. $p^2 + 4p - 2p^2 + 6p$

25. $3m - 7m^2 - 6m + m^2$

26. $8m^2n + 3m^2n$

27. $9ab^2 + 3ab^2$

28. $\dfrac{1}{3}s - \dfrac{1}{2}s$

29. $\dfrac{1}{4}t - \dfrac{1}{2}t$

30. $(3a + 4) + (2a - 5)$

31. $(6x^2 + x) + (2x^2 - x)$

32. $(y^2 - 1) + (2y^2 + 1)$

33. $(7b + 4) + (-8b - 4)$

34. $(3m - 2) - (m + 7)$

35. $(2t - 4) - (5t - 7)$

36. $(5y + 1) - (y^2 + 3y - 5)$

37. $b - (b^2 - 2b + 1)$

38. $(5x^2 + 2x - 5) + (x^2 - 3x + 4) - (3x^2 - 7x + 1)$

39. $(3a + 5) - (a^2 + a + 1) + (a^2 + a + 1)$

40. $(m + n - 1) - (2m + n - 7) - (5m - 7n + 3)$

41. $(7k - 5) - (8k + 5) + (3k + 4) - (6k + 11)$

42. $(9x - 3) - (2x + 7) - (8x + 9) + (x - 11)$

43. $(x^2y - 2xy^2 - 3) - (xy^2 + 3x^2y + 1)$

44. $(5s^2t - 9st^2 + 4st) - (8st^2 + s^2t + 6st)$

45. Subtract $8r^2 - 7r + 3$ from $2r^2 + 9r - 2$.

46. Subtract $5s^2 - 6st + 4t^2 + 7t - 2$ from $5s^2 - 8st + 6s - 9t + 4$.

47. Subtract $x^3y^2 - 5xy^2 + 6x^2y - 11xy$ from $xy^2 - 8x^2y + x^3y^2 + 7xy$.

48. Subtract $9a^2b - 13ab^3 + 7ab$ from $16a^3b - 18ab^2 + 8ab$.

Simplify each polynomial by removing all grouping symbols and performing the indicated operations.

49. $x + 3y - [3x - (4x + y)]$

50. $5a + (2b - a) - [(2b - 4) - a]$

51. $8 - (2x - [3y + (2y - 5x)])$

52. $4a + 3 - [6b - (3 + 5a - 2)]$

53. $5m - [m - (2m + 1)]$

54. $8 - [(1 + 8p) - (5 - 2p)]$

55. $7(3s - 2t) - [s - t - (s - t)]$

56. $-(k + 3) - 4[3k - 2(4 + 5k) - 8]$

57. $3[6r - (r + s)] + 3[-(2r + 4s) + 5(r - s)]$

58. $11x^2 - [6 - (4x - x^2)] - 2[3 - 3(4x - x^2)]$

59. $-3[-x^2 + 2xy - 3y^2] - [-6x^2 + 4xy + 2y^2] + [3 - (5xy + y^2)]$

60. $7[-a^2 - 5ab + 2b^2] - [-3a^2 - 7ab + 6b^2] - 2[1 - 4(ab + b^2)]$

Given $P(x) = x^2 - 1$, $Q(x) = 2x + 3$ and $R(x) = 2x^2 - 3x + 4$, find the following.

61. $P(0)$, $P(2)$, $P(-3)$

62. $Q(1)$, $Q(-3)$, $Q(4)$

63. $P(x) + Q(x)$

64. $P(x) - R(x)$

65. $2Q(x) - 3R(x)$

66. $4P(x) - Q(x)$

67. $P(2) - Q(3) + R(0)$

68. $Q(-3) + P(0) - R(-1)$

Use a calculator to perform the indicated operations.

69. $(-6.85x^2 + 3.241x - 9.173) - (-0.79x^2 + 2.099x + 8.676)$

70. $(0.001y^3 - 17.842y^2 + 1.111y) - (-1.899y^3 - 16.001y^2 + 0.018y)$

71. Evaluate $0.02p^2 + 0.3p - 0.5$ when $p = 0.2$.

72. Evaluate $2a^3 - 7a^2 + 8a - 10$ when $a = 3.1$.

73. A company's cost in dollars for producing and marketing a certain product is given by $C(x) = 50 + 30x$ where x is the number of units produced and marketed. Its revenue in dollars from this product is given by $R(x) = 90x + x^2$ and its profit in dollars is given by $P(x) = R(x) - C(x)$. Find the polynomial representing profit. If the company produces and markets 50 units per day, what is its daily profit?

74. Togs, Incorporated manufactures and sells x pairs of custom-fitted tennis shoes for a profit in dollars of $P(x) = x^2 + 31x - 200$. The total revenue in dollars for x pairs is $R(x) = x^2 + 49x + 600$. If the cost to manufacture and sell x pairs is represented by $R(x) - P(x) = C(x)$, find $C(x)$. Find the cost to manufacture and sell 100 pairs. Find the profit from manufacturing and selling 100 pairs.

Review Exercises

75. Solve $\quad 5a - b = 2$ by substitution.
$\qquad\quad 3a + 3b = 5$

76. Solve $\quad m - 3n = 8$ by elimination.
$\qquad\quad 3m - 2n = 5$

77. Graph the region that is the solution to $\{(x, y) \mid y - x \le 3\} \cap \{(x, y) \mid x + y < 5\}$.

78. Graph the region that is the solution to $\{(x, y) \mid x - y > -6\} \cap \{(x, y) \mid -x - y > 2\}$.

Simplify each of the following.

79. $\dfrac{(3x^{-4}y^{-3})^{-1}}{(-9)^{-1}(x^3y^{-2})}$

80. $\dfrac{[-(10)^4(10^{-2})^2(3^{-3})]^1}{[(-10)^2(-10)^3(9^2)]^0}$

81. $\left(\dfrac{x^{5n-4}y^{2n+3}}{x^{5n-5}y^{2n-5}}\right)^{-2}$

82. $\dfrac{(x^{-2n}y^{3n})^{-2}(x^{4n}y^{-2n})^{-1}}{x^{-n}y^{-2n}}$

4.3   Multiplication of Polynomials

The easiest product to find when multiplying polynomials is that of two monomials. It is found by applying the commutative and associative properties together with the rules for exponents. Consider the product of $3x^2y$ and $2xy^5$.

$$3x^2y \cdot 2xy^5$$

$$= 3 \cdot 2x^2 \cdot x \cdot y \cdot y^5 \qquad \text{Commutative property}$$

$$= (3 \cdot 2)(x^2 \cdot x)(y \cdot y^5) \qquad \text{Associative property}$$

$$= 6x^3y^6 \qquad\qquad\qquad\quad b^m \cdot b^n = b^{m+n}$$

Example 1   Multiply.

(a) $(2rs^2)(8r^7s^9) = (2 \cdot 8)(r \cdot r^7)(s^2 \cdot s^9) = 16r^8s^{11}$

(b) $(-4x^2y^5)(7x^7y^8) = (-4 \cdot 7)(x^2 \cdot x^7)(y^5 \cdot y^8) = -28x^9y^{13}$

(c) $(-4mn)(-3m^2n)(2m^2n^6)$

$$= [(-4)(-3)(2)][m \cdot m^2 \cdot m^2][n \cdot n \cdot n^6]$$

$$= 24m^5n^8 \quad \blacksquare$$

The product of two polynomials, other than two monomials, is found by using the distributive property.

$$a(b + c + d + \cdots + z) = a \cdot b + a \cdot c + a \cdot d + \cdots + a \cdot z$$

Example 2   Find the product of $2x^2$ and $x^3 + 3x^2 + 6$.

$$2x^2(x^3 + 3x^2 + 6)$$

$$= 2x^2 \cdot x^3 + 2x^2 \cdot 3x^2 + 2x^2 \cdot 6 \qquad \text{Distributive property}$$

$$= 2x^5 + 6x^4 + 12x^2 \quad \blacksquare$$

Example 3   Find the product of $-3x^5$ and $x^2 - 2xy - 9$.

In order to understand how the signs of the product are found, remember that subtraction is defined as addition of the additive inverse.

$$-3x^5(x^2 - 2xy - 9)$$

$$= -3x^5[x^2 + (-2xy) + (-9)] \qquad \text{Definition of subtraction}$$

$$= -3x^5(x^2) + (-3x^5)(-2xy) + (-3x^5)(-9) \qquad \text{Distributive property}$$

$$= -3x^7 + (6x^6y) + (27x^5)$$

$$= -3x^7 + 6x^6y + 27x^5 \quad \blacksquare$$

The first two steps in finding the product in Example 3 are usually completed mentally. With this approach all products are separated by + signs.

Example 4 Find the product of $-2c^3$ and $c^2 + 5c - 8$ without rewriting the subtraction as addition.

$$-2c^3(c^2 + 5c - 8)$$
$$= -2c^5 + -10c^4 + +16c^3$$

| The product of $-2c^3$ and c^2 | The product of $-2c^3$ and $5c$ | The product of $-2c^3$ and -8 |

The last step in the product is then rewritten as

$$= -2c^5 - 10c^4 + 16c^3. \quad \blacksquare$$

The product of two binomials is found by using the distributive property twice. To see this, consider the product of $(2x + 5)$ and $(x^2 + 6)$. Remember that $a(b + c) = a \cdot b + a \cdot c$ by the distributive property.

$$\underbrace{(2x + 5)}_{a}(\underbrace{x^2}_{b} + \underbrace{6}_{c})$$

$$= \underbrace{(2x + 5)}_{a} \cdot \underbrace{x^2}_{b} + \underbrace{(2x + 5)}_{a} \cdot \underbrace{6}_{c} \qquad \text{Distribute } (2x + 5) \text{ over } x^2 \text{ and } 6$$

$$= 2x \cdot x^2 + 5 \cdot x^2 + 2x \cdot 6 + 5 \cdot 6 \qquad \text{Distribute } x^2 \text{ and } 6 \text{ over } 2x + 5$$

$$= 2x^3 + 5x^2 + 12x + 30$$

Notice that the next to last line in the product, except for order, is the result of multiplying each term of the second polynomial by each term of the first polynomial. This observation leads to the following rule.

> To find the **product of two polynomials,** multiply each term in the second polynomial by each term in the first polynomial and combine like terms.

Example 5 Multiply $(2x + 3)(x^2 - 6x - 9)$.

Multiply each term in $x^2 - 6x - 9$ by each term in $2x + 3$. **(A term includes the sign before it.)**

$$(2x + 3)(x^2 - 6x - 9) = 2x(x^2) + 2x(-6x) + 2x(-9) + 3(x^2) + 3(-6x) + 3(-9)$$
$$= 2x^3 - 12x^2 - 18x + 3x^2 - 18x - 27$$
$$= 2x^3 - 9x^2 - 36x - 27 \qquad \text{Combine like terms} \quad \blacksquare$$

The product of two binomials can be found when certain patterns are observed. One such pattern, called the **FOIL method,** is illustrated below.

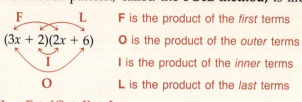

F is the product of the *first* terms

O is the product of the *outer* terms

I is the product of the *inner* terms

L is the product of the *last* terms

$$\text{FOIL} = F + [O + I] + L$$

Example 6   Find the product of each of the following by the FOIL method.

$$\quad\quad\quad\quad\quad\quad\quad F \ + \quad [O \ \ + \ I] \ + L$$

(a) $(x + 2)(x + 3) = x^2 + \quad [3x \ \ + 2x] + 6 = x^2 + \ 5x + 6$

(b) $(2x - 1)(x + 6) = 2x^2 + \ [12x \ \ - \ x] \ - 6 = 2x^2 + 11x - 6$

(c) $(3x - 2)(x - 4) = 3x^2 + [-12x - 2x] + 8 = 3x^2 - 14x + 8$ ∎

Two types of products occur so frequently that the pattern of multiplication should be memorized, symbolically as well as in words. The first of these is the product of the sum and difference of the same two quantities. Consider

$$(a + b)(a - b).$$

The sum   The difference

When the product is found by using the FOIL method,

$$(a + b)(a - b) = a^2 + [-ab + ab] - b^2$$
$$\quad\quad\quad\quad\quad F + \ [O \ \ + \ I] \ + L$$
$$\quad\quad\quad\quad = a^2 - \ b^2, \quad\quad\quad\quad \text{Combine like terms}$$

the product is the difference of the squares of the two quantities. This observation leads to the following rule.

> The product of the sum and difference of the same two quantities is the difference of their squares.
>
> $$(a + b)(a - b) = a^2 - b^2$$

Example 7   Multiply.

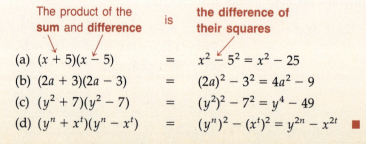

The product of the sum and difference	is	the difference of their squares
(a) $(x + 5)(x - 5)$	=	$x^2 - 5^2 = x^2 - 25$
(b) $(2a + 3)(2a - 3)$	=	$(2a)^2 - 3^2 = 4a^2 - 9$
(c) $(y^2 + 7)(y^2 - 7)$	=	$(y^2)^2 - 7^2 = y^4 - 49$
(d) $(y^n + x^t)(y^n - x^t)$	=	$(y^n)^2 - (x^t)^2 = y^{2n} - x^{2t}$ ∎

With two binomials that are identical, such as $(a + b)(a + b)$ or $(a - b)(a - b)$, a definite pattern also emerges.

$$\text{F} \quad \text{O} \quad \text{I} \quad \text{L}$$

$$(a + b)(a + b) = (a + b)^2 = a^2 + ab + ab + b^2 = a^2 + 2ab + b^2$$
$$(a - b)(a - b) = [a + (-b)]^2 = a^2 - ab - ab + b^2 = a^2 - 2ab + b^2$$

First term Last term The square of the first term Twice the product of the terms The square of the last term

The rule that describes these products is summarized below.

The **square of a binomial** is equal to the square of the first term, plus twice the product of the first and last term, plus the square of the last term.

$$(a + b)^2 = a^2 + 2ab + b^2$$

and

$$(a - b)^2 = a^2 + (-2ab) + b^2 = a^2 - 2ab + b^2$$

▶ *Caution* Be careful to avoid making a common error that may occur when raising a binomial to a power:

Right | *Wrong*
$(a + b)^2 = a^2 + 2ab + b^2$ $(a + b)^2 = a^2 + b^2.$

In general, $(a + b)^n \neq a^n + b^n$; $n \neq 1$.

Example 8 Find the following.

(a) $(x + 3)^2$

$$(x + 3)^2 = x^2 + 2 \cdot 3 \cdot x + 3^2 = x^2 + 6x + 9$$

First term Last term Square of the first term Twice the product of the terms Square of the last term

(b) $(2x - 1)^2$

$$(2x - 1)^2 = (2x)^2 + 2(2x)(-1) + (-1)^2 = 4x^2 - 4x + 1 \quad ■$$

Example 9 Find $(x^n + y^{n+1})^2$.

$$(x^n + y^{n+1})^2 = (x^n)^2 + 2 \cdot x^n \cdot y^{n+1} + (y^{n+1})^2$$
$$= x^{2n} + 2x^n y^{n+1} + y^{2n+2} \qquad (b^m)^n = b^{m \cdot n} \quad ■$$

The rules for multiplying polynomials can be expanded to more complex products. The next three examples illustrate such products.

Example 10   Multiply $(x + 2)(x + 1)(x - 2)$.

First commute the factors to simplify the multiplication.

$$(x + 2)(x - 2)(x + 1)$$
$$= (x^2 - 4)(x + 1) \qquad (x + 2)(x - 2) = x^2 - 4$$
$$= x^3 + x^2 - 4x - 4 \quad \blacksquare$$

The same product would have resulted if the order of multiplication remained unchanged.

Example 11   Multiply $(x - 2)(x + 4)(x - 5)$.

$$(x - 2)(x + 4)(x - 5)$$
$$= (x^2 + 2x - 8)(x - 5) \qquad (x - 2)(x + 4) = x^2 + 2x - 8$$

Now multiply each term in the second polynomial by each term in the first polynomial.

$$= x^3 - 5x^2 + 2x^2 - 10x - 8x + 40$$
$$= x^3 - 3x^2 - 18x + 40 \qquad \text{Combine like terms} \quad \blacksquare$$

Example 12   Multiply $[(2x + 3) + y][(2x + 3) - y]$.

$$[(2x + 3) + y][(2x + 3) - y] \qquad \text{Let } 2x + 3 = a \text{ and } y = b$$
$$= (2x + 3)^2 - y^2 \qquad (a + b)(a - b) = a^2 - b^2$$
$$= 4x^2 + 12x + 9 - y^2 \qquad (a + b)^2 = a^2 + 2ab + b^2 \quad \blacksquare$$

Example 13   Multiply $(3x^2 + 2y + z)(x^2 + 3y - 2z)$.

Multiply each term in the second polynomial by each term in the first polynomial.

$$(3x^2 + 2y + z)(x^2 + 3y - 2z)$$
$$= 3x^4 + 9x^2y - 6x^2z + 2x^2y + 6y^2 - 4yz + x^2z + 3yz - 2z^2$$
$$= 3x^4 + 11x^2y - 5x^2z + 6y^2 - yz - 2z^2 \qquad \text{Combine like terms} \quad \blacksquare$$

Example 14   Multiply and simplify.

$$-x(3x - 4)^2 + 2x(6 - 3x)^2 - (5x^3 + 8)$$
$$= -x\underbrace{(9x^2 - 24x + 16)}_{} + 2x\underbrace{(36 - 36x + 9x^2)}_{} - (5x^3 + 8)$$
$$(a - b)^2 = a^2 - 2ab + b^2$$
$$= -9x^3 + 24x^2 - 16x + 72x - 72x^2 + 18x^3 - 5x^3 - 8 \qquad \text{Distributive property}$$
$$= 4x^3 - 48x^2 + 56x - 8 \quad \blacksquare$$

Example 15   A Norman window is in the form of a rectangle surmounted by a semicircle, as shown in Figure 4.3.

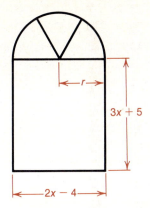

Figure 4.3

(a) Find the area of the window in terms of x.

The area of a rectangle is given by the formula $A_R = L \cdot W$. The area of a semicircle is one-half the area of a circle. Thus

$$A_C = \frac{\pi r^2}{2}$$

where r is the radius. The radius r is one-half the base of the rectangle, or

$$r = \frac{2x - 4}{2} = x - 2.$$

The area of the Norman window is the sum of these two areas.

$$A_R = (2x - 4)(3x + 5) = 6x^2 - 2x - 20$$

$$A_C = \frac{\pi(x - 2)^2}{2} = \frac{\pi}{2}(x^2 - 4x + 4)$$

$$A_W = 6x^2 - 2x - 20 + \frac{\pi}{2}(x^2 - 4x + 4) \qquad \text{Area of window}$$

$$= 6x^2 - 2x - 20 + \frac{\pi x^2}{2} - 2\pi x + 2\pi$$

When the distributive property is used, this becomes

$$= \left(6 + \frac{\pi}{2}\right)x^2 - (2 + 2\pi)x + (2\pi - 20) \qquad \text{Area of the window in terms of } x$$

(b) Find the area if $x = 18$ inches.

To find the area of the window, substitute 18 for x and 3.14 for π.

$$A_W = \left[6 + \frac{3.14}{2}\right](18)^2 - [2 + 2(3.14)](18) + [2(3.14) - 20]$$

$$= 2289.92 \text{ square inches} \quad \blacksquare$$

Exercises 4.3

Multiply each of the following and simplify where possible.

1. $2x^2(3x^3)$
2. $(-4a^3)(2a^4)$
3. $(-7m)(-8m^5)$
4. $(-x^2y^3)^2$
5. $(-p^3q^4)^3$
6. $3^2x(-xy)^3$
7. $(-t^3)^2(t)$
8. $(-2^2xy^2)^2(3x^2y)^2$
9. $-7^2(a^2)^3(-a)^3$
10. $3x(x^2 + 2)$
11. $2y^2(y - 3)$
12. $-6a(3a^2 + 7a - 9)$
13. $-15t^2(-t^2 - 2t - 3)$
14. $6m(m + 7)$
15. $-3s(s^3 - 9)$
16. $-2xy(x^2 + 3 - y^2)$
17. $a^2b(a^2 + 7ab - b^2)$
18. $5u^3v(u^2 + uv - v^3)$
19. $x^3(3x^2 + 7x - 9) - 6x(x^4 - 2x^3 + 8x^2)$
20. $(4k)^2(k^3 + k^2 - k + 1) - (k)^2(3k^3 + 2k^2 - 5k + 6)$
21. $a^n(a^{2n} + 3a^n - 4)$
22. $t^{2n}(3t^{2n} - 4t^n - 6)$
23. $(u + v)(u^2 + uv + v^2)$
24. $(2m + n)(m^2 - mn + n^2)$
25. $(3x^2 - y^2)(2x^3 + x^2y + y^3)$
26. $(2s^2 + t^2)(5s^2 + 6st - t^2)$
27. $(x^n - y^n)(x^{2n} + x^ny^n + y^{2n})$
28. $(x^{2n} + y^{2n})(x^n + xy + y^n)$

Find the product, mentally, of each of the following special types.

29. $(x + 1)^2$
30. $(a + 3)^2$
31. $(2p - 3)(2p - 3)$
32. $(3m - 4)(3m - 4)$
33. $(r + s)(r - s)$
34. $(p^2 + q^2)(p^2 - q^2)$
35. $(3k + 5)(3k - 5)$
36. $(9t - 4)(9t + 4)$
37. $(2h^2 - 9)(2h^2 - 9)$
38. $(3u^2 + 5)(3u^2 + 5)$
39. $(4y^2 - 3)(4y^2 + 3)$
40. $(5m^4 - n^2)(5m^4 + n^2)$
41. $(a^m - b^m)^2$
42. $(u^n + v^n)^2$
43. $(x^{2m} + y^{2m})(x^{2m} - y^{2m})$
44. $(r^{3n} + s^{3n})(r^{3n} - s^{3n})$
45. $[(a + b) - 4][(a + b) + 4]$
46. $[(m - n) - 3][(m - n) + 3]$
47. $[(5p + q) - 5]^2$
48. $[(3x - y) + 2]^2$

Multiply each of the following and simplify.

49. $(x + 3)(x - 1)$
50. $(a + 5)(a - 2)$
51. $(k + 4)(k + 7)$
52. $(3m + 2)(5m - 6)$
53. $(6t + 7)(t + 1)$
54. $(9m + 4)(m + 8)$
55. $(7y - 5)(4y + 8)$
56. $(9a + 4b)(2a - 7b)$
57. $(7r - 5s)(8r + 6s)$
58. $(x^2 - y)(x^2 + 2y)$
59. $(7x^2 - y^2)(3x^2 + 5y^2)$
60. $(m^3 - n^2)(2m^3 + 11n^2)$
61. $(6m^2 + 7n)(11m^2 - 9n)$
62. $(2x + 3)(x^2 - 5x + 6)$
63. $(5p - 3)(2p^2 - 7p + 2)$
64. $(13u - 5)(3u^2 - u + 7)$
65. $(8a - 5b)(3a^2 - 7ab + 2b^2)$
66. $(6s - 7t)(s^2 - 13st + t^2)$
67. $(q^2 - r^2)(q^3 - 3q^2r + 3qr^2 + r^3)$
68. $(m^2 + n^2)(m^3 + 3m^2n - 3mn^2 + n^3)$
69. $(a^n + b^n)(a^{2n} - a^nb^n + b^{2n})$
70. $(x^m - y^m)(x^{2m} + x^my^m + y^{2m})$
71. $(a^2 - 3a + 5)(2a^2 + a - 2)$
72. $(x^2 - 2xy + y^2)(x^2 + 2xy + y^2)$
73. $(3x^2 + 4xy + y^2)(2x^2 - 7xy - 5y^2)$
74. $(8p^2 - 5pq + 2q^2)(3p^2 + pq - 8q^2)$
75. $(x - 3)^2 - (x + 3)^2$
76. $(5x - 2)^2 - (5x + 2)^2$
77. $(3x^n - y^n)^2 - (3x^n + y^n)^2$
78. $(r^n - 2s^n)^2 - (r^n + 2s^n)^2$
79. $(2u + v)(u - v)(2u + 3v)$
80. $(3a - b)(a + b)(3a + 4b)$
81. $(p^{2m+5} - q^{m+3})(p^m - 2p^{m+1}q^{m+1} + q^3)$
82. $(t^{5a-4} + w^{3a+2})(t^{2a+4} + 4tw - 8w^{2a-1})$

A flower garden is divided into 5 areas (A, B, C, D, E) with the dimensions given in the diagram. Find each of the following in terms of x and simplify. (A small box in a corner indicates a right angle.)

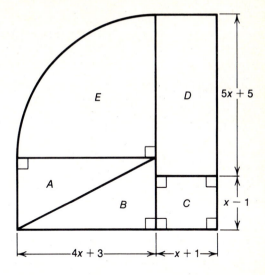

83. $A + B$

84. $D + C$

85. $A + D + E$ (Note: E is one-fourth of a circle.)

86. $A + E - C$

87. $2A + D + C + E$

88. $A + B + C + D$

Use a calculator. Round to two decimal places.

89. If $x = 1.6$ yards, find the area in Exercise 83.

90. If $x = 0.89$ m, find the area in Exercise 84.

4.4 Division of Polynomials

In the last section you learned how to find the product of two or more polynomials. This section will explain methods for dividing polynomials. A monomial can be divided by a monomial by using the rules for exponents that were developed in Section 4.1. For example, $6x^2y^3 \div 3xy^2$ is written as

$$\frac{6x^2y^3}{3xy^2} = \frac{6}{3}x^{2-1}y^{3-2} = 2xy.$$

Division of a polynomial by a monomial is carried out by dividing each term of the polynomial by the monomial. For example, to divide $3x^4 - 18x^2$ by $3x^2$, think of the two terms of the polynomial as $3x^4$ and $-18x^2$. The division would be written as

$$\frac{3x^4 + (-18x^2)}{3x^2} = \frac{3x^4}{3x^2} + \frac{(-18x^2)}{3x^2}$$
$$= x^2 + (-6)$$
$$= x^2 - 6.$$

In general, if a, b, and c represent the terms and $c \neq 0$, then

$$\frac{a}{c} + \frac{b}{c} = \frac{a + b}{c}$$

or $$\frac{a + b}{c} = \frac{a}{c} + \frac{b}{c}.$$

> To **divide a polynomial by a monomial,** divide each term of the polynomial by the monomial.

Example 1   Divide $6x^5 + 9x^2$ by $3x$.

$$\frac{6x^5 + 9x^2}{3x} = \frac{6x^5}{3x} + \frac{9x^2}{3x}$$ Divide each term by 3x

$$= 2x^{5-1} + 3x^{2-1}$$ Divide the coefficients and subtract the exponents

$$= 2x^4 + 3x \quad \blacksquare$$

Example 2   Divide $4x^2y^5 + 18x^5y^7 - 7x^2y$ by $2xy^3$.

$$\frac{4x^2y^5}{2xy^3} + \frac{18x^5y^7}{2xy^3} + \frac{-7x^2y}{2xy^3}$$ Divide each term by 2xy³

$$= 2xy^2 + 9x^4y^4 + \frac{-7xy^{-2}}{2}$$ Divide the coefficients and subtract the exponents

$$= 2xy^2 + 9x^4y^4 - \frac{7x}{2y^2}$$ $b^{-n} = \frac{1}{b^n}$ $\quad \blacksquare$

Notice that the final result is written so as to be free of negative exponents. Also, the quotient of the coefficients is found only if it is an integer.

Example 3   Divide.

$$\frac{-8x^3y - 5xy^4 + x^9y^2z}{-xy^2}$$

$$= \frac{-8x^3y}{-xy^2} + \frac{-5xy^4}{-xy^2} + \frac{x^9y^2z}{-xy^2}$$

$$= 8x^2y^{-1} + (5y^2) + (-x^8z)$$

$$= \frac{8x^2}{y} + 5y^2 - x^8z \quad \blacksquare$$

Example 4   Divide.

$$\frac{x^{3n+4} + x^{n+5}}{x^n} = \frac{x^{3n+4}}{x^n} + \frac{x^{n+5}}{x^n}$$ Divide each term by xⁿ

$$= x^{3n+4-n} + x^{n+5-n}$$ Subtract exponents

$$= x^{2n+4} + x^5 \quad \blacksquare$$

Division of a polynomial by a polynomial is similar to the long division process in arithmetic. The similarities are shown on the facing page.

Divide 552 by 24.

Step 1

Write the problem in standard form.

$24\overline{)552}$

Step 2

24 divides into 55 two times;

$2 \cdot 24 = 48.$

$\begin{array}{r} 2 \\ 24\overline{)552} \\ 48 \end{array}$

Step 3

Subtract and bring down the next term.

$\begin{array}{r} 2 \\ 24\overline{)552} \\ 48\downarrow \\ \hline 72 \end{array}$

Step 4

24 divides into 72 three times;

$3 \times 24 = 72.$

$\begin{array}{r} 23 \\ 24\overline{)552} \\ 48 \\ \hline 72 \\ 72 \\ \hline 0 \end{array}$

The remainder is 0. Since there are no more terms to be brought down, the division is complete.

Step 5

Check by multiplication:

$24 \cdot 23 = 552.$

Divide $x^2 + 7x + 12$ by $x + 4$.

Write the problem in standard form.

$x + 4\overline{)x^2 + 7x + 12}$

x divides into x^2, x times;

$x(x + 4) = x^2 + 4x.$

$\begin{array}{r} x \\ x + 4\overline{)x^2 + 7x + 12} \\ x^2 + 4x \end{array}$

Subtract and bring down the next term.

$\begin{array}{r} x \\ x + 4\overline{)x^2 + 7x + 12} \\ x^2 + 4x \downarrow \\ \hline 3x + 12 \end{array}$

x divides into $3x$, three times;

$3(x + 4) = 3x + 12.$

$\begin{array}{r} x + 3 \\ x + 4\overline{)x^2 + 7x + 12} \\ x^2 + 4x \\ \hline 3x + 12 \\ 3x + 12 \\ \hline 0 \end{array}$

The remainder is 0. Since there are no more terms to be brought down, the division is complete.

Check by multiplication:

$(x + 4)(x + 3) = x^2 + 7x + 12.$

It is interesting to note that when x is replaced by the real number 20, the first problem becomes a special case of the second.

$$x + 3 = 20 + 3 = 23$$
$$x + 4 = 20 + 4 = 24$$
$$x^2 + 7x + 12 = (20)^2 + 7 \cdot 20 + 12 = 552$$

Example 5   Divide $x^2 - 6x - 7$ by $x + 1$.

$$
\begin{array}{r}
x \\
x + 1 \overline{)\, x^2 - 6x - 7\,} \\
\underline{x^2 + x}
\end{array}
$$

x divides into x^2, x times;
$x(x + 1) = x^2 + x$

$$
\begin{array}{r}
x \\
x + 1 \overline{)\, x^2 - 6x - 7\,} \\
\underline{x^2 + x\downarrow} \\
-7x - 7
\end{array}
$$

Subtract and bring down the next term. Remember to subtract, add the additive inverse

$$
\begin{array}{r}
x - 7 \\
x + 1 \overline{)\, x^2 - 6x - 7\,} \\
\underline{x^2 + x} \\
-7x - 7 \\
\underline{-7x - 7}
\end{array}
$$

x divides into $-7x$, -7 times;
$-7(x + 1) = -7x - 7$

The quotient is $x - 7$.

Check: $(x + 1)(x - 7) = x^2 - 6x - 7$. ■

▶ **Caution**   Before carrying out a long division problem, arrange both divisor and dividend in descending powers of the variable.

Example 6   Divide $x^2 + 2x^3 + 6x + 1$ by $x + 2$.

First arrange the terms of the dividend in descending powers of the variable.

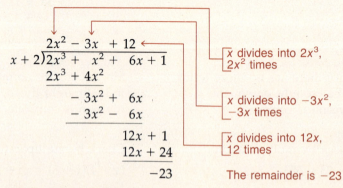

x divides into $2x^3$, $2x^2$ times

x divides into $-3x^2$, $-3x$ times

x divides into $12x$, 12 times

The remainder is -23 ■

The result is written as

$$2x^2 - 3x + 12 + \frac{-23}{x + 2}.$$

Generally, whenever the degree of the remainder is less than the degree of the divisor, the division is complete.

Example 7   Divide $4x^4 - x$ by $2x^2 + 3$.

Notice that $4x^4 - x$ does not contain a third- or second- or zero-degree term. Since these terms may occur in the division process, placeholders with coefficients of zero will be shown in the display.

$$
\begin{array}{r}
2x^2 - 3 \\
2x^2 + 3 \overline{)\,4x^4 + 0x^3 + 0x^2 - x} \\
\underline{4x^4 + 6x^2} \\
-6x^2 - x \\
\underline{-6x^2 - 9} \\
-x + 9
\end{array}
$$

$2x^2$ divides into $4x^4$, $2x^2$ times

$2x^2$ divides into $-6x^2$, -3 times

A second-degree term occurs

Add the additive inverse:
$0x^2 + (-6x^2) = (-6x^2)$

$\leftarrow$ Degree is 1

Since the degree of the remainder is less than the degree of the divisor, the division is complete. The result is written as

$$2x^2 - 3 + \frac{-x + 9}{2x^2 + 3}.$$

Check: $(2x^2 + 3)(2x^2 - 3) + (-x + 9) = 4x^4 - 9 - x + 9$
$$= 4x^4 - x.$$

Divisor · Quotient + Remainder = Dividend ∎

Example 8   Divide $3x^3 + 2x^2 - x + 6$ by $2x - 1$.

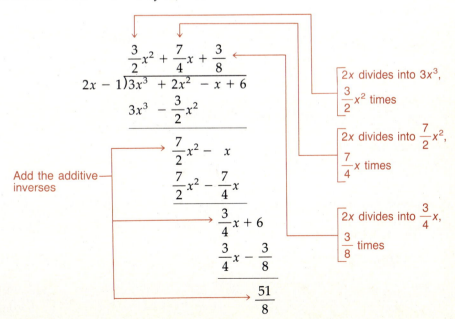

$2x$ divides into $3x^3$, $\frac{3}{2}x^2$ times

$2x$ divides into $\frac{7}{2}x^2$, $\frac{7}{4}x$ times

Add the additive inverses

$2x$ divides into $\frac{3}{4}x$, $\frac{3}{8}$ times

The result is written as

$$\frac{3}{2}x^2 + \frac{7}{4}x + \frac{3}{8} + \frac{\frac{51}{8}}{2x - 1}. \quad \blacksquare$$

Example 9 Suppose a division problem is illustrated as

$$D(x)\overline{)P(x)}^{\displaystyle Q(x)},$$

where $D(x) = x + 7$ and $Q(x) = x - 5$. If the remainder is zero, find $P(x)$.

Since the dividend, $P(x)$, must be the product of the quotient and the divisor,

$$P(x) = (x + 7)(x - 5) = x^2 + 2x - 35. \quad \blacksquare$$

Example 10 Given the dividend $P(y) = 9y^2 + 9y - 3$ and the divisor $D(y) = 3y - 4$, find the quotient $Q(y)$ and the remainder $R(y)$ such that $P(y) = D(y)Q(y) + R(y)$.

$$
\begin{array}{r}
3y \;+\; 7 \\
3y - 4\overline{)9y^2 + 9y - 3} \\
\underline{9y^2 - 12y} \\
21y - 3 \\
\underline{21y - 28} \\
25
\end{array}
$$

$$Q(y) = 3x + 7; \; R(y) = 25 \quad \blacksquare$$

Example 11 Divide $x^{2n} - 3x^n y^n + 2y^{2n}$ by $x^n - y^n$.

$$
\begin{array}{l}
\phantom{x^n - y^n\overline{)}}\; x^n - 2y^n \quad \longleftarrow \quad \text{— } x^n \text{ divides into } x^{2n},\; x^n \text{ times} \\
x^n - y^n\overline{)x^{2n} - \;\;3x^n y^n + 2y^{2n}} \quad \text{— } x^n \text{ divides into } -2x^n y^n,\; -2y^n \text{ times} \\
\phantom{x^n - y^n\overline{)}}\underline{x^{2n} - \;\;\;\;x^n y^n} \quad\longleftarrow\quad \text{— } x^n(x^n - y^n) = x^{2n} - x^n y^n \\
\phantom{x^n - y^n\overline{)xxxx}} -2x^n y^n + 2y^{2n} \\
\phantom{x^n - y^n\overline{)xxxx}} \underline{-2x^n y^n + 2y^{2n}} \quad\longleftarrow\quad \text{— } -2y^n(x^n - y^n) = -2x^n y^n + 2y^{2n} \quad \blacksquare
\end{array}
$$

Exercises 4.4

Divide each of the following. Check each answer.

1. $\dfrac{3a + 6}{3}$

2. $\dfrac{4m + 12}{4}$

3. $\dfrac{15t - 20}{-5}$

4. $\dfrac{36u - 27}{-9}$

5. $\dfrac{4t^2 - 6t + 2}{2}$

6. $\dfrac{12u^2 + 15u - 3}{3}$

7. $\dfrac{64x^3 + 32x^2 - 8x}{8x}$

8. $\dfrac{72y^3 - 54y^2 + 63y}{9y}$

9. $\dfrac{8x^2 + 24x - 32}{-8x}$

10. $\dfrac{55r^2 - 66r + 33}{-11r}$

11. $\dfrac{18a^3 - 22a^2 + 3}{-2a^2}$

12. $\dfrac{39b^3 - 91b^2 - 20}{-13b^2}$

13. $\dfrac{42a^3b^3 + 28a^2b^2}{7a^2b}$

14. $\dfrac{60r^2s^4 - 54rs^2}{6r^2s}$

15. $\dfrac{a^{6m} - a^{4m}}{a^{2m}}$

16. $\dfrac{16t^{9n} - 12t^{6n}}{-4t^{3n}}$

17. $\dfrac{56t^{7n} - 36t^{3n}}{-4t^{-5n}}$

18. $\dfrac{81v^{-2n} + 99v^{5n}}{9v^{-3n}}$

19. $\dfrac{-7a^{11}b^9c^8 + 4a^8b^{10}c^{15}}{a^4b^5c^3}$

20. $\dfrac{15x^{10}y^6z^4 - 3x^8y^{12}z^7}{2x^4y^4z^3}$

21. $\dfrac{-90b^{10}c^8d^5 + 30b^7c^8d^9 + 15b^2cd^4}{-66b^2cd^4}$

22. $\dfrac{-44r^{10}s^8t^{12} - 48r^7s^9t^6 + 28r^3s^8t^5}{-4r^2s^2t^3}$

23. $\dfrac{x^2 + 9x + 8}{x + 8}$

24. $\dfrac{y^2 - 5y + 6}{y - 3}$

25. $(a^2 + 13a + 30) \div (a + 10)$

26. $(m^2 - 3m - 18) \div (m + 3)$

27. $t - 5\overline{)t^2 - 2t - 15}$

28. $s - 7\overline{)s^2 + s - 56}$

29. $\dfrac{x^3 + 2x^2 - 21x + 18}{x - 3}$

30. $\dfrac{m^3 + 5m^2 + 6m + 8}{m + 4}$

31. $(5v^2 - 13v + 6) \div (5v - 3)$

32. $(8k^3 - 6k^2 - 5k + 3) \div (4k + 3)$

33. $7a - 2\overline{)35a^3 + 53a^2 - 19a - 1}$

34. $2y + 3\overline{)-12y^3 - 2y^2 + 21y + 5}$

35. $\dfrac{8x^3 - 1}{2x - 1}$

36. $\dfrac{125y^3 + 1}{5y + 1}$

37. $(15s^4 + 8s^2 + 3s^3 - 1) \div (5s + 1)$

38. $(-5k + 10k^5 - 5k^2 - 4k^3) \div (2k^2 - 3)$

39. $x^n - y^n\overline{)x^{2n} - y^{2n}}$

40. $x^n + y^n\overline{)x^{2n} + 2x^ny^n + y^{2n}}$

41. $x^n - y^n\overline{)x^{2n} - 2x^ny^n + y^{2n}}$

42. $x^n - y^n\overline{)x^{3n} - y^{3n}}$

43. $x^n + y^n\overline{)x^{3n} + y^{3n}}$

44. $x^n + 3y^n\overline{)x^{2n} + x^ny^n - 6y^{2n}}$

45. Given polynomials $P(x) = x^2 + 7x + 13$ and $D(x) = x + 3$, find $Q(x)$ and $R(x)$ such that $P(x) = D(x)Q(x) + R(x)$, where $D(x)$ is the divisor, $P(x)$ is the dividend, $Q(x)$ is the quotient and $R(x)$ is the remainder.

46. Given polynomials $P(y) = 2y^3 + 5y^2 - y - 6$ and $D(y) = y + 2$, find $Q(y)$ and $R(y)$ such that $P(y) = D(y)Q(y) + R(y)$. See problem 45.

47. Given $P(x) = 5x^2 + 7x - 9$ and $D(x) = 4x - 1$, find $Q(x)$ and $R(x)$ such that $P(x) = D(x)Q(x) + R(x)$.

48. Given $P(t) = t^3 - t^2 - 1$ and $D(t) = 5t + 3$, find $Q(t)$ and $R(t)$ such that $P(t) = D(t)Q(t) + R(t)$.

In the figure shown, the area of the rectangle ABCD is $15x^2 + 12x - 3$ *and the area of the trapezoid EBCD is* $10x^2 + 8x - 2$. *Use the formulas for the area of a triangle, rectangle, and trapezoid to complete Exercises 49–52.*

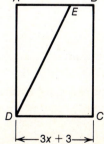

49. Find the length of side BC in terms of x.

50. Find the length of side AE in terms of x.

51. Find the area of triangle AED in terms of x.

52. Find the square of the length of ED in terms of x.

Use a calculator to divide each of the following.

53. $\dfrac{0.06x^3 - 0.07x^2 - 0.03x + 0.01}{0.2x - 0.3}$

54. $\dfrac{0.03y^3 + 0.05y^2 - 0.05y + 0.01}{0.3y - 0.1}$

55. $\dfrac{0.00032s^5 - 1}{0.2s - 1}$

56. $\dfrac{0.81t^2 - 0.09t + 0.83}{0.9t + 1.11}$

Review Exercises

Perform the indicated operations for each of the following.

57. $(6x^3 + 7x^2 - 5x + 2) - (x^3 - 8x^2 - 4x + 3)$

58. $-2(a^2 - 7a + 5) - 3(2a^2 + 8a - 6)$

59. $(3m + 2n)^2$

60. $(7s - 5t)^2$

61. $(11a - 6b)(11a + 6b)$

62. $(5r^2 + 2s^2)(5r^2 - 2s^2)$

63. $(y^2 - 5y + 7)(y^2 - 5y - 7)$

64. $(k^2 + 8k - 3)(k^2 + 8k + 3)$

65. $(3x + 4y)(7x - 9y)$

66. $(7m + 11n)(5m + 6n)$

67. Graph $3x - 5y = 15$ using both intercepts and one other point.

68. Graph $y \leq 3x - 4$.

69. Graph the system $\quad y - x \geq 3$.
$\qquad\qquad\qquad\quad 2y + x \leq 5$

70. Solve the system $5x - 2y = 5$ by graphing.
$\qquad\qquad\qquad x + y = 1$

71. Solve the system $\quad 3x - y = 4$ by elimination.
$\qquad\qquad\qquad\quad 2x + 3y = 5$

4.5   Synthetic Division (Optional)

In certain special cases, long division can be carried out more efficiently by a process called **synthetic division.** For the process to be used, the divisor must be of the form $x + a$ or $x - a$. To see how and why this process works, consider the quotient of $2x^3 + 8x^2 + 11x + 7$ and $x + 2$.

$$
\begin{array}{r}
2x^2 + 4x + 3 \qquad\text{Quotient} \\
x + 2 \overline{)\,2x^3 + 8x^2 + 11x + 7} \quad\text{Dividend} \\
\underline{2x^3 + 4x^2 \qquad\qquad} \\
4x^2 + 11x \qquad \\
\underline{4x^2 + 8x \qquad} \\
3x + 7 \\
\underline{3x + 6} \\
1 \qquad\text{Remainder}
\end{array}
$$

Notice that:

1. The degree of the quotient is one less than the degree of the dividend.

2. The coefficients of the first term in the quotient and dividend are the same.

Keeping the above observations in mind, we could omit all variables and still carry out the division, as shown below.

$$\begin{array}{r} 2+4+\ \ 3 \\ 1+2\overline{\smash{\big)}2+8+\ 11+7} \\ \end{array}$$

    ②$+ 4$

    $\dfrac{}{4+⑪}$

    $\boxed{4}+\ 8$

The circled terms are repetitions of the terms in the dividend and may be omitted

    $3+⑦$

    $\boxed{3}+\ 6$

        1

The boxed terms are repetitions of the terms immediately above them and may be omitted

These omissions yield the following.

$$\begin{array}{r} 2+4+\ 3 \\ 1+2\overline{\smash{\big)}2+8+11+7} \\ 4 \quad\uparrow\quad\uparrow \\ 4 \\ 8 \\ 3 \\ 6 \\ 1 \end{array}$$

All the remaining numbers can be shifted upward into the spaces gained from the omissions

We now have

$$\begin{array}{r} \mathbf{2+4+\ 3} \\ 1+2\overline{\smash{\big)}2+8+11+7} \\ 4+\ 8+6 \\ \hline \mathbf{2+4+\ 3}+\mathbf{1}. \end{array}$$

If the lead term in the quotient "2" were moved to the last row, the first 3 terms would duplicate the quotient

The last number in the last row, 1, is the remainder

This yields

$$\begin{array}{r} 1+2\overline{\smash{\big)}2+8+11+7} \\ 4+\ 8+6 \\ \hline 2+4+\ 3+1. \end{array}$$

Since the first row is repeated in the last row, it can be omitted.

The last row is obtained by subtracting the second row from the first

Finally, since the variable has been omitted, drop the 1 in the divisor and change the sign before 2 to -2. By making this change, every sign in the product will be changed. Each term in the second row will automatically be replaced by its additive inverse so that row 1 and row 2 can be *added* rather than subtracted.

$$\overset{\downarrow \quad \downarrow \quad\ \ \downarrow \quad\ \downarrow}{-2\,\rvert\,2 + 8 + 11 + 7} \qquad \text{Coefficients of the dividend}$$
$$\underline{\quad\ \ -4 - 8 - 6}$$
$$2 + 4 + \ 3 + 1 \qquad \text{The last row is obtained by}$$

Coefficients of the quotient | Remainder

The last row is obtained by **adding** the second row to the first row

The result is

$$2x^2 + 4x + 3 + \frac{1}{x + 2}.$$

Example 1   Use synthetic division to divide $3x^3 + x^2 - 10$ by $x - 2$. Think of $3x^3 + x^2 - 10$ as $3x^3 + x^2 + 0x - 10$.

▶▶ **Caution**   The coefficient of 0 must be present so that like terms will be written in the same column.

Step 1   Write down the coefficients of the dividend. Include a zero for any missing term.

$$\rvert\quad 3 \quad 1 \quad\ \mathbf{0} - 10$$

Step 2   Change the sign of the constant term in the divisor and place it to the left of the coefficients of the dividend.

$$\mathbf{2}\rvert\quad 3 \quad 1 \quad 0 - 10$$

Step 3   Bring down the lead coefficient, 3, from the dividend.

$$\mathbf{2}\rvert\quad 3 \quad 1 \quad 0 - 10$$
$$\downarrow$$
$$\ \overline{\mathbf{3}\qquad\qquad\qquad}$$

Step 4   Multiply 3 by 2 and write the product 6 under the 1. Add to get 7.

$$\mathbf{2}\rvert\quad 3 \quad 1 \quad 0 - 10$$
$$6$$
$$\ \overline{\mathbf{3}\quad 7\qquad\qquad}$$

Step 5   Multiply the 7 by 2 and write the product 14 under 0. Add to get 14.

$$\mathbf{2}\rvert\quad 3 \quad 1 \quad\ \ 0 - 10$$
$$6 \quad 14$$
$$\ \overline{3 \quad 7 \quad 14\qquad}$$

Step 6   Multiply the 14 by 2 and write the product, 28, under -10. Add to get 18.

$$\mathbf{2}\rvert\quad 3 \quad 1 \quad\ \ 0 - 10$$
$$6 \quad 14 \quad 28$$
$$\ \overline{3 \quad 7 \quad 14 \quad 18}\leftarrow\!\text{Remainder is 18}$$

We started with a third-degree dividend. The quotient is a second-degree polynomial. The result is

$$3x^2 + 7x + 14 + \frac{18}{x-2}. \quad \blacksquare$$

Example 2   Divide $x^4 - 12x^2 + 3x + 18$ by $x - 3$.

$$
\begin{array}{r|rrrrr}
3 & 1 & 0 - 12 & 3 & 18 \\
  &   & 3 & 9 - 9 & - 18 \\
\hline
  & 1 & 3 - 3 & - 6 & 0
\end{array}
$$

Fourth-degree

Third-degree

Remainder is zero

The quotient is $x^3 + 3x^2 - 3x - 6$. $\blacksquare$

Example 3   Divide $x^5 - 1$ by $x - 1$.

$$x^5 - 1 = x^5 + 0x^4 + 0x^3 + 0x^2 + 0x - 1$$

$$
\begin{array}{r|rrrrrr}
1 & 1 & 0 & 0 & 0 & 0 - 1 \\
  &   & 1 & 1 & 1 & 1 & 1 \\
\hline
  & 1 & 1 & 1 & 1 & 1 & 0
\end{array}
$$

Fifth-degree

Fourth-degree

Remainder is zero

The quotient is $x^4 + x^3 + x^2 + x + 1$. $\blacksquare$

Example 4   Divide $x^3 + \frac{3}{2}x^2 - 2x + 1$ by $(x + \frac{1}{2})$.

$$
\begin{array}{r|rrrr}
-\dfrac{1}{2} & 1 & \dfrac{3}{2} & - 2 & 1 \\[2mm]
 & & -\dfrac{1}{2} & -\dfrac{1}{2} & \dfrac{5}{4} \\[2mm]
\hline
 & 1 + 1 & -\dfrac{5}{2} & +\dfrac{9}{4}
\end{array}
$$

Change the sign of the constant term in the divisor $(x + \frac{1}{2})$

The result is

$$x^2 + x - \frac{5}{2} + \frac{\dfrac{9}{4}}{x + \dfrac{1}{2}}. \quad \blacksquare$$

A theorem from mathematics (the **remainder theorem**) states that when a polynomial $P(x)$ is divided by $x - a$ the remainder will be $P(a)$.

Example 5   Use synthetic division to find $P(5)$ given that $P(x) = x^3 - 4x^2 - 25$.
$P(5)$ is obtained by dividing $x^3 - 4x^2 - 25$ by $x - 5$.

$$
\begin{array}{r|rrrr}
5 & 1 & -4 & 0 & -25 \\
  &   & 5 & 5 & 25 \\
\hline
  & 1 & 1 & 5 & 0
\end{array}
$$
$P(5)$ is the remainder

The remainder is 0.

$$P(5) = 5^3 - 4 \cdot 5^2 - 25 = 125 - 4 \cdot 25 - 25 = 0$$

$P(5)$ is the remainder. ∎

Example 6   Use synthetic division to find $Q(2)$ if $Q(y) = y^3 - 13y^2 + 40y - 35$.
$Q(2)$ can be found by dividing the polynomial by $y - 2$. The remainder will be $Q(2)$.

$$
\begin{array}{r|rrrr}
2 & 1 & -13 & 40 & -35 \\
  &   & 2   & -22 & 36 \\
\hline
  & 1 & -11 & 18 & 1
\end{array}
$$
Remainder is 1

Since the remainder is 1, $Q(2) = 1$. ∎

Exercises 4.5

Use synthetic division to find the quotient and remainder for each of the following.

1. $(x^2 - 9x + 20) \div (x - 4)$

2. $(y^2 - 11y + 30) \div (y - 6)$

3. $(a^3 + 4a^2 - 7a + 5) \div (a - 2)$

4. $(p^3 - p^2 - 6p + 5) \div (p + 2)$

5. $(3t^3 - t^2 + 8) \div (t + 1)$

6. $(6k^4 - k^2 + 4) \div (k - 3)$

7. $(x^5 - 32) \div (x - 2)$

8. $(y^5 + 32) \div (y + 2)$

9. $(2y^4 - 3y^3 + y^2 - 3y - 6) \div (y - 2)$

10. $(2s^4 + 5s^3 + 6s^2 + 3s - 10) \div (s + 2)$

11. $(r^4 - r^3 + 3r^2 - 7r - 6) \div (r - 2)$

12. $(2m^4 - 3m^3 - m^2 - m - 2) \div (m - 2)$

13. $(a^6 - 3) \div (a - 1)$

14. $(x^6 + 2) \div (x + 1)$

15. $(2t^4 - 7t^3 - t^2 + 7) \div (t - 5)$

16. $(d^4 + 2d^3 - 15d^2 - 32d - 12) \div (d - 4)$

17. $(5y^4 - 18y^2 - 6y + 3) \div (y + 2)$

18. $(2k^3 + 5k^2 - 8k + 6) \div (k + 2)$

19. $(2m^4 - 3m^2 + 5m - 7) \div (m + 3)$

20. $(2x^4 - 10x^2 - 23x + 6) \div (x - 3)$

Use synthetic division to find each of the following.

21. $P(4)$ if $P(x) = x^2 - 5x + 4$

22. $P(-4)$ if $P(y) = y^4 + 3y^3 - y^2 - 2y + 6$

23. $Q(3)$ if $Q(a) = a^4 - a^3 + 2a^2 - 7a$

24. $R(5)$ if $R(t) = 2t^4 - 7t^3 - t^2 + 8$

25. $P(-3)$ if $P(x) = x^3 + 27$

26. $Q\left(-\dfrac{1}{2}\right)$ if $Q(s) = s^3 - 3s + 9$

27. $R(-2)$ if $R(u) = 2u^3 - 4u^2 + u + 1$

28. $P(4)$ if $P(v) = 2v^4 - 5v^3 - 17v^2 + 22v - 6$

29. $Q(-6)$ if $Q(n) = n^5 - 2n^4 - 44n^3 + 24n^2 + 3n - 18$

30. $R\left(\dfrac{1}{2}\right)$ if $R(t) = 2t^4 - 3t^3 - 5t^2 - 13t + 11$

Use synthetic division to find the quotient and remainder in Exercises 31 through 36.

31. $\left(x^2 - \frac{1}{6}x - \frac{1}{6}\right) \div \left(x - \frac{1}{2}\right)$

32. $\left(x^3 - \frac{11}{3}x^2 + \frac{14}{3}x - 2\right) \div \left(x - \frac{1}{3}\right)$

33. $\left(x^5 + 2x^4 + \frac{1}{2}x^3 + x^2 + \frac{5}{4}x + \frac{5}{2}\right) \div (x + 2)$

34. $\left(\frac{1}{2}x^4 - \frac{1}{5}\right) \div (x - 1)$

35. $\left(7x^3 + \frac{1}{2}x - 5\right) \div \left(x - \frac{3}{2}\right)$

36. $\left(\frac{5}{3}x^2 - 8\right) \div \left(x + \frac{1}{4}\right)$

Use long division to find the quotient and remainder in Exercises 37 through 40.

37. $(2x^2 + 13x + 15) \div (4x + 6)$

38. $(5x^2 + 74x - 20) \div (25x - 5)$

39. $(6x^2 + 43x - 34) \div (6x - 5)$

40. $(x^2 - 5x + 1) \div (3x - 1)$

Review Exercises

Perform the indicated operations.

41. $-xy(x^2 + 5x - 7) + 3xy(-5x^2 - 2x + 3) - 4xy(x^2 - x - 1)$

42. $3ac(a^2 + 8a - 9) - 5ac(a^2 - 2a + 4) - ac(-2a^2 + 5a - 7)$

43. $-3[-5 - 2x(x - 7)] + 5[-3x - 7(x^2 - 9)]$

44. $-8t[t - 5(2t - 9)] - [-2 + t(8t - 1)]$

45. $[(3a + 2b) - c][(3a + 2b) + c]$

46. $[(r - 5s) + 3t][(r - 5s) - 3t]$

47. $(3x^2 - 2y)^2$

48. $(7a - 3k^2)^2$

49. $(x^{m+5} - y^{2m-1})(x^{m-1} + 4y^{3m-2})$

50. $(5x^{-m} + 7y^m)(3x^{2m+3} - 5y^{3-m})$

51. $(3x^2 - 2y^2x)^2$

52. $(7a^2bc + 8ab^2c^2)^2$

53. Graph $\{(x, y) \mid y < x\} \cap \{(x, y) \mid x < 4\} \cap \{(x, y) \mid y > 1\}$.

54. Graph $\{(x, y) \mid x + y \geq 1\} \cap \{(x, y) \mid y \geq x - 4\} \cap \{(x, y) \mid y < 3\}$.

4.6   More Applications

The scientific community often deals with very large or very small numbers. They use a special format for writing such numbers, **scientific notation.** When a number is written in scientific notation, it is written as a number between 1 and 10 multiplied by a power of 10. For example, the distance from the earth to the sun is about 93,000,000 miles. In scientific notation this would be written as

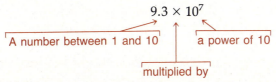

$$9.3 \times 10^7$$

A number between 1 and 10   a power of 10

multiplied by

Recall that when a number is multiplied by 10 the decimal point shifts 1 place to the right. Multiplying by 10^7 is the same as multiplying by 10 seven times. The decimal point would be shifted seven places to the right.

$$9.3 \times 10^7 = 9\,3000000.\qquad \text{seven places}$$

Example 1   Write each of the following in scientific notation.

(a) $927,600,000 = 9.276 \times 10^8$

eight places

(b) $26,000 = 2.6 \times 10^4$ ■

four places

Similarly, very small numbers can be written using negative exponents. For example, the mass of a hydrogen atom is approximately

0.00000000000000000000000017 gram.

This is written as

1.7×10^{-24}.

Dividing by 10 shifts the decimal point 1 place to the left. Since

$$10^{-24} = \frac{1}{10^{24}},$$

multiplying by 10^{-24} is the same as dividing by 10 twenty-four times. The decimal point would be shifted 24 places to the left.

Example 2   Write the following in scientific notation.

(a) $0.000097 = 9.7 \times 10^{-5}$

five places

(b) $0.000854 = 8.54 \times 10^{-4}$ ■

four places

To change a number from scientific notation to **decimal notation,** move the decimal point the number of places indicated by the power of 10. If the exponent is positive, the decimal point is shifted to the right. If the exponent is negative, it is shifted to the left. It may be necessary to add zeros to accomplish the shift.

Example 3   Write each of the following in decimal notation.

(a) $5.78 \times 10^6 = 5,780,000$ Six places right

(b) $3.294 \times 10^9 = 3,294,000,000$ Nine places right

(c) $3.25 \times 10^{-5} = 0.0000325$ Five places left

(d) $8.238 \times 10^{-7} = 0.0000008238$ Seven places left   ■

Example 4   Evaluate

$$\frac{(360,000)(0.00093)}{(0.00009)(310,000)}.$$

Write each number in scientific notation first.

$$\frac{(360,000)(0.00093)}{(0.00009)(310,000)} = \frac{(3.6 \times 10^5)(9.3 \times 10^{-4})}{(9 \times 10^{-5})(3.1 \times 10^5)}$$

$$= \frac{(3.6)(9.3)}{(9)(3.1)} \times 10^{5+(-4)-(-5)-5}$$

$(0.4) = 3.6 \div 9$
$(3) = 9.3 \div 3.1$

$$= (0.4)(3) \times 10^1$$

$$= 1.2 \times 10^1 \qquad \text{Scientific notation for the result}$$

$$= 12 \qquad \text{Decimal notation for the result} \quad \blacksquare$$

Scientific notation is only one of a great number of applications involving exponents. The next two examples show other ways in which exponents are used.

Example 5 The sum of the first n natural numbers is given by the formula

$$1 + 2 + 3 \cdots + n = \frac{n(n + 1)}{2}$$

$$= \frac{n^2 + n}{2}.$$

If $S(n)$ represents the sum, then

$$S(n) = \frac{n^2 + n}{2}.$$

Find the sum of the first 1000 natural numbers, $S(1000)$.

$$S(1000) = \frac{1000^2 + 1000}{2} \qquad \text{Replace } n \text{ with } 1000$$

$$= \frac{1001000}{2}$$

$$= 500,500 \quad \blacksquare$$

Example 6 The volume of a right circular cylinder is given by $V = \pi r^2 h$. Find the volume of a cylinder with $r = 2$ centimeters, and $h = 5$ centimeters. (See Figure 4.4.)

$$V = \pi r^2 h$$

$$V = \pi(2)^2(5) \qquad r = 2, h = 5$$

$$= \pi \cdot 4 \cdot 5$$

$$= 20\pi \text{ cubic centimeters} \quad \blacksquare$$

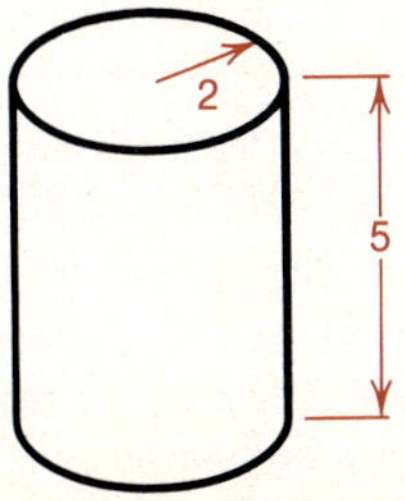

Figure 4.4

Exercises 4.6

Write each of the following in scientific notation.

1. 18,000
2. 3600
3. 0.08
4. 0.00006
5. 43,000,000
6. 1,643,000
7. 0.0000000081
8. 0.0000014

Write each of the following in decimal notation.

9. 8×10^3
10. 6×10^4
11. 1.1×10^{-2}
12. 5.8×10^{-3}
13. 10^3
14. 10^1
15. 10^{-2}
16. 10^{-3}
17. 7.8×10^3
18. 3.5×10^4
19. 1.92×10^{-4}
20. 2.183×10^{-3}

Simplify each of the following. Leave each answer in decimal notation.

21. $(3 \times 10^3)(4 \times 10^{-2})$
22. $(5 \times 10^6)(6 \times 10^{-5})$
23. $(1.1 \times 10^1)(7.0 \times 10^1)$
24. $(1.3 \times 10^4)(5.0 \times 10^{-2})$
25. $(10^8)(10^{-6})(10^3)$
26. $(10^5)(10^{-13})(10^7)$

27. $\dfrac{6.3 \times 10^{-6}}{9.0 \times 10^{-5}}$
28. $\dfrac{8.4 \times 10^{-8}}{1.2 \times 10^{-9}}$
29. $\dfrac{1.56 \times 10^{-17}}{1.30 \times 10^{-16}}$

30. $\dfrac{1.80 \times 10^7}{1.50 \times 10^6}$
31. $\dfrac{75,000,000}{1,500,000}$
32. $\dfrac{3,960,000,000}{12,000,000}$

Write each of the following in scientific notation and simplify. Write the results for Exercises 33–38 in both decimal and scientific notation.

33. $\dfrac{(0.13000)(0.022)(500,000)}{(26,000)(0.000011)(0.0005)}$
34. $\dfrac{(0.48)(3800)(0.025)}{(0.0012)(0.019)(100,000)}$

35. $\dfrac{(20 \times 10^2)^3(6.4 \times 10^{-5})}{4 \times 10^3}$
36. $\dfrac{(80 \times 10^2)^2(3.6 \times 10^{-4})}{(24 \times 10^{-2})^2}$

37. $\dfrac{(21,000)(0.008)(8 \times 10^2)}{(0.000014)(60,000)(2 \times 10^1)}$
38. $\dfrac{(2.4 \times 10^{-3})(0.04)(350,000)}{(1.75 \times 10^2)(0.00001)(2^3 \times 10^{-3})}$

39. The mass of the earth is 6×10^{24} kilograms. If 1 kilogram is 1.1×10^{-3} tons, what is the mass of the earth in tons?

40. The width of the asteroid belt is 2.8×10^8 kilometers. The speed of the space probe Pioneer 10 in passing through this belt was 1.4×10^5 kilometers per hour. How long did it take for Pioneer 10 to pass through this belt?

41. The nematode sea worm is the most plentiful form of sea life. It is estimated that there are 4.0×10^{25} sea worms in the world's oceans. There are about 3.16×10^9 cubic miles of ocean and about 1.10×10^{13} gallons of water per cubic mile. Assuming that the sea worms are uniformly distributed in all the oceans, about how many would there be per gallon?

42. A light year is the distance light travels in a year's time. One light year is about 9.6×10^{12} kilometers. If a star is 220 light years away, how far away is it in kilometers?

43. The amount of money, A, in a bank after n years is given by $A = P(1 + r)^n$ where P is the original amount and r is the rate of interest. To the nearest dollar, how much money would you have after 10 years if you deposited $1000 at 9% interest? Use your calculator.

44. Find the number of one-foot-square tiles that would be needed to cover the surface of the earth. Surface area $= 4\pi r^2$ where r (radius of the earth) is approximately 4000 miles. Let $\pi = 3.1416$. Use your calculator.

45. The sum of the first n odd natural numbers is given by the formula $1 + 3 + 5 + \cdots + (2n - 1) = n^2$. Find the sum, $S(n) = n^2$, of the first 1.0×10^3 odd natural numbers and write your answer in scientific notation.

46. The sum of the first n even natural numbers is given by $2 + 4 + 6 + \cdots + 2n = n^2 + n$. Find the sum, $S(n) = n^2 + n$, of the first 1.5×10^3 even natural numbers and write your answer in scientific notation.

47. The sum of n terms of a pattern of numbers is given by $S(n) = \frac{n}{2}(a_1 + a_n)$ where a_1 is the first number and a_n is the last number in the pattern. Find a_n if $S(n) = 1.01 \times 10^4$, $a_1 = 2.0 \times 10^0$, and $n = 1.0 \times 10^2$.

48. From the formula in Exercise 47 find a_1 if $S(n) = 1.001 \times 10^6$, $n = 1.0 \times 10^3$, and $a_n = 2.0 \times 10^3$.

49. The volume of the silo shown in the figure is given by $V = \pi r^2 h + \frac{2}{3}\pi r^3$. Its dimensions are $r = 1.5 \times 10^1$ feet and $h = 3.2 \times 10^1$ feet. Use $\pi = 3.1416$ to find its volume. (Round the answer to 3 decimal places.)

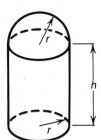

50. The volume of the tower at the right (with square base) is given by $V = w^2 h + \frac{1}{3}w^3$. Its dimensions are $w = 1.2 \times 10^1$ meters and $h = 6.0 \times 10^1$ meters.

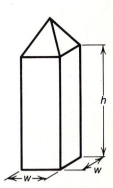

Chapter 4   Summary

[4.1]   Complex expressions can often by simplified by using exponents. An exponential expression consists of two parts, the **base** and the **exponent**.

$$\text{base} \searrow a^n \nwarrow \text{exponent}$$

If a and b are real numbers and m and n are integers, the following rules hold true.

1. $a^m \cdot a^n = a^{m+n}$ **Product Rule**

2. $\dfrac{a^m}{a^n} = a^{m-n}$, $a \neq 0$ **Quotient Rule**

3. $(a^m)^n = a^{m \cdot n}$ **Power Rules**
$(a \cdot b)^n = a^n \cdot b^n$

$\left(\dfrac{a}{b}\right)^n = \dfrac{a^n}{b^n}$, $b \neq 0$

4. $a^{-n} = \dfrac{1}{a^n}$, $a \neq 0$ **Definition of a negative exponent**

5. $a^0 = 1$, $a \neq 0$ **Definition of a zero exponent**

[4.2]   A **polynomial** is the sum of one or more algebraic terms satisfying the following conditions.
   1. All exponents must be whole numbers.
   2. No term may contain a variable in the denominator.
   3. When simplified, no variable may appear under a radical sign.

A polynomial having one term is a **monomial;** one having two terms is a **binomial,** and one having three terms is a **trinomial.**

[4.2] [4.3]   When two polynomials are added or subtracted, the coefficients of **like terms** (same base and exponent) are added by using the distributive property. The variable parts remain unchanged. Polynomials are multiplied by using the distributive property together with the rules for exponents. When the polynomials to be multiplied are binomials, three procedures can be used to carry out the multiplication.

1. **FOIL Method:** Find the products of the first terms (**F**), the outer terms (**O**), the inner terms (**I**), and the last terms (**L**).

2. **Sum and difference:**
   $(a - b)(a + b) = a^2 - b^2$.

3. **Square of a binomial:**
   $(a - b)^2 = a^2 - 2ab + b^2$
   $(a + b)^2 = a^2 + 2ab + b^2$.

[4.4]   To divide a polynomial by a monomial, divide each term of the polynomial by the monomial. A process similar to long division is used to divide a polynomial by a polynomial. If a polynomial in x is to be divided by $x - a$ or $x + a$, **synthetic division** can be used. Synthetic division is a process of division that is dependent upon only the coefficients of the dividend and the constant in the divisor.

[4.5]

[4.6]   Very large or small numbers can be written in exponential form by the use of **scientific notation.** A number in scientific notation is written as a number between 1 and 10 multiplied by a power of 10.

Chapters 3 and 4   Review

1. Graph $2x - 7y = 14$.

2. Solve $\quad x - 3y = 4 \quad$ by graphing.
$\quad\quad\quad 2x - y = -2$

3. Solve $\quad 5x - y = 5 \quad$ by elimination.
$\quad\quad\quad 2x + y = 2$

4. Solve $\quad y = 3x - 1 \quad$ by substitution.
$\quad\quad\quad 2x + y = 4$

5. Graph $y < x + 1$

6. Solve $\quad y > x - 2 \quad$ by graphing.
$\quad\quad\quad y \le -x + 3$

7. Solve $\quad y \le 4 \quad$ by graphing.
$\quad\quad\quad x \ge -1$
$\quad\quad\quad y \ge x$

8. Solve $\quad 3x + 4y = 9$
$\quad\quad\quad 7x - 5y = 7$.

9. Evaluate $(2^{-3})^{-2}$.

10. Simplify $(3ab^2)^{-2}(9a^2b)^{-1}$.

Determine whether each of the following is a polynomial. If it is, (a) write it in standard form, (b) state its degree, (c) state whether it is a monomial, binomial, or trinomial, and (d) find the coefficient of the highest-degree term.

11. $x^2 + 4x^3 - x$

12. $5y - 7y^2 + 6$

13. $k^4 - k^1$

14. $\dfrac{3}{p^3} + 7p$

15. $1 - r^{-5}$

16. $17a - 3a^{-2}$

Simplify each of the following.

17. $(-2x^{-3})(-3x^{-4})^2$

18. $(2a + 5b)^2$

19. $3m^2n(m^2 - 5mn + 6n^3)$

20. $\dfrac{(3x^2)^{-3}(5y^3)^{-2}}{(9x^3)^{-2}(10y^{-3})^2}$

21. $(13r + 3s)(2r - 11s)$

22. $(a + b)(a - b)(a + b)$

23. $\dfrac{39p^3q - 52pq^2 + 91pq}{-13pq}$

24. $(m^3 - m^2 - m + 8) \div (m - 1)$

25. $(8x^2 + 7x - 3) + 2(x^2 + 4)$

26. $6 - \{3m - 3[2n + (5m - 2n)]\}$

27. $(18a^2 + 9a - 36) \div (6a - 7)$

28. Solve
$$
\begin{aligned}
x - 3y + z &= 4 \\
-x + 4y - 4z &= 1 \\
2x - y + 5z &= -3.
\end{aligned}
$$

29. $\dfrac{(0.084)(0.091)(2.0 \times 10^3)}{(7 \times 10^{-3})(1.3 \times 10^1)(4.8 \times 10^3)}$

30. $(2y + 1)(3y^2 - 5y - 2)$

31. Subtract $s^2t - 3st + t^2$ from $-4st - 3s^2t + 2t^2$.

32. A coin machine changes dollar bills into quarters and nickels. If you receive 8 coins from a dollar bill, how many of each type of coin do you receive?

33. Rachel has two concentrations of nitric acid in stock, a 50% solution and an 80% solution. How much of each should she mix to obtain 100 milliliters of a 68% solution?

34. Nett's rowing team can row 6 kilometers down a river in 10 minutes, but it takes the team 15 minutes to row the same distance up the river. How fast can the team row in still water?

35. Rosa's Jewelry Manufacturing makes necklaces and bracelets. The combined number of necklaces (x) and bracelets (y) that can be made in one day is 24, $x + y \leq 24$. The necklace takes one hour of labor and the bracelet $\frac{1}{2}$ hour of labor. The total labor time available per day is 16 hours, $1x + \frac{1}{2}y \leq 16$. Assuming $x \geq 0$ and $y \geq 0$, graph the system of inequalities.

36. If $P(v) = v^3 - 8v$, find $P(-1)$.

37. If $F(x) = 3x^2$ and $R(x) = 4 - x$, find $R(1) - F(-2)$.

38. Marlene accepts a job with a starting salary of $18,000 per year and annual raises of $1100 per year. The formula to find her salary, S, for any given year is $S = a_1 + (n - 1)d$, where a_1 is her beginning salary, n is the number of years worked, and d is the annual raise. What will her salary be 10 years from now?

39. The height, $h(t)$, that an object thrown vertically upward will reach is given by $h(t) = v_0t - 16t^2$, where v_0 is the initial velocity and t is the time in seconds. What will be the projectile's height after 10 seconds if its initial velocity is 3.0×10^4 centimeters per second?

40. The volume of a cylinder with half of a sphere on each end (see the figure) is given by $V = \frac{4}{3}\pi r^3 + \pi r^2h$. Its dimensions are $r = 3.8 \times 10^1$ centimeters and $h = 9.3 \times 10^1$ centimeters. Use $\pi = 3.142$ to find its volume. Round to 3 decimal places.

5

Factoring Polynomials

The process of writing a polynomial as the product of two or more polynomials is helpful in solving equations of degree greater than 1. It also is useful as a means of simplifying fractions involving algebraic expressions. This chapter develops the methods of factoring polynomials and introduces some applications whose solutions depend on factoring.

5.1 Monomial Factors

In Chapter 4 we discussed how to find the product of two polynomials. Factoring is the inverse of this operation. We begin with a polynomial and write it as the product of other polynomials.

The simplest type of factoring associated with polynomials involves removing the **greatest common factor (GCF).** The justification for this type of factoring is based on the distributive property.

$$a(b + c) = ab + ac$$

When the left and right hand members are interchanged, we have

$$ab + ac = a(b + c)$$

Here a is common to both ab and ac and as such is a **common factor** of $ab + ac$.

In the expression

$$6x + 21,$$

the greatest common factor of both $6x$ and 21 is 3.

$$6x + 21 = 3 \cdot 2x + 3 \cdot 7$$
$$= 3(2x + 7) \qquad \text{Distributive property}$$

Example 1 Factor by removing the GCF.

$$8x + 20$$
$$8x + 20 = 4 \cdot 2x + 4 \cdot 5 \qquad \text{4 is the GCF}$$
$$= 4(2x + 5) \qquad \text{Distributive property}$$

Notice that 2 also is a common factor since $2(4x + 10) = 8x + 20$. It is not, however, the GCF. ∎

Example 2   Factor by removing the GCF.

(a) $12y + 9 = 3(4y + 3)$ The GCF is 3

(b) $7x^2 + 14y^3 = 7(x^2 + 2y^3)$ The GCF is 7

(c) $8z + 8 = 8(z + 1)$ The GCF is 8

▶▶ *Caution*   Whenever an entire term is removed as a factor, the remaining factor is 1. The 1 must be included, as in $8z + 8 = 8(z + 1)$.

(d) $6y + 5$ The greatest common factor is 1:
$6y + 5 = 1(6y + 5)$ ■

The greatest common factor frequently involves variables. When such is the case, the GCF will include any common variable with an exponent that is equal to the *smallest exponent on the variable*. In $3x^3 + 2x^2$ the GCF is x^2. There is no common numerical factor other than 1.

$3x^3 + 2x^2 = x^2(3x + 2)$ The smallest exponent is 2

To check, find the product of x^2 and $3x + 2$. The result must be $3x^3 + 2x^2$.

Example 3   Factor by removing the GCF.

(a) $4x^3 + 2x^2 + 8x$

The greatest numerical factor is 2. The variable factor is x. The smallest exponent on x is 1.

$$4x^3 + 2x^2 + 8x = 2x \cdot 2x^2 + 2x \cdot x + 2x \cdot 4$$
$$= 2x(2x^2 + x + 4)$$ Distributive property

(b) $6a^4 + 3a^3 + 12a^2$

The greatest numerical factor is 3. The smallest exponent on a is 2.

$$6a^4 + 3a^3 + 12a^2 = 3a^2(2a^2 + a + 4)$$

(c) $7x^4y + 3x^2y - 15x^3y^3$

The greatest common numerical factor is 1. The smallest exponent on x is 2 and on y is 1. The GCF is $1x^2y$.

$$7x^4y + 3x^2y - 15x^3y^3 = x^2y(7x^2) + x^2y(3) + x^2y(-15xy^2)$$
$$= x^2y(7x^2 + 3 - 15xy^2)$$ ■

When the lead coefficient of a polynomial is negative, it is common practice to seek a common factor that is negative.

Example 4   Factor $-2m^4 + 6m^2 - 4m$.

The greatest numerical factor is 2. The smallest exponent on m is 1. Thus

$$-2m^4 + 6m^2 - 4m = -2m(m^3) + (-2m)(-3m) + (-2m)(2)$$
$$= -2m[m^3 + (-3m) + 2]$$
$$= -2m(m^3 - 3m + 2).$$ ■

Example 5   Remove the indicated factor.

$$6x^2y + \frac{1}{3}x^2 + \frac{2}{9}x^4 = \frac{1}{9}x^2(\qquad)$$

$$= \frac{1}{9}x^2(54y + 3 + 2x^2) \quad \blacksquare$$

Example 6   Factor $5x^{2m-2}y^{m+1} + 10x^{2m}y^{m+3} + 25x^{2m-2}$.

The greatest numerical factor is 5. The smallest exponent on x is $2m-2$. No factor of y is common.

$$5x^{2m-2}y^{m+1} + 10x^{2m}y^{m+3} + 25x^{2m-2}$$
$$= 5x^{2m-2}(y^{m+1}) + 5x^{2m-2}(2x^2y^{m+3}) + 5x^{2m-2}(5) \qquad x^{2m-2}x^2 = x^{2m}$$
$$= 5x^{2m-2}(y^{m+1} + 2x^2y^{m+3} + 5) \quad \blacksquare$$

Example 7   Factor $6x^{m+1} + 3x^{m-1}$.

The greatest numerical factor is 3. The exponent $m + 1$ is 2 greater than the exponent $m - 1$. The smallest exponent on x is $m - 1$.

$$6x^{m+1} + 3x^{m-1} = 3x^{m-1}(2x^2) + 3x^{m-1}(1) \qquad x^{m-1}x^2 = x^{m-1+2} = x^{m+1}$$
$$= 3x^{m-1}(2x^2 + 1) \quad \blacksquare$$

Example 8   Factor by removing the greatest common binomial factor from $3(x + 1)^2 - (x + 1)$.

$$3(x + 1)^2 - (x + 1)$$
$$= (x + 1)[3(x + 1) - 1] \qquad \text{GCF} = (x + 1)$$
$$= (x + 1)[3x + 3 - 1] \qquad \text{Distributive property}$$
$$= (x + 1)[3x + 2] \quad \blacksquare$$

Exercises 5.1

Remove the indicated factor.

1. $3x + 6 = 3(\qquad)$

2. $5a + 10 = 5(\qquad)$

3. $14m - 21 = 7(\qquad)$

4. $15p - 20 = 5(\qquad)$

5. $7s^2 + s = s(\qquad)$

6. $13r^2 - r = r(\qquad)$

7. $-y^2 + 3y = -y(\qquad)$

8. $-c^2 - 11c = -c(\qquad)$

9. $-5x^2 - 10x + 15 = -5(\qquad)$

10. $-16a^2 + 24a - 8 = -8(\qquad)$

11. $-18t^3 + 24t^2 - 12t = -6t(\qquad)$

12. $-28m^3 - 49m^2 + 42m = -7m(\qquad)$

13. $-\frac{1}{5}a^2 + a - \frac{1}{10} = -\frac{1}{10}(\qquad)$

14. $-\frac{1}{2}c^2 + 2c - \frac{1}{4} = -\frac{1}{4}(\qquad)$

15. $\frac{3}{8}p^3 + 2p^2 - \frac{5}{4}p = \frac{3}{8}p(\qquad)$

16. $\frac{5}{7}t^3 - 3t^2 - \frac{1}{5}t = \frac{1}{35}t(\qquad)$

17. $-\frac{4}{9}s^4 + \frac{2}{3}s^3 - \frac{1}{9}s^2 = -\frac{1}{9}s^2(\qquad)$

18. $-\frac{2}{5}y^4 - \frac{1}{10}y^3 + 2y^2 = -\frac{1}{10}y^2(\qquad)$

19. $-\frac{2}{7}r^5 + \frac{3}{4}r^4 - \frac{1}{2}r^3 = -\frac{1}{28}r^3(\qquad)$

20. $-\frac{2}{11}x^5 + \frac{1}{2}x^4 - \frac{1}{3}x^3 = -\frac{1}{66}x^3(\qquad)$

Factor by removing the greatest common monomial factor. If the expression cannot be factored,
write prime.

21. $2x + 10$ **22.** $3x + 21$ **23.** $11y - 55$ **24.** $13y - 39$

25. $15a^2 - 50a$ **26.** $12m^2 - 30m$ **27.** $8 - 16t$ **28.** $7 - 14s$

29. $3a + 9b - 6$ **30.** $7x + 14y - 21$ **31.** $x^2y - xy^2$ **32.** $r^2s - 2rs^2$

33. $12m^2p - 6mp^2 + 18mp$ **34.** $30x^2y - 15xy^2 + 45xy$ **35.** $at + bt - ct$ **36.** $ry - ay + my$

37. $-2x^2 + 10x - 5$ **38.** $-3x^2 + 15x - 2$ **39.** $15r^5s^3 - 24r^2s^4$ **40.** $18a^6b^4 - 30a^3b^2$

41. $5a + 7a^2 - 9a^3$ **42.** $7s^5 - 18s^3 + 24s$ **43.** $28x^4 - 56x^3 - 49x^2$ **44.** $65t^5 - 39t^3 - 91t^2$

45. $8r^5s^3 - 6r^4s^5 - 20r^4s^3$ **46.** $34x^3y^5 - 51x^2y^4 - 102x^2y^6$

47. $48xyz - 64x^2yz^2 - 80xy^2z$ **48.** $44x^2y^2z - 77xy^2z^2 - 99xy^2z$

49. $63a^3c - 81a^2d + 72acd$ **50.** $88ab^2c - 56abc^2 + 80c$

Factor by removing the greatest common binomial factor.

51. $x(x + 1) + 2(x + 1)$ **52.** $y(y - 3) - 2(y - 3)$

53. $a(2a + 1) - (2a + 1)$ **54.** $2t(3t - 4) - (3t - 4)$

55. $m(a + c) + n(a + c)$ **56.** $r(e + f) - t(e + f)$

57. $x(x - y) - y(x - y)$ **58.** $2x(2x - 3y) - 3y(2x - 3y)$

59. $t(t - 1)^2 - (t - 1)$ **60.** $a(a^2 + 1) - (a^2 + 1)$

61. $(m - 5)4 - (m - 5)m$ **62.** $(2y - 3)5 - (2y - 3)3y$

63. $\left(\frac{1}{2} - a\right)\frac{1}{3} + \left(\frac{1}{2} - a\right)a$ **64.** $\left(\frac{1}{4} - m\right)\frac{1}{5} + \left(\frac{1}{4} - m\right)m$

65. $(x - 1)^2 - (x - 1)$ **66.** $(y + 2)^2 - (y + 2)$

67. $(2a + 3)^2 + 2(2a + 3)$ **68.** $(3c - 1)^2 + 5(3c - 1)$

69. $a(a^2 + b)^3 - b(a^2 + b)^2$ **70.** $x(x^2 - 2y)^3 - y(x^2 - 2y)^2$

Factor by removing the greatest common monomial factor.

71. $a^{m+1} + a^m$ **72.** $y^{m+2} + y^m$ **73.** $x^{2m+1} - x^{2m}$ **74.** $t^{3m+1} - t^{3m}$

75. $y^m + y$ **76.** $s^m + s^{m-1}$ **77.** $r^{2m+3} - r^{2m+2}$ **78.** $x^{5m-4} - x^{5m-5}$

Review Exercises

Carry out the indicated operations.

79. $(x + y)(x - y)$ **80.** $(4x^2 + 6x + 11) \div (2x - 3)$

81. $(18x^2y + 6xy^2 - 24xy) \div (-6xy)$ **82.** Use synthetic division. $(a^2 - 7a - 30) \div (a - 10)$

83. $(3m + 2n)^2$ **84.** $(x^m - y^n)(x^m + y^n)$

85. $(a - b)(a^2 + ab + b^2)$ **86.** $(r + s)(r^2 - rs + s^2)$

87. $(x^5 - 32) \div (x - 2)$ **88.** $(-54a^3b^2c - 30a^2b^3c^2 + 18ab^4c) \div (-6ab^2c)$

89. $(a^m + a^n)^2$ **90.** $(x^3 + 3x^2 + 3x + 1) \div (x + 1)$

5.2   Factoring Trinomials I

In this section we will consider factoring trinomials in which the coefficient of the highest degree term is 1. The expressions

$$x^2 + 6x + 5, \quad y^3 - 2y^2 - y, \quad \text{and} \quad m^4 - 3m^2 - 4$$

belong to this group.

A trinomial is often the product of two binomials. Therefore if you try to factor $x^2 + 6x + 5$ it is reasonable to expect that the factors may be two binomials.

To factor $x^2 + 6x + 5$ we want to find two numbers a and b such that

$$(x + a)(x + b) = x^2 + 6x + 5$$

but $(x + a)(x + b) = x^2 + (a + b)x + ab$

This shows that $a \cdot b = 5$ and $a + b = 6$. We are seeking two numbers whose product is **5** and whose sum is **6**. The numbers are **1** and **5**.

$$x^2 + 6x + 5 = (x + 1)(x + 5)$$

The results can be quickly checked by the FOIL method to make sure the sum of the inner and outer products is $6x$.

$$(x + 1)(x + 5) = x^2 + 6x + 5$$

$$x$$
$$5x$$

Example 1   Factor $x^2 + 8x + 12$.

To factor $x^2 + 8x + 12$, find two numbers whose product is **12** and whose sum is **8**.

Product	Sum
$12 \cdot 1 = 12$	$12 + 1 = 13$
$6 \cdot 2 = 12$	$6 + 2 = 8$ ✓
$4 \cdot 3 = 12$	$4 + 3 = 7$

Only 6 and 2 satisfy both conditions. Thus

$$x^2 + 8x + 12 = (x + 2)(x + 6). \quad \blacksquare$$

Example 2   Factor $a^2 - 7a - 18$.

The last term is negative, so the two factors in the product must have opposite signs. We need two numbers whose product is -18 and whose sum is -7. They are -9 and 2.

$$a^2 - 7a - 18 = (a - 9)(a + 2) \quad \blacksquare$$

Example 3   Factor $3y^3 + 12y^2 - 15y$.

The coefficient of the first term is not 1. However, $3y$ is a common factor that can be removed first.

$$3y^3 + 12y^2 - 15y = 3y(y^2 + 4y - 5)$$

Now factor $y^2 + 4y - 5$.

$$3y^3 + 12y^2 - 15y = 3y(y + 5)(y - 1) \quad \blacksquare$$

A polynomial that cannot be factored using real numbers as factors is said to be **prime with respect to the reals.** The polynomial $x^2 + 3x + 9$ is prime, since there are no two real numbers whose product is 9 and whose sum is 3.

Example 4   Factor $y^4 + 15y^2 + 54$.

The polynomial fits the pattern established in Examples 1 through 3. To see this, let $y^2 = a$ so that $y^4 = a^2$.

$$y^4 + 15y^2 + 54 = a^2 + 15a + 54$$
$$= (a + 9)(a + 6)$$

Substituting y^2 for a gives

$$y^4 + 15y^2 + 54 = (y^2 + 9)(y^2 + 6).$$

Since $(y^2 + 9)$ and $(y^2 + 6)$ are prime, the polynomial is factored. $\blacksquare$

Example 5   Factor $x^2(x + 2) - 8x(x + 2) - 20(x + 2)$.

Remove the common factor $x + 2$ first.

$$(x + 2)(x^2 - 8x - 20)$$
$$= (x + 2)(x - 10)(x + 2)$$
$$= (x + 2)^2(x - 10) \qquad (x + 2)(x + 2) = (x + 2)^2 \quad \blacksquare$$

Example 6   Factor $t^2 + \frac{5}{3}t + \frac{4}{9}$.

$$t^2 + \frac{5}{3}t + \frac{4}{9} = \left(t + \frac{1}{3}\right)\left(t + \frac{4}{3}\right) \qquad \frac{1}{3} \cdot \frac{4}{3} = \frac{4}{9} \quad \text{and} \quad \frac{1}{3} + \frac{4}{3} = \frac{5}{3} \quad \blacksquare$$

Example 7   Factor $x^{2n} + 3x^n - 4$.

Think of x^{2n} as $(x^n)^2$.

$$x^{2n} + 3x^n - 4 = (x^n + 4)(x^n - 1) \quad \blacksquare$$

Example 8   Factor $x^{m+2} + 8x^{m+1} + 7x^m$.

x^m is a common factor. Remove it first.

$$x^m(x^2 + 8x + 7) \qquad \begin{aligned} x^m \cdot x^2 &= x^{m+2} \\ x^m \cdot x &= x^{m+1} \end{aligned}$$
$$= x^m(x + 7)(x + 1) \quad \blacksquare$$

▶ **Caution**   When factoring, always remove the greatest common factor first.

Example 9   Factor $-2a^2 - 4a + 70$.

The number -2 is a common factor.

$$-2(a^2 + 2a - 35) \qquad \text{Removing the } (-2) \text{ makes the coefficient of the first term 1}$$

$$-2(a + 7)(a - 5) \quad \blacksquare$$

Exercises 5.2

Factor each of the following trinomials.

1. $x^2 + 3x + 2$
2. $y^2 + 4y + 3$
3. $m^2 + 7m + 6$
4. $a^2 + 9a + 8$
5. $y^2 - 2y + 1$
6. $x^2 - 4x + 3$
7. $p^2 - 5p + 6$
8. $r^2 - 4r + 4$
9. $x^2 - x - 6$
10. $y^2 - y - 12$
11. $t^2 + 2t - 3$
12. $b^2 + 3b - 28$
13. $y^2 - 8y - 20$
14. $x^2 - 5x - 24$
15. $m^2 + m - 30$
16. $a^2 + 2a - 48$
17. $r^2 + 6ra - 16a^2$
18. $m^2 + mn - 72n^2$
19. $a^2 - 7ab - 60b^2$
20. $p^2 - 8pq - 48q^2$

Factor each of the following. Remove any common factors first. If the trinomial is not favorable over the set of real numbers, write **prime**.

21. $2x^2 + 20x + 48$
22. $2y^2 - 12y + 10$
23. $a^2 + 4a + 2$
24. $m^2 + 3m + 1$
25. $5p^2 - 45p + 40$
26. $6r^2 - 30r - 84$
27. $10 - 3s - s^2$
28. $15 - 3x - x^2$
29. $x^3 - 4x^2 - 32x$
30. $y^3 + 9y^2 - 36y$
31. $4a^3 - 16a^2b - 20ab^2$
32. $2x^3 - 2x^2y - 84xy^2$
33. $r^2 + \dfrac{8}{15}r + \dfrac{1}{15}$
34. $s^2 + \dfrac{1}{20}s - \dfrac{1}{20}$
35. $k^2 - \dfrac{1}{6}k - \dfrac{1}{6}$
36. $m^2 - \dfrac{1}{12}m - \dfrac{1}{12}$
37. $y^2 - 7y - 1$
38. $a^2 - 5a - 2$
39. $a^2b^2 - 4ab + 4$
40. $x^2y^2 - 6xy + 9$
41. $-3m^2 - 3m + 36$
42. $-5z^2 - 25z + 70$
43. $-7a^3 - \dfrac{7}{2}a^2 + \dfrac{7}{16}a$
44. $-2r^2 - 2r + \dfrac{2}{4}$
45. $-x^2 + 8x - 12$
46. $-y^2 - 5y + 14$
47. $3st^3 + 33s^2t^2 + 54s^3t$
48. $4a^3b - 4a^2b^2 - 48ab^3$
49. $5p^3q + 10p^2q^2 + 5pq^3$
50. $2x^3y - 12x^2y^2 + 18xy^3$

Factor each polynomial by substitution if needed. See Example 4.

51. $x^4 + 6x^2 + 8$
52. $y^4 + 7y^2 + 12$
53. $a^4 + 8a^2 + 16$
54. $b^4 + 10b^2 + 25$
55. $2m^4 + 18m^2 + 40$
56. $3s^4 + 12s^2 + 12$
57. $-5t^4 - 5t^2 + 150$
58. $-2c^4 + 16c^2 + 130$
59. $-k^3 + \dfrac{11}{30}k^2 - \dfrac{1}{30}k$
60. $-z^3 + \dfrac{10}{21}z^2 - \dfrac{1}{21}z$

Factor each of the following. If there is a common factor, remove it first.

61. $x^2(x + 3) - x(x + 3) - 2(x + 3)$
62. $y^2(y - 2) + 3y(y - 2) + 2(y - 2)$
63. $a^2(a - 3) - a(a - 3) - 6(a - 3)$
64. $z^2(z + 5) + 10z(z + 5) + 25(z + 5)$
65. $t^2(t + 7) + 14t(t + 7) + 49(t + 7)$
66. $s^2(s - 4) - 7s(s - 4) + 12(s - 4)$
67. $k^2\left(k - \dfrac{1}{3}\right) - \dfrac{2}{9}k\left(k - \dfrac{1}{3}\right) + \dfrac{1}{9}\left(k - \dfrac{1}{3}\right)$
68. $c^2\left(c - \dfrac{1}{5}\right) - c\left(c - \dfrac{1}{5}\right) + \dfrac{12}{49}\left(c - \dfrac{1}{5}\right)$

Factor each of the following. If there is a common factor, remove it first.

69. $x^{2m} + 8x^m + 12$
70. $x^{2m} + 9x^m + 20$
71. $y^{2m} - 0.3y^m + 0.02$
72. $a^{2m} - 0.9a^m + 0.2$
73. $c^{m+2} + 11c^{m+1} + 30c^m$
74. $y^{m+2} - 7y^{m+1} + 12y^m$

75. $s^{2m+2} - s^{m+2} - 2s^2$

76. $t^{2m+2} - 5t^{m+2} - 24t^2$

77. $k^{2m+3} - k^{m+3} + 0.21k^3$

78. $z^{2m+3} - 1.3z^{m+3} + 0.42z^3$

Use a calculator to remove the indicated common factor.

79. $0.585y^2 - 0.0715y; \ 0.65y$

80. $0.018x^2 + 0.0204x; \ 0.12x$

81. $-0.1023t^2 - 0.837t; \ -0.93t$

82. $-0.1881a^2 + 0.1767a; \ -0.57a$

5.3   Special Types of Factors and Factoring by Grouping

One of the easiest types of factoring to recognize involves the **difference of two squares.** Recall that

$$(a - b)(a + b) = a^2 - b^2.$$

Applying the symmetric property of equality to this expression gives

$$a^2 - b^2 = (a - b)(a + b).$$

Example 1   Factor.

(a) $x^2 - 25$

$$x^2 - 25 = x^2 - 5^2 = (x - 5)(x + 5)$$

(b) $y^2 - 16$

$$y^2 - 16 = y^2 - 4^2 = (y - 4)(y + 4) \quad ∎$$

Example 2   Factor $m^4 - n^6$.

This expression is the difference of two squares, since m^4 is the square of m^2, and n^6 is the square of n^3.

$$m^4 - n^6 = (m^2)^2 - (n^3)^2 \qquad (b^m)^n = b^{m \cdot n}$$
$$= (m^2 - n^3)(m^2 + n^3) \quad ∎$$

Example 3   Factor $x^2y^4 - z^2$.

$$x^2y^4 - z^2 = (xy^2)^2 - z^2 \qquad (a \cdot b)^m = a^m \cdot b^m$$
$$= (xy^2 - z)(xy^2 + z) \quad ∎$$

Example 4   Factor $s^{2m} - t^{4n}$.

$$s^{2m} - t^{4n} = (s^m)^2 - (t^{2n})^2 \qquad (b^m)^n = b^{m \cdot n}$$
$$= (s^m - t^{2n})(s^m + t^{2n}) \quad ∎$$

The difference of squares method of factoring often can be used more than once in the same problem. For example,

$$x^4 - y^4 = (x^2 - y^2)(x^2 + y^2)$$
$$= (x - y)(x + y)(x^2 + y^2).$$

▶ **Caution**   The **sum** of two squares, such as $x^2 + y^2$, is prime with respect to the real numbers.

Example 5   Factor $x^8 - y^8$.

$$
\begin{aligned}
x^8 - y^8 &= (x^4 - y^4)(x^4 + y^4) \\
&= (x^2 - y^2)(x^2 + y^2)(x^4 + y^4) \\
&= (x - y)(x + y)(x^2 + y^2)(x^4 + y^4) \quad \blacksquare
\end{aligned}
$$

Example 6   Factor $(s + t)^2 - (x + y)^2$.

$$
\underbrace{(s + t)^2}_{a^2} - \underbrace{(x + y)^2}_{b^2} = [\underbrace{(s + t)}_{a} - \underbrace{(x + y)}_{b}][\underbrace{(s + t)}_{a} + \underbrace{(x + y)}_{b}]
$$

$$
= (s + t - x - y)(s + t + x + y) \quad \blacksquare
$$

Two other common types of factoring involving binomials are **the sum and difference of cubes.**

$$
a^3 - b^3 = (a - b)(a^2 + ab + b^2)
$$
$$
a^3 + b^3 = (a + b)(a^2 - ab + b^2)
$$

These factors are easy to justify by multiplication.

$$
\begin{array}{ll}
\begin{array}{r}
a^2 + ab + b^2 \\
a - b \\
\hline
a^3 + a^2b + ab^2 \\
\;\; - a^2b - ab^2 - b^3 \\
\hline
a^3 \qquad\qquad - b^3
\end{array}
&
\begin{array}{r}
a^2 - ab + b^2 \\
a + b \\
\hline
a^3 - a^2b + ab^2 \\
\;\; + a^2b - ab^2 + b^3 \\
\hline
a^3 \qquad\qquad + b^3
\end{array}
\end{array}
$$

Notice that even though the sum of two squares is not factorable, **the sum of two cubes is factorable.**

Example 7   Factor.

(a) $x^3 - 125$

$$
\begin{aligned}
x^3 - 125 &= x^3 - 5^3 \qquad\qquad\qquad a = x, \; b = 5\\
&= \underbrace{(x - 5)}_{(a - b)}\underbrace{(x^2 + 5x + 25)}_{(a^2 + ab + b^2)}
\end{aligned}
$$

(b) $8x^3 + 27$

$$
\begin{aligned}
8x^3 + 27 &= (2x)^3 + 3^3 \qquad\qquad\qquad a = 2x, \; b = 3\\
&= \underbrace{(2x + 3)}_{(a + b)}[\underbrace{(2x)^2 - 3(2x) + 3^2}_{(a^2 - ab + b^2)}]\\
&= (2x + 3)(4x^2 - 6x + 9) \quad \blacksquare
\end{aligned}
$$

▶ **Caution**   Care must be taken not to confuse $a^2 \pm ab + b^2$ with $a^2 \pm 2ab + b^2$, which results from $(a \pm b)^2 = a^2 \pm 2ab + b^2$.

Factoring sometimes involves both the difference of squares and the difference of cubes.

Example 8   Factor $x^6 - y^6$.

Begin by expressing $x^6 - y^6$ as the difference of squares.

$$\begin{aligned} x^6 - y^6 &= (x^3)^2 - (y^3)^2 \qquad\qquad (b^m)^n = b^{m \cdot n}\\ &= (x^3 - y^3)(x^3 + y^3)\\ &= (x - y)(x^2 + xy + y^2)(x + y)(x^2 - xy + y^2) \quad\blacksquare \end{aligned}$$

The expression in Example 8 could also have been written as

$$x^6 - y^6 = (x^2)^3 - (y^2)^3,$$

which would have yielded the factors

$$= (x - y)(x + y)(x^4 - x^2y^2 + y^4).$$

Since the latter expression is not as completely factored as the former, it is not as desirable. In general, if a polynomial can be factored initially in terms of even or odd powers, begin with even powers.

Another type of factoring that is easy to recognize involves the square of a binomial. Recall from Chapter 4 that

$$a^2 + 2ab + b^2 = (a + b)^2$$

and $a^2 - 2ab + b^2 = (a - b)^2$.

Such expressions are called **perfect square trinomials** since they are the squares of binomials. The key to recognizing perfect square trinomials is that the first and last terms, a^2 and b^2, are perfect squares,

$$a^2 \pm 2ab + b^2 = (a \pm b)^2,$$

The square of a   The square of b

and the middle term, $\pm 2ab$, must be twice the product of the terms a and b of the binomial.

Example 9   Factor.

(a) $x^2 + 6x + 9$

$x^2 + 6x + 9 = (x + 3)^2$ $x^2 = (x)^2, \quad 9 = 3^2$
$6x = 2 \cdot 3 \cdot x$

(b) $x^2 - 8x + 16$

$x^2 - 8x + 16 = (x - 4)^2$ $x^2 = (x)^2, \quad 16 = (-4)^2$
$-8x = 2(-4)x \quad\blacksquare$

Example 10   Factor $4x^2 - 20xy + 25y^2$.

$4x^2 - 20xy + 25y^2 = (2x - 5y)^2$ $4x^2 = (2x)^2$
$25y^2 = (5y)^2 \quad\blacksquare$

Example 11   Factor $2x^5 + 36x^4 + 162x^3$.

Remove common factors first.

$$\begin{aligned} 2x^5 + 36x^4 + 162x^3 &= 2x^3(x^2 + 18x + 81)\\ &= 2x^3(x + 9)^2 \quad\blacksquare \end{aligned}$$

One final special type of factoring will be considered in this section. It is called **factoring by grouping.** The name originates from the fact that the way the terms of a polynomial are grouped may determine its factorability. For example, $ax + by + bx + ay$ can be factored if the terms are grouped correctly.

$$ax + by + bx + ay = (ax + bx) + (ay + by) \qquad \text{Commutative and associative properties}$$

$$= x(a + b) + y(a + b) \qquad \text{Distributive property}$$

We now have a common factor, $(a + b)$.

$$= (a + b)(x + y) \qquad \text{Distributive property}$$

The expression can also be factored by grouping the terms containing a and the terms containing b. Try it.

Example 12   Factor $x^2 + 2x + xy + 2y$.

$$x^2 + 2x + xy + 2y = (x^2 + 2x) + (xy + 2y) \qquad \text{Associative property}$$

$$= x(x + 2) + y(x + 2) \qquad \text{Distributive property}$$

$$= (x + 2)(x + y) \qquad \text{Distributive property} \quad \blacksquare$$

In some instances the grouping may involve three or more terms.

Example 13   Factor $x^2 + 2xy + y^2 - z^2$.

$$x^2 + 2xy + y^2 - z^2 = (x^2 + 2xy + y^2) - z^2 \qquad \text{Associative property}$$

$$= (x + y)^2 - z^2 \qquad x^2 + 2xy + y^2 \text{ is a perfect square}$$

The expression can now be factored as the difference of squares.

$$= (x + y - z)(x + y + z) \qquad a^2 - b^2 = (a - b)(a + b) \quad \blacksquare$$

Example 14   Factor $m^2 - mn - m + n$.

$$m^2 - mn - m + n = (m^2 - mn) + (-m + n)$$

$$= m(m - n) - 1(m - n) \qquad \text{When removing a factor of } -1, \text{ change all signs in the remaining factor}$$

$$= (m - n)(m - 1) \quad \blacksquare$$

Example 15   Factor $ax^2 + 4ax + 4a - bx^2 - 4bx - 4b$.

$$ax^2 + 4ax + 4a - bx^2 - 4bx - 4b$$

$$= (ax^2 + 4ax + 4a) + (-bx^2 - 4bx - 4b)$$

$$= a(x^2 + 4x + 4) - b(x^2 + 4x + 4)$$

$$= (a - b)(x^2 + 4x + 4)$$

$$= (a - b)(x + 2)^2 \quad \blacksquare$$

To Factor Polynomials

1. Always remove the greatest common factor first.
2. If the polynomial contains two terms, consider the patterns established for $a^2 - b^2$ and $a^3 \pm b^3$.
3. If the polynomial contains three terms, decide whether or not it is a perfect square trinomial. If not, factor by the methods discussed in Section 5.1.
4. If a polynomial contains four or more terms, factor by grouping.

Exercises 5.3

Factor each difference of squares and simplify.

1. $x^2 - 1$ **2.** $y^2 - 4$ **3.** $2a^2 - 50$ **4.** $3m^2 - 27$

5. $x^2y^2 - 16z^2$ **6.** $r^2s^2 - 49t^2$ **7.** $121p^2 - 144q^2$ **8.** $169m^2 - 100n^2$

9. $a^4 - b^4$ **10.** $18x^4 - 2y^4$ **11.** $3r^{2m} - 3s^{2m}$ **12.** $x^{4m} - 4y^{2m}$

13. $\dfrac{1}{16}a^2b^3 - \dfrac{1}{9}b^5$ **14.** $\dfrac{1}{25}x^3y - \dfrac{1}{4}xy^3$ **15.** $0.64a^3b - 0.49ab^3$ **16.** $0.25m^4n - 0.81n^3$

17. $(a + b)^2 - (a - b)^2$ **18.** $(x - y)^2 - (x + y)^2$ **19.** $(2m - 3n)^2 - (2m + 3n)^2$

20. $(x^2 - 1)^2 - (x^2 + 1)^2$ **21.** $a^8 - b^8$ **22.** $r^8 - s^8$

Factor each sum or difference of cubes.

23. $x^3 - y^3$ **24.** $m^3 - n^3$ **25.** $p^3 + q^3$ **26.** $r^3 + s^3$

27. $27a^3 - 8b^3$ **28.** $64x^3 + 125y^3$ **29.** $x^3y^3 + w^3z^3$ **30.** $p^3q^3 + r^3s^3$

31. $a^{3m} - b^{3m}$ **32.** $c^{3m} + d^{3m}$ **33.** $x^3 + (x - 1)^3$ **34.** $8a^3 + (2a + 1)^3$

35. $(y + 2)^3 - y^3$ **36.** $(m + 1)^3 - m^3$ **37.** $a^6 + b^6$ **38.** $8m^6 - n^6$

39. $x^9 - y^9$ **40.** $r^9 - s^9$ **41.** $(a + b)^3 - (a - b)^3$ **42.** $(c - d)^3 + (c + d)^3$

43. $\dfrac{1}{27}s^3 - \dfrac{1}{8}t^3$ **44.** $\dfrac{1}{64}p^3 - \dfrac{1}{125}q^3$

Factor each perfect square trinomial.

45. $a^2 + 4a + 4$ **46.** $m^2 + 8m + 16$ **47.** $x^2 - 2x + 1$

48. $y^2 - 10y + 25$ **49.** $4x^2 + 12xy + 9y^2$ **50.** $25r^2 + 60rs + 36s^2$

51. $3p^3 - 6p^2q + 3pq^2$ **52.** $5m^3 - 20m^2n + 20mn^2$ **53.** $c^{2m} + 14c^md^n + 49d^{2n}$

54. $w^{2m} + 16w^mz^n + 64z^{2n}$ **55.** $y^4 - 18y^2 + 81$ **56.** $x^4 - 8x^2 + 16$

57. $(a + 2)^2 + 2(a + 2) + 1$ **58.** $(b - 1)^2 - 6(b - 1) + 9$

59. $3m^5 - 30m^3 + 75m$ **60.** $2p^5 - 12p^3 + 18p$

61. $(x - y)^2 - 4(x - y)(a - b) + 4(a - b)^2$ **62.** $(r + s)^2 + 22(r + s)(r - s) + 121(r - s)^2$

63. $\dfrac{1}{16}w^2 + \dfrac{1}{10}wz + \dfrac{1}{25}z^2$ **64.** $\dfrac{1}{49}a^2 + \dfrac{2}{63}ab + \dfrac{1}{81}b^2$

Factor by grouping.

65. $cx + cy + dx + dy$ **66.** $5x + 5y + sx + sy$

67. $a^2 + ab - 2a - 2b$ **68.** $m^2 + mn - 3m - 3n$

69. $mr + 2nr + 2ms + 4ns$

70. $4ac - 6bc + 4ad - 6bd$

71. $x^3 + x^2 - x - 1$

72. $r^3 + r^2 - r - 1$

73. $s^2x^2 - s^2y^2 - t^2x^2 + t^2y^2$

74. $4a^2x^2 - a^2y^2 - 4b^2x^2 + b^2y^2$

75. $a^2 + 2a + 1 - b^2$

76. $x^2 + 4x + 4 - y^2$

77. $y^2 - 9x^2 + 6y + 9$

78. $m^2 - 4n^2 + 2m + 1$

79. $s^2 - 4t^2 - 16t - 16$

80. $25 - y^2 - 4x^2 - 4xy$

81. $6ab - 3bc - 14a + 7c$

82. $8rs + 3t - 8rt - 3s$

83. $8m^3 + 27n^3 + 2m + 3n$

84. $5y - x + 125y^3 - x^3$

85. $128x^3 - 2y^3 + 2y - 8x$

86. $24a^3 - 3b^3 + 3b - 6a$

87. $x^2 + 6x + 9 - y^2 - 4y - 4$

88. $p^2 - 8p + 16 - q^2 + 2q - 1$

89. $ax + bx - cx - ay - by + cy$

90. $dx - ex - fx - dy + ey + fy$

5.4 Factoring Trinomials II; A Test for Factorability

Section 5.2 showed ways to factor trinomials in which the coefficient of the highest-degree term was 1. This section will consider trinomials of the form $ax^2 + bx + c = 0$, $a \neq 0$ or 1, where a, b, c are rational numbers. For example, to factor

$$3x^2 + 7x + 2$$

into the product of two binomials, begin as follows.

$$3x^2 + 7x + 2 = (\qquad)(\qquad).$$

The possible factors of the first term are $3x$ and x. The possible factors of the last term are 2 and 1. *Since all the signs in the trinomial are positive, the signs in the binomials will be positive.* There are two possible ways the factors can be placed in the binomials. Only one will yield the correct middle term.

$$3x^2 + 7x + 2 \overset{?}{=} (3x + 2)(x + 1) \quad \text{Wrong}$$

2x + 3x ≠ 7x

2x
3x

Now exchange the 2 and 1 to see how the middle term is affected.

$$3x^2 + 7x + 2 = (3x + 1)(x + 2) \quad \text{Right}$$

x + 6x = 7x

x
6x

When there is more than one way to factor the first and last term of a trinomial, the number of possible arrangements within the binomial factors increases rapidly.

Example 1   Factor $6x^2 + 11x + 4$.

$$6x^2 = 6x \cdot x \quad \text{or} \quad 3x \cdot 2x$$
$$4 = 4 \cdot 1 \quad \text{or} \quad 2 \cdot 2$$

All signs in the trinomial are positive, so only positive signs can occur in the binomials.

Possible Factors	Middle Term	Conclusion
$(6x + 1)(x + 4)$	$24x + x = 25x$	Wrong
$(3x + 4)(2x + 1)$	$3x + 8x = 11x$	Right

Notice that we did not try other combinations, such as

$$(6x + 4)(x + 1) \quad \text{and} \quad (3x + 2)(2x + 2).$$

Why?

$$(6x + 4)(x + 1) = 2(3x + 2)(x + 1)$$

Since 2 is a factor of this last expression, it would also have to be a factor of $6x^2 + 11x + 4$. The trinomial has no common factors, so the combination is impossible. Thus,

$$6x^2 + 11x + 4 = (3x + 4)(2x + 1). \quad \blacksquare$$

▶ **Caution** In general, if a polynomial has no common factors, then none of its factors can have a common factor.

Example 2 Factor $2y^2 + y - 15$.

$$2y^2 = 2y \cdot y$$
$$15 = 15 \cdot 1 \quad \text{or} \quad 5 \cdot 3$$

Since the sign of the last term in the polynomial is negative, the signs between the terms of the binomial will differ.

Possible Factors	Middle Term	Conclusion
$(2y + 1)(y - 15)$	$-30y + y = -29y$	Wrong
$(2y - 1)(y + 15)$	$30y - y = 29y$	Wrong
$(2y - 3)(y + 5)$	$10y - 3y = 7y$	Wrong
$(2y + 3)(y - 5)$	$-10y + 3y = -7y$	Wrong
$(2y + 5)(y - 3)$	$-6y + 5y = -y$	Wrong
$(2y - 5)(y + 3)$	$6y - 5y = y$	Right

Therefore, $2y^2 + y - 15 = (2y - 5)(y + 3)$. $\quad \blacksquare$

Example 3 Factor $20x^2 - 47x + 21$.

$$20x^2 = 20x \cdot x \quad \text{or} \quad 10x \cdot 2x \quad \text{or} \quad 5x \cdot 4x$$
$$21 = 21 \cdot 1 \quad \text{or} \quad 7 \cdot 3$$

The sign before the last term is positive, so the signs in the binomials will be the same. The sign before the middle term is negative, so both the signs must be negative.

$$20x^2 - 47x + 21 = (5x - 3)(4x - 7) \quad \blacksquare$$

Example 4 Factor $4x^2 + 14x + 12$.
Always remove common factors first.

$$4x^2 + 14x + 12 = 2(2x^2 + 7x + 6)$$
$$= 2(2x + 3)(x + 2) \quad \blacksquare$$

Example 5 Factor $3x^2 + 2xy - 8y^2$.

$$3x^2 + 2xy - 8y^2 = (3x - 4y)(x + 2y) \qquad -4xy + 6xy = 2xy \quad \blacksquare$$

$$-4xy$$
$$6xy$$

Example 6 Factor $4x^{2m} + 7x^m + 3$.
Examine all of the possible factors. All signs of the trinomial are positive, so all signs of the binomial factors must also be positive.

Possible Factors	Middle Term	Conclusion
$(2x^m + 3)(2x^m + 1)$	$6x^m + 2x^m = 8x^m$	Wrong
$(4x^m + 1)(x^m + 3)$	$x^m + 12x^m = 13x^m$	Wrong
$(4x^m + 3)(x^m + 1)$	$4x^m + 3x^m = 7x^m$	Right

Thus $4x^{2m} + 7x^m + 3 = (4x^m + 3)(x^m + 1)$. $\qquad x^m \cdot x^m = x^{2m} \quad \blacksquare$

Example 7 Factor $\frac{1}{9}x^2 - \frac{2}{15}x + \frac{1}{25}$.
Since $\frac{1}{9}x^2 = (\frac{1}{3}x)^2$ and $\frac{1}{25} = (-\frac{1}{5})^2$, the expression may be a perfect square trinomial. Thus $2(\frac{1}{3}x)(-\frac{1}{5})$ must be $-\frac{2}{15}x$.

$$\frac{1}{9}x^2 - \frac{2}{15}x + \frac{1}{25} = \left(\frac{1}{3}x - \frac{1}{5}\right)^2 \quad \blacksquare$$

Example 8 Factor $2r^2 + r + 1$ using integers.
The only possibility to try is $(2r + 1)(r + 1) = 2r^2 + 3r + 1$, which is not correct. Such an expression is said to be prime (not factorable) with respect to the integers. $\blacksquare$

Example 8 showed that some polynomials cannot be factored using integers. When a large number of trials are possible, such a discovery can prove to be very time-consuming. Fortunately, a simple test for factorability has been devised. It is sometimes called **the *ac* test,** since it depends on the coefficients of the trinomial

$$ax^2 + bx + c.$$

ac Test for Factorability
The trinomial $ax^2 + bx + c$ can be factored as the product of two binomials if the following conditions are met:
1. The product ac can be factored into two integers m and n;
2. $m + n = b$.

The test also provides a means to factor the trinomial once its factorability has been determined. If the integers m and n exist, we can substitute their sum for b.

$$ax^2 + (m + n)x + c$$

When the last expression is written as

$$ax^2 + mx + nx + c,$$

factoring by grouping is easily accomplished.

Example 9   Factor $12x^2 + 11x + 2$.

$a = 12, \; c = 2; \; ac = mn = 24$	$b = 11; \; m + n = b$
$m \cdot n = 24$	$m + n = 11$
$24 \cdot 1 = 24$	$24 + 1 = 25$
$12 \cdot 2 = 24$	$12 + 2 = 14$
$\mathbf{8 \cdot 3 = 24}$	$\mathbf{8 + 3 = 11}$ ✓
$6 \cdot 4 = 24$	$6 + 4 = 10$

The third combination, $8 \cdot 3 = 24$, $8 + 3 = 11$, satisfies the requirements of the test. Substitute $8 + 3$ for 11 and break up the middle term.

$$12x^2 + 11x + 2$$
$$12x^2 + (8 + 3)x + 2 \qquad b = 8 + 3$$
$$= 12x^2 + 8x + 3x + 2 \qquad \text{Distributive property}$$
$$= 4x(3x + 2) + 1(3x + 2) \qquad \text{Factor by grouping}$$
$$= (3x + 2)(4x + 1) \quad ■$$

Example 10   Factor $6x^2 - 7x - 5$.

$a = 6, \; c = -5; \; ac = mn = -30$	$b = -7; \; m + n = -7$
$30(-1) = -30$	$30 + (-1) = 29$
$15(-2) = -30$	$15 + (-2) = 13$
$10(-3) = -30$	$10 + (-3) = 7$

Notice in the last trial that only the sign of b is wrong. Change the signs of m and n. Then,

$$(-10)(3) = -30 \quad \text{and} \quad -10 + (3) = -7. \quad ✓$$
$$6x^2 - 7x - 5$$
$$= 6x^2 + (-10 + 3)x - 5 \qquad b = -10 + 3$$
$$= 6x^2 - 10x + 3x - 5 \qquad \text{Break up the middle term}$$
$$= 2x(3x - 5) + 1(3x - 5) \qquad \text{Factor by grouping}$$
$$= (3x - 5)(2x + 1) \quad ■$$

Example 11   Factor $4x^2 - 7x + 9$.

$$ac = mn = 36 \qquad b = -7; \; m + n = -7$$
$$-36(-1) = 36 \qquad -36 + (-1) = -37$$
$$-18(-2) = 36 \qquad -18 + (-2) = -20$$
$$-9(-4) = 36 \qquad -9 + (-4) = -13$$
$$-6(-6) = 36 \qquad -6 + (-6) = -12$$

Since none of the sums is -7, the trinomial is not factorable.   ∎

Exercises 5.4

Factor each polynomial.

1. $2a^2 + 3a + 1$

2. $3p^2 + 4p + 1$

3. $3x^2 + 7x + 2$

4. $4y^2 + 13y + 3$

5. $2s^2 + 7s - 4$

6. $2t^2 + 5t - 6$

7. $12r^2 + 10r - 8$

8. $10x^2 - 5x - 5$

9. $8m^2 - 6m - 5$

10. $4c^2 - 12c - 7$

11. $8x^2 + 26x - 34$

12. $24y^2 + 92y - 16$

13. $4a^2 + 20a + 25$

14. $10d^2 + 13d + 4$

15. $8y^5 + 20y^4 - 100y^3$

16. $6s^4 - 24s^3t + 24s^2t^2$

17. $9c^2 - 24cd - 20d^2$

18. $12x^6 - 25x^5y + 12x^4y^2$

19. $12 + k - 6k^2$

20. $3 + 10b - 8b^2$

21. $2 - 5t - 12t^2$

22. $3 - 16r - 12r^2$

23. $6a^2b^2 - 17ab - 10$

24. $9m^3n + 9m^2n^2 - 4mn^3$

25. $-2k^2x^2 - 14kax^2 - 24a^2x^2$

26. $-3ax^2 - 12axy - 12ay^2$

27. $15y^2 + 49y + 40$

28. $-30t^2 - 17t + 35$

29. $-32x^2 - 4xy + 6y^2$

30. $-27a^2 + 72ab + 60b^2$

31. $a^2b^2 + 5abcd + 6c^2d^2$

32. $x^2y^2 - 5xyuv - 6u^2v^2$

Use the ac test to determine whether each polynomial is factorable. Do not factor.

33. $3x^2 + 4x + 2$

34. $5y^2 + y + 1$

35. $2a^2 - 7a + 9$

36. $6m^2 - 11m + 3$

37. $8p^2 - 6p - 5$

38. $4c^2 + 8c - 5$

39. $6c^2 + 13cd - 8d^2$

40. $4x^2 + 8x - 21$

41. $24 - 26x - 8x^2$

42. $8 + 12x - 36x^2$

Factor each polynomial completely. If it is not factorable write prime.

43. $2x^4 - 16x^2 + 32$

44. $4a^4 + 7a^2 - 2$

45. $3a^4 - 30a^2b^2 + 27b^4$

46. $8x^4 - 74x^2y^2 + 18y^4$

47. $c^6 + 7c^3d^3 - 8d^6$

48. $27m^6 + 26m^3n^3 - n^6$

49. $9x^4 - 85x^2y^2 + 36y^4$

50. $4s^4 - 45s^2t^2 + 81t^4$

51. $98c^4 - 100c^2d^2 + 2d^4$

52. $16x^6 - 130x^3y^3 + 16y^6$

53. $\dfrac{4}{9}y^2 + \dfrac{5}{9}y + \dfrac{1}{6}$

54. $\dfrac{3}{14}a^2 + \dfrac{5}{42}ab - \dfrac{2}{9}b^2$

55. $\dfrac{3}{10}m^2 + \dfrac{1}{60}mn - \dfrac{1}{12}n^2$

56. $\dfrac{2}{15}p^2 + \dfrac{41}{225}pq - \dfrac{1}{15}q^2$

57. $3x^{2m} + 10x^my^n - 8y^{2n}$

58. $9a^{2m} - 6a^mb^n - 8b^{2n}$

59. $12c^{2m} + 23c^md^n - 9d^{2n}$

60. $4r^{2m} + 21r^ms^n - 18s^{2n}$

Factor by any method.

61. $(x + 3)^2 + (x + 3) - 2$

62. $(a - 4)^2 - 4(a - 4) + 3$

63. $24(m + n)^2 + 38(m + n) + 15$

64. $24(t - 2)^2 - 82(t - 2) + 48$

65. $78bx + 24ax + 52by + 16ay$

66. $-33bz + 18ay + 27by - 22az$

67. $4x^2 - 2x - 2xy + y$

68. $5y^2 + y - 5xy - x$

69. $a^2 - 36d^2 - 2ab + b^2$

70. $x^2 - 4z^2 + 6xy + 9y^2$

71. $72xy - 72x^2 - y + x$

72. $27ab - 54b^2 - a + 2b$

73. $30ab + 9b^2 - 16c^2 + 25a^2$

74. $16x^2 - 9z^2 + 40xy + 25y^2$

75. $6(x - y)^2 - 11(x - y)(a + b) - 35(a + b)^2$

76. $6(2x - 3)^2 - 11(2x - 3)(x + 1) - 10(x + 1)^2$

77. $(5y + 1)^2 - 4(5y + 1)(y - 4) - 5(y - 4)^2$

78. $6(a - 1)^2 + 3(a - 1)(a + 1) - 18(a + 1)^2$

5.5 Solving Second-Degree Equations by Factoring

Chapter 2 introduced methods for solving linear (first degree) equations. This section will introduce one method to solve equations of degree greater than 1. A **second-degree equation** in one variable is an equation of the form

$$ax^2 + bx + c = 0, \qquad a, b, c \in R; \quad a \neq 0.$$

Such an equation is called a **quadratic equation.** Some examples are

$$3x^2 + 5x + 7 = 0, \qquad 4x^2 - 9 = 0, \qquad \text{and} \qquad 5x^2 - 25x = 0.$$

If a quadratic equation is factorable, its solution set can be found by applying a theorem called the **zero-factor theorem.**

Theorem 5.1

Zero-Factor Theorem

If a and b represent real numbers and $a \cdot b = 0$, then $a = 0$, $b = 0$, or both a and b are zero.

Although the theorem is stated for real numbers, it is equally true for *complex numbers*, which will be introduced in Section 7.6.

Example 1   Solve $x^2 + 4x + 3 = 0$.

First factor $x^2 + 4x + 3$.

$$(x + 3)(x + 1) = 0$$

By the zero-factor theorem,

$$x + 3 = 0 \qquad \text{or} \qquad x + 1 = 0$$
$$x = -3 \qquad \text{or} \qquad x = -1.$$

To check the solutions, substitute them into the original equation.

$$x^2 + 4x + 3 = 0 \qquad\qquad\qquad x^2 + 4x + 3 = 0$$
$$(-3)^2 + 4(-3) + 3 \stackrel{?}{=} 0 \qquad x = -3 \qquad (-1)^2 + 4(-1) + 3 \stackrel{?}{=} 0$$
$$x = -1$$
$$9 - 12 + 3 \stackrel{?}{=} 0 \qquad\qquad\qquad 1 - 4 + 3 \stackrel{?}{=} 0$$
$$0 \stackrel{\checkmark}{=} 0 \qquad\qquad\qquad\qquad 0 \stackrel{\checkmark}{=} 0$$

The solution set is $\{-3, -1\}$. ∎

Example 2   Solve $2t^2 - 5t = 12$.

First rewrite the equation in the form $ax^2 + bx + c = 0$.

$2t^2 - 5t - 12 = 0$

$(2t + 3)(t - 4) = 0$ Factor first

$2t + 3 = 0$ or $t - 4 = 0$ Zero-factor theorem

$t = -\dfrac{3}{2}$ or $t = 4$

The solution set is $\{-\frac{3}{2}, 4\}$. The check is left to the reader. ■

Example 3   Solve $x^2 - 4 = 0$.

$x^2 - 4 = 0$

$(x - 2)(x + 2) = 0$ Factor first

$x - 2 = 0$ or $x + 2 = 0$

$x = 2$ or $x = -2$

Check: $x^2 - 4 = 0$ $x^2 - 4 = 0$

$(2)^2 - 4 \overset{?}{=} 0$ $x = 2$ $(-2)^2 - 4 \overset{?}{=} 0$

$x = -2$

$4 - 4 \overset{?}{=} 0$ $4 - 4 \overset{?}{=} 0$

$0 \overset{\checkmark}{=} 0$ $0 \overset{\checkmark}{=} 0.$

The solution set is $\{2, -2\}$. ■

Example 4   Solve $4m^2 + 8m = 0$.

$4m^2 + 8m = 0$

$4m(m + 2) = 0$

$4m = 0$ or $m + 2 = 0$

$m = 0$ or $m = -2$

The solution set is $\{0, -2\}$. ■

Example 5   Solve $s^2 - 6s + 9 = 0$.

$s^2 - 6s + 9 = 0$

$(s - 3)(s - 3) = 0$

$s - 3 = 0$ or $s - 3 = 0$

$s = 3$ or $s = 3$

The solution set is $\{3\}$. ■

When an equation has a solution that occurs twice, the equation is said to have a **root** (solution) **of multiplicity 2.** If a root occurs three times, the multiplicity is 3, and so on.

The zero-factor theorem also applies to equations of higher degree than 2.

Example 6 Solve $y^3 + y^2 - 12y = 0$.

$$y^3 + y^2 - 12y = 0$$
$$y(y^2 + y - 12) = 0 \qquad \text{Remove the common factor first}$$
$$y(y + 4)(y - 3) = 0$$
$$y = 0 \quad \text{or} \quad y + 4 = 0 \quad \text{or} \quad y - 3 = 0$$
$$y = 0 \quad \text{or} \qquad y = -4 \quad \text{or} \qquad y = 3$$

The solution set is $\{-4, 0, 3\}$. The check is left to the reader. ■

Example 7 Solve $y^4 - 13y^2 + 36 = 0$.

Let $y^2 = a$. Then $y^4 = a^2$.

$$a^2 - 13a + 36 = 0 \qquad \text{Substitute } a \text{ for } y^2 \text{ and } a^2 \text{ for } y^4$$
$$(a - 9)(a - 4) = 0$$
$$(y^2 - 9)(y^2 - 4) = 0 \qquad y^2 = a$$
$$(y - 3)(y + 3)(y - 2)(y + 2) = 0$$

$$y - 3 = 0 \quad \text{or} \quad y + 3 = 0 \quad \text{or} \quad y - 2 = 0 \quad \text{or} \quad y + 2 = 0$$
$$y = 3 \quad \text{or} \qquad y = -3 \quad \text{or} \qquad y = 2 \quad \text{or} \qquad y = -2$$

The solution set is $\{-3, -2, 2, 3\}$. ■

Example 8 Find the equation with solution set $\{-1, 0, \frac{1}{2}\}$.

If -1, 0, and $\frac{1}{2}$ are solutions, then $x = -1$, or $x = 0$, or $x = \frac{1}{2}$.

Therefore $x + 1 = 0$ or $x = 0$ or $x - \frac{1}{2} = 0$,

so that $(x + 1)(x)\left(x - \frac{1}{2}\right) = 0$

and $x^3 + \frac{1}{2}x^2 - \frac{1}{2}x = 0$

or $2x^3 + x^2 - x = 0$.   Multiply both sides by 2 ■

Many applied problems involve quadratic equations, as shown in Examples 9 and 10.

Example 9 The product of two consecutive even integers is 288. Find the integers.

Let the first even integer $= x$
Let the second even integer $= x + 2$
The product is 288.

$$x(x + 2) = 288$$
$$x^2 + 2x = 288 \qquad \text{Distributive property}$$
$$x^2 + 2x - 288 = 0 \qquad \text{Subtract 288 from each side}$$
$$(x - 16)(x + 18) = 0$$
$$x - 16 = 0 \quad \text{or} \quad x + 18 = 0$$
$$x = 16 \quad \text{or} \quad x = -18$$

There are two possible solutions. If $x = 16$, then $x + 2 = 18$. If $x = -18$, then $x + 2 = -16$. ∎

Example 10   The sum of two numbers is 10. The sum of their squares is 52. Find the numbers.

Let the first number $= n$
Let the second number $= 10 - n$
The sum of their squares is 52.

$$n^2 + (10 - n)^2 = 52$$
$$n^2 + 100 - 20n + n^2 = 52$$
$$2n^2 - 20n + 48 = 0 \qquad \text{Rewrite in the form } ax^2 + bx + c = 0$$
$$2(n^2 - 10n + 24) = 0 \qquad \text{Remove the common factor first}$$
$$2(n - 6)(n - 4) = 0$$

Since $2 \neq 0$,

$$n - 6 = 0 \quad \text{or} \quad n - 4 = 0$$
$$n = 6 \quad \text{or} \quad n = 4.$$

If $n = 6$, then $10 - n = 4$. If $n = 4$, then $10 - n = 6$. The two numbers are 4 and 6. ∎

Exercises 5.5

Solve by factoring. Remember to remove common factors first.

1. $x^2 + 3x = 0$

2. $y^2 + 5y = 0$

3. $a^2 - 3a + 2 = 0$

4. $p^2 - 5p + 4 = 0$

5. $3m^2 - 27 = 0$

6. $7x^2 - 28 = 0$

7. $y^2 - 10y + 21 = 0$

8. $c^2 - c - 72 = 0$

9. $t^2 - 8t - 48 = 0$

10. $s^2 + 4s - 45 = 0$

11. $q^2 - 18q + 81 = 0$

12. $r^2 - 14r + 49 = 0$

13. $4x^2 - 25x - 21 = 0$

14. $2y^2 - 21y + 40 = 0$

15. $10a^2 + a - 24 = 0$

16. $10d^2 - 7d - 6 = 0$

17. $2s^2 - 2s - 12 = 0$

18. $3x^2 - 3x - 60 = 0$

19. $x^3 + 20x^2 + 36x = 0$

20. $a^3 - 13a^2 - 48a = 0$

21. $-96k^3 + 36k^2 + 6k = 0$

22. $-24m^3 - 2m^2 + 15m = 0$

23. $-24p^4 + 10p^3 + 4p^2 = 0$

24. $-9t^4 - 33t^3 + 12t^2 = 0$

25. $(y - 1)(y^2 - 7y + 10) = 0$

26. $(2k + 3)(k^2 + 7k + 12) = 0$

27. $(x^2 - 1)(9x^2 - 24x + 16) = 0$

28. $(p^2 - 9)(21p^2 - 79p - 20) = 0$

29. $\left(x + \dfrac{1}{4}\right)\left(x - \dfrac{1}{3}\right)\left(x - \dfrac{1}{2}\right) = 0$

30. $\left(x - \dfrac{1}{5}\right)\left(x - \dfrac{1}{4}\right)\left(x + \dfrac{3}{7}\right) = 0$

31. $t^2 - \dfrac{2}{3}t + \dfrac{1}{9} = 0$

32. $r^2 - \dfrac{1}{2}r + \dfrac{1}{16} = 0$

33. $a^2 - \dfrac{1}{4}a - \dfrac{1}{8} = 0$

34. $c^2 - \dfrac{2}{15}c - \dfrac{1}{15} = 0$

35. $\dfrac{1}{6}p^2 - \dfrac{13}{36}p + \dfrac{1}{6} = 0$

36. $\dfrac{1}{10}s^2 - \dfrac{3}{20}s + \dfrac{1}{18} = 0$

Determine each equation having the solutions indicated.

37. $\{2, 3\}$ **38.** $\{5, 1\}$ **39.** $\{-3, 0\}$ **40.** $\{-7, 0\}$

41. $\{-1, 0, 2\}$ **42.** $\{-3, 0, 4\}$ **43.** $\left\{5, -\dfrac{1}{2}\right\}$ **44.** $\left\{-6, \dfrac{3}{4}\right\}$

45. $\left\{-\dfrac{1}{3}, 0, \dfrac{2}{7}\right\}$ **46.** $\left\{-\dfrac{1}{2}, 0, \dfrac{4}{5}\right\}$

For each equation, (a) simplify; (b) rewrite in the form $ax^2 + bx + c = 0$; and (c) solve by factoring.

47. $4a^2 + 4a = -1$ **48.** $3c^2 - 10 = 13c$

49. $15y^2 = 13 - 2y$ **50.** $4t^2 = 21 - 8t$

51. $2x(3x - 5) = x + 10$ **52.** $y(3y + 4) = 2 + 3y$

53. $(s + 1)^2 = 16$ **54.** $(k + 2)^2 = 25$

55. $2 + m(m + 7) = m(3 - m)$ **56.** $s(s - 3) - 3 = -s(2 + s)$

57. $3(y^2 - 1) - 5y = 3 - 2y$ **58.** $9a + 12 - 3(a + 4) = a^2 - 16$

59. $2(2x + 11)(x - 4) + 2(x - 2)(x + 4) = 32 + 7(x^2 - 16)$

60. $2y(5y + 2) - 3(2y - 1) = 8y + 23$

Write each of the following as a quadratic equation and solve.

61. The sum of the squares of two consecutive odd positive integers is 290. Find the integers.

62. The product of two consecutive negative integers is 342. Find the integers.

63. If the area of the given rectangle is 2100 square meters, find its dimensions.

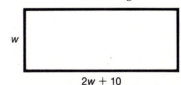

64. The area of a square is numerically 5 more than its perimeter. Find the length of a side.

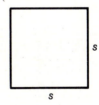

65. The product of two consecutive positive integral multiples of three is 108. Find the integers.

66. The sum of the squares of two consecutive negative integral multiples of 5 is 325. Find the integers. (*Note:* negative integral multiples of five are -5, -10, -15, and so on.)

5.6   More Applications

In this section we will continue to use the five-step process introduced in Section 2.6 for solving word problems. Since the steps are just as applicable to quadratic equations as to linear equations, it is suggested that you reread them before proceeding further.

The possible applications of quadratic equations are far too numerous to provide an example for each. Study the following examples and then apply the knowledge gained to the variety of problems in the exercise set.

Example 1   A square garden is surrounded by a brick walk 2 feet wide. The combined area of the garden and the walk is 196 square feet. Find the length of a side of the garden.

Let a side of the garden be represented by x. The dimensions of the garden plus the walk are $x + 4$ and $x + 4$. (See Figure 5.1.)

Since the area of a square is s^2, we have

$$(x + 4)^2 = 196$$
$$x^2 + 8x + 16 = 196$$
$$x^2 + 8x - 180 = 0$$
$$(x + 18)(x - 10) = 0$$
$$x + 18 = 0 \quad \text{or} \quad x - 10 = 0$$
$$x = -18 \quad \text{or} \quad x = 10.$$

The side of a garden cannot be negative, so we reject -18 as a result. Each side of the garden is 10 feet long.   ■

The **Pythagorean theorem** states that in any right triangle the square of the length of the longest side (**hypotenuse**) is equal to the sum of the squares of the lengths of the other two sides (**legs**). (See Figure 5.2.)

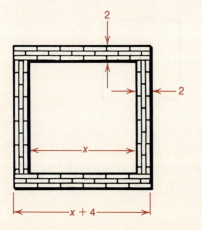

Figure 5.1

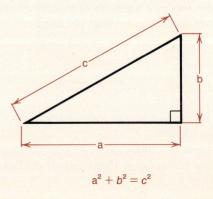

$$a^2 + b^2 = c^2$$

Figure 5.2

Example 2   One side of a right triangle is 7 centimeters longer than the shortest side. The hypotenuse is 8 centimeters longer than the shortest side. Find the length of each side.

Let $\quad x =$ length of the shortest side;
$\quad x + 7 =$ length of the second side;
$\quad x + 8 =$ length of the hypotenuse.

By the Pythagorean theorem,

$$(x + 8)^2 = (x + 7)^2 + x^2 \qquad \text{Substitute for c, a, b}$$
$$x^2 + 16x + 64 = x^2 + 14x + 49 + x^2 \qquad \text{Square each binomial}$$
$$x^2 + 16x + 64 = 2x^2 + 14x + 49$$
$$0 = x^2 - 2x - 15$$
$$0 = (x - 5)(x + 3) \qquad \text{Factor}$$
$$x - 5 = 0 \quad \text{or} \quad x + 3 = 0$$
$$x = 5 \quad \text{or} \qquad x = -3.$$

Reject -3 as a solution. The sides are $x = 5$, $x + 7 = 12$, and $x + 8 = 13$. ■

Example 3   If air resistance is ignored, the position of a projectile thrown vertically upward with a velocity of 96 feet per second is given by $s = -16t^2 + 96t$ (s is its position above ground level). How long will it take for the projectile to strike the ground?

The projectile will strike the ground when $s = 0$.

$$s = -16t^2 + 96t$$
$$0 = -16t^2 + 96t \qquad s = 0$$
$$0 = -16t(t - 6) \qquad \text{Factor}$$
$$-16t = 0 \quad \text{or} \quad t - 6 = 0$$
$$t = 0 \quad \text{or} \qquad t = 6$$

The expression $t = 0$ corresponds to the projectile's position on the ground when it was thrown. The projectile will strike the ground 6 seconds after being thrown. ■

Example 4   A box is to be made by cutting squares from the corners of a square piece of sheet metal and folding up the sides. (See Figure 5.3.) How large a piece must be used if the height of the box is to be 4 inches and its volume 576 cubic inches?

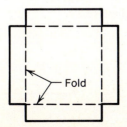
Fold

Figure 5.3

Let x = length of a side of the square piece of sheet metal. If the height is to be 4 inches, a 4-inch square must be cut from each corner. The side of the box is then $x - (2 \cdot 4)$ or $x - 8$ inches. (See Figure 5.4.)

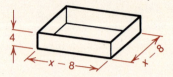

Figure 5.4

$$l \cdot w \cdot h = V$$
$$(x - 8)(x - 8)4 = 576$$
$$(x - 8)(x - 8) = 144 \qquad \text{Divide each side by 4}$$
$$x^2 - 16x + 64 = 144$$
$$x^2 - 16x - 80 = 0$$
$$(x - 20)(x + 4) = 0$$
$$x - 20 = 0 \quad \text{or} \quad x + 4 = 0$$
$$x = 20 \quad \text{or} \quad x = -4$$

The length of a side of the square piece of sheet metal is 20 inches. ■

Exercises 5.6

Write a quadratic equation for each of the following and solve.

1. The square of a number is one-ninth less than two-thirds of the number. Find the number.

2. The square of a positive number less four-thirds of the number is five-ninths. What is the number?

3. The square of twice an integer is eighteen more than six times the integer. Find the integer.

4. The square of three times an integer is three more than six times the integer. What is the integer?

5. The length of a rectangle is 2 inches more than twice the width. The diagonal is 13 inches. Find the dimensions of the rectangle.

6. The length of a rectangle is 1 decimeter more than the width. The diagonal is 5 decimeters. Find the dimensions of the rectangle.

7. At a point 12 yards from the base of a pine tree, the distance to the top of the tree is three more than twice the height. Find the height of the tree.

8. A guy wire is needed to hold a power pole in a vertical position. If the pole is 16 feet high and the guy wire is to be connected to the top of the pole from a point 12 feet from the base of the pole, how long must the guy wire be?

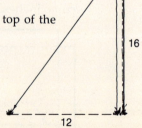

9. Two airplanes flying at right angles to each other pass at noon. How far apart are they at 1:00 P.M. if one is traveling at 300 miles per hour and the other at 400 miles per hour?

10. Louise leaves Newark traveling due north at 45 miles per hour. Johan leaves Newark at the same time traveling due west at 60 miles per hour. How far apart are they one hour later?

11. A 200-square-foot area is to be fenced. Find the dimensions if the length is 10 feet more than the width.

12. Myrna's dining room is three-fourths as wide as it is long. She buys a rug that covers 140 square feet of space. If a one-foot-wide space of floor is left uncovered all around the rug, find the dimensions of the room.

13. Emma has a rectangular backyard with an area of 1000 square feet. If the length is ten feet more than twice the width, find the dimensions.

14. It takes 65 square yards of linoleum to cover the floor in a new building. The length of the rectangular floor is 2 yards less than three times the width. Find the dimensions.

In Exercises 15 to 18 a square is cut out of each corner of a rectangle and the sides are folded up to make a box. (See Example 4 in this section.)

15. A square piece of sheet metal is used to construct a box 4 inches deep, with a volume of 256 cubic inches. Find the length of a side of the sheet metal.

16. A square piece of sheet metal is used to construct a box 2 centimeters deep, with a volume of 162 cubic centimeters. Find the length of a side of the sheet metal.

17. A rectangular sheet of cardboard twice as long as it is wide is used to construct a box 3 inches deep, with a volume of 864 cubic inches. What are the dimensions of the cardboard?

18. A rectangular sheet of cardboard five times as long as it is wide is used to construct a box 5 centimeters deep, with a volume of 1625 cubic centimeters. What are the dimensions of the cardboard?

19. A ball is thrown upward according to the equation $h = -16t^2 + vt$, where h is the height in feet, t is the time in seconds, and v is the initial velocity in feet per second. If the initial velocity is 64 feet per second, how long will it take for the ball to reach a height of 64 feet?

20. If the initial velocity is 64 feet per second, how long will it take for the ball in Exercise 19 to reach a height of 48 feet?

21. Find the lengths of the three sides of a right triangle by the Pythagorean Theorem when the hypotenuse is $a + 2$, and the legs are a and $a + 1$.

22. A 3-inch by 5-inch picture is to be framed as shown (width of the frame is x). If the total area of the frame and picture is 63 square inches, find the width of the frame.

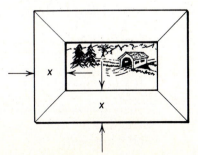

23. The recommended calorie intake, c, for a man (age 20 to 30) is estimated by the equation $c = -5a^2 + 200a + 1200$ where a is his age. If Paul follows this recommendation and consumes 3180 calories per day, what is his age?

24. Joan's height in inches is approximated by the equation $h = a^2 - 4a - 254$, where a is her age. (She is between 15 and 25 years old.) If Joan is five feet, six inches tall, what is the estimate of her age?

25. The concentration of aspirin in the bloodstream t hours after ingestion is estimated by the equation $c = -t^2 + 12t$. Determine the time when the concentration is zero.

26. A hammer is dropped from a tower by a workman. The hammer's distance from the ground is estimated by $s = 25 + 20t - 5t^2$, where s is the height in feet and t is the time in seconds. At what time will the hammer hit the ground?

Chapter 5   Summary

[5.1]   The simplest type of factoring associated with polynomials involves removing the **greatest common factor, GCF.** Once this factor has been removed, the remaining polynomial may sometimes be factored again.

[5.2] [5.3]   Trinomials are factored by recognizing known patterns or by a trial and error process. One easily recognized pattern is the perfect square trinomial.

$$a^2 \pm 2ab + b^2 = (a \pm b)^2$$

[5.3]   Other patterns are the **difference of two squares,**

$$a^2 - b^2 = (a - b)(a + b),$$

and the **sum and difference of two cubes,**

$$a^3 + b^3 = (a + b)(a^2 - ab + b^2)$$
$$a^3 - b^3 = (a - b)(a^2 + ab + b^2).$$

Factoring by grouping is used to factor polynomials of more than three terms.

[5.4]   The factorability of a polynomial of the form $ax^2 + bx + c = 0$ can be determined by the **ac test.** If there are two numbers m and n such that $m \cdot n = ac$ and $m + n = b$, the trinomial is factorable.

A series of steps are followed to make factoring polynomials easier. They are listed below.

1. First remove the *GCF.*

2. If the polynomial contains two terms, consider the patterns $a^2 - b^2$, $a^3 \pm b^3$.

3. If a polynomial contains three terms, decide whether or not it is a perfect square trinomial. If not, factor by trial and error.

4. If a polynomial contains four or more terms, factor by grouping.

[5.5]   Many **quadratic equations** are solved by use of the **zero-factor theorem.** It asserts that for real numbers a and b, if $a \cdot b = 0$, then $a = 0$, $b = 0$, or both a and b are zero.

Chapter 5   Review

Factor by removing the GCF.

1. $27a + 51$

2. $4lm + 6lt - 18lz$

3. $16x(2y - 3) - 10x^2(2y - 3)$

4. $(a - 3)^2 - (a - 3)$

5. $91x^2y^3z - 65xy^2z^2 - 39x^2y^2z^3$

6. $y^{3m+1} - y^{2m}$

Factor each expression completely.

7. $t^2 - 7t - 8$

8. $b^2 - 13b - 30$

9. $s^2 - 64$

10. $2a^3 - 16$

11. $6x^2 - 36x + 30$

12. $16m^2 - 24m + 9$

13. $3r^3s^2 - 3r^2s^3 - 36rs^4$

14. $x^4 - 16y^4$

15. $b^2(b - 3) - 6b(b - 3) + 9(b - 3)$

16. $k^{2m} - 5k^m r^m + 4r^{2m}$

17. $3y^2 + \dfrac{9}{4}y + \dfrac{3}{8}$

18. $s^{3m} - 8t^{3m}$

19. $25a^2 + 60ab + 36b^2$

20. $x^2 + 3x + xy + 3y$

21. $a^2 + 4ab + 4b^2 - c^2$

22. $3k^2 - 2km + 6k - 4m$

23. $6 - 13x + 2x^2$

24. $-7y^2 + 35y + 42$

25. $9k^2 - m^2 - 6m - 9$

26. $64a^6 - b^6$

27. $6ax^2 - 12ax + 6a + bx^2 - 2bx + b$

28. $\dfrac{1}{9}t^2 - \dfrac{1}{6}t + \dfrac{1}{16}$

29. $27s^3 + (2s - 1)^3$

30. $(a - b)^2 - (a + b)^2$

31. $0.49m^2n^2 - 1.21n^4$

32. $x^8 - y^8$

33. $\dfrac{1}{81}a^4 - \dfrac{1}{16}b^4$

34. $(x - 3)^2 - 10(x - 3)(x - 2) - 11(x - 2)^2$

35. $6xy - 4wx + 6yz - 4wz$

36. $10a^3 - 40a^2b + 40ab^2$

37. $s^4 - 9s^2 + 18$

38. $2m^5 - 20m^3 + 50m$

39. $az - bz + cz - at + bt - ct$

40. $a^3 - 8 - (a - 2)$

41. $4xy^3 - 9x^3y + 9x^2y^2$

42. $6k^2 - 32m^2 - 4km$

43. $a^6 - 27b^6 - 26a^3b^3$

44. $18x^{2m} - 4y^{2n} - 21x^m y^n$

Solve each equation by factoring.

45. $x^2 + 3x + 2 = 0$

46. $3y^3 + 2y^2 - 8y = 0$

47. $(k + 4)(k - 4) = 9k - 3(k + 4) + 12$

48. $6a^2 + 13a - 5 = 0$

49. $\dfrac{1}{7}t^2 - \dfrac{9}{21}t - \dfrac{1}{4} = 0$

50. $2b^{m+2} + 14b^{m+1} + 24b^m = 0, \quad m \in N$

Determine the equations with solutions as indicated.

51. $\{-1, 2\}$

52. $\{-3, 0, 1\}$

53. $\left\{-\dfrac{1}{3}, \dfrac{1}{2}\right\}$

54. $\left\{-\dfrac{1}{5}, 0, 7\right\}$

Write each of the following as a quadratic equation and solve.

55. The product of two consecutive real numbers is 72. Find the numbers.

56. A 20-foot-tall tree casts a shadow 15 feet long. How far is it from the tip of the shadow to the top of the tree?

57. Megan has a rectangular pool 30 feet wide and 50 feet long. She decides to put a sidewalk of uniform width around the pool. If the total area including the sidewalk and pool is 1836 square feet, what is the width of the sidewalk?

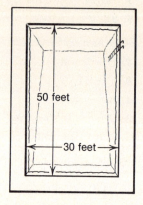

58. Barbara leaves Durango at 3:00 P.M. heading north at an average of 30 miles per hour through the mountains. At 4:00 P.M. Jim leaves Durango heading east at an average of 45 miles per hour. How far apart are they at 5:00 P.M.?

50 feet

30 feet

6

Rational Expressions

Chapter 1 dealt with the set of real numbers and its subsets. One subset was the set of *rational numbers*, which were defined as numbers of the form $\frac{a}{b}$, $b \neq 0$, where a and b are integers.

In this chapter we will consider a particular type of algebraic expression called a *rational expression*. We will also show that the theorems that are used in working with rational expressions are very much like those used for rational numbers.

6.1 Reducing and Building Rational Expressions

A **rational expression** is the quotient of two polynomials in which the denominator cannot equal zero.

> If A and B are polynomials, $B \neq 0$, then $\frac{A}{B}$ is said to be a **rational expression.**

For example,

$$\frac{x + 3}{x - 2}, \quad \frac{y^2 + 3y}{2y - 1}, \quad \frac{y^2 + 3y + 1}{2y^2 + 7y + 9}, \quad \text{and} \quad x^3 + 2 = \frac{x^3 + 2}{1}$$

are rational expressions. Since the denominator of a rational expression cannot be equal to zero, the values of the variable that make it zero must be excluded. To find these values, set the denominator equal to zero and find the solution set for the resulting equation.

Example 1 What values cannot be used as replacements for the variable in each of the following?

(a) $\dfrac{x + 3}{x - 2}$

First, set $x - 2 = 0$, giving $x = 2$. Since $x - 2$ cannot be equal to zero, $x \neq 2$.

(b) $\dfrac{y - 4}{2y - 10}$

Set $2y - 10 = 0$, giving $y = 5$. Since $2y - 10$ cannot be equal to zero, $y \neq 5$.

(c) $\dfrac{a + 2}{a^2 + 3a + 2}$

Set $a^2 + 3a + 2 = 0$.

$(a + 2)(a + 1) = 0$

$a + 2 = 0 \quad$ or $\quad a + 1 = 0$

$a = -2 \quad$ or $\quad\quad a = -1$

Since $a^2 + 3a + 2$ cannot be equal to zero, $a \neq -2, -1$. ∎

Sometimes the denominator of a rational expression contains more than one variable, as in the expression below.

$$\dfrac{x^2y + yz}{x - y}$$

To find the value that must be excluded, set $x - y = 0$.

$x - y = 0$

$x - y + y = 0 + y \quad$ Add y to each side

$x = y$

Thus $x \neq y$.

Example 2   Determine the values of the variable that must be excluded in each rational expression.

(a) $\dfrac{3a - 2}{a^2 - b^2}$

Set $a^2 - b^2 = 0$.

$(a - b)(a + b) = 0 \quad\quad\quad$ Factor

$a - b = 0 \quad$ or $\quad a + b = 0$

$a = b \quad$ or $\quad\quad a = -b$

Thus $a \neq b, -b$. This may be written as $a \neq \pm b$.

(b) $\dfrac{2y + 5}{y^2 + 4}$

Set $y^2 + 4 = 0$. Since $y^2 \geq 0$ for all y, $y^2 + 4 \neq 0$. No real values of y need be excluded. ∎

▶▶ *Caution*   Values that make only the numerator zero need not be excluded:

$$\frac{x-3}{x+6} = \frac{0}{9} = 0 \text{ when } x = 3.$$

Rational expressions are simplified by applying the **fundamental principle of rational expressions,** which states that two rational expressions are equivalent if:

1. **The numerator and denominator are divided by the same nonzero polynomial.**
2. **The numerator and denominator are multiplied by the same nonzero polynomial.**

Theorem 6.1

Fundamental Principle of Rational Expressions

If A, B and C are polynomials, $B \neq 0$, $C \neq 0$, then

$$\frac{A}{B} = \frac{A \cdot C}{B \cdot C} \quad \text{or} \quad \frac{A \cdot C}{B \cdot C} = \frac{A}{B}.$$

Example 3   Reduce to lowest terms.

(a) $\dfrac{3x}{6y} = \dfrac{3x}{3 \cdot 2y} = \dfrac{x}{2y}$ Divide numerator and denominator by 3, $y \neq 0$

(b) $\dfrac{x^2 + 2x}{4x^2 + 8x} = \dfrac{x(x+2)}{4x(x+2)} = \dfrac{1}{4}$ Divide numerator and denominator by $x(x+2)$, $x \neq 0, -2$

(c) $\dfrac{x^2 + 7x + 6}{2x^2 + 9x - 18}$

$\quad = \dfrac{(x+1)(x+6)}{(2x-3)(x+6)} = \dfrac{x+1}{2x-3}$ Divide numerator and denominator by $x + 6$, $x \neq \frac{3}{2}, -6$ ∎

The rational expressions in Example 3 illustrate that the easiest way to discover what polynomial, if any, can be divided out of the numerator and denominator is to factor both of them. Two polynomials having no common factors are said to be **relatively prime.**

Example 4   Reduce to lowest terms.

(a) $\dfrac{a^2 + 12a - 45}{2a^2 - 3a - 35} = \dfrac{(a+15)(a-3)}{(2a+7)(a-5)}$

The numerator and denominator are relatively prime. The rational expression is in lowest terms.

(b) $\dfrac{x^2 - 16}{x^3 - 64} = \dfrac{\text{difference of squares}}{\text{difference of cubes}}$

$= \dfrac{(x - 4)(x + 4)}{(x - 4)(x^2 + 4x + 16)}$

$= \dfrac{x + 4}{x^2 + 4x + 16}$ Divide by $x - 4$, $x \neq 4$

(c) $\dfrac{r - s}{s^2 - r^2} = \dfrac{-(s - r)}{(s - r)(s + r)} = \dfrac{-1}{s + r}$ $s \neq \pm r$

(d) $\dfrac{wy + xy + wz + xz}{(y^2 + z^2)(w^2 - x^2)}$

$= \dfrac{(w + x)(y + z)}{(y^2 + z^2)(w - x)(w + x)}$ Factor by grouping
Difference of squares

$= \dfrac{y + z}{(y^2 + z^2)(w - x)}$ Divide by $w + x$, $w \neq \pm x$ ■

As with addition of rational numbers or fractions, rational expressions also must have the same denominator. When a rational expression is changed to an equivalent rational expression by multiplying numerator and denominator by the same nonzero polynomial, the process is called **building rational expressions.**

Example 5 Use the fundamental principle to write each rational expression so that it has the indicated numerator or denominator.

(a) $\dfrac{x}{x + 3} = \dfrac{?}{2x + 6}$

Since $2x + 6 = 2(x + 3)$, multiply the numerator and denominator of the first rational expression by 2.

$\dfrac{x}{x + 3} = \dfrac{x \cdot 2}{(x + 3)2}$ Multiply by 2

$= \dfrac{2x}{2x + 6}$

(b) $\dfrac{m + 3}{m - 4} = \dfrac{m^2 + m - 6}{?}$

Since $m^2 + m - 6 = (m + 3)(m - 2)$, multiply the numerator and denominator of the first rational expression by $m - 2$.

$\dfrac{m + 3}{m - 4} = \dfrac{(m + 3)(m - 2)}{(m - 4)(m - 2)}$ Multiply by $m - 2$, $m \neq 2$

$= \dfrac{m^2 + m - 6}{m^2 - 6m + 8}$

(c) $\dfrac{x^2 - x - 6}{x^2 - 7x + 12} = \dfrac{?}{x^2 - 6x + 8}$

$$\frac{x^2 - x - 6}{x^2 - 7x + 12} = \frac{(x-3)(x+2)}{(x-3)(x-4)} = \frac{x+2}{x-4}; \; x \neq 3, 4$$

Since $x^2 - 6x + 8 = (x-2)(x-4)$, multiply by $\dfrac{x-2}{x-2}$, $x \neq 2$.

$$\frac{x^2 - x - 6}{x^2 - 7x + 12} = \frac{(x+2)\,(x-2)}{(x-4)\,(x-2)}$$

$$= \frac{x^2 - 4}{x^2 - 6x + 8} \quad \blacksquare$$

Rational expressions often are evaluated for given values of the variable in the same way that polynomials are evaluated.

Example 6   Evaluate $\dfrac{x^2 + 6x + 3}{x^3 - 2x + 4}$ for $x = -1, 3$.

(a) $\dfrac{x^2 + 6x + 3}{x^3 - 2x + 4} = \dfrac{(-1)^2 + 6(-1) + 3}{(-1)^3 - 2(-1) + 4}$ $x = -1$

$$= \frac{1 - 6 + 3}{-1 + 2 + 4} = \frac{-2}{5}$$

(b) $\dfrac{x^2 + 6x + 3}{x^3 - 2x + 4} = \dfrac{3^2 + 6 \cdot 3 + 3}{3^3 - 2 \cdot 3 + 4}$ $x = 3$

$$= \frac{9 + 18 + 3}{27 - 6 + 4} = \frac{30}{25} = \frac{6}{5} \quad \blacksquare$$

Example 7   Evaluate $\dfrac{x^2 - 2xy + y^2}{x^2 - y^2}$ for $x = 2, y = 3$.

$$\frac{x^2 - 2xy + y^2}{x^2 - y^2} = \frac{2^2 - 2 \cdot 2 \cdot 3 + 3^2}{2^2 - 3^2} \qquad x = 2, y = 3$$

$$= \frac{4 - 12 + 9}{4 - 9} = \frac{1}{-5} = \frac{-1}{5}$$

If we reduce first, then

$$\frac{x^2 - 2xy + y^2}{x^2 - y^2} = \frac{(x-y)^2}{(x-y)(x+y)} = \frac{x-y}{x+y}$$

$$= \frac{2 - 3}{2 + 3} \qquad x = 2, y = 3$$

$$= \frac{-1}{5}. \quad \blacksquare$$

Exercises 6.1

Reduce to lowest terms. Assume no denominator is zero.

1. $\dfrac{5x}{15x}$

2. $\dfrac{91y}{13y}$

3. $\dfrac{15xy}{20xyz}$

4. $\dfrac{26ab}{39abc}$

5. $\dfrac{a-b}{b-a}$

6. $\dfrac{s-t}{t-s}$

7. $\dfrac{m^2-n^2}{m-n}$

8. $\dfrac{q^2-r^2}{q+r}$

9. $\dfrac{-17a^2(b-c)}{68a(c-b)}$

10. $\dfrac{-36x^2(y+z)}{54x(y+z)}$

11. $\dfrac{a^2-4a}{a^2-16}$

12. $\dfrac{72a^2-72b^2}{6a^2+12ab+6b^2}$

13. $\dfrac{15r(s-t)(s+t)}{65a^2(s^2-t^2)}$

14. $\dfrac{w^2-3w}{w^2-9}$

15. $\dfrac{x^2y+xy^2}{x^2-y^2}$

16. $\dfrac{p^2-1}{p^2-2p+1}$

17. $\dfrac{(3-2a)(4-a)}{(a-4)(2a-3)}$

18. $\dfrac{(5+b)(2-b)}{(b-2)(b+5)}$

19. $\dfrac{m^2-3m-4}{m^2+3m+2}$

20. $\dfrac{34x^3(x-1)^2(x+5)^3}{17x^3(x-1)^2(x+5)^2}$

Determine which pairs of rational expressions are equivalent. Assume no denominator is zero.

21. $\dfrac{x}{3y},\ \dfrac{x^2y^2}{3xy^3}$

22. $\dfrac{ab}{ac},\ \dfrac{a^2b^2}{a^2c^2}$

23. $\dfrac{st-5}{2s-t},\ \dfrac{3s^2t-15s}{6s^2-3st}$

24. $\dfrac{2m+3}{m-1},\ \dfrac{2m^2+5m+3}{m^2+1}$

25. $\dfrac{3p-5}{2p+1},\ \dfrac{3p^2+p-10}{2p^2+5p+2}$

26. $\dfrac{-3-y}{2y+1},\ \dfrac{9-y^2}{2y^2-5y-3}$

Specify any restrictions on the variable(s) so that each rational expression is defined.

27. $\dfrac{3}{x}$

28. $\dfrac{-5}{a-1}$

29. $\dfrac{3b}{2b^2+3b}$

30. $\dfrac{y(y+4)}{y^3+4y^2}$

31. $\dfrac{p^2-3p+2}{p^2-3p-4}$

32. $\dfrac{m^2-m-2}{m^2-m-6}$

33. $\dfrac{(x-2)(y+3)}{(x-3)(y-1)}$

34. $\dfrac{(r-1)(s+4)}{(s-3)(r+2)}$

35. $\dfrac{3ab}{a-b}$

36. $\dfrac{-18vw}{9(v+w)}$

37. $\dfrac{-xy+2x+3y-6}{xy-x-3y+3}$

38. $\dfrac{xy+2y+3x+6}{xy+2y+4x+8}$

Supply the missing numerator or denominator to make the rational expressions equal. Specify any restrictions on the variable.

39. $\dfrac{5a^2}{8b}=\dfrac{?}{48ab^3}$

40. $\dfrac{-10(xy)^2}{3m^2}=\dfrac{?}{-27m^2xy}$

41. $\dfrac{3}{a+3}=\dfrac{?}{a^2+4a+3}$

42. $\dfrac{5}{x-3}=\dfrac{?}{2x^2-x-15}$

43. $\dfrac{-5p}{p^2-q^2}=\dfrac{?}{p^4-q^4}$

44. $\dfrac{2r}{r-s}=\dfrac{?}{r^2-s^2}$

45. $\dfrac{7}{m-4} = \dfrac{7m+28}{?}$

46. $\dfrac{4a}{2a-b} = \dfrac{8a^2+4ab}{?}$

47. $\dfrac{y+1}{y-1} = \dfrac{?}{3xy-3x+2y-2}$

48. $\dfrac{a-8}{3b+4} = \dfrac{?}{6ab+8a-15b-20}$

49. $\dfrac{-6}{-p+q} = \dfrac{6}{?}$

50. $\dfrac{-(a+3)}{7-a} = \dfrac{a+3}{?}$

Evaluate each rational expression for the given value(s). If a rational expression is not defined for the given value(s), write undefined.

51. $\dfrac{5x-4}{3x+7}$ for $x=2$

52. $\dfrac{7y-2}{5y+1}$ for $y=-1$

53. $\dfrac{a^2-3a+4}{a^2+5a+4}$ for $a=-1$

54. $\dfrac{p^2+6p-1}{p^2-3p+2}$ for $p=3$

55. $\dfrac{13s^2t^3}{s^2+7st+t^2}$ for $s=-1,\ t=0$

56. $\dfrac{m^2-mn+10n^2}{m^2-3mn-10n^2}$ for $m=1,\ n=-4$

Reduce to lowest terms. Some expressions may be in lowest terms already.

57. $\dfrac{6x^2-6x^2y}{18x^3y^2-18x^2y^3}$

58. $\dfrac{15p^3q^2+5pq^2}{10p^2q^2-20p^2q^3}$

59. $\dfrac{x^3-y^3}{x^2+xy+y^2}$

60. $\dfrac{m^3-n^2}{m-n}$

61. $\dfrac{-3a^2+2b+6a-ab}{3a^2-b^2-2ab}$

62. $\dfrac{st+2s+rt+2r}{st+rt+r+s}$

63. $\dfrac{a^2+6a+9}{2a(a+3)-(a+3)^2}$

64. $\dfrac{xy^2-7xy+10x}{x(y-5)^2-x^2(y-5)}$

65. $\dfrac{x^2+y^2}{x^2-y^2}$

66. $\dfrac{p^3-q^3}{p^3+q^3}$

67. $\dfrac{x^3+y^3}{x^2-y^2}$

68. $\dfrac{x^3+xy^3}{x^4+y^4}$

69. $\dfrac{2a^4+a^3-15a^2}{a^4+3a^3}$

70. $\dfrac{3p^4+2p^3-16p^2}{-p^4-7p^3+18p^2}$

6.2 Multiplying and Dividing Rational Expressions

Section 1.2 showed that the set of fractions is a subset of the set of rational numbers, and Section 1.4 reviewed multiplication and division of fractions. In general, if $\frac{a}{b}$ and $\frac{c}{d}$ are fractions, then

$$\frac{a}{b} \cdot \frac{c}{d} = \frac{ac}{bd} \qquad \text{Multiplication of fractions; } b,\ d \neq 0$$

$$\text{and} \quad \frac{a}{b} \div \frac{c}{d} = \frac{a}{b} \cdot \frac{d}{c} = \frac{ad}{bc}. \qquad \text{Division of fractions; } b,\ c,\ d \neq 0$$

These same concepts apply to rational expressions.

Theorem 6.2

To Multiply Two Rational Expressions
If $\frac{A}{B}$ and $\frac{C}{D}$ are rational expressions, then

$$\frac{A}{B} \cdot \frac{C}{D} = \frac{A \cdot C}{B \cdot D}, \qquad B, D \neq 0.$$

Example 1 Multiply $\dfrac{7a}{8b} \cdot \dfrac{3r}{5s}$.

$$\frac{7a}{8b} \cdot \frac{3r}{5s} = \frac{21ar}{40bs} \qquad \text{Multiply numerators} \\ \text{Multiply denominators} \quad \blacksquare$$

Example 2 Multiply $\dfrac{8a^2b^2}{3ab} \cdot \dfrac{6ac}{2ac^2}$ and reduce to lowest terms.

$$\frac{8a^2b^2}{3ab} \cdot \frac{6ac}{2ac^2} = \frac{48a^3b^2c}{6a^2bc^2}$$

$$= \frac{8ab(6a^2bc)}{c(6a^2bc)}$$

$$= \frac{8ab}{c} \qquad \text{Divide numerator and denominator by } 6a^2bc \quad \blacksquare$$

The work in this example could have been shortened if factors in either one of the numerators had been divided out with common factors in either of the denominators.

$$\frac{8a^2b^2}{3ab} \cdot \frac{6ac}{2ac^2} = \frac{8ab}{c}$$

Example 3 Multiply $\dfrac{x + 2}{x^2 + 4x + 3} \cdot \dfrac{x^2 + x - 6}{2x^2 + 5x + 2}$.

$$\frac{x + 2}{x^2 + 4x + 3} \cdot \frac{x^2 + x - 6}{2x^2 + 5x + 2}$$

$$= \frac{x + 2}{(x + 3)(x + 1)} \cdot \frac{(x + 3)(x - 2)}{(2x + 1)(x + 2)} \qquad \text{Factor first}$$

$$= \frac{x - 2}{(x + 1)(2x + 1)} \qquad \text{Divide out common factors} \quad \blacksquare$$

The product of several rational expressions can be found by applying Theorem 6.2 repeatedly.

Example 4 Multiply.

(a) $\dfrac{x^2 - y^2}{x^3 - y^3} \cdot \dfrac{x^2 + 2xy + y^2}{2x + 6y} \cdot \dfrac{x^2 + xy + y^2}{(x + y)^3}$

$= \dfrac{(x - y)(x + y)}{(x - y)(x^2 + xy + y^2)} \cdot \dfrac{(x + y)^2}{2(x + 3y)} \cdot \dfrac{x^2 + xy + y^2}{(x + y)^3}$

$= \dfrac{1}{2(x + 3y)}$

(b) $\dfrac{ax + bx + ay + by}{x^2 - y^2} \cdot \dfrac{cy - cx + dy - dx}{c^2 + 2cd + d^2} \cdot \dfrac{c + d}{a + b}$

$= \dfrac{(a + b)(x + y)}{(x - y)(x + y)} \cdot \dfrac{(c + d)(y - x)}{(c + d)^2} \cdot \dfrac{c + d}{a + b}$

$= \dfrac{y - x}{x - y} = -1$ When two factors differ only by sign, their quotient is -1 ∎

▶ **Caution** Rational expressions are reducible only when the numerator and the denominator have a common factor.

Right	Wrong
$\dfrac{xy + x}{x} = \dfrac{x(y + 1)}{x} = y + 1$	$\dfrac{xy + 1}{x} \neq y + 1$
x is a common factor	x is not a common factor of $xy + 1$

Theorem 6.3

To Divide Two Rational Expressions

If $\dfrac{A}{B}$ and $\dfrac{C}{D}$ are rational expressions, then

$$\dfrac{A}{B} \div \dfrac{C}{D} = \dfrac{A}{B} \cdot \dfrac{D}{C} = \dfrac{A \cdot D}{B \cdot C}, \quad C \neq 0.$$

Example 5 Divide $\dfrac{y + 1}{y} \div \dfrac{y^2 - 2y - 3}{y^2 - 3y}$ and simplify.

$\dfrac{y + 1}{y} \div \dfrac{y^2 - 2y - 3}{y^2 - 3y} = \dfrac{y + 1}{y} \cdot \dfrac{y^2 - 3y}{y^2 - 2y - 3}$

$= \dfrac{y + 1}{y} \cdot \dfrac{y(y - 3)}{(y + 1)(y - 3)}$

$= 1$ ∎

▶ **Caution** When all factors can be divided out, the quotient is 1, not zero.

Example 6   Simplify.

$$\frac{y - x}{x - z} \cdot \frac{(x + y)^2}{x^2 - y^2} \div \frac{x + y}{z - x}$$

$$= \frac{y - x}{x - z} \cdot \frac{(x + y)^2}{(x - y)(x + y)} \cdot \frac{z - x}{x + y}$$

$$= \frac{y - x}{x - z} \cdot \frac{z - x}{x - y}$$

$$= -1(-1) = 1 \qquad\qquad \frac{z - x}{x - z} = -1; \quad \frac{y - x}{x - y} = -1 \quad \blacksquare$$

Example 7   What polynomial will have a quotient of $x^2 + 7x + 10$ when divided by $\frac{x + 3}{x + 5}$?

Recall that a division problem is checked by multiplying the quotient by the divisor when there is no remainder.

Let the polynomial be P.

$$P = \frac{x + 3}{x + 5} \cdot (x^2 + 7x + 10)$$

$$= \frac{x + 3}{x + 5} \cdot (x + 2)(x + 5)$$

$$= (x + 3)(x + 2) = x^2 + 5x + 6$$

The polynomial is $x^2 + 5x + 6$. $\blacksquare$

Exercises 6.2

Write each product as a single fraction and reduce to lowest terms. Assume no denominator is zero.

1. $\dfrac{5}{x} \cdot \dfrac{x^2}{10}$

2. $\dfrac{7}{y} \cdot \dfrac{y^2}{21}$

3. $\dfrac{8p^2}{3a} \cdot \dfrac{9a^2}{4p}$

4. $\dfrac{3b^2}{7c} \cdot \dfrac{35c}{12b}$

5. $\dfrac{3xy}{x - y} \cdot \dfrac{(x - y)^2}{6xy}$

6. $\dfrac{a^2 - a}{a - 1} \cdot \dfrac{a + 1}{a}$

7. $\dfrac{m - 3}{2p^2} \cdot \dfrac{8p^3}{m - 3}$

8. $\dfrac{q - 1}{q + 2} \cdot \dfrac{q^2 - 4}{q^2 - 1}$

Divide and reduce to lowest terms. State any restrictions on the variable(s).

9. $\dfrac{6a}{5b} \div \dfrac{3a}{10b}$

10. $\dfrac{4x}{y} \div \dfrac{12x^2}{y^2}$

11. $3mn \div \dfrac{m}{n}$

12. $15s \div \dfrac{3s}{t}$

13. $\dfrac{p - q}{m - n} \div \dfrac{q - p}{n - m}$

14. $\dfrac{a - b}{c - d} \div \dfrac{b - a}{d - c}$

15. $\dfrac{3y^2 - 12}{6y} \div (y - 2)$

16. $\dfrac{5x^2 - 45}{7x} \div (x + 3)$

Perform the indicated operations and reduce to lowest terms. Assume no denominator is zero.

17. $\dfrac{a^2 - 1}{x + 3} \cdot \dfrac{x + 3}{a^2 + a - 2}$

18. $\dfrac{y^2 - 4}{m - 2} \cdot \dfrac{m^2 - 4}{y^2 - 4y + 4}$

19. $\dfrac{t^2 - t}{t - 3} \div t$

20. $\dfrac{p^2 + p}{p - 4} \div (p + 1)$

21. $\dfrac{a^2 + 7a + 6}{a + 2} \div (a + 6)$

22. $\dfrac{x}{y} \cdot \dfrac{a}{b} \div \dfrac{xb}{ya}$

23. $\dfrac{a^2 b}{p - 3} \cdot \dfrac{(p - 3)^2}{ab} \div a(p - 3)$

24. $\dfrac{(-7y^2 z)^3}{25a^2 x^3} \div \dfrac{49yz}{(-5ax)^2}$

25. $\dfrac{(-3m^2 n)^3}{32x^2 y^3} \cdot \dfrac{xy}{-9m} \div \dfrac{m^5 n^2}{(-2x)^4}$

26. $\dfrac{a^2 - ab}{(b - a)^2} \div \dfrac{a^2}{b^2 - a^2} \cdot (b^2 + a^2)$

27. $\dfrac{w^3 - w^2 x}{wy - w} \div \left(\dfrac{w - x}{y - 1}\right)^2$

28. $\dfrac{a^3 - a^2 b}{ac - a} \div \left(\dfrac{a - b}{c - 1}\right)^2$

29. $\dfrac{r^3 - s^3}{s - r} \cdot \dfrac{(r + s)^2}{r^2 + s^2}$

30. $\dfrac{p^3 + q^3}{q - p} \div \dfrac{(p + q)^2}{p^2 - q^2}$

31. $\dfrac{2a^2 + 9ab + 4b^2}{a^2 + 2ab + b^2} \cdot \dfrac{3a^2 + 2ab - b^2}{3a^2 + 11ab - 4b^2}$

32. $\dfrac{4y^2 + 4y + 1}{2y + 1} \cdot \dfrac{25(3y^2 - 19y + 20)}{5y^2 - 5y - 100}$

33. $(a + 2)(a - 1) \cdot \dfrac{3}{a^2 + a - 2}$

34. $p(p - 3)(p + 4) \cdot \dfrac{-2}{p^2 + p - 12}$

35. $\dfrac{mx + my + 2x + 2y}{6x^2 - 5xy - 4y^2} \div \dfrac{2mx - 4x + my - 2y}{3mx - 6x - 4my + 8y}$

36. $\dfrac{ax - 2a + 2xy - 4y}{ax + 2a - 2xy - 4y} \div \dfrac{ax + 2a + 2xy + 4y}{ax - 2a - 2xy + 4y}$

37. $\dfrac{2p + 1}{3p^2} \cdot \dfrac{5p^2 q}{p - 7} \div \dfrac{4p + 2}{3p - 21}$

38. $\dfrac{r^2 - s^2}{r^2 + rs + s^2} \cdot \dfrac{r^3 - s^3}{r^2 + s^2} \div \dfrac{r^4 - s^4}{r^2 - s^2}$

39. $\dfrac{8(x + 2)}{4x^2 + 16x + 16} \cdot \dfrac{3(x^2 - 4)}{-3x^2 + 192} \div \dfrac{6(x^2 - x - 6)}{7(x + 8)}$

40. $\dfrac{a^2 + 3a - 40}{5a^2 + 39a - 8} \div \dfrac{6a - 3}{a^2 - 25} \div \dfrac{2a^2 + 16a + 30}{8a^2 - 100}$

41. $\dfrac{k^2 - 1}{k^3 + 1} \div \dfrac{k^3 - 1}{k - 1} \cdot (k^2 + k + 1)$

42. $\dfrac{n^2 - 1}{n^3 + 1} \cdot \dfrac{n^3 - 1}{n - 1} \div (n^2 + n + 1)$

43. $\dfrac{x^3 - 3x^2 + 2x - 6}{x + 5} \div \dfrac{x^2 + 2}{x^2 + 7x + 10}$

44. $\dfrac{9y^3 - 18y^2 - 4y + 8}{3y^2 - 4y - 4} \cdot \dfrac{1}{8 - 12y}$

45. $\dfrac{b}{b - a}\left(\dfrac{a^2 - 2ab + b^2}{a^2 - b^2} \div \dfrac{1}{a^2 + 2ab + b^2}\right)$

46. $\dfrac{27x^3 - y^3}{xz + yz} \div \left(\dfrac{9x^2 - y^2}{z^4} \cdot \dfrac{xz^3 - yz^3}{3x^2 - 2xy - y^2}\right)$

Complete each of the following. Assume no denominator is zero.

47. $\dfrac{6a^2b^2}{4wz^2} \div \left(\rule{3cm}{0.4pt}\right) = \dfrac{60a^6b^7}{8w^4z^5}$

48. $\dfrac{3p^3 - 6p^2}{p^2q} \div \left(\rule{3cm}{0.4pt}\right) = \dfrac{-12p}{2 + p}$

49. $\dfrac{x}{x + 1} \cdot \left(\rule{3cm}{0.4pt}\right) = \dfrac{3x^2 + 5x}{x^2 - 1}$

50. $\dfrac{2xy^2}{x + y} \cdot \left(\rule{3cm}{0.4pt}\right) = \dfrac{x + y}{6xy}$

51. $\dfrac{x^2 - 1}{x + 2} \div \left(\rule{3cm}{0.4pt}\right) = x + 1$

52. $(2 - x) \div \left(\rule{3cm}{0.4pt}\right) = \dfrac{-(x + 2)}{x + 3}$

53. $\dfrac{p^2 - p}{p - 3} \cdot \left(\rule{3cm}{0.4pt}\right) = p$

54. $\dfrac{y^2}{2x + 4} \div \dfrac{y}{x + 2} \cdot \left(\rule{3cm}{0.4pt}\right) = \dfrac{x^2}{6y^2}$

Let $M = \dfrac{a}{a + 3}$, $N = \dfrac{a^2}{a^2 - 9}$, and $S = \dfrac{a^3}{(a + 3)^2}$. Perform the indicated operations and reduce to lowest terms. $a \neq \pm 3$.

55. $M \cdot S$

56. $S \div N$

57. $N \div S \cdot M$

58. $S \cdot M \div N$

59. $N \cdot S \cdot M$

60. $N \cdot S \div M$

Perform the indicated operations and reduce to lowest terms. Evaluate by using a calculator. Round answers to 3 decimal places.

61. $\dfrac{3a^2 - 4a - 4}{3a^2 + 14a + 8} \div \dfrac{3a^2 - 10a + 8}{3a^2 + 5a - 12}$ for $a = 1.23$

62. $\dfrac{3y^2 + 2y - 8}{12y^2 - 19y + 4} \cdot \dfrac{-(8y^3 - 27)}{4y^3 + 6y^2 + 9y} \div \dfrac{4y^2 - 4y - 3}{4y^2 - y}$ for $y = 0.881$

63. $\dfrac{2p^2 - 7p + 6}{3p^2 - 10p - 8} \div \dfrac{6p^2 - 5p - 6}{3p - 12}$ for $p = -6.072$

64. $\dfrac{r^2s + s}{2r^2 + 11r + 5} \cdot \dfrac{6r^2 - 17r - 10}{rs^2} \div \dfrac{3r^2 - 4r - 20}{r^2 + 5r}$ for $r = -2.12$, $s = 0.92$

65. Find the dividend D such that the divisor is $\dfrac{6x + 30}{x^2 - 25}$ and the quotient is

$\dfrac{4x^2 - 100}{2x^2 + 16x + 30}$ with a remainder of zero.

66. Find the dividend D such that the divisor is $\dfrac{x^2 - 5x - 14}{x^2 + x - 12}$ and the quotient is

$\dfrac{x^2 + 6x + 8}{x^2 - 10x + 21}$ with a remainder of zero.

6.3 Adding and Subtracting Rational Expressions

Recall that two fractions with the same denominator are added or subtracted by following the definition

$$\frac{a}{c} + \frac{b}{c} = \frac{a+b}{c}, \quad \frac{a}{c} - \frac{b}{c} = \frac{a-b}{c}, \quad c \neq 0.$$

Rational expressions are added by a similar method that is stated below.

Theorem 6.3

If $\dfrac{A}{C}$ and $\dfrac{B}{C}$ are any two rational expressions, then

$$\frac{A}{C} + \frac{B}{C} = \frac{A+B}{C} \quad \text{and} \quad \frac{A}{C} - \frac{B}{C} = \frac{A-B}{C}, \quad C \neq 0.$$

Example 1   Add.

(a) $\dfrac{2x}{3y} + \dfrac{z}{3y}$

Since the denominators are the same, add the numerators.

$$\frac{2x}{3y} + \frac{z}{3y} = \frac{2x+z}{3y}$$

(b) $\dfrac{x^2+3}{x+2} + \dfrac{x^2-2x}{x+2} = \dfrac{2x^2-2x+3}{x+2}$ ■

Example 2   Simplify.

(a) $\dfrac{x^2}{x-2} - \dfrac{4}{x-2}$

$$\frac{x^2}{x-2} - \frac{4}{x-2} = \frac{x^2-4}{x-2} = \frac{(x+2)(x-2)}{x-2}$$

$$= x+2, \quad x \neq 2$$

(b) $\dfrac{x^2-2}{x-3} - \dfrac{2x^2-x}{x-3} + \dfrac{2x^2-10}{x-3}$

▶ **Caution**   When subtracting, be particularly careful to get the signs correct.

$$\frac{x^2 - 2 - (2x^2 - x) + (2x^2 - 10)}{}$$

$$= \frac{x^2 - 2 \;\overset{-}{} 2x^2 \;\overset{+}{} x + 2x^2 - 10}{x - 3} \qquad \text{Watch the signs}$$

$$= \frac{x^2 + x - 12}{x-3} = \frac{(x+4)(x-3)}{x-3} = x+4, \quad x \neq 3 \quad ■$$

When two different rational expressions to be added or subtracted have different denominators, they must be rewritten as equivalent expressions with the same denominator. As in work with fractions, this denominator is called the **least common denominator** (LCD).

> The **least common denominator (LCD)** of two or more rational expressions is the smallest polynomial that is a multiple of each of their denominators.

For example, the LCD of 4 and 6 is 12, since 12 is the smallest number that is a multiple of both 4 and 6.

Example 3   Find the LCD of $\frac{1}{8}$, $\frac{1}{12}$, and $\frac{1}{20}$.
Begin by factoring each denominator.

$$8 = 2 \cdot 2 \cdot 2 = 2^3$$
$$12 = 2 \cdot 2 \cdot 3 = 2^2 \cdot 3$$
$$20 = 2 \cdot 2 \cdot 5 = 2^2 \cdot 5$$

Since the LCD is a multiple of each number it must contain each distinct factor the greatest number of times it occurs in any of the factorizations. The greatest number of times that 2 appears is 3. The factors 3 and 5 each appear once.

$$\text{LCD} = \overbrace{2 \cdot 2 \cdot 2}^{8} \cdot 3 \cdot 5 = 2^3 \cdot 3 \cdot 5 = 120$$

The LCD of 8, 12, and 20 is 120, since it is the smallest number that has each of them as factors. ∎

Example 4   Find the LCD of the denominators $x + 2$, $x^2 - 4$, and $x^2 - x - 6$.
Factor each denominator.

$$x + 2 = x + 2$$
$$x^2 - 4 = (x + 2)(x - 2)$$
$$x^2 - x - 6 = (x + 2)(x - 3)$$

Include each distinct factor.

$$\text{LCD} = (x + 2)(x - 2)(x - 3) \quad ∎$$

To Find the Least Common Denominator of Two or More Denominators
1. Factor each denominator completely.
2. Include in the LCD each distinct factor the greatest number of times it occurs in any denominator.
3. Find, or indicate, the product of the factors in Step 2.

Example 5 Rewrite $\dfrac{2}{x+3}$, $\dfrac{4}{x^2 + 5x + 6}$, and $\dfrac{x}{x+2}$ so that each denominator is the LCD.

$$x + 3 = x + 3$$
$$x^2 + 5x + 6 = (x + 2)(x + 3)$$
$$x + 2 = x + 2$$

The LCD is $(x + 2)(x + 3)$.

$$\frac{2}{x+3} = \frac{2(x+2)}{(x+3)(x+2)}$$

$$\frac{4}{x^2 + 5x + 6} = \frac{4}{(x+3)(x+2)}$$

$$\frac{x}{x+2} = \frac{x(x+3)}{(x+2)(x+3)} = \frac{x(x+3)}{(x+3)(x+2)} \quad\blacksquare$$

Example 6 Add or subtract as indicated.

(a) $\dfrac{3}{x-2} + \dfrac{5}{x+2}$ LCD = $(x-2)(x+2)$

$$\frac{3}{x-2} + \frac{5}{x+2} = \frac{3(x+2)}{(x-2)(x+2)} + \frac{5(x-2)}{(x-2)(x+2)}$$

$$= \frac{3(x+2) + 5(x-2)}{(x-2)(x+2)}$$

$$= \frac{3x + 6 + 5x - 10}{(x-2)(x+2)} \qquad \text{Distributive property}$$

$$= \frac{8x - 4}{(x-2)(x+2)} \qquad \text{Combine terms}$$

$$= \frac{4(x-2)}{(x-2)(x+2)} = \frac{4}{x+2} \qquad \text{Reduce}$$

(b) $\dfrac{4}{a-2} - \dfrac{3}{2-a}$ Notice that $a - 2 = -(2 - a)$

The denominators can be made the same by multiplying the numerator and denominator of either expression by -1.

$$\frac{4}{a-2} - \frac{3}{2-a} = \frac{4}{a-2} - \frac{3(-1)}{(2-a)(-1)}$$

$$= \frac{4}{a-2} - \frac{-3}{a-2} = \frac{4+3}{a-2} = \frac{7}{a-2} \quad \blacksquare$$

Example 7 Simplify

$$\left(\frac{s+2t}{s^2 - 4st + 3t^2} - \frac{s}{s^2 + st - 2t^2} \right) \div \frac{7s + 4t}{s + 2t}.$$

Factor the denominators.

$$\left(\frac{s+2t}{(s-3t)(s-t)} - \frac{s}{(s+2t)(s-t)} \right) \div \frac{7s+4t}{s+2t}$$

The LCD is $(s - 3t)(s - t)(s + 2t)$.

$$\frac{(s+2t)(s+2t) - s(s-3t)}{(s-3t)(s-t)(s+2t)} \div \frac{7s+4t}{s+2t} \qquad \begin{array}{l}\text{Rewrite the expression}\\\text{using the LCD}\end{array}$$

$$= \frac{7st + 4t^2}{(s-3t)(s-t)(s+2t)} \div \frac{7s+4t}{s+2t} \qquad \text{Multiply and collect like terms}$$

$$= \frac{t(7s + 4t)}{(s-3t)(s-t)(s+2t)} \cdot \frac{s+2t}{7s+4t}$$

$$= \frac{t}{(s-3t)(s-t)} \qquad \text{Reduce} \quad \blacksquare$$

Example 8 Combine $\dfrac{x^n}{x^{2n} - y^{2n}} - \dfrac{y^n}{x^n - y^n}$.

Factor the denominators to find the LCD.

$$x^{2n} - y^{2n} = (x^n - y^n)(x^n + y^n).$$

The LCD is $x^{2n} - y^{2n}$, since it contains $x^n - y^n$ as a factor.

$$\frac{x^n}{x^{2n} - y^{2n}} - \frac{y^n}{x^n - y^n} = \frac{x^n}{(x^n - y^n)(x^n + y^n)} - \frac{y^n}{x^n - y^n}$$

$$= \frac{x^n - y^n(x^n + y^n)}{(x^n - y^n)(x^n + y^n)}$$

$$= \frac{x^n - x^n y^n - y^{2n}}{(x^n - y^n)(x^n + y^n)} \qquad y^n \cdot y^n = y^{n+n} = y^{2n} \quad \blacksquare$$

Exercises 6.3

Simplify each of the following expressions and reduce to lowest terms. Assume no denominator is zero.

1. $\dfrac{5x}{7y} + \dfrac{2x}{7y}$

2. $\dfrac{8a}{11b} + \dfrac{3a}{11b}$

3. $\dfrac{6}{a+b} + \dfrac{2}{a+b}$

4. $\dfrac{9}{x-y} + \dfrac{1}{x-y}$

5. $\dfrac{x}{x^2-4} + \dfrac{2}{x^2-4}$

6. $\dfrac{p}{p^2-9} + \dfrac{3}{p^2-9}$

7. $\dfrac{r}{r^2-t^2} - \dfrac{t}{r^2-t^2}$

8. $\dfrac{u}{u^2-v^2} - \dfrac{v}{u^2-v^2}$

9. $\dfrac{m}{m^3-1} - \dfrac{1}{m^3-1}$

10. $\dfrac{a^2}{a^3-b^3} + \dfrac{ab+1}{a^3-b^3}$

Find the least common denominator (LCD) of each of the following denominators. (Only the denominators appear in each exercise.)

11. $6a$, $9a$

12. $13b$, $39b$

13. x^3y, xy^2, x

14. r^2s^2, rs^3, s^2

15. $p-2$, $p-6$

16. $m+3$, $m-3$

17. $q(q+1)$, q^2-q

18. $r^2(r^2-1)$, $r(r+1)$

19. x^2+3x+2, x^2+x-2

20. y^2-4y+4, y^2-3y+2

21. $9b^2+6b+1$, $3b^2+10b+3$

22. $1-2k-3k^2$, $1-9k^2$

23. $2w^2-13w+6$, $2w^2+w-1$, $6+5w-w^2$

24. $-2+3v+2v^2$, $2v^2-7v+3$, $-6-v+v^2$

Simplify and reduce to lowest terms. Assume no denominator is zero.

25. $\dfrac{1}{a} + \dfrac{2}{5a} - \dfrac{3}{2a}$

26. $\dfrac{5}{x} + \dfrac{5}{2x} - \dfrac{5}{3x}$

27. $\dfrac{1}{x^2} - \dfrac{2}{3x} + \dfrac{5}{4x}$

28. $\dfrac{7}{p^2} - \dfrac{3}{2p} + \dfrac{1}{p}$

29. $\dfrac{3a-5}{10a} - \dfrac{3a-4}{15a}$

30. $\dfrac{3w-1}{7w} - \dfrac{w-2}{14w}$

31. $\dfrac{m}{m+3} - \dfrac{m-1}{m-4}$

32. $\dfrac{y-3}{y-2} - \dfrac{y}{y+4}$

33. $\dfrac{7a}{4b} + \dfrac{3b}{8a} - \dfrac{12b^2+56a^2}{32ab}$

34. $\dfrac{4p}{27q} + \dfrac{2q}{9p} - \dfrac{36p^2+54q^2}{54pq}$

35. $\dfrac{2}{m-2} + \dfrac{4}{m+2} + \dfrac{m+6}{4-m^2}$

36. $\dfrac{4}{2a+1} - \dfrac{10a-3}{4a^2-1} - \dfrac{3}{1-2a}$

37. $\dfrac{x^2+1}{x^2-1} - 1$

38. $\dfrac{r^2+9}{r^2-9} - 1$

39. $\dfrac{4k+1}{2k+3} + \dfrac{2}{k}$

40. $\dfrac{2}{1-5v} + \dfrac{3}{4v}$

41. $\dfrac{x+1}{x+2} - \dfrac{x-1}{x-2}$

42. $\dfrac{6}{4-y} - \dfrac{2}{y-4}$

43. $\dfrac{3c}{c^2-c-12} - \dfrac{2c}{c^2-9c+20}$

44. $\dfrac{11-w}{w^2+3w-4} - \dfrac{-22-7w}{w^2+6w+8}$

45. $\dfrac{18s-19}{4s^2+27s-7} - \dfrac{12s-41}{3s^2+17s-28}$

46. $\dfrac{42-22y}{3y^2-13y-10} - \dfrac{21-13y}{2y^2-9y-5}$

47. $ab - \dfrac{a}{b}$

48. $\dfrac{3b}{a} + 2$

49. $m + 1 + \dfrac{1}{m-1}$

50. $3r - 2 + \dfrac{2}{r-1}$

51. $\dfrac{2k+1}{k^2+2k+1} - 1$

52. $2x + \dfrac{xy - y^2}{x+y} - 3y$

Perform the indicated operations and reduce each expression to lowest terms. Assume no denominator is zero.

53. $\left(\dfrac{a^2 - b^2}{u^2 - v^2}\right)\left(\dfrac{av - au}{b - a}\right) + \left(\dfrac{a^2 - av}{u + v}\right)\left(\dfrac{1}{a}\right)$

54. $\left(\dfrac{6r^2}{r^2 - 1}\right)\left(\dfrac{r+1}{3}\right) - \dfrac{2r^2}{r-1}$

55. $\left(\dfrac{m^2 - 5m - 6}{3m}\right)\left(\dfrac{-9m^2}{6 - m}\right) + \dfrac{10m^2 + 10m}{5}$

56. $\left[\dfrac{1}{y^2 - 1} \div \dfrac{1}{y^2 + 1}\right]\left(\dfrac{y^3 + 1}{y^4 - 1}\right) + \dfrac{1}{(y+1)^2(y-1)}$

57. $\left[\dfrac{x^3 - 1}{x^2 + 2x + 1} \div \dfrac{x^3 + 1}{x^2 - 1}\right]\left(\dfrac{x^2 - x + 1}{x^2 - 2x + 1}\right) + \dfrac{-(x^2 - x + 1)}{x^2 + 2x + 1}$

58. $\left[\dfrac{a^3 - 64}{a^2 - 16} \div \dfrac{a^2 - 4a + 16}{a^2 - 4} \div \dfrac{a^2 + 4a + 16}{a^3 + 64}\right] + 4 - a^2$

Perform the indicated operations and reduce to lowest terms. Assume no denominator is zero.

59. $\dfrac{1}{4a^2} + \dfrac{1}{3a^2 + 3a} - \dfrac{1}{3a^2 + 6a + 3}$

60. $\dfrac{3}{12 + 7x + x^2} + \dfrac{x}{x^2 + 5x + 6} + \dfrac{7}{x^2 + 6x + 8}$

61. $\dfrac{2k - 5}{3k^2 - 3k} - \dfrac{k + 5}{k^2 - 1} + \dfrac{4}{k - 1}$

62. $\dfrac{4}{2 - w} - \dfrac{1}{w - 2} - \dfrac{w}{(w - 2)^2} + 2$

63. $\dfrac{-y}{(y - w)(x - w)} + \dfrac{w}{(y - x)(w - y)} - \dfrac{x}{(x - y)(w - x)}$

64. $\dfrac{3}{6x^2 - 5x + 1} - \dfrac{2}{3x^2 - 7x + 2} - \dfrac{1}{2x^2 - 5x + 2}$

65. $\dfrac{a}{(c - a)(a - b)} - \dfrac{b}{(b - a)(c - b)} + \dfrac{c}{(c - b)(a - c)}$

66. $\dfrac{-1}{(r - s)(r - t)} - \dfrac{1}{(s - r)(t - s)} - \dfrac{1}{(t - r)(s - t)}$

67. $\dfrac{x^n}{x^{2n} - y^{2n}} + \dfrac{y^n}{x^n - y^n}$, n a natural number

68. $\dfrac{x^n}{x^{3n} - 1} - \dfrac{1}{x^{2n} - 1}$, n a natural number

6.4 Simplifying Complex Rational Expressions

Complex rational expressions, or complex fractions, are rational expressions in which the numerator or denominator (or both) contain other rational expressions. For example,

$$\frac{\dfrac{x+1}{3}}{\dfrac{x+2}{x}}$$

is a complex rational expression since the numerator and denominator are both rational expressions.

Complex rational expressions are simplified by applying the fundamental principle.

Example 1   Simplify $\dfrac{\dfrac{3}{4}}{\dfrac{2}{5}}$.

To simplify, first find the LCD of $\frac{3}{4}$ and $\frac{2}{5}$, which is 20. Then multiply both the numerator and the denominator by this number.

$$\frac{\dfrac{3}{4}}{\dfrac{2}{5}} = \frac{\dfrac{3}{4} \cdot 20}{\dfrac{2}{5} \cdot 20} \qquad \text{Fundamental principle}$$

$$= \frac{3 \cdot 5}{2 \cdot 4} \qquad \text{Reduce before multiplying}$$

$$= \frac{15}{8} \quad \blacksquare$$

An alternate method for simplifying complex rational expressions is to invert the fraction in the denominator and multiply.

$$\frac{\dfrac{3}{4}}{\dfrac{2}{5}} = \frac{3}{4} \cdot \frac{5}{2} = \frac{15}{8}$$

In this text we will use the method introduced in Example 1.

Example 2   Simplify $\dfrac{\dfrac{x+1}{3}}{\dfrac{x+2}{x}}$.

The LCD of 3 and x is $3x$.

$$\frac{\dfrac{x+1}{3}}{\dfrac{x+2}{x}} = \frac{\left(\dfrac{x+1}{3}\right) \cdot 3x}{\left(\dfrac{x+2}{x}\right) \cdot 3x} \qquad \text{Multiply by the LCD}$$

$$= \frac{(x+1)x}{(x+2)3} \qquad \text{Reduce before multiplying}$$

$$= \frac{x^2 + x}{3x + 6} \quad \blacksquare$$

Example 3   Simplify

$$\frac{\dfrac{1}{x^2} - 2}{3 - \dfrac{2}{x}}.$$

The LCD of all fractions in the numerator and denominator is x^2.

$$\frac{\dfrac{1}{x^2} - 2}{3 - \dfrac{2}{x}} = \frac{\left(\dfrac{1}{x^2} - 2\right) \cdot x^2}{\left(3 - \dfrac{2}{x}\right) \cdot x^2} \qquad \text{Multiply by the LCD}$$

$$= \frac{\dfrac{1}{x^2} \cdot x^2 - 2 \cdot x^2}{3 \cdot x^2 - \dfrac{2}{x} \cdot x^2} \qquad \text{Distributive property}$$

$$= \frac{1 - 2x^2}{3x^2 - 2x} \qquad \text{Reduce and multiply} \quad \blacksquare$$

Example 4   Simplify

$$\frac{\dfrac{1}{a} + \dfrac{1}{b}}{\dfrac{1}{a^2} - \dfrac{1}{b^2}}.$$

The LCD of all fractions in the numerator and denominator is a^2b^2.

$$\frac{\dfrac{1}{a} + \dfrac{1}{b}}{\dfrac{1}{a^2} - \dfrac{1}{b^2}} = \frac{\left(\dfrac{1}{a} + \dfrac{1}{b}\right) \cdot a^2b^2}{\left(\dfrac{1}{a^2} - \dfrac{1}{b^2}\right) \cdot a^2b^2} \qquad \text{Multiply by the LCD}$$

$$= \frac{\dfrac{1}{a} \cdot a^2b^2 + \dfrac{1}{b} \cdot a^2b^2}{\dfrac{1}{a^2} \cdot a^2b^2 - \dfrac{1}{b^2} \cdot a^2b^2}$$

Distributive property

$$= \frac{ab^2 + a^2b}{b^2 - a^2}$$

Reduce and multiply

$$= \frac{ab(b + a)}{(b - a)(b + a)}$$

Factor

$$= \frac{ab}{b - a}$$

Reduce ■

Example 5 Simplify

$$\frac{\dfrac{x - 2}{x - 3} - \dfrac{x - 3}{x - 2}}{\dfrac{2}{x^2 - 5x + 6}}.$$

The LCD is $(x - 2)(x - 3) = x^2 - 5x + 6$.

$$\frac{\dfrac{x - 2}{x - 3} - \dfrac{x - 3}{x - 2}}{\dfrac{2}{x^2 - 5x + 6}} = \frac{\left(\dfrac{x - 2}{x - 3} - \dfrac{x - 3}{x - 2}\right) \cdot (x - 2)(x - 3)}{\left(\dfrac{2}{x^2 - 5x + 6}\right) \cdot (x - 2)(x - 3)}$$

Multiply by the LCD

$$= \frac{(x - 2)^2 - (x - 3)^2}{2}$$

Multiply and reduce

$$= \frac{2x - 5}{2}$$

Expand and collect like terms ■

Example 6 Simplify $\dfrac{x^{-1} + x^{-2}y}{x^{-2} + y^{-1}}$.

Recall that $a^{-n} = \dfrac{1}{a^n}$. Rewrite the expression so that it is free of negative exponents.

$$\frac{\dfrac{1}{x} + \dfrac{y}{x^2}}{\dfrac{1}{x^2} + \dfrac{1}{y}}$$

The LCD is x^2y and the process for simplifying is the same as in Examples 1 through 5. The simplified form is

$$\frac{xy + y^2}{y + x^2}. \quad ■$$

It is interesting to note that if the original expression in Example 6 is multiplied by the LCD x^2y and the rules for exponents are applied, the same result is obtained.

$$\frac{x^{-1} + x^{-2}y}{x^{-2} + y^{-1}} \cdot \frac{x^2y}{x^2y}$$

$$= \frac{x^{-1}(x^2y) + (x^{-2}y)(x^2y)}{x^{-2}(x^2y) + (y^{-1})(x^2y)} \qquad \text{Distributive property}$$

$$= \frac{x^{-1+2}y + x^{-2+2}y^{1+1}}{x^{-2+2}y + x^2y^{-1+1}}$$

$$= \frac{xy + x^0y^2}{x^0y + x^2y^0} = \frac{xy + y^2}{y + x^2} \qquad a^0 = 1$$

Can you discover a rule for finding the LCD without first eliminating any negative exponents?

Example 7 Simplify

$$\frac{a - \dfrac{1}{1 + \dfrac{1}{a}}}{a + \dfrac{1}{a - \dfrac{1}{a}}}.$$

A complex fraction of this type contains complex fractions within complex fractions.

In the numerator $\dfrac{1}{a + \dfrac{1}{a}}$ is a complex fraction, as is $\dfrac{1}{a - \dfrac{1}{a}}$

in the denominator. Simplify them first.

$$\frac{a - \dfrac{1}{1 + \dfrac{1}{a}}}{a + \dfrac{1}{a - \dfrac{1}{a}}} = \frac{a - \dfrac{1 \cdot a}{\left(1 + \dfrac{1}{a}\right) \cdot a}}{a + \dfrac{1 \cdot a}{\left(a - \dfrac{1}{a}\right) \cdot a}} \qquad \text{The LCD of each is } a$$

$$= \frac{a - \dfrac{a}{a + 1}}{a + \dfrac{a}{a^2 - 1}}$$

The problem is now similar to the earlier examples. The result is

$$\frac{a-1}{a}. \quad \blacksquare$$

Exercises 6.4

Simplify. Assume no denominator is zero.

1. $\dfrac{\dfrac{1}{2}}{\dfrac{1}{4}}$

2. $\dfrac{\dfrac{3}{5}}{\dfrac{9}{10}}$

3. $\dfrac{\dfrac{a}{b}}{\dfrac{a^2}{bc}}$

4. $\dfrac{\dfrac{xy}{z}}{\dfrac{x}{yz}}$

5. $\dfrac{\dfrac{1}{4}+\dfrac{1}{3}}{\dfrac{1}{2}-\dfrac{1}{8}}$

6. $\dfrac{\dfrac{3}{5}-\dfrac{4}{9}}{\dfrac{2}{3}+\dfrac{1}{5}}$

7. $\dfrac{2+\dfrac{3}{x}}{x+\dfrac{3}{2}}$

8. $\dfrac{r-\dfrac{1}{r}}{\dfrac{2}{3r}}$

9. $\dfrac{1-\dfrac{1}{3}}{1-\dfrac{1}{9}}$

10. $\dfrac{1-\dfrac{9}{4}}{1-\dfrac{3}{2}}$

11. $\dfrac{1-\dfrac{1}{k}}{1-\dfrac{1}{k^2}}$

12. $\dfrac{1-\dfrac{m^2}{n^2}}{1-\dfrac{m}{n}}$

13. $\dfrac{\dfrac{1}{a}+\dfrac{1}{b}}{\dfrac{1}{a^2}-\dfrac{1}{b^2}}$

14. $\dfrac{\dfrac{1}{x}-\dfrac{1}{y}}{\dfrac{1}{y^2}-\dfrac{1}{x^2}}$

15. $\dfrac{\dfrac{1}{a}-\dfrac{1}{a^2}}{a}$

16. $\dfrac{\dfrac{1}{s}+\dfrac{1}{t}}{st}$

17. $\dfrac{9}{\dfrac{1}{b}-\dfrac{1}{4b}}$

18. $\dfrac{b^2}{\dfrac{1}{b}-\dfrac{1}{b^2}}$

19. $\dfrac{\dfrac{1}{x}+1}{1-\dfrac{1}{x^2}}$

20. $\dfrac{\dfrac{1}{y}-\dfrac{1}{2}}{\dfrac{1}{4}-\dfrac{1}{y^2}}$

21. $\dfrac{v}{1+\dfrac{v+1}{v+1}}$

22. $\dfrac{p}{1-\dfrac{p+1}{p-1}}$

23. $\dfrac{\dfrac{3ab^2}{7cd^2}}{\dfrac{12ab^2}{28cd^2}}$

24. $\dfrac{\dfrac{25m^3n^2}{24cd^2}}{\dfrac{35mn}{18cd^2}}$

25. $\dfrac{\dfrac{3a}{2b}+\dfrac{2a}{3b}}{\dfrac{3a}{2b}+\dfrac{2a}{3b}}$

26. $\dfrac{\dfrac{x}{y}+\dfrac{y}{x}}{\dfrac{x}{y}-\dfrac{y}{x}}$

27. $\dfrac{\dfrac{1}{q}-\dfrac{1}{q-1}}{\dfrac{1}{q}-\dfrac{1}{q-1}}$

28. $\dfrac{\dfrac{3}{r-1}-\dfrac{1}{r}}{\dfrac{1}{r}+\dfrac{3}{r-1}}$

29. $\dfrac{\dfrac{1}{x+1}-\dfrac{1}{x-1}}{\dfrac{1}{1-x^2}}$

30. $\dfrac{\dfrac{4}{w-2}-\dfrac{4}{w+2}}{\dfrac{4}{4-w^2}}$

31. $\dfrac{1+\dfrac{5}{\dfrac{1}{x}+2}}{1-\dfrac{1}{\dfrac{1}{x}+2}}$

32. $\dfrac{1+\dfrac{1}{1-\dfrac{1}{a}}}{1-\dfrac{3}{1-\dfrac{1}{a}}}$

33. $\dfrac{4t^2 - 9}{1 - \dfrac{1-t}{t-2}}$

34. $\dfrac{\dfrac{4}{x^2} - 1}{\dfrac{4}{x^2} - \dfrac{4}{x} + 1}$

35. $\dfrac{6 - \dfrac{9}{b}}{1 - \dfrac{4}{b} + \dfrac{3}{b^2}}$

36. $\dfrac{\dfrac{1+y}{1-y} - \dfrac{1-y}{1+y}}{\dfrac{1}{1-y} - \dfrac{1}{1+y}}$

37. $\dfrac{\dfrac{p-1}{p-2} - \dfrac{p-2}{p-1}}{\dfrac{1}{p-1} - \dfrac{1}{p-2}}$

38. $\dfrac{\dfrac{z}{z-1} - \dfrac{z}{z+1}}{\dfrac{2}{z+1} - \dfrac{1}{z-1}}$

39. $\dfrac{\dfrac{1}{m-4} + \dfrac{1}{m-5}}{\dfrac{1}{m^2 - 9m + 20}}$

40. $\dfrac{\dfrac{1}{k^2 - 7k + 12}}{\dfrac{1}{k-3} + \dfrac{1}{k-4}}$

41. $\dfrac{s + 3 - \dfrac{8}{s-3}}{s + 9 - \dfrac{12s}{s-3}}$

42. $\dfrac{\dfrac{t}{t-1} + 1}{\dfrac{3}{t-2} + 2}$

43. $\dfrac{a^2 + \dfrac{1}{a}}{1 - \dfrac{a}{1 - \dfrac{a}{a-1}}}$

44. $x - \dfrac{1}{1 - \dfrac{1}{1 - \dfrac{1}{1-x}}}$

45. $\dfrac{1 + \dfrac{1}{1 - \dfrac{1}{p}}}{1 - \dfrac{3}{1 - \dfrac{1}{p}}}$

46. $\dfrac{2r + \dfrac{r-1}{2r+1}}{2r - \dfrac{r+3}{2r+1}}$

47. $\dfrac{w - \dfrac{15}{w+2}}{w - 1 - \dfrac{10}{w+2}}$

48. $\dfrac{k - 5 + \dfrac{18}{k+4}}{k + \dfrac{3}{k+4}}$

49. $\left[\dfrac{1}{x+y} - \dfrac{1}{x-y}\right] \div \left[\dfrac{4y}{x^2 + 2xy + y^2}\right]$

50. $\left[\dfrac{a-1}{a+2} + \dfrac{2-a}{a+2}\right] \div \left[\dfrac{a}{a-1} - \dfrac{a-1}{a+2}\right]$

Simplify. Assume no denominator is zero.

51. $\dfrac{3+4}{3^{-1} + 4^{-1}}$

52. $\dfrac{12+6}{12^{-1} + 6^{-1}}$

53. $\dfrac{x^{-1} + y^{-1}}{x^{-1} - y^{-1}}$

54. $\dfrac{a^{-1} + b^{-1}}{(ab)^{-1}}$

55. $(r^{-1} - s^{-1})^{-1} \div \dfrac{1}{(rs)^2}$

56. $\left(\dfrac{m}{n^{-1}} + \dfrac{m^{-1}}{n}\right)^{-1}$

57. $\dfrac{p^{-1}}{p^{-1} + 1}$

58. $\dfrac{2c^{-1} + 6d^{-1}}{4c^{-2} - 4d^{-2}}$

59. $\dfrac{s^{-1} - 2r^{-1}}{r^{-1} + 5s^{-1}}$

60. $\dfrac{2x^{-1} - 3x^{-1}}{x^{-2}}$

61. $\dfrac{6x^{-1} - 3y^{-1}}{12x^{-1} - 6y^{-1}}$

62. $\dfrac{(q^{-2} - t^{-2})^{-1}}{(t^{-1} - q^{-1})^{-1}}$

63. $\left(\dfrac{a^2 - 1}{3} \cdot \dfrac{a}{a^2 + 2a + 1} - \dfrac{4}{a+1}\right) \div \left(a - \dfrac{1}{a+1}\right)$

64. $\left(\dfrac{p^2 - 1}{3} \cdot \dfrac{p}{p^2 + 2p + 1}\right) - \left(\dfrac{4}{p+1} \div p - \dfrac{1}{p+1}\right)$

Review Exercises

Solve each of the following equations or inequalities.

65. $3(x - 3) = 5(x - 4)$

66. $5y - 2(y - 1) \geq 4y + 3$

67. $\begin{cases} x - 2y = 4 \\ x + 3y = 1 \end{cases}$

68. $10a^2 - a - 21 = 0$

69. $|3p - 2| \leq 4$

70. $\dfrac{1}{p} = \dfrac{1}{q} + \dfrac{1}{r}$ for q

6.5 Solving Rational Equations

Section 2.1 discussed the methods used to solve linear equations in one variable. Some of these equations involved rational expressions. They were solved by clearing the equations of fractions by multiplying each side by the least common multiple of the denominators, which we now call the LCD. In this section we will continue to use the same process to solve rational equations with more complex LCDs. We begin with a review problem.

Example 1 Solve $\dfrac{x}{3} + \dfrac{x}{5} = \dfrac{1}{10}$.

The LCD is 30.

$$30\left(\frac{x}{3} + \frac{x}{5}\right) = 30 \cdot \frac{1}{10} \qquad \text{Multiply each side by 30}$$

$$30 \cdot \frac{x}{3} + 30 \cdot \frac{x}{5} = 30 \cdot \frac{1}{10} \qquad \text{Distributive property}$$

$$10x + 6x = 3 \qquad \text{Reduce and multiply}$$

$$16x = 3 \qquad \text{Collect like terms}$$

$$x = \frac{3}{16} \qquad \text{Divide each side by 16}$$

Check: $\dfrac{x}{3} + \dfrac{x}{5} = \dfrac{1}{10}$

$$\frac{\frac{3}{16}}{3} + \frac{\frac{3}{16}}{5} \overset{?}{=} \frac{1}{10} \qquad x = \frac{3}{16}$$

$$\frac{1}{16} + \frac{3}{80} \overset{?}{=} \frac{1}{10}$$

$$\frac{8}{80} \overset{?}{=} \frac{1}{10}$$

$$\frac{1}{10} \overset{\checkmark}{=} \frac{1}{10}.$$

The solution set is $\{\frac{3}{16}\}$. ■

Recall from Section 2.2 that one should always note any restrictions on the variable before solving an equation.

Example 2   Solve $\dfrac{3}{x-2} = \dfrac{2}{x+2}$, $x \neq 2, -2$.

Multiply each side by the LCD, $(x-2)(x+2)$.

$$(x-2)(x+2) \cdot \left(\frac{3}{x-2}\right) = (x-2)(x+2) \cdot \left(\frac{2}{x+2}\right)$$

$$3(x+2) = 2(x-2)$$

$$3x + 6 = 2x - 4$$

$$3x - 2x = -4 - 6 \qquad \text{Subtract 2x and 6}$$

$$x = -10$$

Check: $\dfrac{3}{x-2} = \dfrac{2}{x+2}$

$$\frac{3}{-10-2} \overset{?}{=} \frac{2}{-10+2} \qquad x = -10$$

$$\frac{3}{-12} \overset{?}{=} \frac{2}{-8}$$

$$\frac{-1}{4} \overset{\checkmark}{=} \frac{-1}{4}.$$

The solution set is $\{-10\}$. ∎

In Example 2 the solution was obtained by multiplying both sides by an expression involving variables. This sometimes results in false solutions, as the next example shows.

Example 3   Solve $\dfrac{x}{x-4} - 6 = \dfrac{4}{x-4}$.

First note that $x \neq 4$, then multiply each side by the LCD, $x - 4$.

$$(x-4) \cdot \left(\frac{x}{x-4} - 6\right) = (x-4) \cdot \left(\frac{4}{x-4}\right)$$

$$x - 6(x-4) = 4$$

$$x - 6x + 24 = 4$$

$$-5x = -20$$

$$x = 4$$

Since $x \neq 4$, the equation has no solution. The solution set is ∅. ∎

Example 4   Solve $\dfrac{3}{k-1} - \dfrac{2}{k-2} = \dfrac{4}{k^2 - 3k + 2}$.

$$k^2 - 3k + 2 = (k-1)(k-2) \qquad \text{LCD} = (k-1)(k-2); \quad k \neq 1, 2$$

$$(k - 1)(k - 2) \cdot \left(\frac{3}{k - 1} - \frac{2}{k - 2} \right) = (k - 1)(k - 2) \cdot \left(\frac{4}{k^2 - 3k + 2} \right)$$

$$3(k - 2) - 2(k - 1) = 4$$

$$3k - 6 - 2k + 2 = 4$$

$$k - 4 = 4$$

$$k = 8$$

The solution set is {8}. The check is left to the reader. ■

Literal equations often involve rational expressions that are more complicated than those found in Chapter 2.

Example 5   Solve $\dfrac{x}{b + c} + \dfrac{y}{b - c} = \dfrac{2}{b + c}$ for b.

The LCD is $(b + c)(b - c), \quad b \neq c, -c$.
Multiply each side by the LCD.

$$(b + c)(b - c) \cdot \left(\frac{x}{b + c} + \frac{y}{b - c} \right) = (b + c)(b - c) \cdot \left(\frac{2}{b + c} \right)$$

$$x(b - c) + y(b + c) = 2(b - c)$$

$$bx - cx + by + cy = 2b - 2c$$

Now isolate all terms involving b on one side of the equation.

$$bx + by - 2b = -2c - cy + cx$$

$$b(x + y - 2) = -c(2 + y - x) \qquad \text{Factor}$$

$$b = \frac{-c(2 + y - x)}{x + y - 2} \qquad \begin{array}{l}\text{Divide by the} \\ \text{coefficient of } b\end{array} \quad ■$$

The next example illustrates an operation called *finding the inverse of a function*, which will be covered in Section 9.9.

Example 6   Given $y^2 = \dfrac{x + 2}{x + 3}$, interchange x and y, then solve for y.

$$y^2 = \frac{x + 2}{x + 3}$$

$$x^2 = \frac{y + 2}{y + 3} \qquad \text{Interchange } x \text{ and } y$$

$$x^2(y + 3) = y + 2 \qquad \text{Multiply by } y + 3$$

$$x^2 y + 3x^2 = y + 2 \qquad \text{Distribute}$$

$$3x^2 - 2 = y - x^2 y \qquad \text{Isolate } y$$

$$3x^2 - 2 = y(1 - x^2) \qquad \text{Factor}$$

$$\frac{3x^2 - 2}{1 - x^2} = y \qquad \text{Divide by } (1 - x^2) \quad ■$$

Exercises 6.5

Solve each of the following. State any restrictions on the variable. Check your answers.

1. $\dfrac{x}{2} - 4 = x$

2. $a + 5 = \dfrac{a}{3}$

3. $\dfrac{3p}{4} = \dfrac{5p}{7} + 2$

4. $\dfrac{5t}{6} = \dfrac{1}{12} + \dfrac{2t}{3}$

5. $\dfrac{m}{2} + \dfrac{4m}{3} = 11$

6. $\dfrac{s}{6} + \dfrac{3s}{5} = 2$

7. $\dfrac{7}{4}x - \dfrac{1}{2}x = 5$

8. $\dfrac{3}{2}y - \dfrac{1}{3}y = 7$

9. $\dfrac{5k - 3}{7} = \dfrac{3k - 1}{5}$

10. $\dfrac{4b - 5}{9} = \dfrac{2b - 1}{5}$

11. $\dfrac{8}{r} - \dfrac{1}{3} = \dfrac{5}{r}$

12. $\dfrac{7}{q} + \dfrac{1}{4} = \dfrac{9}{q}$

13. $\dfrac{1}{6} + \dfrac{5}{6y} = \dfrac{6}{7y}$

14. $\dfrac{3}{4} - \dfrac{3}{4y} = \dfrac{-4}{5y}$

15. $\dfrac{x + 1}{x + 3} = 3$

16. $\dfrac{p - 2}{p - 6} = 5$

17. $\dfrac{2s + 3}{3s + 2} = \dfrac{5}{6}$

18. $\dfrac{2k - 3}{4k - 5} = \dfrac{2}{5}$

19. $\dfrac{3}{m - 2} = \dfrac{3}{m - 4}$

20. $\dfrac{5}{t - 4} = \dfrac{5}{t - 2}$

21. $\dfrac{1}{5} + \dfrac{4}{x} = \dfrac{16}{3} - \dfrac{1}{5x}$

22. $\dfrac{7}{8} - \dfrac{1}{n} = \dfrac{1}{2n} - \dfrac{1}{4}$

23. $\dfrac{x^2 - 8x + 16}{x^2 - 16} = 1$

24. $\dfrac{y^2 + 12y + 36}{y^2 + 2y - 24} = 1$

25. $\dfrac{1}{p + 3} - \dfrac{2}{p - 1} = 0$

26. $\dfrac{4}{b + 3} - \dfrac{3}{b - 4} = 0$

27. $\dfrac{5}{s - 1} + \dfrac{3}{s} = \dfrac{4}{s - 1}$

28. $\dfrac{2}{q} - \dfrac{4}{q + 3} = \dfrac{9}{q + 3}$

29. $\dfrac{6}{3a - 7} = \dfrac{9}{3a - 7} - \dfrac{5}{6a}$

30. $9 - \dfrac{4}{2m + 5} = \dfrac{-3}{2m + 5}$

Determine whether the given value of the variable is a solution to the equation.

31. $\dfrac{-3}{x + 3} = 5 - \dfrac{1}{x - 2}$, $x = -3$

32. $\dfrac{2}{a - 7} = 1 - \dfrac{1}{a + 2}$, $a = -2$

33. $\dfrac{3p - 2}{2p - 5} - 2 = \dfrac{4}{2p - 5}$, $p = 4$

34. $\dfrac{2}{3k + 8} - 5 = \dfrac{7}{3k + 8}$, $k = -3$

35. $\dfrac{6r}{4r + 1} + \dfrac{8}{4r + 1} = 5$, $r = \dfrac{1}{4}$

36. $\dfrac{10}{2t + 3} - 8 = \dfrac{12t}{2t + 3}$, $t = -\dfrac{1}{2}$

37. $\dfrac{3}{5y + 2} - \dfrac{2}{2y - 1} = \dfrac{8y + 5}{10y^2 - y - 2}$, $y = \dfrac{1}{2}$

38. $\dfrac{7}{m^2 - 4} - 3 = \dfrac{4m - 1}{m + 2}$, $m = 2$

Solve each literal equation for the specified variable. Assume no denominator is zero.

39. $p = \dfrac{A}{1 + rt}$ for A

40. $s = \dfrac{a - r1}{1 - r}$ for r

41. $\dfrac{2}{x} + \dfrac{4}{x + a} = \dfrac{-6}{a - x}$ for x

42. $\dfrac{1}{b - x} - \dfrac{1}{x} = \dfrac{-2}{b + x}$ for x

43. $\dfrac{s - vt}{t^2} = -16$ for v

44. $A = p\left[1 + \dfrac{r}{n}\right]$ for r

45. $\dfrac{1}{p} = \dfrac{1}{f} + \dfrac{1}{F}$ for f

46. $h = \dfrac{v^2}{2g} + \dfrac{p}{c}$ for p

47. $ab - 3c + 3ac = 2c$ for c

48. $k^2 - ky + 1 = m(p - m)$ for p

Interchange x and y and then solve for y. Assume no denominator is zero.

49. $y^3 = \dfrac{2x - 3}{3x - 1}$

50. $y^2 + 3 = \dfrac{x - 1}{x - 3}$

51. $y^2 - 2y = \dfrac{x - 2}{x + 1}$

52. $y^2 - 2y + 1 = \dfrac{x + 9}{x}$

Solve each of the following. State any restrictions on the variable.

53. $\dfrac{a + 1}{2a^2 + 7a - 4} - \dfrac{a}{2a^2 - 7a + 3} = \dfrac{1}{a^2 + a - 12}$

54. $\dfrac{2p}{6p^2 + 7p - 3} - \dfrac{p - 3}{3p^2 + 11p - 4} = \dfrac{5}{2p^2 + 11p + 12}$

55. $\dfrac{r - 2}{r^2 - 10r + 9} - \dfrac{3r - 12}{r^2 - 4r - 45} = \dfrac{1 - 2r}{r^2 + 4r - 5}$

56. $\dfrac{y - 6}{y^2 + 3y - 28} + \dfrac{y + 8}{y^2 + 16y + 63} = \dfrac{2y - 5}{y^2 + 5y - 36}$

57. $\dfrac{k + 2}{5k^2 - 27k - 18} + \dfrac{2 - k}{4k^2 - 29k + 30} = \dfrac{-(k + 1)}{20k^2 - 13k - 15}$

58. $\dfrac{2 - q}{30 - 29q + 4q^2} + \dfrac{q + 1}{15 + 13q - 20q^2} = \dfrac{q + 2}{5q^2 - 27q - 18}$

Review Exercises

Rewrite each of the following as an algebraic expression.

59. The sum of a number and its reciprocal.

60. The portion of a job done in one hour if it takes seven hours to do the whole job.

61. The distance traveled in one hour if you traveled 600 miles in x hours.

62. The time it takes to travel 725 kilometers at y kilometers per hour.

63. The rate of interest if N dollars earns 126 dollars simple interest in one year.

64. The part of a tank filled by a pipe in 3 hours if the tank can be filled by the pipe in x hours.

6.6   More Applications

Applications involving rational expressions are numerous and varied. Many of the applications unfortunately require techniques and knowledge beyond the scope of this course. For this reason we will include only a few special types. As you study the examples and solve the problems in the exercise set remember the step-by-step process outlined in the earlier chapters.

Number Problems

Example 1   If a number is added to its reciprocal the sum is $\frac{13}{6}$. Find the number.

Let n represent the number. Its reciprocal is $\frac{1}{n}$.

A number   added to   its reciprocal   is

$$n \quad + \quad \frac{1}{n} \quad = \quad \frac{13}{6}$$

The LCD is $6n$.

$$6n\left(n + \frac{1}{n}\right) = 6n \cdot \frac{13}{6}$$

$$6n^2 + 6 = 13n$$

$$6n^2 - 13n + 6 = 0$$

$$(3n - 2)(2n - 3) = 0$$

$$3n - 2 = 0 \quad \text{or} \quad 2n - 3 = 0$$

$$n = \frac{2}{3} \quad \text{or} \quad n = \frac{3}{2}$$

If the number is $\frac{2}{3}$, its reciprocal is $\frac{3}{2}$. If the number is $\frac{3}{2}$, its reciprocal is $\frac{2}{3}$. There are two solutions.   ■

Example 2   The denominator of a fraction is 3 more than the numerator. If 1 is subtracted from both numerator and denominator, the resulting fraction has a value of $\frac{2}{3}$. Find the original fraction.

Let the numerator of the fraction be x. The denominator is $x + 3$. The original fraction is $\frac{x}{x+3}$. When 1 is subtracted from both the numerator and denominator, the resulting fraction is

$$\frac{x - 1}{x + 3 - 1} = \frac{x - 1}{x + 2}.$$

Thus $$\frac{x - 1}{x + 2} = \frac{2}{3}.$$

$$3(x - 1) = 2(x + 2) \qquad \text{The LCD is } 3(x + 2)$$

$$3x - 3 = 2x + 4$$

$$x = 7$$

The original fraction is $\frac{x}{x+3} = \frac{7}{10}$.

Check: $$\frac{7 - 1}{10 - 1} = \frac{6}{9} = \frac{2}{3}.$$ ■

Distance-Rate-Time

When setting up an equation to solve a distance-rate-time problem remember that the formula $D = r \cdot t$ can be written in two alternate forms:

$$t = \frac{D}{r} \quad \text{and} \quad r = \frac{D}{t}.$$

Example 3   A motorboat can travel 10 miles per hour upstream and 16 miles per hour downstream. If Juan rents the boat for 4 hours, how far can he go upstream before heading back?

Let x = number of miles he can travel upstream. Here $t = \frac{D}{r}$.

	Time	Rate	Distance
Upstream	$\frac{x}{10}$	10	x
Downstream	$\frac{x}{16}$	16	x

Since Juan rented the boat for 4 hours, the sum of the times is 4.

$$\frac{x}{10} + \frac{x}{16} = 4$$

$$8x + 5x = 320 \qquad \text{LCD} = 80$$

$$13x = 320$$

$$x = 24\frac{8}{13} \text{ miles}$$

Juan can travel $24\frac{8}{13}$ miles upstream. ■

Example 4   An airplane can fly 150 miles against the wind in the same time it takes to fly 180 miles with the wind. If the wind speed is 15 miles per hour, how fast would the plane have been moving in still air?

When the plane moves against the wind its speed will be slowed by an amount equal to the wind speed. When it flies with the wind its speed will be increased by an amount equal to the wind speed.

Let x = speed in still air
$x - 15$ = speed against the wind
$x + 15$ = speed with the wind

	Time	Rate	Distance
Against wind	$\dfrac{150}{x - 15}$	$x - 15$	150
With wind	$\dfrac{180}{x + 15}$	$x + 15$	180

Since the two times are equal,

$$\frac{150}{x - 15} = \frac{180}{x + 15}$$

$$150(x + 15) = 180(x - 15)$$

$$150x + 2250 = 180x - 2700$$

$$4950 = 30x$$

$$165 = x.$$

The plane would have been moving at 165 miles per hour in still air. ■

Work Problems

Example 5   Randy and Todd contract to paint 10 identical tract homes. Randy can paint a house in 6 days with power equipment. When Todd joins him without power equipment they can paint a house in 4 days. How long would it take Todd to paint a house, working alone?

Let x = number of days Todd requires working alone. If Randy takes 6 days to paint a house, he can paint $\frac{1}{6}$ of it in one day.

	Total Time in Days	Amount in 1 day
Randy	6	$\frac{1}{6}$
Todd	x	$\frac{1}{x}$
Together	4	$\frac{1}{4}$

What they can do together in one day, $\frac{1}{4}$, must equal the sum of what they can do individually in one day.

$$\frac{1}{6} + \frac{1}{x} = \frac{1}{4}$$

$$2x + 12 = 3x$$

$$12 = x$$

Todd would require 12 days working alone. ■

Example 6   Two pipes are used to fill a swimming pool. The larger pipe could fill the pool in 12 hours and the smaller pipe in 16 hours. How long would it take to fill the pool if both pipes were used?

Let x = the number of hours for both pipes to fill the pool.

	Total Time in Hours	Amount in 1 Hour
Larger pipe	12	$\dfrac{1}{12}$
Smaller pipe	16	$\dfrac{1}{16}$
Together	x	$\dfrac{1}{x}$

$$\frac{1}{12} + \frac{1}{16} = \frac{1}{x}$$
$$4x + 3x = 48$$
$$7x = 48$$
$$x = \frac{48}{7} = 6\frac{6}{7}$$

Together the pipes would fill the pool in $6\frac{6}{7}$ hours. ■

Exercises 6.6

Write each of the following as an equation and solve.

Number Problems

1. A number added to seven times its reciprocal is -8. Find the number.

2. A number added to twenty times its reciprocal is 9. Find the number.

3. A natural number decreased by 27 times its reciprocal is 6. Find the natural number.

4. A natural number decreased by 30 times its reciprocal is 1. Find the natural number.

5. The denominator of a certain fraction is 6 more than the numerator. If the numerator is increased by 7 and the denominator decreased by 3, the value of the new fraction is $\frac{5}{4}$. Find the original fraction.

6. The denominator of a certain fraction is 4 more than the numerator. If 3 is subtracted from the numerator and added to the denominator, the value of the new fraction is $\frac{4}{9}$. Find the original fraction.

7. What number can be subtracted from numerator and denominator of the fraction $\frac{8}{11}$ so that the value of the new fraction will be $\frac{3}{5}$?

8. If a certain number is subtracted from the numerator, and twice the number is subtracted from the denominator of the fraction $\frac{19}{30}$, the value of the new fraction is $\frac{2}{3}$. What is the number?

9. What number must be subtracted from the numerator and denominator of the fraction $\frac{12}{7}$ so that the value of the new fraction will be $\frac{5}{3}$?

10. What number must be added to the numerator and denominator of the fraction $\frac{5}{7}$ so that the value of the new fraction will be $\frac{5}{3}$?

Distance-Rate-Time Problems

11. Two students leave Portland at the same time heading for Seattle. One, in a sports car, travels at 65 miles per hour. The second, on a motorcycle, travels at 55 miles per hour. How long will it be before they are 55 miles apart?

12. Dave and Fran leave home at the same time going in opposite directions. Fran travels at 64 kilometers per hour and Dave travels at 54 kilometers per hour. How far apart will they be after 30 minutes?

13. The Paytons leave El Paso heading east at 45 miles per hour. Their daughter Joy leaves El Paso one half hour later traveling east at 60 miles per hour. How long does it take for her to catch up with them? How far does Joy have to go?

14. An airplane travels 1260 miles in the same time that a car travels 420 miles. If the rate of the car is 120 miles per hour less than the rate of the airplane, find the rate of each.

15. San Antonio and Baton Rouge are 460 miles apart. Penny leaves San Antonio at 8:00 A.M. for Baton Rouge. Bill leaves Baton Rouge for San Antonio at 10:00 A.M. traveling 20 miles per hour faster than Penny. If they meet 210 miles from Baton Rouge, how fast is each traveling?

16. By increasing his usual average speed 20 kilometers per hour, Tim finds that he can cut his time on a 240-kilometer drive by two hours. Find his usual average speed.

17. Janet can row a canoe 2 miles per hour faster than Linda. If Janet rows 5 miles in the same amount of time that Linda rows 4 miles, how fast can Janet row her canoe?

18. Rex can walk 11 miles in the same amount of time that Larry can walk 8 miles. Rex walks one mile per hour faster than Larry. How fast can Rex walk?

19. Darwin's boat goes 15 miles per hour in still water. If it takes him as much time to travel 6 miles downstream as it does to travel 4 miles upstream on the Missouri River, what is the speed of the current?

20. Madalyn's plane travels at a speed of 350 miles per hour in still air. One day she travels 160 miles against the wind in the same time it takes to travel 190 miles with the wind. What is the speed of the wind?

Work Problems

21. A painter can prepare and paint a room in 10 hours. His son requires 18 hours to do the same job. How long will it take them to paint the room working together?

22. Marcie can prepare a manual in 6 hours. Joanie can prepare it in 4 hours. How long would it take them to prepare the manual working together?

23. Gary can spade his yard in 6 hours, but with his son helping it only takes them 4 hours. How long would it take his son to spade the yard working alone?

24. A pool can be filled by a pipe in 10 hours. If a second pipe is also opened it only takes 4 hours to fill the pool. How long would it take to fill the pool using only the second pipe?

25. It takes Mac $\frac{2}{3}$ as long as it takes Bennie to paint a compact car. If together Mac and Bennie take 12 hours, how long would it take Bennie working alone?

26. A hot tub can be filled by the cold water pipe in 21 minutes and by the hot water pipe in 28 minutes. How long would it take to fill the hot tub using both pipes?

27. A bathtub can be filled with cold water in 9 minutes and then drained in 12 minutes. If the cold water is then turned on and the drain left open by mistake, how long will it take to fill the bathtub?

28. A vat can be filled by an inlet pipe in 8 hours. How long will it take to drain the vat, if when both the inlet and drain pipes are opened the vat is filled in 24 hours?

29. Lewis can clean the house in 12 hours. After Lewis and Minnie have both been cleaning for one hour they are joined by their daughter Rachel and they complete the cleaning in 3 more hours. If it takes Minnie 10 hours to clean the house alone, how long would it take Rachel?

30. Bob can clean the kitchen in 45 minutes. Karen can clean it in 30 minutes. If both start cleaning and 5 minutes later their daughter Kelsey joins them, it takes 7 minutes more to finish. How long would it take Kelsey to clean the kitchen?

Chapter 6   Summary

[6.1]   A **rational expression** is the quotient of two polynomials in which the denominator is not zero. When working with a rational expression, one should first note and exclude any values of the variable that make the denominator zero.

A rational expression has many equivalent forms. Many of the forms can be discovered by use of the **fundamental principle of rational expressions:**

$$\frac{A}{B} = \frac{A \cdot C}{B \cdot C} \qquad B \neq 0, \; C \neq 0.$$

The fundamental principle asserts that when the numerator and denominator of a rational expression are multiplied or divided by the same nonzero polynomial, an equivalent rational expression results. If a rational expression has no common factors in the numerator and denominator, the numerator and denominator are said to be **relatively prime.** When this is the case the rational expression is in lowest terms.

[6.2]   The product of two rational expressions

$$\frac{A}{B} \text{ and } \frac{C}{D}, \quad B, D, \neq 0, \quad \text{is } \frac{A \cdot C}{B \cdot D}.$$

The quotient of the same two rational expressions is

$$\frac{A \cdot D}{B \cdot C}.$$

[6.3]   The sum of two rational expressions

$$\frac{A}{C} \text{ and } \frac{B}{C} \text{ is } \frac{A + B}{C}, \quad C \neq 0, \quad \text{and their difference is } \frac{A - B}{C}.$$

If their denominators are not the same they must first be rewritten in terms of their **LCD (least common denominator).** The LCD is found by factoring each denominator completely and forming the product of each distinct factor, the greatest number of times it occurs in any denominator.

[6.4]   A **complex rational expression** (complex fraction) is one in which the numerator or denominator (or both) contain other rational expressions. Complex rational expressions are simplified by using the fundamental principle. Both numerator and denominator of the original complex rational expression are multiplied by the LCD of all the denominators within the expression.

[6.5]   To solve an equation involving rational expressions, first note any restrictions on the variable. Then clear the equation of fractions by multiplying each side of the equation by the LCD. Each solution should be checked against the excluded values.

There are numerous applications involving rational expressions. They include number problems, distance-rate-time problems, work problems, and many that are beyond the scope of this book.

Chapters 5 and 6   Review

Factor each of the following.

1. $a^2 - 81$

2. $6x^2y + 12xy^2 + 6y^3$

3. $20t^2 - 23t - 21$

4. $2m^3 + 16$

5. $\dfrac{1}{16}s^3t - \dfrac{1}{10}s^2t^2 + \dfrac{1}{25}st^3$

6. $p^4 - 24p^2 - 25$

Solve each equation by factoring.

7. $v^2 - 5v + 4 = 0$

8. $6x^2 + 7x - 3 = 0$

9. $3a^4 - 39a^2 + 36 = 0$

10. $-35p + 42 + 7p^2 = 0$

Perform the indicated operations and simplify.

11. $\dfrac{7x}{3y} \cdot \dfrac{15y}{21x}$

12. $\dfrac{a^2 - 9}{b - 3} \cdot \dfrac{b^2 - 9}{a^2 - 6a + 9}$

13. $\left(\dfrac{m^3 - 27}{n + 2} \cdot \dfrac{n^2 - 4}{m^2 + 3m + 9} \right) \div (n + 2)$

14. $\left(\dfrac{2}{3q + 1} - \dfrac{5}{3q - 1} \right) \div \dfrac{1}{9q^2 - 1}$

15. $\dfrac{a - b}{d - c} + \dfrac{a - b}{c - d}$

16. $\left[\dfrac{27x^3y^2}{16x^2z^2} \div \dfrac{3x^2}{8z^2} \right] \cdot \left[\dfrac{yz}{9x} \right]$

17. $3s - \dfrac{rs - s^2}{r + s} - 2r$

18. $1 - \dfrac{-(-1 - 2p)}{p^2 - 2p + 1}$

19. $\dfrac{1}{u^2 - v^2} + \dfrac{1}{u - v} - \dfrac{1}{u + v}$

20. $5x - 2 - \dfrac{2 - x}{x - 3}$

Simplify.

21. $\dfrac{x + \dfrac{1}{2}}{1 + \dfrac{1}{x}}$

22. $\dfrac{p - \dfrac{1}{p}}{\dfrac{5}{3p}}$

23. $\dfrac{\dfrac{1}{y} + \dfrac{1}{y - 1}}{\dfrac{1}{y} - \dfrac{1}{y - 1}}$

24. $\dfrac{\dfrac{9}{9 - x^2}}{\dfrac{3}{x - 3} - \dfrac{3}{x + 3}}$

25. $\dfrac{-\dfrac{1}{t^2} - 1}{\dfrac{5}{t^2} - \dfrac{5}{t} - 10}$

26. $\dfrac{x^{-1} - y^{-1}}{x^{-1} + y^{-1}}$

27. $\left[\dfrac{1}{a+b} + \dfrac{1}{a-b}\right] \div \left[\dfrac{1}{a+b} - \dfrac{1}{a-b}\right]$

28. $\left[\dfrac{x^2 - y^2}{a^2 - b^2}\right] \div \left[\dfrac{x-y}{a+b}\right]$

Solve each of the following. Check your answers.

29. $\dfrac{x}{3} - \dfrac{x}{2} = 5$

30. $\dfrac{1}{x} - \dfrac{3}{x} = 2$

31. $\dfrac{x-3}{x-4} = -2$

32. $\dfrac{3b-2}{2b+1} = -1$

33. $\dfrac{1}{4} + \dfrac{3}{s} = \dfrac{7}{3} - \dfrac{1}{5s}$

34. $\dfrac{t^2 - 7t + 12}{t^2 - 16} = 1$

35. $\dfrac{x^2 - 3x + 2}{x^2 - 8} = 2$

36. $\dfrac{a^2 + 16a + 3}{a^2 - 5} = 3$

37. $\dfrac{1}{x^2 - 2} - \dfrac{1}{x^2 - 3x + 1} = 0$

38. $\dfrac{m}{m^2 - 9} - \dfrac{1}{m+2} = 0$

39. $\dfrac{6}{y} - \dfrac{4}{y+2} = \dfrac{9}{y+2}$

40. $\dfrac{7}{q} - \dfrac{3}{q-1} = \dfrac{5}{q-1}$

Write each of the following as a rational equation and solve.

41. Three times an integer decreased by 5 times its reciprocal is 2. Find the integer.

42. The denominator of a number is 4 more than the numerator. If both the numerator and denominator are increased by 5, the result is $\frac{7}{9}$. Find the number.

43. Mort leaves Minot, Maine at 11:00 A.M. heading south at an average of 56 miles per hour. Helen leaves Minot at 11:30 A.M. the same day to overtake Mort traveling at an average of 64 miles per hour. How long does it take Helen to overtake Mort? How far will they have traveled?

44. Todd runs a one-mile track in 5 minutes. Rhonda runs the track in 7 minutes. If Todd and Rhonda start at opposite ends of the track and run until they meet, how long does it take?

45. Because of fog Ray decreases his usual speed by 20 miles per hour for a 240-mile trip. He finds that this increases his usual time by 2 hours. What is his usual speed?

46. A tank can be filled by an inlet pipe in 7 hours. It takes 10 hours to drain the tank. How long will it take to fill the tank if the inlet pipe is opened and the drain pipe is left half open?

7

Exponents and Radicals

Chapter 4 examined the laws of exponents as they apply to integers. We began by defining the exponent as a natural number that indicated the number of times the base was to be used as a factor. It soon became necessary to give meaning to zero as an exponent:

$$b^0 = 1,$$

and to negative integers as exponents:

$$b^{-n} = \frac{1}{b^n}.$$

In this chapter we will examine exponential expressions such as

$$b^{m/n},$$

where the exponent m/n is a rational number, $n \neq 0$. We will show how the rules of exponents are used when the exponents represent any rational numbers.

7.1 Rational Exponents

If the rules of exponents are to hold for exponents that are rational numbers, we will have to give meaning to expressions such as

$$4^{1/2} \quad \text{or} \quad (-27)^{1/3}.$$

To discover the meaning, consider

$$(4^{1/2})^2 = 4^{(1/2)\cdot 2} = 4 \qquad \text{Recall that } (b^m)^n = b^{m \cdot n}$$

This implies that $4^{1/2}$ is a number whose square is 4. In the same manner

$$[(-27)^{1/3}]^3 = (-27)^{3/3} = -27,$$

so $(-27)^{1/3}$ must be a number whose cube is -27.

Recall from elementary algebra that $\sqrt{4} = 2$ and $\sqrt[3]{-27} = -3$. Thus, $4^{1/2}$ is another way of indicating a square root of 4 and $(-27)^{1/3}$ another way of indicating a cube root of -27. In general,

$$(b^{1/n})^n = b^{n/n} = b, \quad b \geq 0, \ n \in N,$$

so $b^{1/n}$ indicates an nth root of b.

There may be more than one nth root of a number. The number 4 has two real square roots, 2 and -2, since $2^2 = 4$ and $(-2)^2 = 4$. The number -27 has only one real cube root, -3. In this book we will use the following interpretation.

Finding the nth Root of b

| n even | $b \geq 0$; $b^{1/n}$ is the positive nth root of b or zero | $b < 0$; $b^{1/n}$ is not a real number |
| n odd | $b \geq 0$; $b^{1/n}$ is the positive nth root of b or zero | $b < 0$; $b^{1/n}$ is the negative nth root of b |

Example 1   Find the root of each real number.

(a) $4^{1/2} = 2$ since $2^2 = 4$

(b) $8^{1/3} = 2$ since $2^3 = 8$

(c) $(-64)^{1/3} = -4$ since $(-4)^3 = -64$

(d) $(-16)^{1/2}$ is not a real number.   ■

▶ **Caution**   Be particularly careful to use grouping symbols correctly when writing the sign of a number in a problem involving roots.

$(-9)^{1/2}$ is not a real number

$-9^{1/2} = -1 \cdot 9^{1/2} = -1 \cdot 3 = -3$

Example 2   Find the root of each real number.

(a) $-4^{1/2} = -2$, since $-2^2 = -4$

(b) $(-16)^{1/4}$ is not a real number.

(c) $(-32)^{1/5} = -2$ since $(-2)^5 = -32$ ■

What interpretation is given to $b^{m/n}$? If the laws of exponents are to hold, then

$$b^{m/n} = (b^{1/n})^m = (b^m)^{1/n}. \qquad b \geq 0 \text{ when } n \text{ is even}$$

For example,

$$8^{2/3} = (8^{1/3})^2 = 2^2 = 4$$
$$\text{and} \qquad 8^{2/3} = (8^2)^{1/3} = (64)^{1/3} = 4.$$

In general it is better to treat $b^{m/n}$ as $(b^{1/n})^m$ since less effort is usually required to find the root first.

Example 3   Simplify.

(a) $(64^{5/6}) = (64^{1/6})^5 = 2^5 = 32$ $64^{1/6} = 2$

(b) $(81)^{3/4} = (81^{1/4})^3 = 3^3 = 27$ $81^{1/4} = 3$

(c) $(0.125)^{2/3} = [(0.125)^{1/3}]^2 = [0.5]^2 = 0.25$ ■

We now turn our attention to rational exponents in general. The laws of exponents hold as long as we do not apply them to even roots of negative numbers.

Example 4   Simplify each of the following.

(a) $9^{-1/2} = \dfrac{1}{9^{1/2}} = \dfrac{1}{3}$ $b^{-n} = \dfrac{1}{b^n}$

(b) $64^{-1/6} = \dfrac{1}{64^{1/6}} = \dfrac{1}{2}$

(c) $x^{1/2} \cdot x^{1/3} = x^{1/2+1/3} = x^{5/6}, \; x > 0$ $b^m b^n = b^{m+n}$

(d) $\dfrac{x^{2/3}}{x^{3/5}} = x^{2/3-3/5} = x^{1/15}$ $\dfrac{b^m}{b^n} = b^{m-n}$

(e) $(xy)^{2/3} = x^{2/3}y^{2/3}$ $(ab)^m = a^m b^m$

(f) $\left(\dfrac{a^{2/3}}{b^{1/2}}\right)^{4/5} = \dfrac{a^{(2/3)(4/5)}}{b^{(1/2)(4/5)}} = \dfrac{a^{8/15}}{b^{2/5}}$ $\left(\dfrac{a}{b}\right)^m = \dfrac{a^m}{b^m}$ ∎

Example 5   Simplify.

(a) $\dfrac{9a^{-2/3}}{3a^{-7/3}} = 3a^{-2/3-(-7/3)} = 3a^{-2/3+7/3} = 3a^{5/3}$

(b) $3x^{2/3}(4x^{7/3}) = 12x^{2/3+7/3} = 12x^3$

(c) $x^{1/3}(x^{1/2} + x^{2/3}) = x^{1/3+1/2} + x^{1/3+2/3} = x^{5/6} + x$ ∎

Example 6   Factor $a^{2/3} - a^{3/5}$.

Write each rational exponent in terms of the LCD of the exponents.

$$a^{2/3} - a^{3/5} = a^{10/15} - a^{9/15}$$

The greatest common factor is $a^{9/15}$ or $a^{3/5}$.

$$a^{2/3} - a^{3/5} = a^{9/15}(a^{1/15} - 1) = a^{3/5}(a^{1/15} - 1)$$ ∎

Example 7   Simplify $\dfrac{x^{2/3}y^{-1/2}z^{1/4}}{x^{1/3}y^{3/2}z^{3/4}}$.

Applying the laws of exponents, we have

$$x^{2/3-1/3}y^{-1/2-3/2}z^{1/4-3/4} = x^{1/3}y^{-2}z^{-1/2}$$

$$= \dfrac{x^{1/3}}{y^2 z^{1/2}}.$$ ∎

Exercises 7.1

Assume all variables in the exercises for this section represent positive real numbers.

Simplify each of the following.

1. $25^{1/2}$ 2. $36^{1/2}$ 3. $27^{1/3}$

4. $8^{1/3}$ 5. $16^{-1/4}$ 6. $81^{-1/4}$

7. $4^{3/2}$

8. $9^{3/2}$

9. $(-27)^{1/3}$

10. $(-64)^{1/3}$

11. $27^{2/3}$

12. $125^{2/3}$

13. $(8)^{1/3} \cdot (8)^{2/3}$

14. $(5)^{1/2} \cdot (5)^{3/2}$

15. $(3)^{3/2} \cdot (3)^{-1/2}$

16. $\left(\dfrac{9}{4}\right)^{-1/2}$

17. $\left(\dfrac{8}{27}\right)^{2/3}$

18. $\left(\dfrac{25}{36}\right)^{3/2}$

19. $\left(\dfrac{1}{32}\right)^{-1/5}$

20. $\left(\dfrac{1}{81}\right)^{-1/4}$

21. $\left(\dfrac{4}{49}\right)^{3/2}$

Simplify each of the following. Express each answer using only positive exponents.

22. $(0.00243)^{2/5}$

23. $(0.0256)^{-1/4}$

24. $(-0.125)^{-1/3}$

25. $x^{1/2} \cdot x^{1/2}$

26. $y^{3/4} \cdot y^{1/4}$

27. $p^{3/5} \cdot p^{1/5}$

28. $q^{4/3} \cdot q^{-1/3}$

29. $(2t^{1/3})(5t^{2/3})$

30. $(7s^{3/2})(2s^{-1/2})$

31. $(a^4)^{1/2}$

32. $(c^3)^{4/3}$

33. $6m^{2/3} \cdot 3m^{1/6}$

34. $2x^{1/4} \cdot x^{-1/3}$

35. $3k^{-5/2} \cdot k^{1/3}$

36. $9y^{-3/4} \cdot 4y^{1/3}$

37. $\dfrac{a^{3/4}}{a^{1/4}}$

38. $\dfrac{p^{5/8}}{p^{-3/8}}$

39. $\dfrac{6r^{-2/3}}{3r^{-5/3}}$

40. $\dfrac{15s^{-1/4}}{5s^{-5/4}}$

41. $\left(\dfrac{x^{1/2}}{x^{1/4}}\right)^4$

42. $\left(\dfrac{m^{3/2}}{m^{-1/2}}\right)^{1/2}$

43. $\left(\dfrac{17y^{8/3}}{34y^{-1/3}}\right)^{1/3}$

44. $\left(\dfrac{18a^{-2/5}}{27a^{3/5}}\right)^{-1}$

45. $(18x^{4/3}y^{7/3}) \div 54x^{1/3}y^{1/3}$

46. $(63a^{9/5}b^{7/5}) \div (7a^{4/5}b^{2/5})$

47. $(16^{2/3}p^{5/9}q^{-1/3}) \div (64^{1/3}p^{-2/7}q^{4/3})$

48. $(-91s^{-1/3}t^{-2/5}) \div (13s^{1/3}t^{2/5})$

49. $x^{1/2}(x^{1/2} + x^{1/4})$

50. $a^{1/3}(a^{2/3} + a^{1/4})$

51. $k^{-1/2}(k^{3/2} - k^{1/2})$

52. $p^{-1/4}(p^{5/4} - p^{1/4})$

53. $3t^{-2/3}(5t^{1/4} - 2t^{3/2})$

54. $7m^{-3/4}(2m^{-1/3} + 3m^{1/2})$

55. $(x^{1/2} - y^{1/2})(x^{1/2} + y^{1/2})$

56. $(a^{1/2} + b^{1/2})(a^{1/2} - b^{1/2})$

57. $(a^{1/2} + b^{1/4})^2$

58. $(y^{1/2} - y^{1/4})^2$

59. $(x^{1/3} + y^{1/3})(x^{2/3} - x^{1/3}y^{1/3} + y^{2/3})$

60. $(p^{1/3} - q^{1/3})(p^{2/3} + p^{1/3}q^{1/3} + q^{2/3})$

61. $(a - 1)^{2/3}[(a - 1)^{1/3} - (a - 1)^{-2/3}]$

62. $(s - t)^{-2/3}[(s - t)^{2/3} + (s - t)^{5/3}]$

Factor each of the following as indicated.

63. $p^{5/4} = p^{1/4}(\qquad)$

64. $y^{-3/4} = y^{1/4}(\qquad)$

65. $x^{1/2} + x^{-1/2} = x^{-1/2}(\qquad)$

66. $t^{1/3} - t^{-1/3} = t^{-1/3}(\qquad)$

67. $a - a^{1/3} = a(\qquad)$

68. $k - k^{1/5} = k(\qquad)$

69. $x - y = (x^{1/2} + y^{1/2})(\qquad)$

70. $a - b = (a^{1/2} - b^{1/2})(\qquad)$

71. $8^{2/3}p^{3/4} - 4^{1/2}p^{-1/2} = 2p^{-1/2}(\qquad)$

72. $9^{3/2}s^{-1/3} - 9^{1/2}s^{4/3} = 3s^{-1/3}(\qquad)$

73. $(x - y) = (x^{1/3} - y^{1/3})(\qquad)$

74. $(a + b) = (a^{1/3} + b^{1/3})(\qquad)$

Simplify each of the following. Express each answer using only positive exponents.

75. $\dfrac{x^5 y^0 z^{1/4}}{x^{1/3} y^{1/2} z^{2/3}}$

76. $\dfrac{a^{2/3} b^{-1/2} c^{1/4}}{a^{2/3} b^{1/2} c^{-3/4}}$

77. $\dfrac{-25^{3/2} r^{3/5} s^2 t^{1/3}}{5^2 r^{1/4} s^{1/3} t^{1/4}}$

78. $\dfrac{(9u)^{1/2}(v)^{1/3}(25w)^{3/2}}{(8u)^{1/3}(9v)^{3/2}(125w)^{1/3}}$

79. $[(x^{-2/3}y^{1/4}z^{1/5})^{-1/2}]^{-3/2}$

80. $[(m^{-1/4})^{1/3}(9p)^{-3/2}(27q)^{1/3})^{-2}]^{1/3}$

81. $\dfrac{9^{-2}a^{-1/3}b^{4/3}}{(4a^{2/3}b^{-4/3})^{1/2}}$

82. $\left(\dfrac{16x^{3/4}y^{-3/5}}{24x^0y^{3/5}}\right)^{-1/3}$

83. $\left(\dfrac{r^{1/2}s^{-2}}{t^{1/2}}\right)^2 \cdot \left(\dfrac{r^{-1/2}s^0}{t^{-4}}\right)^{-2}$

84. $\left(\dfrac{mn^{1/2}}{p^{1/2}q^{3/2}}\right)^{-4} \cdot \left(\dfrac{m^2n^2}{p^2q^4}\right)^{3/2}$

85. $(x^{-1/4}y^{-1/3})^{-1} \cdot (x^{-3/2})^2 \cdot (y^{-5/3})^{-2}$

86. $(a^{1/4}b^{-3/8})^{-1} \cdot (a^{-3/4})^{-4} \cdot (b^{-2/3})^4$

Simplify. Assume m and n are positive rational numbers.

87. $(y^{n+1})^2$

88. $(x^{m-1})^3$

89. $(t^{2m})^{3/2}$

90. $\left(\dfrac{s^{3m}}{s^m}\right)^{1/3}$

91. $\left(\dfrac{x^{5n-1}}{x^{3n-2}}\right)^{1/2}$

92. $\left(\dfrac{y^{2n-3}}{y^{n-1}}\right)^{1/3}$

93. $\left(\dfrac{x^{2m}}{y^{2n}}\right)^{1/3} \cdot \left(\dfrac{y^{3n}}{x^{3m}}\right)^{1/2}$

94. $\left(\dfrac{a^{2n}}{b^{-3/2}}\right)^{1/3} \cdot \left(\dfrac{a^{2/3}}{b^{1/4}}\right)^n$

Use a calculator to evaluate each of the following. Round each answer to 4 decimal places.

95. $5^{1/2}$

96. $6^{1/2}$

97. $3^{1/3}$

98. $4^{1/3}$

99. $3^{-1/2}$

100. $7^{-1/3}$

101. $9^{3/4}$

102. $100^{2/3}$

103. $(0.0124)^{1/3}$

104. $(1.2826)^{1/2}$

105. $(-0.893)^{-1/3}$

106. $(-0.101)^{-1/3}$

7.2 Rational Exponents and Radicals

In Section 7.1 the symbol $b^{1/n}$ was used to indicate the *n*th root of *b*. Another symbol that is used to indicate a root of a number is the radical symbol, shown below.

$$\sqrt{}$$

The symbol $\sqrt{b}$ means *the square root of b*. Since every positive number has two square roots, one positive and one negative, $\sqrt{b}$ will be used to indicate the **positive** or **principal square root** of *b*.

> The **principal square root** of any positive number *b*, written $\sqrt{b}$, is that positive number *a* such that $a^2 = b$.

Example 1   Find each square root.

(a) $\sqrt{25} = 5$, since $5^2 = 25$

(b) $\sqrt{36} = 6$, since $6^2 = 36$

(c) $\sqrt{9} = 3$, since $3^2 = 9$ ■

When the root of a number is other than a square root, an **index** is used to indicate it. For example, in

$$\sqrt[3]{b}$$

the index is 3, which indicates that the cube root of b is to be found. The symbol

$$\sqrt[n]{b}$$

means that the **nth root** of b is to be found. The quantity that appears under the radical sign is called the **radicand.**

Index ─── ── Radical symbol

$\sqrt[n]{b}$ ←── Radicand

There is a direct relation between radical notation and the rational exponent.

$$\sqrt{4} = 4^{1/2}$$ The index is not shown for square roots;
$$\sqrt[3]{-27} = (-27)^{1/3}$$ it is understood to be 2
$$\sqrt[5]{36} = 36^{1/5}$$
$$\sqrt[4]{a^3} = (a^3)^{1/4} = a^{3/4}$$
$$(\sqrt[7]{x})^9 = (x^{1/7})^9 = x^{9/7}$$

The Relationship Between Rational Exponents and Radicals

$$b^{m/n} = (b^{1/n})^m = (\sqrt[n]{b})^m$$

and $$b^{m/n} = (b^m)^{1/n} = \sqrt[n]{b^m}$$, $b \geq 0$ if n is even

---

Example 2 Rewrite in radical form.

(a) $3^{1/2} = \sqrt{3}$ The index is not shown for square roots

(b) $5^{1/4} = \sqrt[4]{5}$

(c) $7^{2/3} = \sqrt[3]{7^2}$

(d) $(8^{1/4})^2 = (\sqrt[4]{8})^2 = \sqrt[4]{8^2}$ ∎

Example 3 Rewrite using rational exponents.

(a) $\sqrt{5} = 5^{1/2}$

(b) $\sqrt[3]{3^2} = (3^2)^{1/3} = 3^{2/3}$

(c) $(\sqrt[3]{x^3 + y^3}) = (x^3 + y^3)^{1/3}$

(d) $(\sqrt{(x + y)^2} = [(x + y)^2]^{1/2} = (x + y)^{2/2} = x + y$ ∎

Example 4 Find the indicated roots.

(a) $\sqrt{36} = 6$ $6^2 = 36$

(b) $\sqrt[3]{8} = 2$ $2^3 = 8$

(c) $\sqrt[3]{a^6 b^9} = a^2 b^3$ $(a^2 b^3)^3 = a^6 b^9$

(d) $\sqrt[4]{625 x^8} = 5 x^2$ $(5 x^2)^4 = 625 x^8$

(e) $\sqrt[3]{(a + b)^9} = (a + b)^3$ $\qquad$ $[(a + b)^3]^3 = (a + b)^9$

(f) $\sqrt{x^2 + 4x + 4} = \sqrt{(x + 2)^2} = x + 2$ ■

In Examples 1 through 4 the variables in the radicand represented positive quantities whenever the index was even. The following example shows why the restriction was made.

$$\sqrt{b^2} = b \qquad\qquad\qquad\qquad \sqrt{b^2} = b$$
$$\text{if } b = 3 \qquad\qquad\qquad\qquad \text{if } b = -3$$
$$\sqrt{3^2} = 3 \qquad\qquad\qquad\qquad \sqrt{(-3)^2} \overset{?}{=} -3$$
$$\sqrt{9} = 3 \qquad\qquad\qquad\qquad \sqrt{9} \overset{?}{=} -3$$

$\qquad\quad 3 = 3$ $\quad$ True statement $\qquad\qquad 3 = -3$ $\quad$ False statement

In order for $\sqrt{b^2} = b$ to be true for any real number b, two cases must be considered:

$$\sqrt{b^2} = b \text{ if } b \geq 0$$

and $\qquad \sqrt{b^2} = -b \text{ if } b < 0.$

Since $|b| = b$ if $b \geq 0$ and $|b| = -b$ if $b < 0$, both cases can be defined as follows.

$$\sqrt{b^2} = |b| \textbf{ for all } b \in R$$

No problem exists in the event the index is odd. For example,

$$\sqrt[3]{3^3} = 3 \qquad \text{and} \qquad \sqrt[3]{(-3)^3} = -3$$
$$\sqrt[3]{27} = 3 \qquad\qquad\qquad \sqrt[3]{-27} = -3$$
$$3 \overset{\checkmark}{=} 3 \qquad\qquad\qquad -3 \overset{\checkmark}{=} -3.$$

Example 5 Find the indicated roots.

(a) $\sqrt{x^2} = |x|$

(b) $\sqrt{(x + 2)^2} = |x + 2|$

(c) $\sqrt[5]{x^5} + \sqrt{x^2} = x + |x|$ ■

Radicals can be simplified. Simplification requires a theorem involving radicals in which we assume the roots exist.

Theorem 7.1

Properties of Radicals

$$\sqrt[n]{a \cdot b} = \sqrt[n]{a}\sqrt[n]{b} \quad \text{or} \quad (a \cdot b)^{1/n} = a^{1/n}b^{1/n}, \; a \text{ and } b \text{ not both} < 0, \; n \text{ even}$$

$$\sqrt[n]{\frac{a}{b}} = \frac{\sqrt[n]{a}}{\sqrt[n]{b}} \quad \text{or} \quad \left(\frac{a}{b}\right)^{1/n} = \frac{a^{1/n}}{b^{1/n}}$$

To illustrate Theorem 7.1, consider the following.

$$\sqrt{144} = \sqrt{9 \cdot 16} = \sqrt{9} \cdot \sqrt{16} = 3 \cdot 4 = 12 \qquad \sqrt{144} = 12$$

$$\text{and} \qquad \sqrt{\frac{144}{9}} = \frac{\sqrt{144}}{\sqrt{9}} = \frac{12}{3} = 4 \qquad \qquad \sqrt{\frac{144}{9}} = \sqrt{16} = 4$$

For a radical to be in simplest form, the following conditions must be satisfied.

1. **No exponent on a factor in the radicand is greater than or equal to the index.**
2. **No radicand contains a fraction.**

Example 6   Simplify each of the following radicals.

(a) $\sqrt{150} = \sqrt{25 \cdot 6} = \sqrt{5^2 \cdot 6} = 5\sqrt{6} \qquad \sqrt[n]{ab} = \sqrt[n]{a}\sqrt[n]{b}$

(b) $\sqrt[4]{a^7 b^6}, \; a, \, b > 0$

This radical is not in simplest form because the exponents in the radicand are greater than the index.

$$\sqrt[4]{a^7 b^6} = \sqrt[4]{a^4 a^3 b^4 b^2}$$

$$= \sqrt[4]{a^4}\sqrt[4]{b^4}\sqrt[4]{a^3 b^2} \qquad \sqrt[n]{ab} = \sqrt[n]{a}\sqrt[n]{b}$$

$$= ab\sqrt[4]{a^3 b^2}$$

(c) $\sqrt[3]{(x+y)^7} = \sqrt[3]{[(x+y)^2]^3 (x+y)}$

$$= (x+y)^2 \sqrt[3]{x+y} \qquad \sqrt[3]{b^3} = b, \; b = (x+y)^2$$

(d) $\sqrt{\dfrac{25}{36}} = \dfrac{\sqrt{25}}{\sqrt{36}} = \dfrac{5}{6} \qquad \dfrac{\sqrt[n]{a}}{\sqrt[n]{b}} = \sqrt[n]{\dfrac{a}{b}}$

(e) $\sqrt{\dfrac{1}{3}}$

This radical is not in simplest form because the radicand contains a fraction. In order to eliminate the fraction in the radicand the denominator must be made into a perfect square.

$$\sqrt{\frac{1}{3}} = \sqrt{\frac{1}{3} \cdot \frac{3}{3}} \qquad \qquad \text{Fundamental property of fractions}$$

$$= \sqrt{\frac{3}{3^2}} = \frac{\sqrt{3}}{\sqrt{9}} = \frac{\sqrt{3}}{3}$$

(f) $\sqrt[3]{\dfrac{1}{4}} = \sqrt[3]{\dfrac{1}{2^2}}$

Since $\sqrt[3]{2^3} = 2$, we multiply the radicand by $\frac{2}{2}$.

$$\sqrt[3]{\frac{1}{2^2}} = \sqrt[3]{\frac{1}{2^2} \cdot \frac{2}{2}} = \sqrt[3]{\frac{2}{2^3}} = \frac{\sqrt[3]{2}}{\sqrt[3]{2^3}} = \frac{\sqrt[3]{2}}{2}$$

(g) $\sqrt[3]{0.008} = \sqrt[3]{8 \times 10^{-3}}$ Scientific notation

$= \sqrt[3]{8} \cdot \sqrt[3]{10^{-3}}$ Property 1

$= 2 \times 10^{-1}$

$= 0.2$ ∎

▶ *Caution* Remember that $\sqrt{x^2 + y^2} \neq x + y$. If the two expressions were equal, the following would be a true statement.

$$\sqrt{25} = \sqrt{9 + 16} = \sqrt{3^2 + 4^2} = 3 + 4 = 7$$

We know that $\sqrt{25}$ is 5, not 7.

Exercises 7.2

Rewrite using radical notation.

1. $6^{1/2}$
2. $3^{1/2}$
3. $11^{1/3}$
4. $17^{1/4}$
5. $x^{2/3}$
6. $y^{3/5}$
7. $p^{-1/3}$
8. $r^{-1/5}$
9. $6t^{-3/7}$
10. $7k^{-2/3}$
11. $(5y)^{1/3}$
12. $(13s)^{2/3}$

Rewrite using rational exponents. Express all answers using positive exponents only.

13. $\sqrt{7}$
14. $\sqrt{15}$
15. $\sqrt[3]{2}$
16. $\sqrt[3]{5}$
17. $\sqrt[3]{x^2}$
18. $\sqrt[5]{k^4}$
19. $\sqrt[7]{y^3}$
20. $\sqrt[7]{s^5}$
21. $\sqrt[3]{(5)^{-2}}$
22. $\sqrt[3]{(3)^{-2}}$
23. $\sqrt[3]{5x^2}$
24. $\sqrt[3]{4y^2}$
25. $\sqrt[3]{7^3}$
26. $\sqrt[5]{13^5}$
27. $\sqrt[5]{(3^{-2})(m^3)}$
28. $\sqrt[3]{(7^{-2})(k^2)}$
29. $\sqrt{x + y}$
30. $\sqrt{a - b}$
31. $\sqrt[3]{p^3 + q^3}$
32. $\sqrt[3]{x^3 - y^3}$

Find the indicated roots. Assume all variables represent positive real numbers.

33. $\sqrt{4}$
34. $\sqrt{25}$
35. $\sqrt[3]{64}$
36. $\sqrt[3]{27}$
37. $\sqrt{x^2}$
38. $\sqrt{y^2}$
39. $\sqrt[3]{-8}$
40. $\sqrt[3]{-125}$
41. $\sqrt[4]{16p^4}$
42. $\sqrt[4]{81s^4}$
43. $\sqrt{16t^4}$
44. $\sqrt{36s^6}$
45. $\sqrt[3]{r^6t^9}$
46. $\sqrt[5]{m^{10}p^{15}}$
47. $\sqrt[5]{3^{-5}x^5y^{-5}}$
48. $\sqrt[3]{5^{-3}k^6p^{-6}}$
49. $\sqrt[3]{(x + y)^3}$
50. $\sqrt{(a + b)^2}$
51. $\sqrt{(p + q)^4}$
52. $\sqrt[3]{(m + n)^6}$

Simplify each of the following. Assume all variables represent positive real numbers and no denominator is zero.

53. $\sqrt{8}$
54. $\sqrt{32}$
55. $\sqrt{20}$
56. $\sqrt{12}$
57. $\sqrt{\dfrac{27}{16}}$

58. $\sqrt{\dfrac{18}{25}}$
59. $\sqrt{\dfrac{128}{50}}$
60. $\sqrt{\dfrac{243}{12}}$
61. $\sqrt{x^3}$
62. $\sqrt{y^5}$

63. $\sqrt{p^3q}$
64. $\sqrt{r^5s}$
65. $\dfrac{1}{2}\sqrt{72}$
66. $\dfrac{1}{3}\sqrt{45}$
67. $\dfrac{1}{5}\sqrt{250}$

68. $\dfrac{5}{4}\sqrt{128}$
69. $\sqrt{\dfrac{225}{400}}$
70. $\sqrt{\dfrac{98}{18}}$
71. $\sqrt{\dfrac{x^3}{y^2}}$
72. $\sqrt{\dfrac{a^5}{b^4}}$

73. $\sqrt{0.04}$
74. $\sqrt{0.09}$
75. $\sqrt{9 \cdot 10^4}$
76. $\sqrt{4 \cdot 10^6}$
77. $\sqrt{63ab^2}$

78. $\sqrt{250xy^3}$ **79.** $\sqrt[3]{24a^7}$ **80.** $\sqrt[3]{32n^5}$ **81.** $\sqrt[4]{32p^9}$ **82.** $\sqrt[5]{96t^7}$

83. $\sqrt[6]{r^6s^7t^{13}}$ **84.** $\sqrt[5]{x^8y^7z^6}$ **85.** $\sqrt{\dfrac{1}{3^3}}$ **86.** $\sqrt{\dfrac{1}{2^3}}$ **87.** $\sqrt[3]{\dfrac{2}{3^5}}$

88. $\sqrt[3]{\dfrac{3}{5^4}}$ **89.** $\sqrt{\dfrac{x}{12}}$ **90.** $\sqrt{\dfrac{y}{18}}$ **91.** $\sqrt[3]{\dfrac{1}{16}}$ **92.** $\sqrt[3]{\dfrac{1}{36}}$

93. $\sqrt[4]{\dfrac{1}{32}}$ **94.** $\sqrt[4]{\dfrac{1}{27}}$ **95.** $\sqrt[3]{\dfrac{2}{100}}$ **96.** $\sqrt[3]{\dfrac{2}{250}}$ **97.** $\sqrt{\dfrac{3}{x^3}}$

98. $\sqrt{\dfrac{7}{y^3}}$ **99.** $\sqrt[3]{\dfrac{x}{a^7}}$ **100.** $\sqrt[4]{\dfrac{t}{t^6}}$

101. $\sqrt{x^2 + 2x + 1}$ **102.** $\sqrt{y^2 + 6y + 9}$ **103.** $\sqrt{p^2 + 4p + 4}$

104. $\sqrt{s^2 + 22s + 121}$ **105.** $\sqrt{x^2 + 2xy + y^2}$ **106.** $\sqrt{u^2 + 10uv + 25v^2}$

107. $\sqrt{\dfrac{a^2 + 12a + 36}{b^2 + 14b + 49}}$ **108.** $\sqrt{\dfrac{(w + y)^3}{w^2 + 2wy + y^2}}$

Simplify each of the following. All variables represent real numbers and no denominator is zero. Use absolute value as needed.

109. $\sqrt{x^2}$ **110.** $\sqrt{v^4}$ **111.** $\sqrt{25k^4}$ **112.** $\sqrt{36p^2}$

113. $\sqrt[3]{-108y^5}$ **114.** $\sqrt[3]{-24w^4}$ **115.** $\sqrt{(s^2)^{-2}}$ **116.** $\sqrt{(t^2)^{-4}}$

117. $\sqrt{\dfrac{98}{r^2}}$ **118.** $\sqrt{\dfrac{125}{x^4}}$ **119.** $\sqrt[5]{\dfrac{-64x}{x^6}}$ **120.** $\sqrt[5]{\dfrac{-96a^2}{a^7}}$

Use a calculator to evaluate. Round answers to three decimal places.

121. $\sqrt[3]{6.823}$ **122.** $\sqrt[5]{-8.141}$ **123.** $\sqrt{16.328}$

124. $\sqrt{29.079}$ **125.** $(1.82^{1/2})^{1/3}$ **126.** $(2.92^{1/3})^{1/2}$

7.3 Arithmetic Operations Involving Radicals

Radicals can be added, subtracted, multiplied, and divided if certain conditions are met. If two or more radicals are to be added or subtracted they must have the same index and the same radicand. Once this condition is met the sum or difference is found by adding or subtracting the coefficients.

Example 1   Simplify each of the following.

(a) $\sqrt{6} + 3\sqrt{6} = (1 + 3)\sqrt{6} = 4\sqrt{6}$ Distributive property

(b) $2\sqrt[3]{a} + 4\sqrt[3]{a} - 7\sqrt[3]{a} = (2 + 4 - 7)\sqrt[3]{a}$
$$= -\sqrt[3]{a}$$

(c) $a\sqrt[7]{b^2} + c\sqrt[7]{b^2} + 3a\sqrt[7]{b^2} = (a + c + 3a)\sqrt[7]{b^2}$
$$= (4a + c)\sqrt[7]{b^2} \quad\blacksquare$$

When the radicands and indexes are not exactly the same, they often can be made the same by using the properties of radicals.

Example 2   Add $\sqrt{32} + \sqrt{50}$.

First rewrite the two radicals in simplest radical form to see whether or not the radicands are the same.

$$\sqrt{32} + \sqrt{50} = \sqrt{16 \cdot 2} + \sqrt{25 \cdot 2}$$
$$= 4\sqrt{2} + 5\sqrt{2} \qquad \sqrt[n]{ab} = \sqrt[n]{a}\sqrt[n]{b}$$
$$= 9\sqrt{2} \quad \blacksquare$$

To Add Two Radicals

1. If the radicand and the index are the same, add their coefficients, simplify, and multiply by the radical.
2. If the radicand and index are *not* the same, use the properties of radicals to simplify each as much as possible. Combine any like terms generated by the method of Step 1.

In the remainder of this section assume that all variables represent positive real numbers.

Example 3   Simplify.

(a) $\sqrt{5} + \sqrt{20} - \sqrt{125} = \sqrt{5} + \sqrt{4 \cdot 5} - \sqrt{25 \cdot 5}$
$$= \sqrt{5} + 2\sqrt{5} - 5\sqrt{5}$$
$$= -2\sqrt{5}$$

(b) $\sqrt[3]{8x^5} - x\sqrt[3]{27x^2} = \sqrt[3]{2^3 x^3 x^2} - x\sqrt[3]{3^3 x^2}$
$$= 2x\sqrt[3]{x^2} - 3x\sqrt[3]{x^2}$$
$$= -x\sqrt[3]{x^2}$$

(c) $\dfrac{-3\sqrt{a^3 b^2}}{4} + \dfrac{ab\sqrt{a}}{2} - \dfrac{5ab\sqrt{a}}{3}$

$$= \dfrac{-3ab\sqrt{a}}{4} + \dfrac{ab\sqrt{a}}{2} - \dfrac{5ab\sqrt{a}}{3}$$

$$= \dfrac{-9ab\sqrt{a} + 6ab\sqrt{a} - 20ab\sqrt{a}}{12} \qquad \text{The LCD is 12}$$

$$= \dfrac{-23ab\sqrt{a}}{12}$$

(d) $\dfrac{-a + \sqrt{c^2 - 4d}}{2} + \dfrac{-a - \sqrt{c^2 - 4d}}{2}$

$$= \dfrac{-a + \sqrt{c^2 - 4d} - a - \sqrt{c^2 - 4d}}{2}$$

$$= \dfrac{-2a}{2} = -a \quad \blacksquare$$

Expressions involving radicals are multiplied or factored using the same methods used for polynomials. Recall from the laws of exponents and the definition of $a^{1/n}$ that

$$\sqrt[n]{a}\sqrt[n]{b} = a^{1/n} \cdot b^{1/n} = (a \cdot b)^{1/n} = \sqrt[n]{ab}.$$

Example 4   Find each of the following products.

(a) $\sqrt{2}\sqrt{8} = \sqrt{2 \cdot 8} = \sqrt{16} = 4$

(b) $3(2 + \sqrt{5}) = 6 + 3\sqrt{5}$ Distributive property

(c) $\sqrt{5}(\sqrt{3} + \sqrt{2}) = \sqrt{5}\sqrt{3} + \sqrt{5}\sqrt{2}$

$= \sqrt{15} + \sqrt{10}$ $\sqrt[n]{a}\sqrt[n]{b} = \sqrt[n]{ab}$

(d) $(\sqrt{x} + y)(\sqrt{x} - y) = (\sqrt{x})^2 - y^2$ $(a + b)(a - b) = a^2 - b^2$

$= x - y^2$ $(\sqrt{x})^2 = (x^{1/2})^2 = x$

(e) $(\sqrt{3} + \sqrt{5})^2 = (\sqrt{3})^2 + 2\sqrt{3}\sqrt{5} + (\sqrt{5})^2$

$= 3 + 2\sqrt{15} + 5$

$= 8 + 2\sqrt{15}$

(f) $(3\sqrt{2} - \sqrt{3})(\sqrt{6} - \sqrt{2})$

$= 3\sqrt{2}\sqrt{6} - 3(\sqrt{2})^2 - \sqrt{3}\sqrt{6} + \sqrt{3}\sqrt{2}$

$= 3\sqrt{12} - 3 \cdot 2 - \sqrt{18} + \sqrt{6}$

$= 6\sqrt{3} - 6 - 3\sqrt{2} + \sqrt{6}$ $\sqrt{12} = 2\sqrt{3}, \; \sqrt{18} = 3\sqrt{2}$ ■

Exponents can be particularly useful when simplifying radicals and finding their products.

Example 5   Use exponents to carry out the indicated operations.

(a) $\sqrt{x}\sqrt[3]{x} = x^{1/2}x^{1/3} = x^{1/2+1/3} = x^{5/6} = \sqrt[6]{x^5}$

(b) $\sqrt[4]{9} = \sqrt[4]{3^2} = 3^{2/4} = 3^{1/2} = \sqrt{3}$

(c) $\sqrt[3]{a}(\sqrt[6]{ab} + \sqrt[3]{ab^2}) = a^{1/3}[a^{1/6}b^{1/6} + a^{1/3}b^{2/3}]$

$= a^{1/3+1/6}b^{1/6} + a^{2/3}b^{2/3}$

$= a^{3/6}b^{1/6} + a^{2/3}b^{2/3}$

$= \sqrt[6]{a^3b} + \sqrt[3]{a^2b^2}$ ■

Example 6   Factor each of the following.

(a) $\sqrt[3]{ab} + 3\sqrt[3]{a} = \sqrt[3]{a}\sqrt[3]{b} + 3\sqrt[3]{a}$

$= \sqrt[3]{a}(\sqrt[3]{b} + 3)$

(b) $x + y$ as the sum of two cubes

Recall that $a^3 + b^3 = (a + b)(a^2 - ab + b^2)$.

$x + y = (\sqrt[3]{x})^3 + (\sqrt[3]{y})^3$

$= (\sqrt[3]{x} + \sqrt[3]{y})[(\sqrt[3]{x})^2 - \sqrt[3]{x}\sqrt[3]{y} + (\sqrt[3]{y})^2]$

$= (\sqrt[3]{x} + \sqrt[3]{y})(\sqrt[3]{x^2} - \sqrt[3]{xy} + \sqrt[3]{y^2})$

(c) $x - y$ as the difference of squares

$$x - y = (\sqrt{x})^2 - (\sqrt{y})^2$$
$$= (\sqrt{x} - \sqrt{y})(\sqrt{x} + \sqrt{y}) \quad \blacksquare$$

Example 7   Is $\dfrac{-1 + \sqrt{5}}{2}$ a solution to the equation $x^2 + x - 1 = 0$?

$$\left(\frac{-1 + \sqrt{5}}{2}\right)^2 + \left(\frac{-1 + \sqrt{5}}{2}\right) - 1 \overset{?}{=} 0 \qquad \text{Substitute } \frac{-1 + \sqrt{5}}{2} \text{ for } x$$

$$\frac{1 - 2\sqrt{5} + 5}{4} + \frac{-1 + \sqrt{5}}{2} - 1 \overset{?}{=} 0$$

$$1 - 2\sqrt{5} + 5 - 2 + 2\sqrt{5} - 4 \overset{?}{=} 0 \qquad \text{Multiply both sides by 4}$$

$$0 \overset{\checkmark}{=} 0$$

The equation is true, so $\dfrac{-1 + \sqrt{5}}{2}$ is a solution. $\blacksquare$

Exercises 7.3

Combine like radicals.

1. $2\sqrt{3} + 5\sqrt{3}$

2. $7\sqrt{2} + 6\sqrt{2}$

3. $\sqrt[3]{5} - 6\sqrt[3]{5}$

4. $11\sqrt[3]{7} - \sqrt[3]{7}$

5. $5x\sqrt{10} - x\sqrt{10}$

6. $8y\sqrt{13} + 2y\sqrt{13}$

7. $\sqrt[4]{3} + 2\sqrt[4]{2} - \sqrt[4]{3}$

8. $\sqrt[4]{2} - 3\sqrt[4]{5} + 3\sqrt[4]{2}$

9. $6x\sqrt{y} - 5x\sqrt{y}$

10. $3m\sqrt{2p} - m\sqrt{p} + \sqrt{2p}$

11. $u\sqrt[3]{3v} - u\sqrt[3]{v} + 2u\sqrt[3]{v}$

12. $7r\sqrt[3]{st} - 5\sqrt[3]{st} - r\sqrt[3]{st}$

Simplify and combine like radicals. Assume all variables represent positive real numbers.

13. $\sqrt{2} + \sqrt{8}$

14. $\sqrt{3} - \sqrt{27}$

15. $3\sqrt[3]{2} + 5\sqrt[3]{16}$

16. $\sqrt[3]{5} - \sqrt[3]{40}$

17. $\sqrt{5} + \sqrt{45} - \sqrt{80} + \sqrt{180}$

18. $\sqrt{3} - \sqrt{12} + 3\sqrt{8} + \sqrt{200}$

19. $\sqrt{75} - \sqrt{125} - 2\sqrt{108} + \sqrt{320}$

20. $\sqrt[5]{-64} + 3\sqrt[5]{2}$

21. $-7\sqrt[4]{162} + 3\sqrt[4]{32}$

22. $\sqrt{27} + \sqrt{288} - \sqrt{48} - \sqrt{162}$

23. $6\sqrt{32} - \sqrt{98} + 3\sqrt{96} - 2\sqrt{54}$

24. $\sqrt{7x^3} + 2x\sqrt{7x}$

25. $-\sqrt{27u^3} + 3u\sqrt{3u}$

26. $-5p\sqrt[3]{8p^5} + 3p^2\sqrt[3]{27p^2}$

27. $-v\sqrt[4]{2v^5} + v^2\sqrt[4]{32v}$

28. $\sqrt{\dfrac{7}{4}} + \dfrac{\sqrt{7}}{2}$

29. $\sqrt{\dfrac{5}{9}} - \dfrac{-2\sqrt{5}}{3}$

30. $7\sqrt{98} - (\sqrt{32} - 2\sqrt{18})$

31. $\sqrt{175} - (\sqrt{28} - \sqrt{63})$

32. $\dfrac{\sqrt{36a^2b^4c}}{2} - \dfrac{b\sqrt{16a^2b^2c}}{3}$

33. $\dfrac{y\sqrt{18xyz}}{5} + \dfrac{\sqrt{32xy^3z}}{4}$

34. $7\sqrt{300a^3} - 2a\sqrt{75a}$

35. $6\sqrt{600u^5} - 2u^2\sqrt{150u}$

36. $-7c\sqrt[3]{375cd^5} + 6d\sqrt[3]{81c^4d^2}$

37. $-5z\sqrt[3]{81y^5z^2} + 4y\sqrt[3]{24y^2z^5}$

38. $-5a\sqrt{75b^3} + \sqrt{18a^2b} + 7b\sqrt{108a^2b} - a\sqrt{50b}$

39. $3\sqrt{24x^3y} - y\sqrt{27x} + 11x\sqrt{54x^2y} - y\sqrt{48x}$

40. $\dfrac{-b + \sqrt{b^2 - 4ac}}{2a} + \dfrac{-b - \sqrt{b^2 - 4ac}}{2a}$

41. $\dfrac{c - \sqrt{c^2 - 4ad}}{2a} + \dfrac{c + \sqrt{c^2 - 4ad}}{2a}$

Reduce to lowest terms.

42. $\dfrac{6 + 2\sqrt{3}}{2}$
43. $\dfrac{14 + 10\sqrt{5}}{2}$
44. $\dfrac{30 - 15\sqrt{2}}{5}$

45. $\dfrac{18 - 21\sqrt{7}}{3}$
46. $\dfrac{-16 + 18\sqrt{3}}{-2}$
47. $\dfrac{-91 + 63\sqrt{11}}{-7}$

Multiply and reduce to simplest radical form. Assume all variables represent positive real numbers.

48. $2(3 + \sqrt{5})$
49. $3(5 - \sqrt{6})$
50. $\sqrt{2}(7 - \sqrt{2})$

51. $\sqrt{11}(3 - \sqrt{11})$
52. $3\sqrt{2}(\sqrt{3} - 2)$
53. $6\sqrt{3}(2\sqrt{2} - \sqrt{3})$

54. $\sqrt{x}(\sqrt{x} + \sqrt{y})$
55. $\sqrt{u}(\sqrt{v} - \sqrt{u})$
56. $\sqrt{ab}(\sqrt{a} - \sqrt{b})$

57. $(\sqrt{2} + \sqrt{3})(\sqrt{2} - \sqrt{3})$
58. $(\sqrt{5} - \sqrt{2})(\sqrt{5} + \sqrt{2})$

59. $\sqrt[3]{6} \cdot \sqrt[3]{36}$
60. $3\sqrt[3]{9} \cdot \sqrt[3]{9}$

61. $2\sqrt[3]{98x} \cdot \sqrt[3]{14x} \cdot \sqrt[3]{2x}$
62. $5\sqrt[4]{8a^3} \cdot 2\sqrt[4]{2a^2} \cdot \sqrt[4]{2^4 a^3}$

63. $(\sqrt{2} + \sqrt{5})^2$
64. $(\sqrt{7} - \sqrt{3})^2$

65. $(\sqrt{a + 2} - 1)^2$
66. $(\sqrt{c - 3} + 1)^2$

67. $(3\sqrt{5} - 2\sqrt{2})(3\sqrt{5} + 2\sqrt{2})$
68. $(7\sqrt{3} - 2\sqrt{6})(7\sqrt{3} + 2\sqrt{6})$

69. $\dfrac{-b + \sqrt{b^2 - 4ac}}{2a} \cdot \dfrac{-b - \sqrt{b^2 - 4ac}}{2a}$
70. $\dfrac{c - \sqrt{c^2 - 4ad}}{2a} \cdot \dfrac{c + \sqrt{c^2 - 4ad}}{2a}$

71. $(\sqrt[3]{a} + \sqrt[3]{b})(\sqrt[3]{a^2} - \sqrt[3]{ab} + \sqrt[3]{b^2})$
72. $(\sqrt[3]{s} - \sqrt[3]{t})(\sqrt[3]{s^2} + \sqrt[3]{st} + \sqrt[3]{t^2})$

Factor each of the following as indicated.

73. $3\sqrt{5} + 3 = 3(\quad\quad)$
74. $14\sqrt{2} - 7 = 7(\quad\quad)$

75. $3\sqrt{5} - x\sqrt{5} = \sqrt{5}(\quad\quad)$
76. $2\sqrt{12} + \sqrt{3} = \sqrt{3}(\quad\quad)$

77. $\sqrt[3]{xy} - \sqrt[3]{x} = \sqrt[3]{x}(\quad\quad)$
78. $3\sqrt[3]{81} - 10\sqrt[3]{6} = \sqrt[3]{3}(\quad\quad)$

79. $\sqrt[5]{xy} - \sqrt[5]{xz} = \sqrt[5]{x}(\quad\quad)$
80. $\sqrt[5]{ab} - \sqrt[5]{ac} = \sqrt[5]{a}(\quad\quad)$

81. Factor $a - b$ as the difference of squares.
82. Factor $m - n$ as the difference of squares.

83. Factor $r - s$ as the difference of cubes.
84. Factor $p + q$ as the sum of cubes.

Determine whether or not the given value is a solution to the equation.

85. $x^2 + x - 4 = 0$; $\dfrac{-1 - \sqrt{17}}{2}$
86. $y^2 - 2y - 4 = 0$; $1 + \sqrt{5}$

87. $3a^2 - a - 5 = 0$; $\dfrac{1 - \sqrt{61}}{3}$
88. $5v^2 - 2v - 1 = 0$; $\dfrac{1 - \sqrt{6}}{2}$

89. $p^2 + 9p + 1 = 0$; $\dfrac{-9 + \sqrt{77}}{2}$
90. $c^2 + 7c + 9 = 0$; $\dfrac{-7 + \sqrt{13}}{2}$

91. $5x^2 + 7x - 3 = 0$; $\dfrac{-7 - \sqrt{108}}{12}$
92. $8y^2 + 3y - 11 = 0$; $\dfrac{-3 - \sqrt{359}}{16}$

7.4 Division of Radicals; Rationalizing

Section 7.2 listed two conditions that had to be met for an expression involving radicals to be in simplest form. In this section we add two additional conditions.

A Radical Is in Simplest Form If:

1. No exponent on a factor in the radicand is greater than or equal to the index.
2. No radicand contains a fraction.
3. No fraction contains a radical in the denominator.
4. The index and the exponent in the radicand have no common factors.

The third condition is another way of considering the division of expressions involving radicals. Recall that

$$\sqrt[n]{\dfrac{a}{b}} = \dfrac{\sqrt[n]{a}}{\sqrt[n]{b}}.$$

This property can often be used to eliminate a radical from the denominator.

Example 1   Eliminate the radical in the denominator and simplify.

(a) $\dfrac{6\sqrt[3]{x^7}}{3\sqrt[3]{x^4}} = 2\sqrt[3]{\dfrac{x^7}{x^4}} = 2\sqrt[3]{x^3} = 2x$ $\sqrt[n]{b^n} = b$

(b) $\dfrac{4\sqrt{x}}{8\sqrt{x^3}} = \dfrac{1}{2}\sqrt{\dfrac{x}{x^3}} = \dfrac{1}{2}\sqrt{\dfrac{1}{x^2}} = \dfrac{1}{2x}$

(c) $\dfrac{\sqrt[5]{64x^7y^6}}{\sqrt[5]{2x^2y}} = \sqrt[5]{\dfrac{64x^7y^6}{2x^2y}} = \sqrt[5]{32x^5y^5} = 2xy$ ■

In Example 1 we used the second property of radicals to eliminate the radical from the denominator. The fundamental property of fractions also can be used to eliminate a radical from the denominator. When the denominator of a fraction is changed from an irrational number to a rational number, the process is called **rationalizing the denominator.**

Example 2   Simplify.

Irrational⟶ $\dfrac{3}{\sqrt{5}} = \dfrac{3}{\sqrt{5}} \cdot \dfrac{\sqrt{5}}{\sqrt{5}}$ Fundamental property

Rational⟶ $= \dfrac{3\sqrt{5}}{5}$ $\sqrt{5} \cdot \sqrt{5} = 5$ ■

Example 3   Simplify.

$$\dfrac{\sqrt{2}}{\sqrt{7}} = \dfrac{\sqrt{2}}{\sqrt{7}} \cdot \dfrac{\sqrt{7}}{\sqrt{7}} = \dfrac{\sqrt{14}}{7}$$ ■

To eliminate a radical from the denominator of a fraction, simplify first, then multiply numerator and denominator by an expression that will make the exponents in the radicand of the denominator equal to the index.

Example 4   Simplify.

(a) $\dfrac{2xy}{\sqrt[3]{y}}$

Since the index is 3, multiply by $\sqrt[3]{y^2}$.

$$\frac{2xy}{\sqrt[3]{y}} \cdot \frac{\sqrt[3]{y^2}}{\sqrt[3]{y^2}} = \frac{2xy\sqrt[3]{y^2}}{\sqrt[3]{y^3}} \qquad \sqrt[n]{a}\sqrt[n]{b} = \sqrt[n]{ab}$$

The exponent and index are equal

$$= \frac{2xy\sqrt[3]{y^2}}{y} \qquad \sqrt[3]{y^3} = y$$

$$= 2x\sqrt[3]{y^2} \qquad \text{Reduce}$$

(b) $\dfrac{6}{\sqrt[4]{a^5b^3}} = \dfrac{6}{a\sqrt[4]{ab^3}}$ Simplify first

$$= \frac{6}{a\sqrt[4]{ab^3}} \cdot \frac{\sqrt[4]{a^3b}}{\sqrt[4]{a^3b}}$$

$$= \frac{6\sqrt[4]{a^3b}}{a\sqrt[4]{a^4b^4}} \qquad \text{The exponent and index are equal}$$

$$= \frac{6\sqrt[4]{a^3b}}{a^2b} \quad \blacksquare$$

Example 5   Simplify.

(a) $\dfrac{\sqrt{x-y}}{\sqrt{x+y}} = \dfrac{\sqrt{x-y}}{\sqrt{x+y}} \cdot \dfrac{\sqrt{x+y}}{\sqrt{x+y}} = \dfrac{\sqrt{x^2-y^2}}{x+y}$

(b) $\sqrt[4]{\dfrac{7}{x}} = \sqrt[4]{\dfrac{7}{x} \cdot \dfrac{x^3}{x^3}} = \dfrac{\sqrt[4]{7x^3}}{x}$

(c) $\dfrac{\sqrt{3}}{\sqrt[4]{9}} = \dfrac{\sqrt{3}}{\sqrt[4]{3^2}} = \dfrac{\sqrt{3}}{\sqrt{3}} = 1$ The index and exponent have a common factor, 2

(d) $\sqrt{3} - \dfrac{2}{\sqrt{3}} = \dfrac{\sqrt{3}\sqrt{3} - 2}{\sqrt{3}}$ The LCD is $\sqrt{3}$

$$= \frac{3-2}{\sqrt{3}} = \frac{1}{\sqrt{3}} = \frac{\sqrt{3}}{3}$$

(e) $\dfrac{\sqrt{2}}{\sqrt{a}} - \dfrac{\sqrt{3}}{\sqrt{2}} + \sqrt{a} = \dfrac{\sqrt{2}\sqrt{2} - \sqrt{3}\sqrt{a} + \sqrt{a}\sqrt{2a}}{\sqrt{2a}}$ The LCD is $\sqrt{a}\sqrt{2}$ or $\sqrt{2a}$

$$= \frac{2 - \sqrt{3a} + a\sqrt{2}}{\sqrt{2a}}$$

$$= \frac{(2 - \sqrt{3a} + a\sqrt{2})}{\sqrt{2a}} \cdot \frac{\sqrt{2a}}{\sqrt{2a}}$$

$$= \frac{2\sqrt{2a} - a\sqrt{6} + 2a\sqrt{a}}{2a} \quad \blacksquare$$

When the denominator of a fraction contains a sum or difference, more thought is required to choose the expression needed to rationalize it. Recall that

$$(a - b)(a + b) = a^2 - b^2,$$
$$(a - b)(a^2 + ab + b^2) = a^3 - b^3$$

and $$(a + b)(a^2 - ab + b^2) = a^3 + b^3.$$

If $(\sqrt{x} - \sqrt{y})$ is multiplied by $(\sqrt{x} + \sqrt{y})$ the product is $(\sqrt{x})^2 - (\sqrt{y})^2 = x - y$; $x, y \geq 0$. If the irrational number $1 + \sqrt{7}$ is multiplied by the irrational number $1 - \sqrt{7}$, the product is $1^2 - (\sqrt{7})^2 = 1 - 7 = -6$, a rational number. The expressions $1 + \sqrt{7}$ and $1 - \sqrt{7}$ are said to be **conjugates** of each other.

> The **conjugate** of a binomial involving square roots is a binomial of the same terms with the opposite sign between them.

Example 6   Simplify.

(a) $\dfrac{2}{\sqrt{5} - \sqrt{2}} = \dfrac{2}{\sqrt{5} - \sqrt{2}} \cdot \dfrac{\sqrt{5} + \sqrt{2}}{\sqrt{5} + \sqrt{2}}$ Multiply by the conjugate $\sqrt{5} + \sqrt{2}$

$$= \frac{2(\sqrt{5} + \sqrt{2})}{5 - 2} = \frac{2(\sqrt{5} + \sqrt{2})}{3}$$

(b) $\dfrac{\sqrt{3} + \sqrt{6}}{\sqrt{2} + \sqrt{3}} = \dfrac{\sqrt{3} + \sqrt{6}}{\sqrt{2} + \sqrt{3}} \cdot \dfrac{\sqrt{2} - \sqrt{3}}{\sqrt{2} - \sqrt{3}}$

$$= \frac{\sqrt{3}\sqrt{2} - \sqrt{3}\sqrt{3} + \sqrt{6}\sqrt{2} - \sqrt{6}\sqrt{3}}{2 - 3}$$

$$= \frac{\sqrt{6} - 3 + 2\sqrt{3} - 3\sqrt{2}}{-1} = -(\sqrt{6} - 3 + 2\sqrt{3} - 3\sqrt{2}) \quad \blacksquare$$

Example 7   Simplify $\dfrac{1}{\sqrt[3]{x} + \sqrt[3]{y}}$.

To eliminate the radicals from the denominator, multiply by $\sqrt[3]{x^2} - \sqrt[3]{xy} + \sqrt[3]{y^2}$.

$$\frac{1}{\sqrt[3]{x} + \sqrt[3]{y}} \cdot \frac{\sqrt[3]{x^2} - \sqrt[3]{xy} + \sqrt[3]{y^2}}{\sqrt[3]{x^2} - \sqrt[3]{xy} + \sqrt[3]{y^2}}$$

$$= \frac{\sqrt[3]{x^2} - \sqrt[3]{xy} + \sqrt[3]{y^2}}{x + y} \quad \blacksquare$$

Exercises 7.4

Rationalize the denominator and simplify if possible. Assume all variables represent positive real numbers.

1. $\dfrac{1}{\sqrt{2}}$

2. $\dfrac{1}{\sqrt{3}}$

3. $\dfrac{6}{\sqrt{3}}$

4. $\dfrac{4}{\sqrt{2}}$

5. $\dfrac{15}{\sqrt{5}}$

6. $\dfrac{48}{\sqrt{6}}$

7. $\dfrac{-28}{\sqrt{7}}$

8. $\dfrac{-44}{\sqrt{11}}$

9. $\sqrt{\dfrac{1}{3}}$

10. $\sqrt{\dfrac{3}{7}}$

11. $-\sqrt{\dfrac{5}{6}}$

12. $-\sqrt{\dfrac{8}{13}}$

13. $\dfrac{3\sqrt[3]{5}}{\sqrt[3]{3}}$

14. $\dfrac{10\sqrt[3]{2}}{\sqrt[3]{5}}$

15. $\dfrac{-6\sqrt[3]{7}}{5\sqrt[3]{6}}$

16. $\dfrac{-11\sqrt[3]{5}}{10\sqrt[3]{3}}$

17. $\dfrac{x}{\sqrt[4]{x^2}}$

18. $\dfrac{y}{\sqrt[4]{y^2}}$

19. $\dfrac{2a}{\sqrt[3]{a}}$

20. $\dfrac{4p}{\sqrt[3]{2p}}$

21. $\dfrac{x+y}{\sqrt{x+y}}$

22. $\dfrac{s-t}{\sqrt{s-t}}$

23. $\sqrt[3]{\dfrac{3}{r}}$

24. $\sqrt[3]{\dfrac{a}{b}}$

25. $\sqrt[4]{\dfrac{m}{n}}$

26. $\sqrt[4]{\dfrac{5}{k}}$

27. $\dfrac{u}{\sqrt[5]{u}}$

28. $\dfrac{wz}{\sqrt[5]{wz}}$

29. $\sqrt[3]{2} \div \sqrt[3]{4}$

30. $\sqrt[3]{5} \div \sqrt[3]{25}$

31. $\sqrt[4]{3x} \div \sqrt[4]{27x^3}$

32. $\sqrt[4]{2a^2} \div \sqrt[4]{32a^5}$

33. $16\sqrt[5]{w^3} \div \sqrt[5]{2w^6}$

34. $27\sqrt[5]{k^4} \div \sqrt[5]{27k^7}$

Perform the indicated operations and simplify if possible. Assume all variables represent positive real numbers.

35. $\dfrac{6}{\sqrt{3}} + 4\sqrt{3}$

36. $\dfrac{3}{\sqrt{2}} - \sqrt{2}$

37. $\sqrt{5} - \dfrac{1}{\sqrt{5}}$

38. $\sqrt{a} - \dfrac{a}{\sqrt{a}}$

39. $\sqrt{3p} + \dfrac{p}{\sqrt{3p}}$

40. $\sqrt{5u} - \dfrac{u}{\sqrt{5u}}$

41. $\dfrac{4}{\sqrt{2}} + 2 - \dfrac{3}{\sqrt{3}} - \dfrac{5}{\sqrt{3}}$

42. $\dfrac{3a}{\sqrt{a}} + \dfrac{5}{\sqrt{2}} - 7\sqrt{2} + \dfrac{2a}{\sqrt{a}}$

43. $\dfrac{6w}{\sqrt{2w}} - \dfrac{9\sqrt{3}}{\sqrt{2}} + \dfrac{3w}{5\sqrt{2w}} - \dfrac{3\sqrt{6}}{2}$

44. $\dfrac{7\sqrt{5}}{\sqrt{3}} - \dfrac{2\sqrt{xy}}{\sqrt{x}} + \dfrac{6\sqrt{15}}{2} + \dfrac{3y}{\sqrt{y}}$

45. $\dfrac{x}{\sqrt{x^2+1}} - \dfrac{\sqrt{x^2+1}}{x}$

46. $\dfrac{3a}{\sqrt{a^2-1}} + \dfrac{\sqrt{a^2-1}}{2a}$

Find the quotient and simplify where possible. Assume all variables represent positive real numbers.

47. $\dfrac{15\sqrt{24} - 6\sqrt{12}}{3\sqrt{3}}$

48. $\dfrac{20\sqrt{10} - 45\sqrt{6}}{6\sqrt{2}}$

49. $\dfrac{6a\sqrt{5} + 8\sqrt{2a^2}}{a\sqrt{2}}$

50. $\dfrac{40t\sqrt{7} - 45\sqrt{15t^2}}{t\sqrt{3}}$

51. $\dfrac{3 + \sqrt{x}}{\sqrt{x}}$

52. $\dfrac{5 - \sqrt{p}}{\sqrt{2p}}$

53. $\dfrac{\sqrt{5w} + \sqrt{6z}}{\sqrt{30wz}}$

54. $\dfrac{\sqrt{7r} - \sqrt{3s}}{\sqrt{21rs}}$

Rationalize the denominators and simplify where possible. Assume all variables represent positive real numbers.

55. $\dfrac{\sqrt{3}}{4 - \sqrt{7}}$

56. $\dfrac{\sqrt{2}}{4 + \sqrt{6}}$

57. $\dfrac{-1}{4\sqrt{3} - \sqrt{5}}$

58. $\dfrac{-1}{2\sqrt{3} - 5\sqrt{2}}$

59. $\dfrac{-2\sqrt{3}}{4\sqrt{3} - 7}$

60. $\dfrac{6\sqrt{5}}{2\sqrt{6} - 5}$

61. $\dfrac{3 - \sqrt{2}}{3 + \sqrt{2}}$

62. $\dfrac{4 + \sqrt{5}}{4 - \sqrt{5}}$

63. $\dfrac{-(3 + \sqrt{5})}{7 - 5\sqrt{2}}$

64. $\dfrac{-(5 + \sqrt{3})}{2\sqrt{5} - 4\sqrt{3}}$

65. $\dfrac{\sqrt{6} - \sqrt{3}}{\sqrt{6} + 2\sqrt{3}}$

66. $\dfrac{\sqrt{2} - \sqrt{6}}{3\sqrt{2} + \sqrt{6}}$

67. $\dfrac{x + y}{\sqrt{x} + \sqrt{y}}$

68. $\dfrac{a - b}{\sqrt{a} - \sqrt{b}}$

69. $\dfrac{3\sqrt{u} - \sqrt{v}}{2\sqrt{u} + \sqrt{v}}$

70. $\dfrac{6\sqrt{k} - 5\sqrt{k}}{3\sqrt{k} + 5\sqrt{k}}$

71. $\dfrac{9\sqrt{xy} - \sqrt{ab}}{9\sqrt{xy} + \sqrt{ab}}$

72. $\dfrac{6\sqrt{2r} - \sqrt{5s}}{\sqrt{2r} + \sqrt{5s}}$

73. $\dfrac{1}{\sqrt[3]{a} - \sqrt[3]{b}}$

74. $\dfrac{1}{\sqrt[3]{x} + \sqrt[3]{y}}$

75. $\dfrac{1}{\sqrt{m + n} - \sqrt{p}}$

76. $\dfrac{1}{\sqrt{u - v} + \sqrt{w}}$

77. $\dfrac{1}{\sqrt[3]{s^2} + \sqrt[3]{st} + \sqrt[3]{t^2}}$

78. $\dfrac{1}{\sqrt[3]{x^2} - \sqrt[3]{xy} + \sqrt[3]{y^2}}$

Rationalize each denominator and simplify where possible.

79. $\dfrac{a^{1/2} + a^{2/3}}{a^{1/3}}, \quad a \neq 0$

80. $\dfrac{x^{3/5} - x^{1/3}}{x^{2/3}}, \quad x \neq 0$

81. $\dfrac{x^{1/3}y^{1/3} - x^{1/5}y^{1/5}}{(xy)^{2/5}}, \quad x, y \neq 0$

82. $\dfrac{r^{1/7}s^{1/3} - r^{1/5}s^{3/5}}{r^{3/5}s^{2/7}}, \quad r, s \neq 0$

83. $\dfrac{(x - y)^{2/3} + (x - y)^{4/3}}{(x - y)^{2/3}}, \quad x \neq y$

84. $\dfrac{(w^2 - z^2)^{1/3} - (w^2 - z^2)^{5/3}}{(w^2 - z^2)^{1/3}}, \quad w \neq \pm z$

Rationalize each numerator.

85. $\dfrac{\sqrt{3}}{3\sqrt{2} + 9\sqrt{6}}$

86. $\dfrac{2\sqrt{5}}{5 + 10\sqrt{3}}$

87. $\dfrac{\sqrt{3} - \sqrt{2}}{\sqrt{3} + \sqrt{2}}$

88. $\dfrac{\sqrt{5} + \sqrt{7}}{\sqrt{5} - \sqrt{7}}$

89. $\dfrac{2\sqrt{3} - 3\sqrt{2}}{2\sqrt{3} + 3\sqrt{2}}$

90. $\dfrac{5\sqrt{6} - 7\sqrt{5}}{3\sqrt{6} + 4\sqrt{5}}$

Review Exercises

Solve each equation by factoring.

91. $a^2 - 4a - 5 = 0$

92. $5t^2 + 4t - 1 = 0$

93. $4w^2 + 12w + 9 = 0$

94. $y^2 - 2y - 3 = 12$

95. $5k^2 - 10k - 15 = 0$

96. $3m^3 - 3m = 0$

97. $\dfrac{1}{2}t^2 + \dfrac{1}{2}t - 3 = 0$

98. $a^2 + \dfrac{1}{3}a - \dfrac{1}{2} = \dfrac{1}{2}a - \dfrac{1}{3}$

99. $3(4 - 2s)^2 + (s + 2)^2 = 37$

100. $\left[\dfrac{1}{2}p - 1\right]^2 + 4 = 4p - 11$

7.5   Equations Involving Radicals

Equations involving radicals are solved by using the **power property of equality.**

If A and B are two algebraic expressions with $A = B$, then
$$A^n = B^n, \quad n \in N.$$

When $n = 2$, the property is called the **squaring property.** For example,

$3 = 3$
$3^2 = 3^2$ Squaring property
$9 = 9.$

When $n = 3$ the property is called the **cubing property.**

$2 = 2$
$2^3 = 2^3$ Cubing property
$8 = 8$

Using the power property to solve an equation can result in a false solution to the equation. Consider what happens when both sides of the equation $x = 2$ are squared.

$x = 2$
$x^2 = 2^2$ Squaring property
$x^2 = 4$

The new equation has two solutions, but the original equation had only one solution.

$$x^2 = 4$$
$$x^2 - 4 = 0$$
$$(x - 2)(x + 2) = 0$$
$$x - 2 = 0 \quad \text{or} \quad x + 2 = 0$$
$$x = 2 \quad \text{or} \quad x = -2$$

Notice that squaring both sides has introduced the solution $x = -2$, which does not satisfy the original equation. Such a solution is said to be **extraneous.** The introduction of extraneous solutions when applying the power property of equality necessitates that we *check all solutions* found in this manner.

Theorem 7.2

When the power property of equality is applied to an equation, the solution set of the original equation will always be a subset of the solution set of the new equation.

Theorem 7.2 states that all, part, or none of the solutions to an equation solved by the power property may be extraneous.

Example 1   Solve $\sqrt{x+2} = 4$.

To eliminate the radical, square both sides.

$$(\sqrt{x+2})^2 = 4^2$$
$$x + 2 = 16$$
$$x = 14$$

Check: $\sqrt{14 + 2} \overset{?}{=} 4 \qquad x = 14$
$$\sqrt{16} \overset{?}{=} 4$$
$$4 \overset{\checkmark}{=} 4.$$

The solution set is {14}.   ■

Example 2   Solve $\sqrt{x-2} = 2 - x$.

$$(\sqrt{x-2})^2 = (2-x)^2 \qquad \text{Squaring property}$$
$$x - 2 = 4 - 4x + x^2$$
$$0 = x^2 - 5x + 6$$
$$0 = (x-2)(x-3)$$
$$x - 2 = 0 \quad \text{or} \quad x - 3 = 0$$
$$x = 2 \quad \text{or} \qquad x = 3$$

Check: $\sqrt{2-2} \overset{?}{=} 2 - 2 \qquad\qquad \sqrt{3-2} \overset{?}{=} 2 - 3$
$$\sqrt{0} \overset{?}{=} 0 \qquad\qquad\qquad \sqrt{1} \overset{?}{=} -1$$
$$0 \overset{\checkmark}{=} 0 \qquad\qquad\qquad 1 \neq -1$$

Since only the 2 checks, the solution set is {2}. Note that {2} $\subseteq$ {2, 3}, as is asserted by Theorem 7.2.   ■

Example 3   Solve $\sqrt[4]{1-x} = 4$.

To eliminate the radical, raise both sides to the 4th power.

$$(\sqrt[4]{1-x})^4 = 4^4$$
$$1 - x = 256$$
$$-x = 255$$
$$x = -255.$$

The check is left to the reader.   ■

Some equations involving radicals require that the power property be applied more than once to eliminate the radicals.

Example 4   Solve $\sqrt{3x+1} = \sqrt{x} - 1$.

Square both sides.

$$(\sqrt{3x+1})^2 = (\sqrt{x} - 1)^2$$
$$3x + 1 = x - 2\sqrt{x} + 1 \qquad (a-b)^2 = a^2 - 2ab + b^2$$

Now isolate the term involving the radical on one side of the equation and square again.

$$2x = -2\sqrt{x}$$
$$(2x)^2 = (-2\sqrt{x})^2 \qquad \text{Squaring property}$$
$$4x^2 = 4x$$
$$4x^2 - 4x = 0$$
$$4x(x - 1) = 0$$
$$4x = 0 \quad \text{or} \quad x - 1 = 0$$
$$x = 0 \quad \text{or} \qquad x = 1$$

Check: $\sqrt{3 \cdot 0 + 1} \stackrel{?}{=} \sqrt{0} - 1 \qquad \sqrt{3 \cdot 1 + 1} \stackrel{?}{=} \sqrt{1} - 1$

$$\sqrt{1} \stackrel{?}{=} -1 \qquad\qquad\qquad \sqrt{4} \stackrel{?}{=} \sqrt{1} - 1$$
$$1 \neq -1 \qquad\qquad\qquad\qquad 2 \neq 0$$

The solution set is $\emptyset$.   ■

Example 5   Solve for u. Note any restrictions on the variable first.

$$\frac{2}{\sqrt{u + 1}} = \sqrt{u + 1} + \sqrt{u + 2}$$

We know that $\sqrt{u + 1}$ will be a real number if $u \geq -1$, and $\sqrt{u + 2}$ will be a real number if $u \geq -2$. The intersection of these sets is $u \geq -1$. Division by zero would result if $u = -1$. Thus u must be greater than -1.

$$2 = u + 1 + (\sqrt{u + 2})(\sqrt{u + 1}) \qquad \text{LCD} = \sqrt{u + 1}$$
$$2 = u + 1 + (\sqrt{u^2 + 3u + 2})$$

To solve for u, isolate the radical.

$$-u + 1 = \sqrt{u^2 + 3u + 2}$$
$$u^2 - 2u + 1 = u^2 + 3u + 2 \qquad \text{Squaring property}$$
$$-5u = 1$$
$$u = -\frac{1}{5}$$

The solution checks. The solution set is $\{-\frac{1}{5}\}$.   ■

Exercises 7.5

Solve each of the following. Check each answer. Write $\emptyset$ if no real number solution exists.

1. $\sqrt{x} = 2$

2. $\sqrt{y} = 3$

3. $\sqrt{a} = -1$

4. $\sqrt{p} = -16$

5. $\sqrt{w + 1} = 5$

6. $\sqrt{s - 1} = 2$

7. $\sqrt{2t - 3} = 1$

8. $\sqrt{3k + 1} = 1$

9. $\sqrt{2y - 1} = -2$

10. $\sqrt{5m - 1} = 2$

11. $\sqrt{3x - 1} = 2$

12. $\sqrt{2m + 1} = 3$

13. $\sqrt{2u - 4} = -4$

14. $\sqrt{4v - 5} = -5$

15. $\sqrt{1 - 3s} = 6$

16. $\sqrt{2 - t} = 1$

17. $-\sqrt{3y} = -5$

18. $-\sqrt{2w} = -7$

19. $\sqrt[3]{x} = 2$

20. $\sqrt[3]{t} = 1$

21. $\sqrt[3]{2s} = -1$

22. $\sqrt[3]{5k} = -2$

23. $\sqrt[5]{-a} = 2$

24. $\sqrt[4]{-x} = 2$

25. $\sqrt[4]{2v} = -1$ **26.** $\sqrt[4]{3y} = -2$ **27.** $\sqrt[4]{-x} = 3$ **28.** $\sqrt[4]{1+y} = 3$

29. $-\sqrt[3]{2a-1} = -2$ **30.** $-\sqrt[3]{3c+2} = -3$ **31.** $\sqrt{x+1} = \sqrt{3}$ **32.** $\sqrt{t-2} = \sqrt{3}$

33. $\sqrt{3w-1} = \sqrt{w}$ **34.** $\sqrt{4s-3} = \sqrt{s}$ **35.** $\sqrt{7y+3} = \sqrt{2y+5}$

36. $\sqrt{u-4} = \sqrt{2u-17}$ **37.** $\sqrt{3x+2} = \sqrt{x+4}$ **38.** $\sqrt{v+4} = -\sqrt{2v+1}$

39. $-\sqrt{3t+5} = \sqrt{t+2}$ **40.** $\sqrt[3]{k-1} = \sqrt[3]{2k}$ **41.** $\sqrt[3]{3a-5} = \sqrt[3]{a}$

42. $\sqrt[3]{7w+2} = -\sqrt[3]{2w}$ **43.** $\sqrt[3]{7m-3} = -\sqrt[3]{m}$ **44.** $\sqrt{x+5} = \sqrt{x-7}$

45. $\sqrt{9v-2} = \sqrt{9v-1}$ **46.** $\sqrt{4k-5} = 2\sqrt{k}$ **47.** $\sqrt{25u+3} = 5\sqrt{u}$

48. $\sqrt{36s-5} = 6\sqrt{s}$ **49.** $2\sqrt[3]{k} = \sqrt[3]{8k+2}$ **50.** $\sqrt[3]{16m} = 3\sqrt[3]{2m+3}$

51. $\sqrt[3]{5s+1} = \sqrt[3]{2s+1}$ **52.** $\sqrt[4]{k-1} = 2\sqrt[4]{k-16}$

Each of the following becomes a factorable quadratic equation. Solve and check each answer. Exclude answers that result in even roots of negative numbers.

53. $u - 5 = \sqrt{u-3}$ **54.** $\sqrt{p-4} = 4 - p$ **55.** $m = 1 + \sqrt{m+11}$

56. $4 - k = -\sqrt{k+2}$ **57.** $\sqrt{a}\sqrt{a-5} = 6$ **58.** $\sqrt{v}\sqrt{v-6} = 4$

59. $\sqrt{x-6} = \dfrac{4}{\sqrt{x}}$ **60.** $\sqrt{y-9} = \dfrac{6}{\sqrt{y}}$ **61.** $\sqrt{2y-1} = 2y - 3$

62. $\sqrt{t-1} - 2 = t - 3$ **63.** $\sqrt{t+2} - 1 = \sqrt{2t-10}$ **64.** $\sqrt{s} = \sqrt{2s+1} - 1$

65. $\sqrt{p} = p - 6$ **66.** $\sqrt{y} = y - 8$ **67.** $\sqrt[3]{v^2 - 1} = 2$

68. $\sqrt[3]{x^2 - 9} = 3$ **69.** $\sqrt[4]{y^2 - 8} = 1$ **70.** $\sqrt[4]{m^2 - 19} = 3$

7.6   Complex Numbers

In the earlier sections of this chapter we considered expressions such as $\sqrt{4}, \sqrt[4]{16}$, and the general expression $\sqrt[n]{b}$. When n was even, b was restricted to the set of positive real numbers. In this section we will consider numbers such as $\sqrt{-1}$.

$$\sqrt{-1} \neq 1 \text{ since } 1 \cdot 1 = 1 \neq -1$$
$$\sqrt{-1} \neq -1 \text{ since } (-1)(-1) = 1 \neq -1$$

Since the square of any real number is positive or zero, $\sqrt{-1}$ cannot be a real number.

In order to give meaning to the square root of a negative number, mathematicians introduce the imaginary unit, i, which is defined as follows.

$$\sqrt{-1} = i \quad \text{or} \quad i^2 = -1$$

Assuming the rules for exponents hold for the number "i", we have

$$i = \sqrt{-1}$$
$$i^2 = -1$$
$$i^3 = i^2 \cdot i = -1 \cdot i = -i$$
$$i^4 = (i^2)^2 = (-1)^2 = 1$$

$$i^5 = i^4 \cdot i = 1 \cdot i = i$$
$$i^6 = (i^2)^3 = (-1)^3 = -1$$
$$i^7 = i^6 \cdot i = -1 \cdot i = -i$$
$$i^8 = (i^4)^2 = 1^2 = 1$$

Only four results occur when i is raised to a positive integer power: i, -1, $-i$, 1. Since $i^4 = 1$, it is easy to find i^n for any positive integer.

Example 1   Find i^{67}.

Since $67 = 4 \cdot 16 + 3$,

$$i^{67} = (i^4)^{16} \cdot i^3 = (1)^{16} \cdot i^3 = -1. \qquad i^4 = 1, \quad i^3 = -i \quad \blacksquare$$

To find the value of i raised to a positive integer power, divide the integer power by 4 and use the remainder as a power of i. Since $i^4 = 1$, i to the remainder power is equivalent to i to the original power.

Example 2   Simplify.

(a) $i^{29} = (i^4)^7 \cdot i^1 = (1)^7 \cdot i^1 = i$ The remainder is 1

(b) $i^{322} = (i^4)^{80} \cdot i^2 = (1)^{80} \cdot i^2 = -1$ The remainder is 2

(c) $i^{123} = (i^4)^{30} \cdot i^3 = (1)^{30} \cdot i^3 = -i$ The remainder is 3

(d) $i^{80} = (i^4)^{20} \cdot i^0 = (1)^{20} \cdot i^0 = 1$ The remainder is 0 $\blacksquare$

Properties of radicals can now be used to simplify expressions such as $\sqrt{-4}$, $\sqrt{-9}$, or $\sqrt{-45}$. Recall that $\sqrt{ab} = \sqrt{a}\sqrt{b}$. If we extend this property to cover negative radicands, then

$$\sqrt{-4} = \sqrt{4(-1)} = \sqrt{4}\sqrt{-1} = 2i$$
$$\sqrt{-9} = \sqrt{9(-1)} = \sqrt{9}\sqrt{-1} = 3i$$
$$\sqrt{-45} = \sqrt{9 \cdot 5(-1)} = \sqrt{9}\sqrt{5}\sqrt{-1} = 3\sqrt{5}i.$$

The process outlined above is generally shortened as follows.

$$\sqrt{-b} = \sqrt{b}\,i, \quad b > 0$$

▶ **Caution** $\sqrt[n]{a}\sqrt[n]{b} = \sqrt[n]{a \cdot b}$ *does not apply* if a and b are both negative. The radicals must be converted to the i form first.

Right	*Wrong*
$\sqrt{-9}\sqrt{-4} = 3i \cdot 2i$	$\sqrt{-4}\sqrt{-9} = \sqrt{(-4)(-9)}$
$\qquad = 6i^2$	$\qquad = \sqrt{36}$
$\qquad = -6 \quad i^2 = -1$	$\qquad = 6$

Numbers such as bi, where b is a real number, are a subset of a larger set called the **complex numbers.**

A **complex number** is any number of the form $a + bi$ where a, $b \in R$.

A complex number consists of a real part and an imaginary part.

$$a + bi$$

Real part   Imaginary part

Thus, $3 + 2i$, $4 - 3i$, $5 + 6i$, and $2 - \sqrt{-3} = 2 - \sqrt{3}i$ are all complex numbers. The set of real numbers is a subset of the set of complex numbers, since real numbers are just complex numbers in which the imaginary part is zero.

$$2 = 2 + 0i$$
Real numbers $3 = 3 + 0i$ Complex form
$$5 = 5 + 0i$$

If the real part is zero, the numbers are said to be **pure imaginary numbers.**

$$2i = 0 + 2i$$
Pure imaginary $3i = 0 + 3i$ Complex form
numbers $5i = 0 + 5i$

The properties of real numbers stated in Chapter 1 hold for the complex numbers as well. We now define equality, addition, and multiplication of complex numbers.

1. **Equality** $a + bi = c + di$ if and only if $a = c$ and $b = d$
2. **Addition** $a + bi + (c + di) = (a + c) + (b + d)i$
3. **Multiplication** $(a + bi)(c + di) = (ac - bd) + (ad + bc)i$

The definition of equality states that *two complex numbers are equal only when their real parts are equal and their imaginary parts are equal.* This definition is often used to solve equations involving complex numbers.

Example 3   Given $2x + yi = 4 + 7i$, solve for x and y.

$2x = 4$, $y = 7$ The real parts, $2x$ and 4, are equal
$x = 2$ The imaginary parts, y and 7, are equal   ■

Example 4   Solve $(3 + 2i)x + (4 - 2i)y = 2 + 6i$.
First rewrite the equation so the parts can be identified.

$3x + 2xi + 4y - 2yi = 2 + 6i$ Distributive property
$(3x + 4y) + (2x - 2y)i = 2 + 6i$

(a) $3x + 4y = 2$ The real parts are equal

(b) $2x - 2y = 6$ The imaginary
parts are equal

The equations (a) and (b) form a system of equations that can be solved by any of the methods of Chapter 3. The solution is $x = 2$, $y = -1$.   ■

To **add two complex numbers,** treat them as binomials and add the sum of their real parts to i times the sum of their imaginary parts.

Example 5   Add.

(a) $(4 + 6i) + (3 - 2i)$

The real parts are 4 and 3. The imaginary parts are 6 and -2.

$$(4 + 6i) + (3 - 2i) = (4 + 3) + (6 - 2)i$$
$$= 7 + 4i$$

(b) $(8 + 6i) + (3 + 4i) - (5 + 2i)$

Remove symbols of grouping.

$$8 + 6i + 3 + 4i - 5 - 2i$$
$$= \quad (8 + 3 - 5) \quad + \quad (6 + 4 - 2)i$$

$\qquad\qquad$ Add the real parts $\quad$ Add the imaginary parts

$$= 6 + 8i \quad \blacksquare$$

To **multiply two complex numbers,** treat them as binomials and simplify any powers of i that may occur.

Example 6   Multiply $(3 + 5i)(6 + 3i)$.

$$(3 + 5i)(6 + 3i) = 18 + 9i + 30i + 15i^2$$
$$\qquad\qquad\qquad\quad \uparrow \quad\, \uparrow \quad\, \uparrow \qquad \uparrow$$
$$\qquad\qquad\qquad\quad F \quad O \quad I \qquad L$$
$$= 18 + 9i + 30i - 15 \qquad i^2 = -1$$
$$= 3 + 39i \quad \blacksquare$$

Example 7   Expand $(2 + i)^3$.

$$(2 + i)^3 = (2 + i)^2(2 + i)$$
$$= (4 + 4i + i^2)(2 + i) \qquad (a + b)^2 = a^2 + 2ab + b^2$$
$$= (3 + 4i)(2 + i) \qquad\quad i^2 = -1, \quad 4 - 1 = 3$$
$$= 6 + 11i + 4i^2 \qquad\qquad \text{FOIL}$$
$$= 2 + 11i \quad \blacksquare$$

The process used to divide two complex numbers is similar to the process used to rationalize the denominator. To use this process, observe the following property.

The product of two complex conjugates is a real number.

$$(a + bi)(a - bi) = a^2 - b^2i^2 = a^2 + b^2$$

Example 8   Divide $\dfrac{6 + i}{3 + 4i}$. Write the answer in the form $a + bi$.

Multiply numerator and denominator by the conjugate of the denominator.

$$\dfrac{6 + i}{3 + 4i} = \dfrac{6 + i}{3 + 4i} \cdot \dfrac{3 - 4i}{3 - 4i} \qquad \text{Fundamental property}$$

$$= \dfrac{18 - 21i - 4i^2}{9 + 16}$$

$$= \dfrac{22 - 21i}{25} = \dfrac{22}{25} - \dfrac{21}{25}i$$

To show that the quotient is correct, we need only show that the product of the quotient and the divisor is equal to the dividend. That is, if $\frac{a}{b} = c$, then $b \cdot c = a$.

$$\left(\dfrac{22}{25} - \dfrac{21i}{25}\right)(3 + 4i) = \dfrac{66}{25} + \dfrac{88i}{25} - \dfrac{63i}{25} - \dfrac{84i^2}{25}$$

$$= \dfrac{150}{25} + \dfrac{25i}{25}$$

$$\overset{\checkmark}{=} 6 + i \quad \blacksquare$$

Example 9   Divide $\dfrac{\sqrt{5} - \sqrt{-3}}{\sqrt{-2} + \sqrt{3}}$. Write the answer in the form $a + bi$.

First write $\sqrt{-2}$ as $\sqrt{2}i$ and $\sqrt{-3}$ as $\sqrt{3}i$.

$$\dfrac{\sqrt{5} - \sqrt{3}i}{\sqrt{2}i + \sqrt{3}}$$

Now multiply the numerator and denominator by the conjugate of the denominator, $\sqrt{2}i - \sqrt{3}$.

$$\dfrac{\sqrt{5} - \sqrt{3}i}{\sqrt{2}i + \sqrt{3}} \cdot \dfrac{\sqrt{2}i - \sqrt{3}}{\sqrt{2}i - \sqrt{3}} = \dfrac{(\sqrt{6} - \sqrt{15}) + (3 + \sqrt{10})i}{-2 - 3}$$

$$= \dfrac{\sqrt{15} - \sqrt{6}}{5} - \dfrac{(3 + \sqrt{10})i}{5} \quad \blacksquare$$

Exercises 7.6

Simplify.

1. i^2
2. $-i^2$
3. $-i^3$
4. i^3
5. i^4

6. $-i^4$
7. $-i^5$
8. i^5
9. i^{20}
10. i^{28}

11. i^{101}
12. i^{97}
13. i^{-8}
14. i^{-4}

Perform the indicated operations and write each answer in the form a + bi.

15. $(2 + i) + (3 - 2i)$

16. $(5 - i) + (2 + 7i)$

17. $(6 - 11i) - (2 - i)$

18. $(5 - 3i) - (-1 - 2i)$

19. $(-15 - 2i) - (-3 - 2i)$

20. $(-6 + 5i) - (-6 + 3i)$

21. $(-6 + i) - 2(3 + i)$

22. $-8(1 - i) + 4(3 - 2i)$

23. $i(3 + 4i)$

24. $2i(1 - i)$

25. $(1 + i)(1 - i)$

26. $(3 - i)(3 + i)$

27. $(5 - 7i)(6 - 11i)$

28. $(3 - 13i)(5 + 2i)$

29. $(6 - 5i)^2$

30. $(7 + 2i)^2$

31. $(2 - i)^3$

32. $(1 + i)^3$

33. $\dfrac{5}{i}$

34. $\dfrac{7}{2i}$

35. $\dfrac{4}{1 + i}$

36. $\dfrac{6}{1 - i}$

37. $\dfrac{2 + i}{3 - i}$

38. $\dfrac{5 - i}{2 + i}$

39. $\dfrac{3 + 4i}{5 - 2i}$

40. $\dfrac{6 + 5i}{3 + 7i}$

41. $(2 + 3i)^2 - 5(2 - i) + 4$

42. $(3 - 5i)^2 + 4(7 - 2i) + 6$

Write each of the following in terms of i and simplify.

43. $\sqrt{-4}$

44. $\sqrt{-9}$

45. $\sqrt{-25}$

46. $\sqrt{-16}$

47. $\sqrt{-36}$

48. $\sqrt{-81}$

49. $\sqrt{-32}$

50. $\sqrt{-50}$

51. $\sqrt{-98}$

52. $-2\sqrt{-250}$

53. $-3\sqrt{-200}$

54. $-5 + \sqrt{-2}$

55. $-3 + 5\sqrt{-27}$

56. $-\sqrt{-8} + \sqrt{-18}$

57. $\sqrt{-45} + \sqrt{-20} - \sqrt{-5}$

58. $2 + \sqrt{-5} - 3\sqrt{-80}$

59. $\sqrt{-3} - 5\sqrt{-12} + 2\sqrt{-48}$

60. $\sqrt{-2} - 5\sqrt{-18} + 6\sqrt{-50} - \sqrt{-8}$

61. $7\sqrt{-8} + 5\sqrt{-12} - 6\sqrt{-18} + 2\sqrt{-27}$

62. $11\sqrt{-3} + \sqrt{-75} - \sqrt{-16} + \sqrt{-25}$

63. $\dfrac{1}{\sqrt{-3}}$

64. $\dfrac{1}{\sqrt{-8}}$

65. $\dfrac{3}{\sqrt{-27}}$

66. $\dfrac{2}{1 - \sqrt{-3}}$

67. $\dfrac{3}{2 + \sqrt{-3}}$

68. $\dfrac{5}{7 - \sqrt{-2}}$

69. $\dfrac{4 - \sqrt{-5}}{1 - \sqrt{-1}}$

70. $\dfrac{1 + \sqrt{-2}}{3 - \sqrt{-3}}$

71. $\dfrac{3 - 2\sqrt{-50}}{4 + 3\sqrt{-8}}$

72. $\dfrac{11 - \sqrt{-2}}{\sqrt{-3} - \sqrt{-2}}$

73. $\dfrac{6 - \sqrt{-7}}{\sqrt{-5} + \sqrt{-2}}$

74. $\dfrac{2 - \sqrt{-3}}{\sqrt{-3} + \sqrt{-7}}$

Solve each of the following for x and y.

75. $3x + 5yi = 9 + 10i$

76. $x - 3yi = -5 + 2i$

77. $(3 + 2i)x - (1 - i)y = 4i$

78. $(2 - i)x - (3 - i)y = 7i$

79. $(4x + y - 24)i = 2y + 7 - 3x$

80. $(-3x + 18 - y)i + 17 = 6y - 7x$

81. $3x + (y - 2)i - (3y - 8)i = 5 - 2x$

82. $5 - 4i - 2x + 1 = (3y + 2)i$

83. $9i + 5 - 9ix = 4y + 4iy + 3x$

84. $2ix + 9 - 13i - 9y = -4x - 6iy$

Perform the indicated operations, simplify and write each answer in the form a + bi.

85. $\left(-\dfrac{1}{2} - \dfrac{\sqrt{3}}{2}i\right)^3$

86. $\left(-\dfrac{1}{2} + \dfrac{\sqrt{3}}{2}i\right)^3$

87. $\left(\dfrac{1}{\sqrt{2}} - \dfrac{1}{\sqrt{2}}i\right)^4$

88. $\left(\dfrac{1}{\sqrt{2}} + \dfrac{1}{\sqrt{2}}i\right)^4$

Determine whether or not the complex number is a solution to the given equation.

89. $2a^2 - 4a + 3 = 0$; $1 + \dfrac{\sqrt{2}}{2}i$

90. $5x^2 + 2 = 2x$; $\dfrac{1 - 3i}{5}$

91. $2y^2 + y + 1 = 0$; $\dfrac{-1 + \sqrt{7}i}{4}$

92. $p^2 + 3 = 3p$; $\dfrac{3 - \sqrt{2}i}{2}$

93. $m^2 + 7 = -4m$; $2 + \sqrt{3}i$

94. $t^2 + 19 = 6t$; $5 + \sqrt{10}i$

7.7   More Applications

Many areas of applied mathematics use formulas that involve radicals. It is not necessary that you know where the formulas are applied in order to be able to work with them. We begin with a problem about numbers.

Example 1   What number subtracted from its square root yields a difference of -12?

Let n represent the number. Its square root is $\sqrt{n}$.

$$\sqrt{n} - n = -12$$
$$\sqrt{n} = n - 12 \qquad \text{Isolate the radical}$$
$$(\sqrt{n})^2 = (n - 12)^2 \qquad \text{Squaring property}$$
$$n = n^2 - 24n + 144$$
$$0 = n^2 - 25n + 144$$
$$0 = (n - 16)(n - 9) \qquad \text{Factor}$$
$$n = 16 \quad \text{or} \quad n = 9$$

Check in the words of the problem.

$$\sqrt{16} = 4; \quad 4 - 16 \overset{\checkmark}{=} -12$$
$$\sqrt{9} = 3; \quad 3 - 9 \neq -12$$

The number is 16.   ∎

Example 2   In statistics the calculated value of the z score is given by the formula

$$z = \frac{P - p}{\sqrt{\dfrac{pq}{n}}}.$$

Solve the formula for n.

$$z^2 = \frac{(P - p)^2}{\left(\sqrt{\dfrac{pq}{n}}\right)^2} \qquad \begin{array}{l} \text{Square both sides to} \\ \text{eliminate the radical} \end{array}$$

$$z^2 = \frac{(P - p)^2}{\dfrac{pq}{n}}$$

Now simplify the complex fraction on the right.

$$z^2 = \frac{n(P - p)^2}{pq}$$

$$z^2pq = n(P - p)^2 \qquad \text{Clear of fractions}$$

$$\frac{z^2pq}{(P - p)^2} = n \qquad \text{Divide by } (P - p)^2 \quad \blacksquare$$

Example 3 A point moves in such a way that its position at any time is given by $\sqrt{(x + 1)^2} + y = 4$. Solve the equation for y in terms of x.

$$\sqrt{(x + 1)^2} + y = 4$$

$$(x + 1)^2 + y = 16 \qquad \text{Squaring property}$$

$$y = 16 - (x + 1)^2 \qquad \text{Subtract } (x + 1)^2 \text{ from each side} \quad \blacksquare$$

Example 4 The formula for the radius of a circle inscribed in a triangle is given by

$$r = \sqrt{\frac{(s - a)(s - b)(s - c)}{s}},$$

where $s = \frac{1}{2}(a + b + c)$. Find the radius of the circle in Figure 7.1.

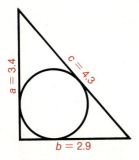

Figure 7.1

$$s = \frac{1}{2}(a + b + c) = \frac{1}{2}(10.6) = 5.3 \text{ inches}$$

Substituting this result in the formula yields

$$r = \sqrt{\frac{(5.3 - 3.4)(5.3 - 2.9)(5.3 - 4.3)}{5.3}}$$

$$= \sqrt{\frac{(1.9)(2.4)(1)}{5.3}}.$$

Using a calculator, we find $r = 0.9$ to the nearest tenth. $\blacksquare$

Exercises 7.7

Solve each of the following as indicated. Round answers to 2 decimal places as needed.
Let $\pi = 3.1416$.

1. A number exceeds twice its square root by three. Find the number.

2. The standard deviation a of a theoretical binomial probability distribution is given by $a = \sqrt{npq}$, where n is the number in the sample, p is the probability of the event occurring and q is the probability of the event not occurring. Solve for p.

3. The number n required in a sample to determine an error less than or equal to E is given by

$$\sqrt{n} = \frac{M \cdot D}{E}.$$

   Solve for M.

4. The pooled estimate D for standard deviation, where n is one sample size, N is the second sample size, s is the standard deviation of one sample, and S is the standard deviation of the second sample, is given by

$$D = \sqrt{\frac{(n-1)s + (N-1)S}{N+n-2}}.$$

   Solve the equation for S.

5. The area A of a regular octagon of side s is given by

$$A = \frac{s^2(4 + 2\sqrt{2})}{\sqrt{2}}.$$

   Find the area of the given octagon.

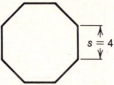

$s = 4$

6. The length of a side s of a square with diagonal D is given by

$$s = \sqrt{\frac{D^2}{2}}.$$

   Find the length of a side of the given square.

$D = 12$ s

7. The frequency with which a ball on the end of a steel rod will oscillate is given by

$$f = \frac{1}{2\pi}\sqrt{\frac{K}{m}}.$$

   Find the frequency of oscillation f of a ball of mass $m = 0.0625$ when K, the steel rod constant, is 102.

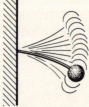

8. The lateral surface area L of a cone is given by $L = \pi r\sqrt{h^2 + r^2}$. Find the lateral surface area of the given figure.

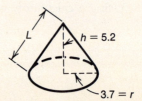

L $h = 5.2$

$3.7 = r$

9. The radius r of a sphere is given by the equation

$$r = \sqrt[3]{\frac{3V}{4\pi}}.$$

Find the radius of a sphere if its volume V is 18.23 cu. in.

10. The volume V of a frustrum of a pyramid is given by

$$V = \frac{h(A + a + \sqrt{Aa})}{3},$$

where A, the area of the lower base is 6.04 square inches; a, the area of the upper base is 4.74 square inches, and the height h is 3.96 inches. Find the volume.

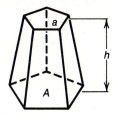

11. The radius of a circle circumscribed about a triangle with sides $a = 12.1$ inches; $b = 14.5$ inches; and $c = 9.6$ inches, is given by

$$r = \frac{abc}{4\sqrt{s(s - a)(s - b)(s - c)}}$$

where $s = \frac{1}{2}(a + b + c)$. Find the radius r.

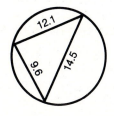

12. In the equation

$$I^{2/3} = \frac{V}{10^2},$$

I is current in amperes and V is voltage in volts. Find V.

13. In the theory of relativity the formula

$$m = \frac{m_0}{\sqrt{1 - v^2/c^2}}$$

gives the mass of an object at a speed v when its mass at rest is m_0 and c is the velocity of light. If $v = 0$, show that $m = m_0$.

14. The velocity V of sound in gases is given by

$$V = c\sqrt{\frac{T}{M}}.$$

Find the velocity of sound in air if $M = 28$, $T = 293$ and $c = 3.41$ meters per second.

15. A string stretched with a tension $t = 50$ newtons produces the note C with a frequency $f = 256$ cycles/second. The frequency is given by the proportion

$$\frac{f}{F} = \frac{\sqrt{t}}{\sqrt{T}}.$$

Here f and F are the frequencies, t and T are the tensions. Find the tension T required to produce a frequency of $F = 512$ cycles/second.

16. The equation of a simple parabola is found by $\sqrt{(x - a)^2 + y} = x + a$. Solve this equation for y.

17. The path of a projectile thrown horizontally from a point y feet above the ground with a velocity v feet per second is a parabola whose equation is

$$x = \sqrt{\frac{-2v^2 \cdot y}{g}}.$$

Solve this equation for y.

18. In a sphere the surface area is given by $S = 4\pi r^2$ and the volume by $V = (4/3)\pi r^3$. Solve for V in terms of S and r.

19. The side x of a cube of volume V is given by $x = V^{1/3}$ and the surface area by $S = 6x^2$. Solve for S in terms of V.

Chapter 7   Summary

[7.1]   The properties of exponents as they apply to integers also apply to rational exponents. The expression $b^{1/2}$ means $\sqrt{b}$ and $b^{1/3}$ means $\sqrt[3]{b}$. In general, $b^{1/n}$ means the nth root of b. If n is even, b must be positive or zero for the expression to represent a real number.

[7.2]   Another way to indicate a root of a number is to use a radical symbol. The symbol $\sqrt{b}$ means the square root of b; $\sqrt[3]{b}$ means the cube root of b, and the $\sqrt[n]{b}$ means the nth root of b. A radical consists of three parts: the **index**, the **radical**, and the **radicand.**

$$\text{Index} \longrightarrow \sqrt[n]{b} \longleftarrow \text{Radical}$$
$$\text{Radicand}$$

The principal square root of a positive number b is written as $\sqrt{b}$. It indicates the number a such that $a^2 = b$. The square root of the square of a number is the absolute value of the number.

$$\sqrt{b^2} = |b|$$

Radicals are simplified by using the theorem indicating

$$\sqrt[n]{a \cdot b} = \sqrt[n]{a}\sqrt[n]{b} \quad \text{and}$$

$$\sqrt[n]{\frac{a}{b}} = \frac{\sqrt[n]{a}}{\sqrt[n]{b}}.$$

A radical is in **simplest form** if four conditions are satisfied.

[7.2]   1. No exponent on a factor in the radicand is greater than or equal to the index.

2. No radicand contains a fraction.

[7.4]   3. No fraction contains a radical in the denominator.

4. The index and the exponent in the radicand have no common factors.

[7.3]   Radicals with the same index and radicand can be added and subtracted. If they are not the same, the properties of radicals often can be used to simplify them. If it is helpful, radicals can be written in terms of rational exponents and the rules for exponents used in the simplification process.

[7.4]   The fundamental property of fractions is used to eliminate radicals from the denominator of a fraction. Both numerator and denominator are multiplied by a quantity sufficiently large to make the exponents on the radicand equal to the index. The process is called **rationalizing the denominator.** When the denominator containing radicals is a binomial, its **conjugate** is used in the rationalizing process. The conjugate of $a + b$ is $a - b$.

[7.5]   Equations involving radicals are simplified and solved by the **power property of equality.** When it is used to eliminate a square root it is called the **squaring property;** when it is used to eliminate a cube root it is called the **cubing property.**

[7.6]   The number $i = \sqrt{-1}$ is called the imaginary unit. When i is raised to positive integer powers it takes on one of the four values, i, -1, $-i$, or 1. A number of the form $a + bi$, $a, b \in R$ is called a complex number. Addition and multiplication of complex numbers are carried out in the same way that binomials such as $ax + by$ and $cx + dy$ are added or multiplied. Complex numbers are divided by rationalizing the denominator. If the denominator is a binomial, it is rationalized by multiplying by its conjugate. The conjugate of $a + bi$ is $a - bi$, and $(a + bi)(a - bi) = a^2 + b^2$, which is a real number.

Chapter 7   Review

Simplify each of the following. Write all answers with positive exponents.

1. $(0.0081)^{-3/4}$

2. $2^{-1/2} + 3^{-1/2}$

3. $\dfrac{5^{3/4} \cdot 5^{2/3}}{5^{1/3} \cdot 5^{-5/2}}$

4. $\dfrac{x^{-5/2} \cdot x^{5/3}}{x^{7/4}}$

5. $(16a^{-8/9}b^{4/3})^{3/4}$

6. $(a + b)^{-1/2}$

7. $(x + 3)(x - 2)^{-1/2} + (x - 2)^{1/2}$

8. $2a^{-1/2} + 3b^{-1/2}$

9. $(m^{1/2} + n^{1/2})(m^{1/2} - n^{1/2})$

10. $(s^{1/2} + s^{-1/2})^2$

Simplify each of the following. Assume all variables represent positive real numbers.

11. $\sqrt[4]{64t^4}$

12. $\sqrt{400}$

13. $\sqrt[3]{-32}$

14. $\sqrt{-100}$

15. $\sqrt{n^2 + 8n + 16}$

16. $-\sqrt[4]{256k^7}$

17. $(\sqrt{3} + \sqrt{2})(\sqrt{3} - \sqrt{2})$

18. $\sqrt{2} + 5\sqrt{72} + 4\sqrt{50} - \sqrt{128}$

19. $\sqrt{x^6 + x^4}$

20. $(3\sqrt{5} + 4\sqrt{2})(\sqrt{5} - 3\sqrt{2})$

21. $(4\sqrt{3} + 5\sqrt{6})^2$

22. $2\sqrt{x}(7\sqrt{x} + \sqrt{2x} - \sqrt{4x})$

23. $(\sqrt{3} + i)^3$

24. $3\sqrt[3]{x^5y} + (x/y)\sqrt[3]{125x^2y^4} + (3/y)\sqrt[3]{-x^8y}$

25. $(\sqrt{a} + \sqrt{a - 1})(\sqrt{a} - \sqrt{a - 1})$

26. $(\sqrt{x + 1} - 1)^2$

27. $\dfrac{\sqrt{6}}{\sqrt{10}}$

28. $\sqrt[3]{\dfrac{4x^2y^7}{9xy^8}}$

29. $\dfrac{\sqrt{3a} - \sqrt{7b}}{\sqrt{21ab}}$

30. $\dfrac{\sqrt{2} - \sqrt{3}}{2\sqrt{3} + \sqrt{2}}$

31. $\dfrac{3k}{\sqrt[3]{2k^2}}$

32. $\dfrac{4a}{\sqrt[3]{2a}}$

33. $\dfrac{5 + 3\sqrt{x}}{5 - 3\sqrt{x}}$

34. $\dfrac{\sqrt{a + b} + \sqrt{a}}{\sqrt{a + b} - \sqrt{a}}$

Solve each of the following. Check all solutions. Write Ø if no real number solution exists. Exclude answers that would result in even roots of negative numbers.

35. $\sqrt{a+1} = 3$

36. $\sqrt{3y+2} = 5$

37. $\sqrt[3]{t+1} = -2$

38. $\sqrt[4]{x} = -1$

39. $\sqrt{s+5} - 2\sqrt{s} = -\sqrt{s-3}$

40. $\sqrt{r+7} = \sqrt{r+2} - \sqrt{4r+17}$

41. $\sqrt{6m-5} - 5 = 0$

42. $\sqrt{x^2 - 5x + 2} = x + 3$

43. $\sqrt{y} \cdot \sqrt{y-7} = 2\sqrt{2}$

44. $\sqrt{2a+3} \cdot \sqrt{3a-1} + 2 = 2$

45. $\dfrac{3}{\sqrt{p+5}} = \sqrt{p+7} + \sqrt{p+5}$

46. $\dfrac{k+2}{\sqrt{k+1}} = \sqrt{k+5}$

Solve for x and y.

47. $2iy - y + 19 = ix + 6x + 12i$

48. $(-3 + 5i)y + 6 + 19i = (4 + 3i)x$

Simplify each of the following.

49. $\sqrt{-9} + \sqrt{-36} - \sqrt{-81}$

50. $-4i(\sqrt{-8} + \sqrt{-50})$

51. $\dfrac{3 + 2i}{2 - i}$

52. $\dfrac{4 - \sqrt{2}i}{3 + \sqrt{2}i}$

53. $(5 - \sqrt{2}i)^2$

54. $(2 - i)(3 - i)(2 + i)$

55. $(3 - \sqrt{2}i)^{-2}$

56. $(\sqrt{3} - i)^3$

8

Quadratic Equations

8.1 Solving Quadratic Equations by the Square Root Method

In Section 5.5 we solved equations of degree greater than one by using the zero-factor theorem. As the name implies, this theorem can be used to find the solution set of a factorable equation. The next two examples review this method.

Example 1   Solve $6x^3 - 17x^2 + 12x = 0$.

$$x(6x^2 - 17x + 12) = 0 \qquad \text{\color{red}{Remove common factor first}}$$
$$x(3x - 4)(2x - 3) = 0$$
$$x = 0 \quad \text{or} \quad 3x - 4 = 0 \quad \text{or} \quad 2x - 3 = 0 \qquad \text{\color{red}{Zero-factor theorem}}$$
$$x = 0 \quad \text{or} \qquad x = \frac{4}{3} \quad \text{or} \qquad x = \frac{3}{2}$$

The solution set is $\left\{0, \frac{4}{3}, \frac{3}{2}\right\}$. ■

Example 2   Solve $x^2 - 9 = 0$ by factoring.

$$(x - 3)(x + 3) = 0 \qquad \text{\color{red}{$a^2 - b^2 = (a - b)(a + b)$}}$$
$$x - 3 = 0 \quad \text{or} \quad x + 3 = 0$$
$$x = 3 \quad \text{or} \qquad x = -3$$

The solution set is $\{3, -3\}$. ■

A second means of solving quadratic equations, **the square root method,** often is easier to use than factoring. Consider again the equation in Example 2: $x^2 - 9 = 0$. Adding 9 to each side gives the equation $x^2 = 9$. If we take the square root of each side, we have

$$x^2 = 9$$
$$|x| = 3. \qquad \text{\color{red}{$\sqrt{x^2} = |x|$}}$$

Since $|3| = 3$ and $|-3| = 3$, the solution set is $\{3, -3\}$ as before. In practice the operation above is shortened to

$$x^2 = 9$$
$$\sqrt{x^2} = \pm\sqrt{9}$$
$$x = \pm 3.$$

Since we knew there would be a positive and a negative solution, the sign $\pm$ was used to represent them. The expression ±3 is read "positive or negative 3."

Example 3   Solve $x^2 - 45 = 0$.

This equation is not readily factorable. We will use the square root method.

$$x^2 = 45$$
$$\sqrt{x^2} = \pm\sqrt{45} \qquad \text{Take the square root of each side}$$
$$x = \pm3\sqrt{5} \qquad \sqrt{45} = \sqrt{9 \cdot 5} = 3\sqrt{5}$$

The solution set is $\{3\sqrt{5},\ -3\sqrt{5}\}$. ■

Example 4   Solve $(2a - 1)^2 = -4$ by the square root method.

$$(2a - 1)^2 = -4$$
$$\sqrt{(2a - 1)^2} = \pm\sqrt{-4} \qquad \text{Take the square root of each side}$$
$$2a - 1 = \pm2i$$
$$2a = 1 \pm 2i \qquad \text{Add 1 to each side}$$
$$a = \frac{1 \pm 2i}{2} \qquad \text{Divide each side by 2}$$

When the two solutions are written in $a + bi$ form, the solution set is $\left\{\frac{1}{2} + i,\ \frac{1}{2} - i\right\}$.

Check: $\left[2\left(\frac{1}{2} + i\right) - 1\right]^2 = -4 \qquad x = \frac{1}{2} + i$
$$[1 + 2i - 1]^2 = -4$$
$$4i^2 = -4$$
$$-4 = -4 \qquad i^2 = -1$$

The check of the other solution is left to the reader. ■

Example 5   Solve $y^4 - 64 = 0$.

Begin by factoring into quadratic factors.

$$(y^2 - 8)(y^2 + 8) = 0$$
$$y^2 - 8 = 0 \quad \text{or} \quad y^2 + 8 = 0 \qquad \text{Zero-factor theorem}$$

Each of these can be solved by the square root method.

$$y^2 - 8 = 0 \qquad\qquad y^2 + 8 = 0$$
$$y^2 = 8 \qquad\qquad\quad y^2 = -8$$
$$y = \pm2\sqrt{2} \qquad\qquad y = \pm2\sqrt{2}i$$

The solution set is $\{2\sqrt{2},\ -2\sqrt{2},\ 2\sqrt{2}i,\ -2\sqrt{2}i\}$. ■

Example 6   Solve each of the following by the square root method. Simplify all solutions.

(a) $A^3 = 3r^2$ for r

$$\frac{A^3}{3} = r^2$$

$$\pm\sqrt{\frac{A^3}{3}} = r$$

$$\pm\frac{A}{3}\sqrt{3A} = r$$

The solution set is $\left\{-\frac{A}{3}\sqrt{3A}, \frac{A}{3}\sqrt{3A}\right\}$.

(b) $\dfrac{tx^2}{s} = f$ for x

$$tx^2 = sf$$

$$x^2 = \frac{sf}{t}$$

$$x = \pm\frac{\sqrt{sft}}{t} \quad \text{or} \quad \pm\frac{1}{t}\sqrt{sft}$$

The solution set is $\left\{-\dfrac{\sqrt{sft}}{t}, \dfrac{\sqrt{sft}}{t}\right\}$.   ■

When the solution set to an equation is known it is not difficult to find the equation. To find the equation whose solution set is {5, −2}, proceed as follows:

$$x = 5 \quad \text{or} \quad x = -2$$
$$x - 5 = 0 \quad \text{or} \quad x + 2 = 0$$
$$(x - 5)(x + 2) = 0 \qquad \text{Zero-factor theorem}$$
$$x^2 - 3x - 10 = 0. \qquad \text{Multiply}$$

A third-degree equation such as the result in Example 7 is called a **cubic equation.**

Example 7   Find an equation whose solution set is $\{2, 3 + \sqrt{2}i, 3 - \sqrt{2}i\}$.

$$x = 2 \quad \text{or} \qquad\qquad x = 3 + \sqrt{2}i \quad \text{or} \qquad\qquad x = 3 - \sqrt{2}i$$
$$x - 2 = 0 \quad \text{or} \quad x - 3 - \sqrt{2}i = 0 \qquad \text{or} \quad x - 3 + \sqrt{2}i = 0$$
$$(x - 2)(x - 3 - \sqrt{2}i)(x - 3 + \sqrt{2}i) = 0$$
$$(x - 2)(x^2 - 6x + 11) = 0$$
$$x^3 - 8x^2 + 23x - 22 = 0 \quad ■$$

Exercises 8.1

Solve each of the following equations by factoring. Remember to remove common factors first.

1. $x^2 - 6x + 8 = 0$

2. $y^2 + y - 2 = 0$

3. $2a^2 + 2 = 5a$

4. $2k^2 - k = 0$

5. $2t^2 - 5t - 3 = 0$

6. $15m^2 - 23m - 28 = 0$

7. $6s^2 - 33s + 45 = 0$

8. $24p^3 - 14p^2 - 20p = 0$

9. $\dfrac{1}{6}w^2 - \dfrac{11w}{36} + \dfrac{1}{12} = 0$

10. $\dfrac{1}{4}x^3 - \dfrac{1}{9}x = 0$

11. $15r^2 + 46r + 35 = 0$

12. $q^2 - 81 = 0$

13. $0.015a^2 - 0.019ab + 0.006b^2 = 0$

14. $0.02y^2 + 0.05yz - 0.07z^2 = 0$

Solve each of the following by the square root method.

15. $x^2 = 16$

16. $y^2 = 9$

17. $z^2 = -4$

18. $w^2 = -25$

19. $a^2 - 81 = 0$

20. $b^2 - 121 = 0$

21. $v^2 = 10^{-6}$

22. $m^2 = 10^4$

23. $4n^2 - 1 = 0$

24. $16t^2 - 169 = 0$

25. $x^2 = 7$

26. $y^2 = 5$

27. $2a^2 = 3$

28. $3c^2 = 11$

29. $50p^2 = 32$

30. $49k^2 = 98$

31. $(2a - 1)^2 = 5$

32. $(3q - 2)^2 = 7$

33. $(4m + 3)^2 = 1$

34. $(7u - 1)^2 = 13$

35. $(2v + 3)^2 + 3 = 0$

36. $(9d - 5)^2 + 7 = 0$

37. $(7x - 5)^2 + 4 = 0$

38. $(y - 11)^2 + 36 = 0$

39. $(x - a)^2 = b^2$

40. $(y - c)^2 = p^2$

41. $81k^2 + 25 = 0$

42. $64q^2 + 225 = 0$

Write the quadratic or cubic equation having the given solution set.

43. $\{2, 3\}$

44. $\{1, -4\}$

45. $\{0, -1, 5\}$

46. $\{0, 3, -2\}$

47. $\left\{\dfrac{1}{2}, \dfrac{1}{3}\right\}$

48. $\left\{-\dfrac{1}{2}, \dfrac{1}{5}\right\}$

49. $\{\pm 2\sqrt{2}\}$

50. $\{\pm 3\sqrt{3}\}$

51. $\{\pm 3i\}$

52. $\{\pm 2i\}$

53. $\{2 \pm \sqrt{5}\}$

54. $\{1 \pm 2\sqrt{3}\}$

55. $\{1 \pm 2i\}$

56. $\{3 \pm 5i\}$

57. $\{1 \pm \sqrt{3}i\}$

58. $\{4 \pm \sqrt{5}i\}$

59. $\{2a, 3b\}$

60. $\{6c, -7d\}$

61. $\left\{\dfrac{-b \pm \sqrt{b^2 - 4ac}}{2a}\right\}$

62. $\left\{\dfrac{-c \pm \sqrt{c^2 - 4ad}}{2a}\right\}$

Write each of the following in the form $(ax + b)^2 = c$ and solve by the square root method. Simplify all solutions.

63. $a^2 + 2a + 1 = 5$

64. $p^2 - 4p + 4 = 3$

65. $m^2 - 6m + 9 = 7$

66. $k^2 + 12k + 36 = 1$

67. $4x^2 + 12x + 9 = 8$

68. $25y^2 + 60y + 36 = 12$

69. $49s^2 - 42s + 9 = 32$

70. $121w^2 - 44w + 4 = 27$

71. $36t^2 - 108t + 81 = -18$

72. $144r^2 - 168r + 49 = -12$

73. $64u^2 - 112u + 49 = -50$

74. $49q^2 + 98q + 49 = -72$

Solve each of the following. Factor first, then solve by the square root method.

75. $x^4 - 81 = 0$

76. $y^4 - 16 = 0$

77. $16a^4 - 64 = 0$

78. $15s^4 - 135 = 0$

79. $p^4 - b^4 = 0, \quad b > 0$

80. $t^4 - c^4 = 0, \quad c > 0$

81. $m^4 - 5m^2 + 6 = 0$

82. $6r^4 + 13r^2 + 6 = 0$

Solve each of the following for the indicated variable.

83. $A = \pi r^2$, $r > 0$

84. $V = \dfrac{1}{3}\pi r^2 h$, $r > 0$

85. $S = \dfrac{1}{2}gt^2$, $t > 0$

86. $a^2 + b^2 = c^2$, $a > 0$

87. $A = P(1 + r)^2$, $r > 0$

88. $S = h - r^2$, $r > 0$

Solve each of the following for x.

89. $x^2 - c = 0$

90. $dx^2 = b$

91. $\dfrac{rx^2}{s} = t$

92. $\dfrac{ax^2}{b} = c$

93. $(ax - b)^2 = c$

94. $(cx + d)^2 = d$

95. $(mx + n)^2 = n^2$

96. $(3rx + 2s)^2 = s^2$

8.2 Solving Quadratic Equations by Completing the Square

In Section 8.1 we solved equations such as $(x + 2)^2 = 6$ by taking the square root of each side. In this section we will learn to solve equations that are not easily factorable but can be written in the form

$$(ax + b)^2 = c,$$

which can be solved by the square root method. Consider the equation

$$x^2 + 6x - 8 = 0.$$

Adding 8 to each side of the equation gives

$$x^2 + 6x = 8.$$

The left side of the equation is not a perfect square trinomial but can be made into one by adding 9 to each side.

$$x^2 + 6x + 9 = 8 + 9$$
$$(x + 3)^2 = 17$$
$$x + 3 = \pm\sqrt{17}$$
$$x = -3 \pm \sqrt{17}$$

The solution set is $\{-3 + \sqrt{17},\ -3 - \sqrt{17}\}$.

The process that was used to change $x^2 + 6x$ into $x^2 + 6x + 9$ is called **completing the square.**

How did we know to add 9 to create a perfect square trinomial? Consider the square of the binomial

$$(x + a)^2 = x^2 + 2ax + a^2.$$

If the coefficient of the second degree term is 1, the quantity necessary to complete the square is the square of one half of the coefficient of the first-degree term.

$$x^2 + 2ax + a^2 = (x + a)^2$$

$\frac{1}{2}$ of 2a squared $\frac{1}{2}$ of 2a

Example 1   Complete the square.

$$\begin{array}{ccc} \text{I} & \text{II} & \text{III} \end{array}$$

(a) $x^2 + 4x + 4 = (x + 2)^2$

(b) $x^2 - 10x + 25 = (x - 5)^2$

(c) $x^2 + 12x + 36 = (x + 6)^2$

Each quantity in column II is the square of $\frac{1}{2}$ the coefficient in column I. Each quantity in column III is $\frac{1}{2}$ the coefficient in column I.   ■

In general, to complete the square for the expression $x^2 + Bx$, add the quantity $\left(\frac{B}{2}\right)^2$.

$$x^2 + Bx + \left(\frac{B}{2}\right)^2 = \left(x + \frac{B}{2}\right)^2.$$

Example 2   Solve $p^2 + 2p + 5 = 0$ by completing the square.

$$\begin{array}{ll} p^2 + 2p = -5 & \text{Subtract 5 from each side} \\ p^2 + 2p + 1 = -5 + 1 & \text{Complete the square} \\ (p + 1)^2 = -4 & \text{Write as the square of a binomial} \\ p + 1 = \pm 2i & \text{Take the square root of each side} \\ p = -1 \pm 2i & \end{array}$$

The solution set is $\{-1 + 2i, -1 - 2i\}$.   ■

Example 3   Solve $3t^2 + 4t - 5 = 0$.

First write the equation as $3t^2 + 4t = 5$. To complete the square, the lead coefficient must be 1, so divide each side by 3.

$$t^2 + \frac{4}{3}t = \frac{5}{3}$$

$$t^2 + \frac{4}{3}t + \frac{4}{9} = \frac{5}{3} + \frac{4}{9} \qquad \left(\frac{2}{3}\right)^2 = \frac{4}{9}$$

$$\left(t + \frac{2}{3}\right)^2 = \frac{5}{3} + \frac{4}{9} \qquad \tfrac{1}{2} \text{ of } \tfrac{4}{3} \text{ is } \tfrac{2}{3}$$

$$\left(t + \frac{2}{3}\right)^2 = \frac{19}{9} \qquad \text{The LCD is 9}$$

$$t + \frac{2}{3} = \pm \frac{\sqrt{19}}{3}$$

$$t = -\frac{2}{3} \pm \frac{\sqrt{19}}{3}$$

The solution set is

$$\left\{ -\frac{2}{3} + \frac{\sqrt{19}}{3}, \ -\frac{2}{3} - \frac{\sqrt{19}}{3} \right\}. \quad ■$$

To solve an equation by completing the square, use the five steps below.

> **To Solve a Quadratic Equation by Completing the Square**
> *Step 1* Isolate all terms involving the variable on one side of the equation.
> *Step 2* If the coefficient of the second-degree term is not 1, divide both sides by the coefficient.
> *Step 3* Add the square of one-half of the coefficient of the first-degree term to each side of the equation.
> *Step 4* Express the perfect square trinomial from Step 3 as the square of a binomial.
> *Step 5* Take the square root of each side and solve the resulting equation.

Some of the processes used to solve equations in earlier sections and chapters result in equations that can be solved by completing the square.

Example 4   Solve for x:

$$\frac{x}{x^2 + 3x + 2} - \frac{x + 3}{x + 2} = \frac{7}{x + 1}; \quad x \neq -1, \, -2.$$

$$x - (x + 3)(x + 1) = 7(x + 2) \qquad \text{LCD} = (x + 1)(x + 2)$$

When simplified, this becomes

$$x^2 + 10x + 17 = 0.$$

Completing the square yields

$$(x + 5)^2 = 8$$
$$x = -5 \pm 2\sqrt{2}.$$

The solution set is $\{-5 + 2\sqrt{2}, \, -5 - 2\sqrt{2}\}$. ∎

Example 5   Rewrite $x^2 + 4x + y^2 + 6y = 12$ in terms of the squares of binomials.

$$(x^2 + 4x + 4) + (y^2 + 6y + 9) = 12 + 13 \qquad \text{Addition property of equality}$$
$$\text{(13 added to each side)}$$
$$(x + 2)^2 + (y + 3)^2 = 25 \quad ∎$$

Exercises 8.2

Complete the square in each of the following, then factor and write as the square of a binomial.

1. $a^2 + 2a + $ _____
2. $b^2 + 4b + $ _____
3. $x^2 - 6x + $ _____
4. $y^2 - 8y + $ _____
5. $r^2 + 10r + $ _____
6. $t^2 + 12t + $ _____
7. $v^2 - 14v + $ _____
8. $u^2 - 16u + $ _____

Solve by completing the square. Simplify all solutions.

9. $x^2 - 2x - 3 = 0$

10. $w^2 - 4w - 5 = 0$

11. $a^2 + 8a + 12 = 0$

12. $c^2 + 6c + 5 = 0$

13. $k^2 - k - 2 = 0$

14. $m^2 - m - 6 = 0$

15. $t^2 - 2t + 3 = 0$

16. $w^2 - 4w + 8 = 0$

17. $2s^2 + 3s - 1 = 0$

18. $3r^2 + r - 1 = 0$

19. $3y^2 - 5y + 3 = 0$

20. $x^2 - 9x - 19 = 0$

21. $a^2 + \dfrac{3a}{2} - \dfrac{1}{2} = 0$

22. $b^2 + \dfrac{1b}{2} - 1 = 0$

23. $4w^2 + 8w + 1 = 0$

24. $2c^2 + 5c - 1 = 0$

25. $2v^2 + 7v + 6 = 0$

26. $3t^2 + 4t + 2 = 0$

27. $3a^2 + 7a + 3 = 0$

28. $3q^2 - 8q + 3 = 0$

29. $5m^2 + 7m = 0$

30. $7v^2 - 9v = 0$

31. $t^2 - \dfrac{4t}{3} + \dfrac{1}{9} = 0$

32. $s^2 + \dfrac{7s}{15} - \dfrac{4}{15} = 0$

Solve by any method.

33. $w^2 + 11w + 30 = 0$

34. $n^2 - 5n - 24 = 0$

35. $(2a - 3)^2 = -11$

36. $(3t + 4)^2 = -1$

37. $x^2 + x + 1 = 0$

38. $y^2 - y + 1 = 0$

39. $p^4 - 625 = 0$

40. $s^4 - 256 = 0$

41. $6c^2 - c - 12 = 0$

42. $35u^2 + 2u - 24 = 0$

43. $3m^2 - 5m - 2 = 0$

44. $60v^2 + v - 10 = 0$

Solve each of the following. Check all solutions.

45. $x + \sqrt{x + 5} = 7$

46. $p = 1 + \sqrt{13 - p}$

47. $\sqrt{6 + 6a} = 3 + \sqrt{a + 4}$

48. $\sqrt{t + 8} = 4 - \sqrt{2t - 1}$

Solve each of the following. State any restrictions on the variables.

49. $\dfrac{1}{x + 2} = 3x - 1$

50. $\dfrac{2a + 1}{3} = \dfrac{1}{a - 1}$

51. $\dfrac{5}{v + 1} - 3 = \dfrac{4}{2v - 1}$

52. $\dfrac{1}{2w - 5} = 1 - \dfrac{1}{1 - w}$

Write each of the following as a quadratic equation and solve by any method.

53. The sum of twice the square of an integer and three times the integer is 9. Find the integer.

54. The sum of three times the square of an odd integer and twice the integer is 33. Find the odd integer.

55. The larger of two numbers exceeds the smaller by 2. Find the two numbers if the sum of their squares is 74.

56. The smaller of two numbers is 4 less than the larger. Find the numbers if the sum of their squares is 58.

57. The sum of the reciprocals of two consecutive integers is $\frac{7}{12}$. Find the integers.

58. A natural number added to its reciprocal is $\frac{122}{11}$. Find the number.

59. The length of a rectangle exceeds its width by 8 feet. Find the dimensions if the area is 48 square feet.

60. The length of a rectangle is 4 centimeters shorter than twice the width. Find the dimensions if the area is 96 square centimeters.

61. The base of a triangle is 2 meters longer than twice the altitude. Find the dimensions if the area is 12 square meters.

62. The second side of a right triangle is 7 units longer than the first side and the third side is one unit longer than the second side. Use the Pythagorean theorem to find the lengths of the three sides.

Complete the square of both x and y in each of the following. Rewrite as binomials squared.

63. $x^2 + 2x + y^2 + 4y = 4$

64. $x^2 + 6x + y^2 + 10y = 2$

65. $x^2 - 4x + y^2 - 12y = 7$

66. $x^2 - 14x + y^2 - 6y = 6$

8.3   The Quadratic Formula

In the last section we saw that any quadratic equation can be solved by the method of completing the square. In Section 5.5 we used the equation $ax^2 + bx + c = 0$, $a \neq 0$, to represent any quadratic equation. When this equation is solved by completing the square, the solution set provides a formula that can be used to solve any quadratic equation.

$$ax^2 + bx + c = 0$$

$$ax^2 + bx = -c \qquad \text{Subtract } c \text{ from each side}$$

$$x^2 + \frac{b}{a}x = -\frac{c}{a} \qquad \text{Divide each side by } a$$

To complete the square we must add the square of one-half the coefficient of x, $\frac{b}{a}$, to each side.

$$\frac{1}{2}\left(\frac{b}{a}\right) = \frac{b}{2a}; \quad \left(\frac{b}{2a}\right)^2 = \frac{b^2}{4a^2}$$

$$x^2 + \frac{b}{a}x + \frac{b^2}{4a^2} = \frac{b^2}{4a^2} - \frac{c}{a}$$

$$\left(x + \frac{b}{2a}\right)^2 = \frac{b^2 - 4ac}{4a^2} \qquad \begin{array}{l}\text{Factor as the square of a}\\ \text{binomial; LCD} = 4a^2\end{array}$$

$$x + \frac{b}{2a} = \pm\sqrt{\frac{b^2 - 4ac}{4a^2}} \qquad \text{Take the square root of each side}$$

$$x + \frac{b}{2a} = \pm\frac{\sqrt{b^2 - 4ac}}{2a} \qquad \sqrt{4a^2} = 2a$$

$$x = \frac{-b}{2a} \pm \frac{\sqrt{b^2 - 4ac}}{2a} \qquad \text{Subtract } \frac{b}{2a} \text{ from each side}$$

$$x = \frac{-b \pm \sqrt{b^2 - 4ac}}{2a} \qquad \text{LCD} = 2a$$

The solution set is $\left\{\dfrac{-b + \sqrt{b^2 - 4ac}}{2a}, \dfrac{-b - \sqrt{b^2 - 4ac}}{2a}\right\}.$

When the solution set is written in the form

$$x = \frac{-b \pm \sqrt{b^2 - 4ac}}{2a}, \quad a \neq 0,$$

it is called the **quadratic formula.**

Theorem 8.1   **Quadratic Formula**

The solution set of the equation $ax^2 + bx + c = 0$, $a \neq 0$, is given by the formula

$$x = \frac{-b \pm \sqrt{b^2 - 4ac}}{2a}.$$

When the sign $\pm$ is used in the context above, it is read "plus or minus" rather than "positive or negative." The quadratic formula is stated in terms of the variable x, but it is equally true for any other variable. To solve a quadratic equation, identify the values of a, b, and c, and substitute them into the formula.

▶▶ *Caution*   To identify a, b, and c, first write the equation in *standard form*:

$ax^2 + bx + c = 0.$

Example 1   Solve $3x^2 + 6x + 1 = 0$ by the formula.

First identify a, b, and c.

$$ax^2 + bx + c = 0$$
$$\downarrow \quad\quad \downarrow \quad\quad \downarrow$$
$$3x^2 + 6x + 1 = 0$$

$$a = 3, b = 6, c = 1$$

$$x = \frac{-b \pm \sqrt{b^2 - 4ac}}{2a}$$

$$x = \frac{-6 \pm \sqrt{6^2 - 4 \cdot 3 \cdot 1}}{2 \cdot 3} \qquad \text{Substitute for } a, b, \text{ and } c$$

$$x = \frac{-6 \pm \sqrt{36 - 12}}{6}$$

$$= \frac{-6 \pm \sqrt{24}}{6}$$

$$= \frac{-6 \pm 2\sqrt{6}}{6} \qquad \sqrt{24} = 2\sqrt{6}$$

$$= \frac{2(-3 \pm \sqrt{6})}{2 \cdot 3} \qquad \text{Factor}$$

$$= \frac{-3 \pm \sqrt{6}}{3} \qquad \text{Reduce}$$

The solution set is

$$\left\{ \frac{-3 + \sqrt{6}}{3}, \frac{-3 - \sqrt{6}}{3} \right\}. \quad ■$$

The quadratic formula can be used even if an equation is factorable. Since factoring usually requires less effort, first check to see if the equation can be factored easily.

Example 2   Solve $2x^2 - 4x = -3$.

Think of $2x^2 - 4x = -3$ as $2x^2 + (-4)x + 3 = 0$ to find the values for a, b, and c. Here $a = 2$, $b = -4$, and $c = 3$.

$$x = \frac{-b \pm \sqrt{b^2 - 4ac}}{2a}$$

$$= \frac{-(-4) \pm \sqrt{(-4)^2 - 4(2)(3)}}{2 \cdot 2}$$

$$= \frac{4 \pm \sqrt{16 - 24}}{4}$$

$$= \frac{4 \pm \sqrt{-8}}{4}$$

$$= \frac{4 \pm 2\sqrt{2}i}{4}$$

$$= 2 \pm \sqrt{2}i \qquad\qquad \text{2 is a common factor}$$

The solution set is $\{2 + \sqrt{2}i, \, 2 - \sqrt{2}i\}$. ∎

Notice in Example 2 that when one solution to a quadratic equation with real coefficients is of the form $a + bi$, the other solution is its **conjugate,** $a - bi$.

The quadratic formula also applies to second-degree equations whose coefficients involve imaginary numbers. We will consider only one case, in which both a and c are imaginary.

Example 3   Solve $2im^2 - 3m + 4i = 0$.

Here $a = 2i$, $b = -3$, and $c = 4i$.

$$m = \frac{-(-3) \pm \sqrt{(-3)^2 - 4(2i)(4i)}}{2(2i)}$$

$$= \frac{3 \pm \sqrt{9 - 32i^2}}{4i}$$

$$= \frac{3 \pm \sqrt{41}}{4i} \qquad\qquad i^2 = -1$$

$$= \frac{(3 \pm \sqrt{41})}{4i} \cdot \frac{i}{i} \qquad\qquad \text{Rationalize the denominator}$$

$$= \frac{(3 \pm \sqrt{41})i}{4i^2}$$

$$= \frac{-(3 \pm \sqrt{41})i}{4} \qquad\qquad i^2 = -1$$

The solution set is

$$\left\{\frac{-(3 + \sqrt{41})i}{4}, \frac{-(3 - \sqrt{41})i}{4}\right\}. \quad \blacksquare$$

The radicand from the quadratic formula, $b^2 - 4ac$, allows one to determine the nature of the solutions to a quadratic equation without actually solving the equation. As such, $b^2 - 4ac$ is called the **discriminant** of the quadratic equation. To see how the discriminant is used, consider the two solutions to the quadratic equation, which we will designate as

$$x_1 = \frac{-b + \sqrt{b^2 - 4ac}}{2a} \quad \text{and} \quad x_2 = \frac{-b - \sqrt{b^2 - 4ac}}{2a}.$$

If $b^2 - 4ac = 0$, then

$$x_1 = \frac{-b + 0}{2a} = \frac{-b}{2a} \quad \text{and} \quad x_2 = \frac{-b - 0}{2a} = \frac{-b}{2a}.$$

There is only one solution of multiplicity two (see Section 5.5), namely $\frac{-b}{2a}$. If $b^2 - 4ac > 0$, then the radicand will be positive and, although the solutions will be real numbers, they will be unequal. If $b^2 - 4ac < 0$ the radicand will be negative. In this case the solutions will be complex conjugates. These observations are outlined below.

Discriminant	Nature of Solutions	Example
$b^2 - 4ac = 0$	One real solution of multiplicity two	$x^2 + 6x + 9 = 0$ The solution set is $\{-3\}$.
$b^2 - 4ac > 0$	Two real solutions	$x^2 - 4x + 3 = 0$ The solution set is $\{1, 3\}$.
$b^2 - 4ac < 0$	Complex conjugates	$x^2 + x + 1 = 0$ The solution set is $\left\{\frac{-1}{2} + \frac{\sqrt{3}i}{2}, \frac{-1}{2} - \frac{\sqrt{3}i}{2}\right\}$.

Example 4 Determine the nature of the solutions to each of the following.

(a) $3p^2 + 4p + 7 = 0$

$$b^2 - 4ac = 4^2 - 4 \cdot 3 \cdot 7 = -68 < 0$$

The solutions are complex conjugates.

(b) $4t^2 - 3t - 2 = 0$

$$b^2 - 4ac = (-3)^2 - 4 \cdot 4 \cdot (-2) = 41 > 0$$

There are two real solutions.

(c) $4m^2 - 20m + 25 = 0$

$$b^2 - 4ac = (-20)^2 - 4 \cdot 4 \cdot 25 = 0$$

There is one solution of multiplicity two. $\quad \blacksquare$

As a final topic in this section we will consider the relationship between the solutions of a quadratic equation.

Theorem 8.2

If $x_1 = \dfrac{-b + \sqrt{b^2 - 4ac}}{2a}$ and $x_2 = \dfrac{-b - \sqrt{b^2 - 4ac}}{2a}$ are solutions to the quadratic equation $ax^2 + bx + c = 0$, then

1. $x_1 + x_2 = \dfrac{-b}{a}$ 2. $x_1 \cdot x_2 = \dfrac{c}{a}$.

We will prove the first part of the theorem.

$$x_1 + x_2 = \frac{-b + \sqrt{b^2 - 4ac}}{2a} + \frac{-b - \sqrt{b^2 - 4ac}}{2a}$$

$$= \frac{-b + \sqrt{b^2 - 4ac} - b - \sqrt{b^2 - 4ac}}{2a}$$

$$= \frac{-2b}{2a} = \frac{-b}{a}$$

The second part is left as Exercise 91.

These relationships can be used to check the solutions to a quadratic equation.

Example 5   Solve and check $3x^2 + 6x + 1 = 0$.
From Example 1 the solution set is

$$\left\{ \frac{-3 + \sqrt{6}}{3}, \frac{-3 - \sqrt{6}}{3} \right\}.$$

Check: $x_1 + x_2 = \dfrac{-b}{a} = -2$ and $x_1 \cdot x_2 = \dfrac{c}{a} = \dfrac{1}{3}$.

Let $x_1 = \dfrac{-3 + \sqrt{6}}{3}$ and $x_2 = \dfrac{-3 - \sqrt{6}}{3}$.

$$x_1 + x_2 = \frac{-3 + \sqrt{6} + (-3 - \sqrt{6})}{3} = \frac{-6}{3} = -2$$

$$x_1 \cdot x_2 = \left(\frac{-3 + \sqrt{6}}{3} \right)\left(\frac{-3 - \sqrt{6}}{3} \right) = \frac{9 - 6}{9} = \frac{1}{3} \quad \blacksquare$$

Example 6   Find an equation with solutions $2 + i$ and $2 - i$.
Consider what happens when both sides of $ax^2 + bx + c = 0$ are divided by a.

$$ax^2 + bx + c = 0$$

$$x^2 + \frac{b}{a}x + \frac{c}{a} = 0$$

$$x^2 + \frac{b}{a}x + \frac{c}{a} = 0$$

The negative of the sum of $x_1 + x_2$

The product of x_1 and x_2

$$x_1 + x_2 = 2 + i + 2 - i = 4 = \frac{-b}{a}.$$

Thus $\dfrac{b}{a} = -4$.

$$x_1 \cdot x_2 = (2 + i)(2 - i) = 5 = \frac{c}{a}$$

The equation is $x^2 - 4x + 5 = 0$. ∎

Exercises 8.3

Solve each of the following by using the quadratic formula. Write in standard form first, if necessary.

1. $x^2 - 6x + 8 = 0$
2. $a^2 - a - 2 = 0$
3. $p^2 + 8p + 15 = 0$
4. $y^2 - 7y - 18 = 0$
5. $2w^2 - 7w - 4 = 0$
6. $3s^2 - 10s + 3 = 0$
7. $3u^2 + 3u - 6 = 0$
8. $4m^2 + m - 3 = 0$
9. $2q^2 + 5q - 12 = 0$
10. $v^2 + v + 1 = 0$
11. $y^2 + 3y + 4 = 0$
12. $x^2 - x + 1 = 0$
13. $3a^2 + 2a + 2 = 0$
14. $c^2 - 7c - 2 = 0$
15. $3n^2 + 6 = -19n$
16. $2d^2 + 5d = 12$
17. $2k^2 + 3k = 2$
18. $t^2 + 2t = -11$
19. $-4v^2 + 3 = 2v$
20. $5r^2 + 10r = 1$
21. $c(c - 1) = 5$
22. $u(u - 8) = 6$
23. $x(x + 3) = 1$
24. $12p(3p + 2) = 1$
25. $\dfrac{z^2}{7} + \dfrac{z}{4} = -4$
26. $\dfrac{p^2}{5} + 3 = \dfrac{p}{4}$
27. $\dfrac{a^2}{4} + 3a = 5$
28. $\dfrac{w^2}{2} + 3w = 1$
29. $x^2 + 12 = 0$ (Hint: $b = 0$.)
30. $y^2 - 16 = 0$
31. $p^2 - 18 = 0$
32. $k^2 + 28 = 0$
33. $s^2 + 5s = 0$ (Hint: $c = 0$.)
34. $w^2 - 7w = 0$
35. $5m^2 + 3m = 0$
36. $4w^2 - 9w = 0$
37. $(x - 2)^2 = -8$
38. $(2c - 3)^2 = -5$
39. $(a - 2)^2 = 15$
40. $(4p + 1)^2 = 1$
41. $4iv^2 + 7i = 12v$
42. $2it^2 + 5i = 4t$
43. $5ik^2 + 4k - 5i = 0$
44. $2iy^2 - 5i = y$

Use the discriminant to determine the nature of the solutions for each of the following.

45. $x^2 - 3x + 2 = 0$
46. $y^2 - 7y - 8 = 0$
47. $a^2 + a + 1 = 0$
48. $c^2 - c - 3 = 0$
49. $3v^2 - 4v + 2 = 0$
50. $b^2 - 4b + 1 = 0$
51. $4k^2 + 8k + 1 = 0$
52. $2p^2 - p + 2 = 0$
53. $9q^2 - 24q + 16 = 0$
54. $16t^2 + 40t + 25 = 0$
55. $2h^2 + 5h + 4 = 0$
56. $12d^2 + d - 6 = 0$

Find all real values of K for which the solutions to the given equation are as indicated.

57. $a^2 + 6a + K = 0$; two real solutions

58. $p^2 - 7p + K = 0$; two real solutions

59. $p^2 + 8p + K = 0$; one solution

60. $c^2 - 10c + K = 0$; one solution

61. $Km^2 - m - 1 = 0$; complex conjugate solutions

62. $Ky^2 - y - 1 = 0$; complex conjugate solutions

63. $Kd^2 - 2d + 1 = 0$; complex conjugate solutions

64. $Kt^2 + 6t + 5 = 0$; complex conjugate solutions

65. $Kv^2 - 3v + 8 = 0$; two real solutions

66. $Kr^2 - 3r - 2 = 0$; two real solutions

Write the quadratic equation whose solutions x_1 and x_2 satisfy the following conditions. Use Theorem 8.2.

67. $x_1 = 5$, $x_2 = -3$

68. $x_1 = -9$, $x_2 = -6$

69. $x_1 = \dfrac{2 + i}{3}$, $x_2 = \dfrac{2 - i}{3}$

70. $x_1 = \dfrac{1 + \sqrt{3}}{2}$, $x_2 = \dfrac{1 - \sqrt{3}}{2}$

71. $x_1 = \sqrt{7}$, $x_2 = -\sqrt{7}$

72. $x_1 = 5 + \sqrt{2}i$, $x_2 = 5 - \sqrt{2}i$

Solve by any method. Check Exercises 75 through 78 by Theorem 8.2.

73. $(2x + 5)^2 = -11$

74. $(3a - 7)^2 = -15$

75. $25k^2 - 20k + 4 = 0$

76. $8t^2 - 10t + 3 = 0$

77. $15 - 7v - 4v^2 = 0$

78. $5 - 3z^2 + 2z = 0$

79. $(8x + 19)x = 27$

80. $(24u + 94)u = 25$

81. $36 + \dfrac{60}{s} = \dfrac{-25}{s^2}$

82. $\dfrac{10c + 19}{c} = \dfrac{15}{c^2}$

83. $m - 16 = \dfrac{105}{m}$

84. $25v - 40 + \dfrac{16}{v} = 0$

85. $\dfrac{2x + 11}{2x + 8} = \dfrac{3x - 1}{x - 1}$

86. $\dfrac{2r - 5}{2r + 1} + \dfrac{6}{2r - 3} = \dfrac{7}{4}$

87. $\sqrt{10 - x} - 8 = 2x$

88. $\sqrt{16 - p} + 4 = p$

89. $\sqrt{5b + 1} = 1 + \sqrt{3b}$

90. $6\sqrt{h} - 3 = \sqrt{10h - 1}$

91. Prove part 2 of Theorem 8.2.

8.4 Equations Involving Quadratics

The solutions to many equations that are not quadratic require, in part, the techniques used to solve quadratic equations. For example, consider

$$x^3 - 1 = 0.$$

In Section 5.3 it was shown that $x^3 - 1$ is factored as the difference of two cubes:

$$(x - 1)(x^2 + x + 1) = 0.$$

When the zero-factor theorem is applied, we have

$$x - 1 = 0 \quad \text{or} \quad x^2 + x + 1 = 0.$$

The solution to the first equation, $x - 1 = 0$, is $x = 1$. The second equation, $x^2 + x + 1 = 0$, is quadratic and can be solved by the formula.

$$x^2 + x + 1 = 0$$

$$x = \frac{-1 \pm \sqrt{1-4}}{2}$$

$$x = \frac{-1 \pm \sqrt{3}i}{2}$$

The solution set is

$$\left\{1, \frac{-1+\sqrt{3}i}{2}, \frac{-1-\sqrt{3}i}{2}\right\}.$$

Notice that the equation is of degree 3 and that it has three solutions. The following theorem relates the number of solutions of a polynomial equation to its degree.

Theorem 8.3   A polynomial of degree n has at most n distinct solutions.

For example, a third-degree equation can have one, two, or three distinct solutions. If the equation has only two distinct solutions, one solution must be of multiplicity two. If it has one distinct solution it is of multiplicity three.

Example 1   Solve $x^4 - 3x^2 + 2 = 0$.

The equation is of degree 4. It can have at most four solutions.

$$x^4 - 3x^2 + 2 = 0$$
$$(x^2 - 2)(x^2 - 1) = 0$$
$$x^2 = 2 \quad \text{or} \quad x^2 = 1$$
$$x = \pm\sqrt{2} \quad \text{or} \quad x = \pm 1$$

The solution set is $\{-1, 1, \sqrt{2}, -\sqrt{2}\}$. ∎

Example 2   Solve $p^6 - 64 = 0$.

The equation is of degree 6. It can have at most six distinct solutions. The equation $p^6 - 64$ can be factored as the difference of squares.

$$(p^3)^2 - 8^2 = 0$$
$$(p^3 - 8)(p^3 + 8) = 0$$

Each of these can be factored further.

$$(p - 2)(p^2 + 2p + 4)(p + 2)(p^2 - 2p + 4) = 0$$
$$p - 2 = 0 \quad \text{or} \quad p^2 + 2p + 4 = 0 \quad \text{or} \quad p + 2 = 0 \quad \text{or} \quad p^2 - 2p + 4 = 0$$

Two solutions are 2 and -2. When the quadratic factors are solved by using the formula, the solution set is found to be $\{2, -2, 1 + \sqrt{3}i, 1 - \sqrt{3}i, -1 + \sqrt{3}i, -1 - \sqrt{3}i\}$. ∎

When an equation is not quadratic it can often be converted to quadratic form by an appropriate substitution.

Example 3   Solve $s^{1/2} + 4s^{1/4} - 5 = 0$.

To write the equation in quadratic form, let $m = s^{1/4}$. Then

$$m^2 = (s^{1/4})^2 = s^{1/2}.$$

Substituting for $s^{1/4}$ and $s^{1/2}$ yields the following.

$$m^2 + 4m - 5 = 0$$

$$(m + 5)(m - 1) = 0$$

$$m + 5 = 0 \quad \text{or} \quad m - 1 = 0$$

$$m = -5 \quad \text{or} \quad m = 1$$

Now "back substitute" $s^{1/4}$ for m.

$$s^{1/4} = -5 \quad \text{or} \quad s^{1/4} = 1$$

$$s = (-5)^4 \quad \text{or} \quad s = 1^4 \qquad \text{Power property, raise each side}$$
$$\text{to the fourth power}$$

$$s = 625 \qquad\qquad s = 1$$

Since the power property was used, the solutions must be checked.

Check: $(625)^{1/2} + 4(625)^{1/4} - 5 \overset{?}{=} 0$ $\qquad (1)^{1/2} + 4(1)^{1/4} - 5 \overset{?}{=} 0$

$\qquad\qquad\qquad 25 + 4(5) - 5 \overset{?}{=} 0 \qquad\qquad 1 + 4 \cdot 1 - 5 \overset{?}{=} 0$

$\qquad\qquad\qquad\qquad\qquad 40 \neq 0 \qquad\qquad\qquad\qquad 0 \overset{\checkmark}{=} 0$

The solution set is $\{1\}$.   ■

Example 4   Solve $2x^{-2} + 5x^{-1} + 3 = 0$.

Let $m = x^{-1}$ then $m^2 = (x^{-1})^2 = x^{-2}$

$$2m^2 + 5m + 3 = 0 \qquad \text{Substitute } m = x^{-1}$$

$$(2m + 3)(m + 1) = 0$$

$$2m + 3 = 0 \quad \text{or} \quad m + 1 = 0$$

$$m = -\frac{3}{2} \quad \text{or} \quad m = -1$$

Now back substitute for m.

$$x^{-1} = \frac{-3}{2} \quad \text{or} \quad x^{-1} = -1$$

$$\frac{1}{x} = \frac{-3}{2} \quad \text{or} \quad \frac{1}{x} = -1 \qquad x^{-1} = \frac{1}{x}$$

Clear of fractions and solve for x.

$$2 = -3x \quad \text{or} \quad x = -1$$

$$x = -\frac{2}{3}$$

Check to verify that the solution set is $\{-1, \frac{-2}{3}\}$.   ■

Example 5 Solve $(y - 2)^4 + (y - 2)^2 - 2 = 0$.

This equation is of degree 4. There are at most four distinct solutions. Let $m = (y - 2)^2$; then $m^2 = (y - 2)^4$.

$$m^2 + m - 2 = 0 \quad \text{Substitute } m = (y - 2)^2$$
$$(m + 2)(m - 1) = 0$$
$$m + 2 = 0 \quad \text{or} \quad m - 1 = 0$$
$$m = -2 \quad \text{or} \quad m = 1$$

Thus $(y - 2)^2 = -2$ or $(y - 2)^2 = 1$. Each equation can be solved by the square root method. The solution set is $\{1, 3, 2 + \sqrt{2}i, 2 - \sqrt{2}i\}$. ■

Example 6 Two men working together complete a job in four days. If each had worked on the job alone the slower worker would have required six days longer than the faster worker. How long would the faster worker take to do the job alone?

Let the faster worker's time to do the job be d days. The slower worker would take $d + 6$ days.

	Fractional Part Done in One Day	Fractional Part Done in Four Days
Faster worker	$\dfrac{1}{d}$	$\dfrac{4}{d}$
Slower worker	$\dfrac{1}{d + 6}$	$\dfrac{4}{d + 6}$

The complete job includes the fractional parts done by each worker in four days. Thus,

$$\frac{4}{d} + \frac{4}{d + 6} = 1.$$
$$4(d + 6) + 4d = d(d + 6)$$
$$d^2 - 2d - 24 = 0$$
$$(d - 6)(d + 4) = 0$$
$$d = 6 \quad \text{or} \quad d = -4 \quad \text{Reject } -4$$

The faster worker would take six days. ■

Exercises 8.4

Factor and solve.

1. $x^4 - 3x^2 + 2 = 0$
2. $k^4 - 5k^2 + 4 = 0$
3. $p^4 - 6p^2 + 9 = 0$
4. $w^4 - 8w^2 + 16 = 0$
5. $y^4 - 3y^2 - 54 = 0$
6. $m^4 - 13m^2 + 36 = 0$
7. $t^5 + 5t^3 + 6t = 0$
8. $r^5 + r^3 - 56r = 0$
9. $2z^4 - 30z^2 + 112 = 0$
10. $3x^4 + 30x^2 + 63 = 0$
11. $a^3 - 8 = 0$
12. $c^3 - 27 = 0$
13. $s^3 + 1 = 0$
14. $n^3 + 64 = 0$
15. $x^4 + 125x = 0$
16. $y^4 + y = 0$
17. $6k^4 - 6k = 0$
18. $5z^4 - 320z = 0$

Factor as the difference of squares, then solve the resulting equations.

19. $p^6 - 1 = 0$ **20.** $r^6 - 64 = 0$ **21.** $3k^6 - 3 = 0$

22. $7m^6 - 7 = 0$ **23.** $64x^6 - 729 = 0$ **24.** $3y^7 - 192y = 0$

Use an appropriate substitution to change each of the following into a quadratic equation and solve.

25. $t^{1/3} - 4t^{1/6} + 3 = 0$ **26.** $y + 2y^{1/2} - 3 = 0$ **27.** $m^{2/3} - m^{1/3} - 12 = 0$

28. $a^{4/3} + 7a^{2/3} + 12 = 0$ **29.** $2p^{2/5} + 5p^{1/5} + 2 = 0$ **30.** $r^{1/2} - 5r^{1/4} + 6 = 0$

31. $(x^2 - x)^2 - 4(x^2 - x) - 12 = 0$ **32.** $(z^2 + 2z)^2 - (z^2 + 2z) - 6 = 0$

33. $4(t^2 + 1)^2 - 7(t^2 + 1) = 2$ **34.** $4(m^2 - 1)^2 - 12(m^2 - 1) = -8$

35. $(2r^2 - 1)^2 + 11(2r^2 - 1) = -30$ **36.** $(k^2 + 2)^2 + 13(k^2 + 2) + 12 = 0$

37. $6q^{-2} - 5q^{-1} - 6 = 0$ **38.** $3s^{-2} - 11s^{-1} = 20$

Solve each of the following. State any restrictions on the variables.

39. $\dfrac{2}{x - 1} + \dfrac{5}{x - 7} = \dfrac{4}{3x - 1}$ **40.** $\dfrac{1}{a} + \dfrac{3}{a - 1} - \dfrac{2}{a + 1} = 0$

41. $\dfrac{1}{2} - \dfrac{1}{p - 1} = \dfrac{6}{p^2 - 1}$ **42.** $\dfrac{m}{m - 2} - 1 + m = \dfrac{1 - m}{2}$

43. $\dfrac{2a + 7}{a^2 - 5a + 4} - \dfrac{a - 8}{a^2 - 4a + 3} = \dfrac{a}{a^2 - 7a + 12}$ **44.** $\dfrac{k - 7}{k + 4} - \dfrac{k + 3}{k} = \dfrac{3}{k}$

Solve each of the following and check all solutions.

45. $\sqrt[3]{x^2 - x - 6} - 2 = 0$ **46.** $\sqrt{2z + 1} - \sqrt{z} = 1$

47. $a^2 - 6a - \sqrt{a^2 - 6a} - 3 = 3$ **48.** $\dfrac{4 - x}{\sqrt{x^2 - 8x + 32}} = \dfrac{3}{5}$

(*Hint:* use an appropriate substitution.)

49. $\sqrt{x} - \dfrac{2}{\sqrt{x}} = 1$ **50.** $\sqrt{\sqrt{x + 16} - \sqrt{x}} = 2$

Rewrite each of the following as a quadratic equation and solve.

51. Working together, two cranes can unload a ship in four hours. The slower crane working alone requires six hours more time than the faster crane to unload the ship. How long would it take each crane alone to unload the ship?

52. A roofer and his helper can finish a roofing job in four hours. Working alone, the roofer can finish the job in six hours less time than his helper. How long would it take each working alone to finish the job?

53. Computers A and B together can complete a data processing job in two hours. Working alone, computer A can do the job in three hours less time than B. How long would it take each computer working alone to complete the job?

54. Lydia and Myrna can do an advertising layout in six days together. Working alone, Myrna takes sixteen days longer than Lydia to do the layout. How long would it take for each person working alone to do the layout?

55. Beverly flies a distance of 600 miles. She could fly the same distance in 30 minutes less time by increasing her average speed by 40 miles per hour. Find her average speed.

56. Christi travels 40 miles in 45 minutes and then completes her trip of 50 more miles at the same rate. How long did it take her to travel the last 50 miles?

57. Troy traveled 36 kilometers down a river and back in 8 hours. If the rate of the boat in still water is 12 kilometers per hour, what is the rate of the current?

58. Bob and Maria start at the same time from the same place and travel along roads that are perpendicular to each other. Maria travels four miles per hour faster than Bob. At the end of two hours they are 40 miles apart. How fast is each one traveling?

8.5   Quadratic Inequalities

A **quadratic inequality** is one that can be written in the form

$$ax^2 + bx + c < 0$$

or in any alternate form where the inequality symbol is $>$, $\leq$, or $\geq$.

Quadratic inequalities are solved by a variety of methods. One of the easier methods is based on the following observations.

1. If $a \cdot b > 0$, ($a \cdot b$ is positive), then a and b have the same sign.

2. If $a \cdot b < 0$, ($a \cdot b$ is negative), then a and b have opposite signs.

For example, to solve the inequality $x^2 - x - 12 > 0$, first factor to write it in the form $a \cdot b > 0$.

$$(x - 4)(x + 3) > 0$$

Observe that when $x = 4$, the first factor, $x - 4$, is 0. Values greater than 4 make the factor positive and values less than 4 make it negative. On a number line this appears as Figure 8.1.

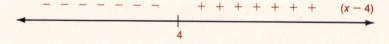

Figure 8.1

The negative signs to the left of 4 indicate that the factor is negative for $x < 4$, and the positive signs to the right of 4 indicate that the factor is positive for $x > 4$. Such a line is called a **sign graph.** Similarly, $x + 3$ is positive for $x > -3$ and negative for $x < -3$. The sign graph for $x + 3$ is shown in Figure 8.2.

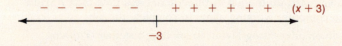

Figure 8.2

To find the values for which $(x - 4)(x + 3) > 0$, construct the sign graphs of the factors on the same line, as in Figure 8.3.

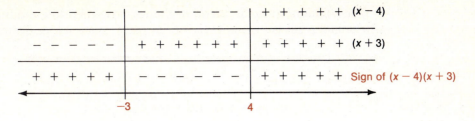

Figure 8.3

The product, $(x - 4)(x + 3)$, will be positive if x is greater than 4 or if x is less than -3. The solution set is the union of the regions where the product is positive. The solution set is

$$(-\infty, -3) \cup (4, \infty).$$

Example 1   Solve $x^2 - 6x + 5 \leq 0$.

$(x - 5)(x - 1) \leq 0$ Factor first

Now construct a sign graph showing both factors, as in Figure 8.4.

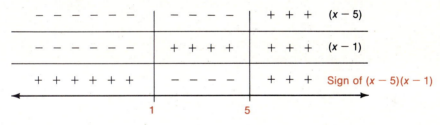

Figure 8.4

The product must be negative or zero. This value occurs for x between 1 and 5 inclusive. The solution set is [1, 5]. ■

Sign graphs can be used to solve inequalities involving polynomials of degree greater than 2 as well as those involving quotients of polynomials.

Example 2   Solve $y(y + 2)(y - 5) > 0$ and graph the solution set.
Construct a sign graph showing each factor, as in Figure 8.5.

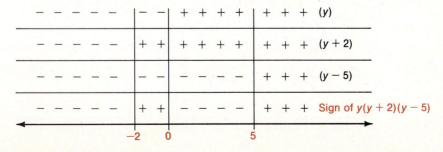

Figure 8.5

The solution set is the union of the regions where the product is positive. The solution set is $(-2, 0) \cup (5, \infty)$. Its graph is shown in Figure 8.6. ■

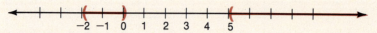

Example 3   Solve $\dfrac{x+2}{x-3} \le 2$, $x \ne 3$.

The tendency here is to clear the inequality of fractions by multiplying each side by $x - 3$. Since we don't know whether $x - 3$ is positive or negative, however, we don't know whether to leave the inequality sign as is or to reverse its direction. To avoid this difficulty, proceed as follows.

$$\frac{x+2}{x-3} \le 2$$

$$\frac{x+2}{x-3} - 2 \le 0 \qquad \text{Subtract 2 from each side}$$

$$\frac{x+2-2(x-3)}{x-3} \le 0 \qquad \text{LCD} = x - 3$$

$$\frac{x+2-2x+6}{x-3} \le 0$$

$$\frac{-x+8}{x-3} \le 0$$

In Figure 8.7 note that values of x greater than 8 make the factor $-x + 8$ negative.

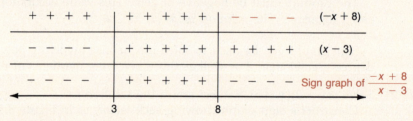

Figure 8.7

The solution set is $(-\infty, 3) \cup [8, \infty)$. ■

Example 4   Solve $\dfrac{2x}{x-2} \le \dfrac{x}{x-3}$, $x \ne 2, 3$.

$$\frac{2x}{x-2} - \frac{x}{x-3} \le 0 \qquad \text{Subtract } \frac{x}{x-3} \text{ from each side}$$

$$\frac{x^2 - 4x}{(x - 2)(x - 3)} \le 0 \qquad \text{LCD} = (x - 2)(x - 3)$$

$$\frac{x(x - 4)}{(x - 2)(x - 3)} \le 0$$

The sign graph is shown in Figure 8.8.

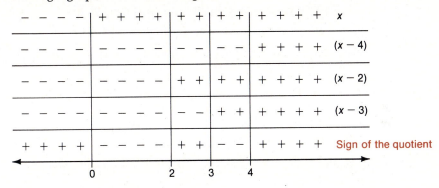

Figure 8.8

The solution set is $[0, 2) \cup (3, 4]$. ∎

Example 5   Solve and graph $-7 < x^2 - 8 \le 8$.

$$-7 < x^2 - 8 \le 8$$
$$1 < x^2 \le 16 \qquad\qquad\qquad \text{Add 8 to each part}$$

If $-4 \le x < -1$,   then $1 < x^2 \le 16$.
If $1 < x \le 4$,   then $1 < x^2 \le 16$.

The solution set is $[-4, -1) \cup (1, 4]$. The graph is shown in Figure 8.9. ∎

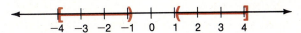

Figure 8.9

Example 6   A ball thrown vertically upward with an initial velocity of 96 feet per second reaches a height of $h = 96t - 16t^2$ after t seconds.

(a) When will the ball be on the ground?
   The ball will be on the ground when $h = 0$.

$$0 = 96t - 16t^2$$
$$0 = 16t(6 - t)$$
$$16t = 0 \quad \text{or} \quad 6 - t = 0$$
$$t = 0 \qquad\qquad t = 6$$

The ball is on the ground when $t = 0$ or $t = 6$.

(b) At what time is the ball more than 80 feet from the ground?

The ball will be more than 80 feet from the ground when $h > 80$. Thus

$$96t - 16t^2 > 80.$$

$$0 > 16t^2 - 96t + 80$$

$$0 > 16(t^2 - 6t + 5)$$

Dividing each side by 16 and factoring yields

$$0 > (t - 5)(t - 1) \quad \text{or} \quad (t - 5)(t - 1) < 0.$$

The reader should construct the sign graph to verify that the solution set is the interval $(1, 5)$. ∎

Exercises 8.5

Solve each of the following and graph as indicated.

1. $y(y + 1) < 0$

2. $p(p + 2) < 0$

3. $(a + 1)(a - 2) < 0$; graph

4. $(x + 3)(x - 1) < 0$; graph

5. $(w - 2)(w + 3) \leq 0$

6. $(h + 4)(h - 1) \leq 0$

7. $v(v - 3) > 0$; graph

8. $t(t - 4) > 0$; graph

9. $(m + 5)(m - 3) > 0$

10. $(2b - 1)(b + 6) > 0$

11. $(3c - 1)(2c + 3) \geq 0$; graph

12. $(4y + 3)(2y - 5) \geq 0$; graph

13. $(1 - x)(2 + x) \geq 0$

14. $(3 - a)(4 + a) \geq 0$

Factor each of the following and solve.

15. $h^2 - h - 12 < 0$

16. $r^2 + 4r + 3 < 0$

17. $a^2 + 7a + 6 \leq 0$

18. $x^2 - 5x - 6 \leq 0$

19. $y^2 + 8y + 16 > 0$

20. $t^2 + 6t + 9 > 0$

21. $u^2 - 9u - 22 \geq 0$

22. $p^2 - 7p - 18 \geq 0$

23. $w^2 - 4w + 4 \leq 0$

24. $h^2 - 10h + 25 \leq 0$

25. $r^2 - 9 < 0$

26. $v^2 - 16 < 0$

27. $6b^2 + 7b > 0$

28. $5c^2 - 6c > 0$

Solve each of the following.

29. $\dfrac{c}{c + 2} < 0$

30. $\dfrac{m}{m - 3} < 0$

31. $\dfrac{1}{a + 4} \leq 1$

32. $\dfrac{1}{w - 2} \leq 1$

33. $\dfrac{t + 3}{t + 2} < 1$

34. $\dfrac{v + 5}{v + 2} < 2$

35. $\dfrac{w + 2}{w + 4} \geq 3$

36. $\dfrac{s + 2}{s + 9} \geq 1$

37. $\dfrac{b - 1}{b - 5} \leq 2$

38. $\dfrac{2v + 6}{v - 1} \geq 5$

39. $\dfrac{4}{h - 3} > \dfrac{-2}{h + 11}$

40. $\dfrac{9}{3y + 4} > \dfrac{1}{2y + 1}$

41. $\dfrac{2x}{x + 2} \leq \dfrac{-2x}{x - 5}$

42. $\dfrac{2p}{p - 2} > \dfrac{3p}{4p + 1}$

Solve each of the following and graph as indicated.

43. $a(a - 1)(a + 2) \leq 0$

44. $x(x + 1)(x - 2) \leq 0$

45. $(2y - 3)(3y - 1)y \geq 0$; graph

46. $(p - 5)(2p + 1)p \geq 0$; graph

47. $(3w - 4)(w + 1)w < 0$

48. $(t - 1)(2t + 5)t < 0$

49. $-2 < c^2 - 3 < 13$

50. $2 < h^2 + 1 < 5$

51. $-3 < m^2 - 4 < 5$; graph

52. $6 < k^2 + 5 < 9$; graph

53. $11 < s^2 + 7 < 16$

54. $8 < x^2 - 1 < 15$

Solve each of the following as indicated.

55. Lisa's business shows a profit or loss, P, per hour if she works t hours of an 8-hour day according to the formula $P = 15t - 3t^2$.
    (a) For what hours t during the day is her profit greater than zero?
    (b) For what values of t does her profit P become less than zero (a loss)?
    (c) How many hours should she work per day if her profit is to be at least $12.00 per hour?

56. The height h of a certain projectile above the ground is given by $h = 32t - 16t^2$, where t is the time in seconds.
    (a) For what range of values of t is the projectile above ground level?
    (b) At what time t does the projectile strike the ground?
    (c) For what range of values of t is the projectile less than 16 feet above the ground?

57. A ball is thrown directly upward from ground level at an initial velocity of 40 feet per second and attains a height h of $h = 40t - 16t^2$ after t seconds.
    (a) During what time interval is the ball at a height of at least 16 feet?
    (b) At what time will the ball fall back to the ground?
    (c) During what time interval is the ball less than 16 feet above the ground?

58. A manufacturer of solar heaters finds that when x units are made and sold, the profit (in thousands of dollars) is given by $P = x^2 - 50x - 5000$, $0 \le x \le 200$.
    (a) For what values of x will the firm show a loss?
    (b) For what values of x will the firm break even?
    (c) For what values of x will the firm show a profit?

59. The tensile strength S, in pounds per square inch, of a new plastic varies with the temperature according to $S = 500 + 600t - 20t^2$, where t is the temperature in degrees, $0 \le t < 50$.
    (a) For what temperature range is $S > 4500$?
    (b) For what temperature range is $S \le 4500$?
    (c) For what temperature, to the nearest degree, is the tensile strength 120 pounds per square inch?

60. A realtor charges the seller of a house 7% commission on the selling price. Suppose that the seller wants to clear at least $60,000.00 after paying the commission and that similar houses are selling for $70,000.00 or less.
    (a) What is the range of selling prices for the house?
    (b) If the house is sold, what is the range of the realtor's commission?

Review Exercises

Solve each of the following and write all solutions using interval notation.

61. $|a - 3| < 3$

62. $|5t + 1| < 11$

63. $|4x - 1| > 1$

64. $|3y + 6| \le 12$

65. $|3a - 1| \le 8$

66. $|2v - 6| > 4$

67. $|3y + 2| \ge -1$

68. $\left| \frac{1}{3} - h \right| < \frac{2}{3}$

69. $\left| \frac{5 - c}{3} \right| > 4$

70. $\left| \frac{2x + 1}{3} \right| \le 5$

71. $\left| \frac{3w - 2}{4} \right| < 1$

72. $\left| \frac{2k + 1}{3} \right| < 0$

8.6   More Applications

Applications of quadratic equations and inequalities are too numerous for this book to include an example of each type. The following examples and exercises cover a few of them. Since quadratic equations often have more than one solution, it is essential that the solutions be checked in the words of the problem. Remember to follow the five-step process outlined earlier.

Example 1   The length of a rectangle exceeds its width by 2 meters. If the area of the rectangle is 224 square meters, find its dimensions.

Let w = the width and

$w + 2$ = the length.

The area of a rectangle is $A = lw$.

$$w(w + 2) = 224$$
$$w^2 + 2w = 224$$
$$w^2 + 2w - 224 = 0$$
$$(w + 16)(w - 14) = 0$$
$$w + 16 = 0 \quad \text{or} \quad w - 14 = 0$$
$$w = -16 \quad \text{or} \quad w = 14$$

Since the width cannot be negative, $w = 14$ and $w + 2 = 16$.

Check: The length, 16, is 2 more than the width, 14. Since their product is 224, the solution checks.   ■

Example 2   A group of students charter a bus for a trip to Disneyland for $720.00. If they can persuade 12 more students to go, each student's cost will decrease by $2.00. How many students are in the group?

Let x = the number of students in the group.

	Number of Students	Total Cost	Cost Per Student
Original group	x	720	$\dfrac{720}{x}$
Increased group	$x + 12$	720	$\dfrac{720}{x + 12}$

Since the cost per student in the larger group decreases by $2.00 per student, we have

$$\frac{720}{x} - \frac{720}{x + 12} = 2.$$
$$720(x + 12) - 720x = 2x(x + 12)$$
$$8640 = 2x^2 + 24x$$
$$0 = x^2 + 12x - 4320$$
$$0 = (x - 60)(x + 72)$$
$$x = 60 \quad \text{or} \quad x = -72$$

The number of students must be greater than zero. There are 60 students in the group. ∎

Example 3   Two buses, an express and a local, leave Pismo Beach for Los Angeles, a distance of 200 miles. The express makes the trip in one hour less time and travels an average of 10 miles per hour faster. Determine the rate (speed) of each bus.

Let x = the rate of the local bus and

$x + 10$ = the rate of the express bus.

The distance for both buses is 200 miles, and $t = \frac{D}{r}$.

	t	r	D
Local bus	$\dfrac{200}{x}$	x	200
Express bus	$\dfrac{200}{x + 10}$	$x + 10$	200

The local bus takes one hour longer. Thus

$$\frac{200}{x} - \frac{200}{x + 10} = 1.$$

$$200(x + 10) - 200x = x(x + 10)$$

$$0 = x^2 + 10x - 2000$$

$$0 = (x - 40)(x + 50)$$

$$x - 40 = 0 \quad \text{or} \quad x + 50 = 0$$

$$x = 40 \quad \text{or} \quad x = -50$$

We reject $x = -50$. The local bus averages 40 miles per hour and the express bus 50 miles per hour. The check is left to the reader. ∎

Exercises 8.6

Solve each of the following.

1. The area of a triangular plate is 20 square meters. Find the base and height if the base is 6 meters longer than the height.

2. Determine the base and altitude of a triangle whose altitude exceeds the base by three feet, and whose area is 14 square feet.

3. What are the dimensions of a rectangular piece of metal whose length is 50 inches more than its width and whose area is 1875 square inches?

4. A rectangular plot of grass is 10 yards long and 4 yards wide. New strips of grass of equal width are to be planted at one end and one side. Find what width of strip is necessary to double the area of the original plot.

5. The hypotenuse of a right triangle is 13 centimeters long. Find the lengths of the sides of the triangle if one side is 7 centimeters longer than the other.

6. A rectangular field 250 meters long and 180 meters wide has a concrete walkway of uniform width as its border. If there are 2616 square meters of concrete in the walkway, what is the width of the walkway?

7. Mel and Nett working together can mow and water a lawn in three hours less time than Mel alone. Nett working alone takes one hour longer than Mel. Determine the time it takes for them to do the work together.

8. Mindy can review a set of books in 7 hours less time than Jill. Together they can review the books in 9 hours less time than Mindy alone. Find the time they take together to review the books.

9. Rico can row 20 miles downstream and back in 7.5 hours. If his rate of rowing in still water is six miles per hour, what is the rate of the stream?

10. Celia finds that a trip of 70 miles covered at her normal speed can be decreased by 15 minutes if she travels 5 miles per hour faster. What is her normal speed?

11. Two fuel lines of different sizes working together can fill a tank in 4 hours. If the larger line fills the tank in 4 hours less time than the smaller line, how much time is needed for the larger line to fill the tank?

12. A gardener sets 180 plants in rows. Each row contains the same number of plants. If there were 40 more plants in each row, he would need 6 less rows. How many rows are there?

13. The members of the chemistry department at a college agreed to contribute equal amounts of money to make up a scholarship fund of $280.00. Since then, three new members of the department have been hired; as a result, each member's share has been reduced by $30.00. How many members are there now?

14. Dave has scores on his first three tests of 65, 78, and 84. What range of scores on his fourth test will give him an average of less than 80 and no less than 70 for the four tests?

15. Sandra's scores on her first four tests in English are 68, 76, 83, and 63. What range of scores on a fifth test will give her an average less than 80 but no less than 70 for the five tests?

16. A rectangular solar collector is to have a height of 1.5 meters, but its length is still to be determined. What is the range of values for this length if the collector provides 400 watts per square meter, and if it must provide between 2000 and 3500 watts?

17. How far is the horizon to the nearest mile from a plane that is 31,680 feet high (6 miles)? (See the figure.)

18. How far is the horizon to the nearest mile from a plane that is 8 miles high? (See Exercise 17.)

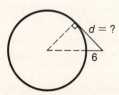

Radius of the
earth is
4000 miles.

19. What should the radius of the inner circle be so that the area of the inner circle is equal to the area between the circles?

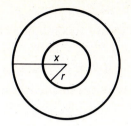

20. A Norman window (a semicircle on a rectangle) is as tall as it is wide, with an area of 16 square feet. How wide is the window to the nearest hundredth?

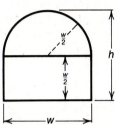

21. Matt bought several cartons of large eggs for $11.20. If he had bought extra-large eggs, which cost 10 cents more per carton, he would have had 2 less cartons of eggs for the same amount of money. How much did he pay for a carton of large eggs?

22. Nancy bought $28.00 worth of pizzas for a party. If she had bought the pizzas at a different pizza parlor, which charges 50 cents less per pizza, she would have had one more pizza for the same amount of money. How many pizzas did she buy?

23. One pipe can fill a tank in 5 hours less than another pipe. Together both pipes can fill the tank in 6 hours. How long does it take each pipe alone to fill the tank?

24. A drain takes 3 minutes longer to empty a sink full of water than it takes for the cold water faucet to fill the empty sink. If it takes 6 minutes to fill the sink with the drain accidentally left open, how long does it take to fill the empty sink with the drain closed?

25. Andy's boat travels 15 miles per hour in still water. It takes him 3 hours more to travel 60 miles up a river than it does to travel 60 miles down the river. What is the speed of the current?

26. A helicopter takes one hour longer to travel 200 miles against a 25-mile-per-hour wind than it does to travel the same distance with the wind. What is the speed of the helicopter in still air to the nearest hundredth?

Chapter 8   Summary

[8.1]   Quadratic equations can be solved by using the **zero-factor theorem,** which was discussed in Section 5.5. After factoring an equation, each factor is set equal to zero. The solutions to the resulting equations form the solution set.

The **square root method** gives another way to solve quadratic equations. To use this method, the side of the equation involving the variable must be a perfect square.

If $x^2 = a$, then $x = \pm\sqrt{a}$.

Results obtained in this manner may be real or imaginary numbers.

[8.2]   A third method for solving quadratic equations is called **completing the square.** To use this method, isolate all terms involving the variable on one side of the equation. Next, a quantity sufficient to make this side a perfect square must be added to each side of the equation so that the solution set can be found by the square root method. To complete the square, the coefficient of the second-degree term must be 1. The quantity to be added to each side is the square of one-half of the coefficient of the first-degree term.

To complete the square of $x^2 + Bx$,   add $\left(\dfrac{B}{2}\right)^2$.

[8.3]   When the method of completing the square is applied to the general quadratic equation

$ax^2 + bx + c = 0$,

the result is called the **quadratic formula:**

$$x = \frac{-b \pm \sqrt{b^2 - 4ac}}{2a}.$$

The quadratic formula provides a means of determining the nature of the solutions without actually solving the equation. The tool used to do this is called the **discriminant** of the equation. The discriminant, $b^2 - 4ac$, is the radicand of the quadratic formula.

If $b^2 - 4ac = \mathbf{0}$, there is one solution of **multiplicity two.**

If $b^2 - 4ac < \mathbf{0}$, the solutions are **complex conjugates.**

If $b^2 - 4ac > \mathbf{0}$, there are **two real solutions.**

[8.4]   Many equations can be put into quadratic form after a suitable substitution. Once the substitution has been made the equation can be solved by one of the four methods listed above.

If a polynomial is of degree n it can have at most n distinct solutions. When r solutions represent the same number, the equation is said to have a solution of multiplicity r.

[8.5]   Inequalities are solved by means of **sign graphs.** To use a sign graph, first write the inequality in factored form, then determine the regions in which each factor is positive or negative and plot these regions on a number line. The region satisfying the original inequality with respect to the product or quotient of these factors is the solution set.

[8.6]   To solve an applied problem involving quadratics, write a quadratic equation or inequality describing the conditions of the problem, then find the solutions by one of the above methods.

Chapters 7 and 8   Review

Simplify. Write all answers with positive exponents.

1. $(x^{1/2} - y^{1/2})(x^{1/2} + y^{1/2})$

2. $(x^{1/3} + y^{1/3})(x^{2/3} - x^{1/3}y^{1/3} + y^{2/3})$

3. $5a^{-1/2} - 6b^{-1/2}$

4. $(81a^{-8/3}b^{-4/3})^{3/4}$

5. $\left(\dfrac{a^{3n}}{b^{6n}}\right)^{-3/n}$

6. $\left(\dfrac{x^{3n-4}}{y^{n-2}}\right)^{4/n}$

Simplify. State any restrictions on the variables.

7. $\sqrt[3]{-8x^3}$

8. $\sqrt{y^6}$

9. $\sqrt[3]{(a+b)^9}$

10. $\sqrt{\dfrac{150}{6}}$

11. $\sqrt{\dfrac{x^3}{x}}$

12. $\sqrt{\dfrac{u^5}{p^4}}$

13. $\sqrt[4]{\dfrac{2}{27}}$

14. $\sqrt[4]{\dfrac{3x}{y}}$

15. $\dfrac{\sqrt[3]{4}}{\sqrt[3]{6}}$

16. $\sqrt{27} + \sqrt{12} - \sqrt{75}$

17. $y\sqrt{18yz} - \sqrt{32xy^3z}$

18. $(\sqrt{3}+\sqrt{5})(\sqrt{3}-\sqrt{5})$

19. $(\sqrt{6} - \sqrt{2})^2$

20. $\dfrac{\sqrt{3}+2}{\sqrt{2}}$

21. $\dfrac{\sqrt{5}-\sqrt{3}}{\sqrt{3}+\sqrt{5}}$

Solve. Check all solutions.

22. $\sqrt{x+3} = 5$

23. $\sqrt{t-4} = -\sqrt{3t+1}$

24. $\sqrt{x^2 - 5x + 4} = x$

25. $3 = \sqrt{x+1}\sqrt{x-1}$

26. $\sqrt[3]{y^3 - 7} = 1$

27. $x - 3 = -\sqrt{x+1}\sqrt{x+2}$

Perform the indicated operations and simplify.

28. i^{37}

29. $(1 + i)^2$

30. $(3 + 4i) - (5 - i)$

31. $(\sqrt{-8} + 3)(2i - \sqrt{-18})$

32. $(2 - i)^3$

33. $(4 + 3i)^2$

34. $\dfrac{2 + 3i}{2 - 3i}$

35. $\dfrac{6 + 7i}{5 - 3i}$

36. $\dfrac{5 - 2\sqrt{-50}}{4 + 3\sqrt{-8}}$

37. $\dfrac{3 - \sqrt{-5}}{\sqrt{-5} + 2}$

Solve for x and y.

38. $(3 - 4i)x - (2 - i)y = 7i$

39. $(x + y - 3)i + 7x + 2y = 5$

Solve by the square root method.

40. $(3a - 5)^2 = 16$

41. $(x - 7)^2 = 6$

42. $(7w - 5)^2 = -8$

43. $(2t + 11)^2 = -27$

Solve by completing the square.

44. $y^2 + 4y + 3 = -1$

45. $3s^2 + 7s - 8 = 0$

46. $6c^2 - c + 1 = 0$

47. $2h^2 + h + 4 = 0$

Solve by the quadratic formula.

48. $x^2 - 5x + 4 = 0$

49. $2w^2 + 3w - 5 = 0$

50. $3k^3 + k^2 + 3k = 0$

51. $3(m - 1)^2 = 5$

Use an appropriate substitution, if needed, and solve.

52. $x^3 - 8 = 0$

53. $t^4 - 6t^2 - 7 = 0$

54. $a^{1/3} - 7a^{1/6} + 6 = 0$

55. $p^{-2} - 11p^{-1} + 30 = 0$

56. $\left(\dfrac{u^2 + 3}{u}\right)^2 - 6\left(\dfrac{u^2 + 3}{u}\right) + 8 = 0$

57. $2r^6 - 5r^3 + 2 = 0$

Solve. Write all solutions in interval notation.

58. $(x - 5)(x + 1) < 0$

59. $h^2 + 3h - 4 \geq 0$

60. $\dfrac{1}{2a + 1} < \dfrac{9}{3a + 4}$

9

Relations, Functions, and Graphs

One of the most fundamental concepts in mathematics is the relationship between two or more quantities. The distance one travels in a given period of time is directly related to the average speed maintained. The area of a circle is directly related to the length of its radius, and the profit a merchant receives when selling an article depends in part on how much he paid the distributor for it. Mathematicians refer to such relationships as **relations.** This chapter is devoted to considering relations and a special subset of them called **functions.**

9.1 Relations and Functions

Suppose that a person averages 50 miles per hour on a trip between two cities. In one hour he travels 50 miles and in three hours 150 miles. The relationship between time and distance can be expressed using ordered pairs, as shown below.

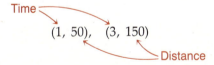

In fact, a relation is defined in terms of ordered pairs.

A **relation** is a set of ordered pairs.

For example, the set $A = \{(-1, 2), (2, 3), (3, 4), (5, 6)\}$ is a relation. The set $B = \{(x, y) \mid y = 2x + 3\}$ also is a relation. Ordered pairs in B can be obtained by assigning any real number value to x and computing the corresponding y-value. Thus $(0, 3)$, $(1, 5)$, $(-2, -1)$ and $(\frac{1}{2}, 4)$ all belong to B. The relation described in A has a **finite** number of elements. The one in B has an **infinite** number of elements—one corresponding to each real number replacement for x.

The set of all first elements in the ordered pairs of a relation is called its **domain.** The set of all second elements is called its **range.** In set A above, the domain is $\{-1, 2, 3, 5\}$ and the range is $\{2, 3, 4, 6\}$.

Example 1   Determine the domain and range of $C = \{(-2, 7), (4, -2), (3, -2)\}$.
The domain is $\{-2, 3, 4\}$. The range is $\{-2, 7\}$.   ■

To determine the domain and range of a relation whose ordered pairs are defined by equations such as

$$y = \frac{2x + 1}{x - 1} \quad \text{or} \quad y = \sqrt{x - 3},$$

one must determine the values that x and y cannot assume. These values must then be excluded. Such values generally fall into one of the following categories:

(1) Values that would make the denominator zero;
(2) Values that would require taking an even root of a negative number.

Unless otherwise specified, consider all variables to represent real numbers.

Example 2   Find the domain and range of each of the following.

(a) $\{(x, y) \mid y = 2x + 4, x \in N\}$
If x is replaced by any real number, y will be defined. The value of y will be zero when $x = -2$. If $x > -2$, y will be positive and if $x < -2$, y will be negative. Thus,

Domain $= \{x \mid x \in N\}$, and
Range $= \{y \mid y \in N\}$.

(b) $\left\{(x, y) \mid y = \dfrac{2x + 1}{x - 1}\right\}$

If $x = 1$, $y = \frac{3}{0}$, which is undefined. Any other value of x is permissible. Thus

Domain $= \{x \mid x \neq 1\}$.

The easiest way to determine the range is to solve the equation for x.

$$y = \frac{2x + 1}{x - 1}$$

$y(x - 1) = 2x + 1$ Clear of fractions
$xy - 2x = 1 + y$ Isolate x
$x(y - 2) = 1 + y$ Factor

$$x = \frac{1 + y}{y - 2}$$

From the last equation it is easy to see that $y \neq 2$. Thus

Range $= \{y \mid y \neq 2\}$.

(c) $\{(x, y) \mid y = \sqrt{x - 3}\}$
If $x < 3$, the radicand will be negative. The square root of a negative number is not real.

Domain $= \{x \mid x \geq 3\}$

Since y is the positive square root of $x + 3$, the smallest value it could have is zero ($y = 0$ when $x = 3$). Thus,

Range $= \{y \mid y \geq 0\}$. ■

A **function** is a special type of relation in which certain restrictions are placed on the ordered pairs.

> A **function** is a correspondence or rule that pairs each element in the domain with exactly one element in the range.

The relations $A = \{(1, 2), (3, 4), (5, 6)\}$ and $B = \{(1, 2), (3, 4), (5, 4)\}$ are functions, because each element in the domain corresponds to one element in the range. The lines in Figure 9.1 show the correspondence.

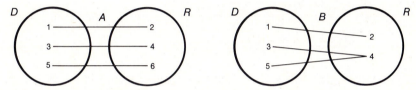

Figure 9.1

The relation $C = \{(1, 2), (2, 3), (2, 7)\}$ is not a function, since the element 2 in the domain corresponds to two elements, 3 and 7, in the range. (See Figure 9.2.)

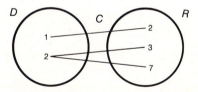

Figure 9.2

The equation $y = x^2$ describes a function, since each value of x corresponds to exactly one value of y. The equation $x = y^2$ does not describe a function, since for $x = 1$, $y = 1$ or $y = -1$. Each value of x does not correspond to exactly one value of y.

Functions are often represented by the letters f, g, or h. For example,

$f = \{(x, y) \mid y = 2x + 4\}$

indicates that the set of ordered pairs describes a function. The notation $f(x)$, read "function of x" or "f of x," is called **functional notation.** For example,

$f = \{(x, y) \mid y = 2x + 4\}$ in functional notation is $f(x) = 2x + 4$, where $f(x)$ illustrates that y depends both on the function and the value of x.

To find the value of f when $x = 1$, substitute 1 for each x in $f(x) = 2x + 4$.

$$f(1) = 2 \cdot 1 + 4 = 6$$

When $x = -3$,

$$f(-3) = 2(-3) + 4 = -2.$$

When $x = 0$,

$$f(0) = 2 \cdot 0 + 4 = 4.$$

Example 3   Rewrite each of the following sets using functional notation.

(a) $g = \{(x, y) \mid y = 3x^2 + 2x\}$
$$g(x) = 3x^2 + 2x$$

(b) $h = \left\{(x, y) \mid y = \dfrac{1}{x + 2}\right\}$

$$h(x) = \dfrac{1}{x + 2} \quad \blacksquare$$

Example 4   Given $g(x) = x^2 + 2x + 1$, find $g(1)$, $g(-2)$, and $g(a + 3)$.

$$g(1) = 1^2 + 2 \cdot 1 + 1 = 4 \qquad \text{Replace } x \text{ with } 1$$
$$g(-2) = (-2)^2 + 2(-2) + 1 = 1 \qquad \text{Replace } x \text{ with } -2$$
$$g(a + 3) = (a + 3)^2 + 2(a + 3) + 1 \qquad \text{Replace } x \text{ with } a + 3$$
$$= a^2 + 8a + 16 \quad \blacksquare$$

Example 5   Given $f(x) = x + 3$, find $\dfrac{f(x + h) - f(x)}{h}$.

To find $f(x + h)$, replace x with $x + h$.

$$f(x + h) = (x + h) + 3$$
$$\frac{f(x + h) - f(x)}{h} = \frac{(x + h) + 3 - (x + 3)}{h}$$
$$= \frac{x + h + 3 - x - 3}{h}$$
$$= 1 \quad \blacksquare$$

Example 6   A company that sells parts for repairing alternators has determined that the price in dollars per unit for selling x units is $P(x) = 50 - (0.3)x$, $1 \le x \le 20$. Find the price per unit when 4 units are sold and the price per unit when 20 units are sold.

When 4 units are sold,

$$P(4) = 50 - (0.3)4 = \$48.80.$$

When 20 units are sold,

$$P(20) = 50 - (0.3)20 = \$44.00. \quad \blacksquare$$

New functions are often created by **composition.** If f represents the function $f(x)$ and g the function $g(x)$, then $(f \circ g)(x) = f(g(x))$ is called the **composite** of f with g.

Example 7   Find each of the following if $f(x) = x^2 + x$ and $g(x) = 2x + 3$.

(a) $f(g(2))$

$$g(2) = 2 \cdot 2 + 3 = 7$$
$$f(g(2)) = f(7) = 7^2 + 7 = 56$$

(b) $g(f(2))$

$$f(2) = 2^2 + 2 = 6$$
$$g(f(2)) = g(6) = 2 \cdot 6 + 3 = 15$$

(c) $f(g(x))$

$$f(g(x)) = f(2x + 3)$$
$$= (2x + 3)^2 + (2x + 3)$$
$$= 4x^2 + 14x + 12 \quad \blacksquare$$

When the graph of a relation is known, it is easy to determine whether or not the relation is a function. If a vertical line intersects the graph of the relation in more than one point, the graph is not the graph of a function. This is called the **vertical line test.** Figure 9.3 illustrates the test.

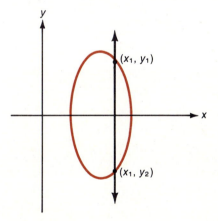

Figure 9.3

Since the x-coordinate is the same for both points on the vertical line $x = x_1$, there are two values of y for x_1. The relation is not a function.

Exercises 9.1

Determine whether or not each of the following relations is a function. Find the domain and range of each relation. Unless noted otherwise, $x \in R$.

1. $\{(1, 3), (2, 7), (3, 8), (-1, 2)\}$

2. $\{(-3, -7), (-2, 5), (0, 1), (2, 3)\}$

3. $\{(-4, 16), (0, -8), (2, -8), (6, 16)\}$

4. $\{(-5, -12), (-2, 3), (0, 3), (3, -12)\}$

5. $\{(3, 7), (2, -5), (3, -7), (6, -5)\}$

6. $\{(0, 6), (6, 0), (0, -6), (-6, 0)\}$

7. $\{(-1, 5), (0, -3), (2, -1), (-3, 9)\}$

8. $\{(0, 0), (2, 0), (3, 2), (5, 3)\}$

9. $\{(2, 1), (4, 3), (3, 4), (2, 0)\}$

10. $\{(1, 0), (-1, 8), (0, 1), (-1, 0)\}$

11. $\{(x, y)|y = 2x - 3, x \in N\}$

12. $\{(x, y)|y = -2x + 7, x \in W\}$

13. $\{(x, y)|y = |x|, -3 < x < 3\}$

14. $\{(x, y)|y = -|x|, -4 < x < 2\}$

15. $\{(x, y)|y = \sqrt{x + 2}\}$

16. $\{(x, y)|y = \sqrt{3 + x}\}$

17. $\left\{(x, y)|y = \dfrac{1}{x - 1}\right\}$

18. $\left\{(x, y)|y = \dfrac{2}{x + 3}\right\}$

19. $\{(x, y)|y = \sqrt{x^2 - 4}, x < 3\}$

20. $\{(x, y)|y = \sqrt{x^2 + 1}, x < 1\}$

21. $\left\{(x, y)|y = \dfrac{x - 4}{x + 3}\right\}$

22. $\left\{(x, y)|y = \dfrac{2x - 1}{x - 4}\right\}$

23. $\{(x, y)|y = -\sqrt{x + 3}\}$

24. $\{(x, y)|y = -\sqrt{2 - x}\}$

25. $\{(x, y)|y = \sqrt{25 - x^2}\}$

26. $\{(x, y)|y = \sqrt{36 - x^2}\}$

Find the indicated function values for each of the following.

27. $f(x) = x - 3; f(1), f(-1), f(3)$

28. $g(x) = 3x - 4; g(1), g(0), g(4/3)$

29. $h(x) = x^2 - 1; h(0), h(-1), h(1)$

30. $F(x) = x^2 - 4; F(0), F(2), F(-2)$

31. $g(x) = \sqrt{16 - x^2}; g(4), g(-4), g(0)$

32. $h(x) = \sqrt{x^2 + 5}; h(0), h(2), h(-2)$

33. $f(x) = 3x^2 + x - 8; f(0), f(2), f(3), f(-1)$

34. $F(x) = 6x^2 - 25x + 14; F(2/3), F(7/2), F(0), F(-1)$

35. $h(x) = 3x - 2; h(a), h(3b), h(-2c), h(t^2)$

36. $g(x) = 7x - 3; g(b), g(x + b), g(x - c), g(c^2 - 1)$

37. $f(x) = ax^2 + bx + c; f\left(\dfrac{-b}{2a}\right), f\left(\dfrac{b}{2a}\right), f\left(\dfrac{-b + \sqrt{b^2 - 4ac}}{2a}\right)$

38. $F(x) = \dfrac{x}{3x - 2}; F(0), F(2/3), F(t - 3)$

39. $f(x) = x^2 + 2; f(r), f(s), f(r + s), f(r) + f(s)$

40. $g(x) = x^2 - 9; g(a), g(b), g(a + b), g(a) + g(b)$

Find $\dfrac{f(x + h) - f(x)}{h}$ *for each of the following functions.*

41. $f(x) = x$

42. $f(x) = x - 5$

43. $f(x) = 2x + 7$

44. $f(x) = 3x - 2$

45. $f(x) = -5x + 4$

46. $f(x) = -8x + 1$

47. $f(x) = x^2 + 1$

48. $f(x) = x^2 - 2$

49. $f(x) = -2x^2 - 5$

50. $f(x) = -3x^2 + 2$

Use the vertical line test to determine whether or not each of the following is a function.

51.

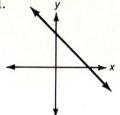

52.

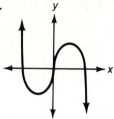

53.

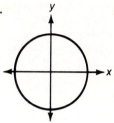

54.

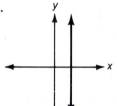

55.

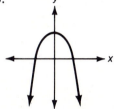

56.

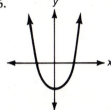

57.

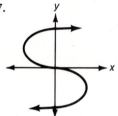

58.

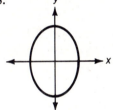

Each of the following defines a function. Solve as indicated.

59. The value $V(t)$ in dollars of factory equipment depreciates with time t in years according to the function

$$V(t) = 15{,}000 - 1000t.$$

(a) What is the value of the equipment after 15 years?
(b) What is the value of the equipment when new ($t = 0$)?
(c) At what time t would the equipment have lost one-half of its value?

60. The cost $C(x)$ in dollars of producing x pens is given by

$$C(x) = \frac{1}{3}x + 212.$$

(a) What is the cost of producing 30 pens?
(b) How many pens could be produced for a cost of $272.00?
(c) How many pens should be produced so that the cost for each pen is no more than 74¢?

61. A company finds its daily cost function to be $C(n) = 100 + 50n$ and its revenue function to be $R(n) = 102n - n^2$, where n is the number of units produced and sold, and cost and revenue are in dollars.

(a) Find the profit function $P(n)$ where $P(n) = R(n) - C(n)$.
(b) Find the production values where the profit is zero.

(c) For what production values does the company make money?

(d) For what production values does the company lose money?

(e) If this were your company, how many units would you choose to produce and sell daily?

62. A tractor manufacturer finds his daily cost function for producing n tractors to be $C(n) = 180 + 272n$. The total revenue from the sale of the tractors is $R(n) = 320n - 3n^2$, $n < 20$.

(a) Find the profit function $P(n)$ where $P(n) = R(n) - C(n)$.

(b) Find the production values (n) where the profit is zero.

(c) For what production values does the company make money?

(d) For what production values does the company lose money?

(e) If this were your company, how many tractors would you try to produce and sell each day?

9.2 Linear Functions

Section 3.2 developed the concept of linear equations. The graphs of the linear equations are straight lines. Unless a line is vertical, a vertical line cannot intersect it in more than one point. With the exception of vertical lines, then, all linear equations can also be classified as linear functions. This section will show how the equations of linear functions are derived.

Consider once again a straight line passing through the fixed point (x_1, y_1) and the general point (x, y). From Chapter 3 we know that the slope is

$$m = \frac{y - y_1}{x - x_1}.$$

When this equation is cleared of fractions by multiplying both sides by $x - x_1$, we have

$$y - y_1 = m(x - x_1).$$

This is called the **point-slope form** of the equation of a line. It can be used to find the equation of a line if one point on the line and the slope of the line are known.

Example 1   Find the equation of the line passing through (2, 3) with $m = \frac{1}{2}$.

$$y - y_1 = m(x - x_1)$$

$$y - 3 = \frac{1}{2}(x - 2) \qquad \text{(x_1, y_1) = (2, 3); } m = \frac{1}{2}$$

$$2y - 6 = x - 2 \qquad \text{Clear of fractions}$$

or $\qquad 0 = x - 2y + 4$ ■

When an equation is written in the form $ax + by + c = 0$, it is said to be in **general form.**

Example 2   Find the equation of the line through $(3, 2)$ and $(-5, 4)$.
First compute the slope of the line. Let $(x_1, y_1) = (3, 2)$ and $(x_2, y_2) = (-5, 4)$. Then

$$m = \frac{y_2 - y_1}{x_2 - x_1} = \frac{4 - 2}{-5 - 3} = \frac{2}{-8} = \frac{-1}{4}.$$

The equation is

$$y - 2 = -\frac{1}{4}(x - 3). \qquad (x_1, y_1) = (3, 2)$$

$$4y - 8 = -x + 3$$

When written in the general form, this becomes

$$x + 4y - 11 = 0. \quad \blacksquare$$

It should be noted that $(-5, 4)$ could have been used to find the equation in Example 2. The equation $y - 4 = -\frac{1}{4}[x - (-5)]$ simplifies to the same general equation.

When the known point on a line is the y-intercept, a special form for the equation of a line results. For example, if the line passes through $(0, b)$ with slope m, then

$$y - y_1 = m(x - x_1)$$
$$y - b = m(x - 0) \qquad (x_1, y_1) = (0, b)$$
$$y - b = mx$$

or $y = mx + b.$

This latter form is known as the **slope-intercept form** of the equation of a line.

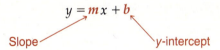

$$y = mx + b$$

Slope   y-intercept

When the equation $y = mx + b$ is written as $f(x) = mx + b$, it is called a **linear function.**

Recall from Section 3.2 that a vertical line has undefined slope. The point-slope form and the slope-intercept form cannot be used to find its equation. A horizontal line has zero slope. This does not present a problem.

Example 3   Find the equation of the line with slope $\frac{2}{3}$ and y-intercept 2. Sketch its graph.

$$y = mx + b$$

$$y = \frac{2}{3}x + 2 \qquad m = \frac{2}{3},\ b = 2$$

The graph is shown in Figure 9.4.   ■

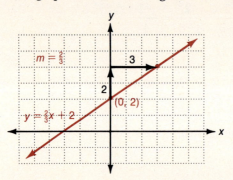

Figure 9.4

Example 4   Find the slope and y-intercept of the line $-3x - 4y - 6 = 0$.

Write the equation of the line in the slope-intercept form by solving for y.

$$-3x - 4y - 6 = 0$$

$$-4y = 3x + 6$$

$$y = \frac{-3}{4}x - \frac{3}{2} \qquad \text{Divide by } -4$$

The slope is $\frac{-3}{4}$, and the y-intercept is $-\frac{3}{2}$.   ■

Example 5   Find the equation of the line through $(3, 2)$ and perpendicular to the line $2y + 4x = 3$.

The slope-intercept form for $2y + 4x = 3$ is

$$y = -2x + \frac{3}{2}.$$

Its slope is -2. The slope of a perpendicular line is $\frac{1}{2}$. (See Section 3.2.) The equation of the line is

$$y - 2 = \frac{1}{2}(x - 3).$$

In slope-intercept form this is written

$$y = \frac{1}{2}x + \frac{1}{2} \quad \text{or} \quad f(x) = \frac{1}{2}x + \frac{1}{2}. \quad ■$$

Example 6   Find the equation of the horizontal line through the intersection of the lines $3x + 2y = 4$ and $2x + 3y = 7$.

Since the line is horizontal, its y-value is constant. Thus we need to find the y-coordinate of the intersection of the lines. Using the method of determinants, we have

$$y = \frac{\begin{vmatrix} 3 & 4 \\ 2 & 7 \end{vmatrix}}{\begin{vmatrix} 3 & 2 \\ 2 & 3 \end{vmatrix}} = \frac{21 - 8}{9 - 4} = \frac{13}{5}.$$

The equation is $y = \frac{13}{5}$. As a review, the reader should verify the solution for y by the substitution and elimination methods.   ■

Summary of Linear Equations (Functions)

General form	$ax + by + c = 0$
Point-slope form	$y - y_1 = m(x - x_1)$
	Point (x_1, y_1); slope $= m$
Slope-intercept form	$y = mx + b$
	Slope $= m$; y-intercept $= b$
Horizontal line	$y = $ constant
	Slope is zero.
Vertical line	$x = $ constant
	Slope is undefined.

Exercises 9.2

Write the equation of the line passing through the given point and having the given slope.
Express each equation in general form.

1. $(0, 2)$, $m = 1$

2. $(0, 3)$, $m = 2$

3. $(4, -1)$, $m = -3$

4. $(3, -2)$, $m = -1$

5. $(-1, -2)$, $m = \frac{1}{2}$

6. $(-3, 0)$, $m = \frac{2}{3}$

7. $(-5, 4)$, $m = \frac{-2}{5}$

8. $(3, -3)$, $m = \frac{-1}{3}$

9. $(4, 4)$, $m = 0$

10. $(3, 1)$, $m = 0$

11. $(2, 5)$, m is undefined

12. $(4, -2)$, m is undefined

13. $(-3, 4)$, parallel to $3x - y = 1$

14. $(1, 0)$, parallel to $x - y = 2$

15. $(4, 0)$, perpendicular to $y = -x$

16. $(0, 3)$, perpendicular to $y = x$

17. $(6, 2)$, parallel to $y = 5$

18. $(1, -1)$, parallel to $y = -2$

19. $(3, 2)$, perpendicular to $x = \frac{-1}{2}$

20. $(2, -5)$, perpendicular to $x = \frac{3}{4}$

Write the equation, in general form, of the line passing through the indicated points.

21. $(1, 6)$, $(5, -2)$

22. $(6, 1)$, $(3, -4)$

23. $(2, -3)$, $(-4, 2)$

24. $(-1, 7)$, $(-2, 1)$

25. $(3, 4)$, $(5, 4)$

26. $(-2, 3)$, $(6, 3)$

27. $(-5, 2)$, $(-5, 0)$ **28.** $(3, 4)$, $(3, -1)$ **29.** $\left(6, \frac{1}{2}\right)$, $\left(3, \frac{1}{3}\right)$

30. $\left(\frac{1}{2}, 2\right)$, $\left(\frac{-1}{4}, 3\right)$ **31.** $\left(\frac{3}{5}, \frac{2}{3}\right)$, $\left(\frac{-1}{2}, \frac{-1}{4}\right)$ **32.** $\left(\frac{-1}{6}, \frac{3}{5}\right)$, $\left(\frac{1}{2}, \frac{-1}{3}\right)$

Write the equation of the line, in $f(x) = mx + b$ form, with the given slope and y-intercept.

33. $m = 2$, $b = 5$ **34.** $m = 1$, $b = -3$ **35.** $m = \frac{-1}{3}$, $b = \frac{1}{2}$

36. $m = \frac{-1}{4}$, $b = -4$ **37.** $m = 0$, $b = -5$ **38.** $m = 0$, $b = 3$

For Exercises 39–56, write the equation of the line, in general form, that satisfies the given conditions.

39. x-intercept 3, y-intercept does not exist

40. y-intercept -2, x-intercept does not exist

41. x-intercept -4, y-intercept 1

42. x-intercept 2, y-intercept -3

43. x-intercept 0, y-intercept 0, parallel to $2x - y = 4$

44. x-intercept 1, y-intercept 1, parallel to $y = -x$

45. x-intercept -5, perpendicular to $y = -2x + 7$

46. x-intercept -1, perpendicular to $3x + y = -4$

47. y-intercept 2, parallel to the line containing $(3, 1)$ and $(-2, 4)$

48. y-intercept -1, parallel to the line containing $(0, 2)$ and $(4, -1)$

49. Passes through $(0, 3)$, perpendicular to $3x - 4y = 1$

50. Passes through $(2, -1)$, perpendicular to $5x + 3y = 7$

51. x-intercept 3, slope $= \frac{1}{2}$ **52.** x-intercept -2, slope $= \frac{1}{3}$

53. y-intercept -1, slope $= \frac{-2}{5}$ **54.** y-intercept 4, slope $= \frac{1}{2}$

55. Passes through the origin, undefined slope   **56.** Passes through the origin, zero slope

57. Do the points $(0, 4)$, $(1, 6)$ and $(-3, -2)$ lie on a straight line?

58. Do the points $(3, 0)$, $(0, \frac{12}{5})$ and $(2, \frac{1}{2})$ lie on a straight line?

59. Write the equations of the lines forming the sides of a triangle with vertices $A(1, 2)$, $B(3, 1)$, and $C(3, 8)$.

60. Write the equations of the lines forming the sides of a triangle with vertices $A(3, -1)$, $B(5, 7)$, and $C(-3, 7)$.

61. Write the equations of the four lines forming the sides of a trapezoid $ABCD$ with vertices $A(-4, 1)$, $B(3, 0)$, $C(4, 1)$, and $D(-2, 3)$.

62. Write the equations of the four lines forming the sides of a parallelogram $ABCD$ with vertices $A(-2, -1)$, $B(4, 1)$, $C(4, 5)$, and $D(-2, 3)$.

Write the equation of a line having the given characteristics.

63. Passes through the point of intersection of the lines $3x - 2y + 10 = 0$ and $4x + 3y - 7 = 0$ and through the point $(2, 1)$

64. Passes through the point of intersection of the lines $2x - 5y = 3$ and $x - 3y = 7$ and is perpendicular to $4x + y = 1$

65. Is perpendicular to the line $2x + 7y = 3$ at its point of intersection with the line $3x - 2y = 8$

66. Passes through the point of intersection of the lines $3x - 5y + 9 = 0$ and $4x + 7y - 28 = 0$ and is parallel to $2x + 3y - 5 = 0$

9.3 The Distance Formula and Circles

This section will show how to find the equation of a circle. First, we need to discover a formula that can be used to find the distance between any two points on a rectangular coordinate system. To do this, we must use the **Pythagorean theorem.**

Pythagorean Theorem

In a right triangle the square of the hypotenuse (longest side) is equal to the sum of the squares of the other two sides (legs).

$$c^2 = a^2 + b^2$$

Figure 9.5 shows a right triangle with legs a and b and hypotenuse c.

Let $P_1 = (x_1, y_1)$ and $P_2 = (x_2, y_2)$ be any two points on a rectangular coordinate system. Let P_3 be the point where a horizontal line through P_1 and a vertical line through P_2 intersect to form a right triangle. (See Figure 9.6.)

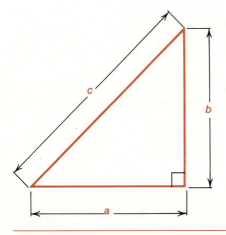

Figure 9.5

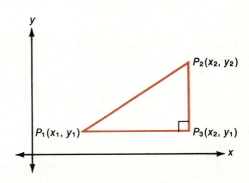

Figure 9.6

Since the line through (x_1, y_1) is horizontal, P_1 and P_3 will have the same y-coordinate, y_1. The line through P_2 is vertical, so P_2 and P_3 have the same x-coordinate, x_2. The horizontal distance from P_1 to P_3 is the difference in

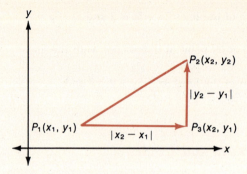

Figure 9.7

their x-coordinates, $|x_2 - x_1|$. Similarly, the vertical distance from P_2 to P_3 is the difference of their y-coordinates, $|y_2 - y_1|$. See Figure 9.7.

The absolute value is used because distance (length of a line segment) is positive. The expressions $|x_2 - x_1|$ and $|x_1 - x_2|$ represent the same length.

If we let d represent the distance from P_1 to P_2, then by the Pythagorean theorem,

$$d^2 = (x_2 - x_1)^2 + (y_2 - y_1)^2.$$

The absolute value symbols are not necessary since the square of any real number is positive. The distance formula is derived by taking the square root of each side.

Theorem 9.1

Distance Formula

The distance between two points (x_1, y_1) and (x_2, y_2) in a plane is given by

$$d = \sqrt{(x_2 - x_1)^2 + (y_2 - y_1)^2}.$$

Example 1   Find the distance between the points $(3, 4)$ and $(-2, 6)$.

Let $(x_1, y_1) = (-2, 6)$ and $(x_2, y_2) = (3, 4)$. Substituting into the distance formula yields

$$d = \sqrt{[3 - (-2)]^2 + [4 - 6]^2}$$
$$= \sqrt{5^2 + (-2)^2}$$
$$= \sqrt{29}. \quad \blacksquare$$

We are now ready to find the equation of a circle with a given radius and a given center point.

A **circle** is the set of points in a plane equidistant from a fixed point called the center.

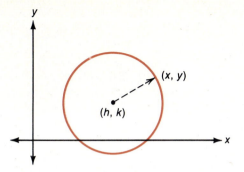

Figure 9.8

The distance of each point in the set from the center is called the **radius** of the circle.

Consider a circle whose center is located at the point (h, k) and whose radius is r, as in Figure 9.8.

If (x, y) is any point on the circle, the distance from (x, y) to the center (h, k) is r. By the distance formula,

$$\sqrt{(x - h)^2 + (y - k)^2} = r.$$

Both sides are then squared to remove the radical, giving the following equation.

> The **equation of a circle** with center at (h, k) and radius r is
>
> $$(x - h)^2 + (y - k)^2 = r^2.$$

If the center of the circle is at the origin, $h = k = 0$, the equation becomes

$$x^2 + y^2 = r^2.$$

Example 2 Find the equation of a circle with center at $(3, -2)$ and radius 4.

$$h = 3, \quad k = -2, \quad r = 4$$
$$(x - h)^2 + (y - k)^2 = r^2$$
$$(x - 3)^2 + (y - (-2))^2 = 4^2$$
$$(x - 3)^2 + (y + 2)^2 = 16 \quad \blacksquare$$

Example 3 Find the equation of the circle with center at $(2, 4)$, tangent to the x-axis. Sketch its graph.

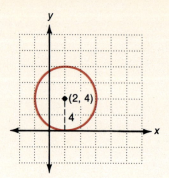

Figure 9.9

If the circle is tangent to the x-axis, its center will be 4 units from the axis. Its radius is 4. The graph is shown in Figure 9.9.

$$(x - 2)^2 + (y - 4)^2 = 4^2$$
$$(x - 2)^2 + (y - 4)^2 = 16 \quad \blacksquare$$

The equation of a circle is often given in the form

$$x^2 + y^2 + Dx + Ey + F = 0.$$

For example, $x^2 + y^2 + 4x - 6y - 12 = 0$ is a circle with center at $(-2, 3)$ and radius 5. To see this, complete the square with respect to both x and y.

$$x^2 + y^2 + 4x - 6y - 12 = 0$$
$$x^2 + 4x + y^2 - 6y = 12$$
$$x^2 + 4x + 4 + y^2 - 6y + 9 = 12 + 13 \qquad \text{Add 13 to each side}$$
$$(x + 2)^2 + (y - 3)^2 = 25$$

We have $h = -2, \quad k = 3, \quad r^2 = 25$, giving $r = 5$. $\quad \blacksquare$

Example 4   Write $(x - 4)^2 + (y - 2)^2 = 4$ in $x^2 + y^2 + Dx + Ey + F = 0$ form.

$$(x - 4)^2 + (y - 2)^2 = 4$$
$$x^2 - 8x + 16 + y^2 - 4y + 4 = 4$$
$$x^2 + y^2 - 8x - 4y + 16 = 0 \quad \blacksquare$$

If one knows the coordinates of both of the endpoints of a line segment forming the diameter of a circle, the equation of the circle can be found once the center is known. The center would lie at the midpoint of the diameter. Suppose that $P_1(x_1, y_1)$ and $P_1(x_2, y_2)$ are two points in a plane and $P_m(x_m, y_m)$ is the midpoint of the line joining them, as in Figure 9.10.

From geometry, we know that triangle P_1P_mS is congruent to triangle P_mP_2T. Thus segment P_1S is half as long as segment P_1P_3, or

$$P_1S = \frac{P_1P_3}{2} = \frac{x_2 - x_1}{2},$$

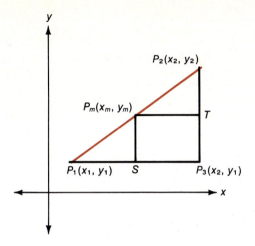

Figure 9.10

which is positive since $x_2 > x_1$. The coordinate x_m can be found as follows:

$$x_m = x_1 + \frac{x_2 - x_1}{2}$$

or $\quad x_m = \frac{x_1 + x_2}{2}.$

The reader should verify that $y_m = \frac{y_1 + y_2}{2}.$

These results show that the coordinates of the midpoint of a line segment can be found by averaging the coordinates of the endpoints.

> The **midpoint** of the line joining (x_1, y_1) to (x_2, y_2) is given by
>
> $$(x_m, y_m) = \left(\frac{x_1 + x_2}{2}, \frac{y_1 + y_2}{2} \right).$$

Example 5 Find the midpoint of the line segment joining $(2, 6)$ to $(-4, 5)$.

$$x_m = \frac{x_1 + x_2}{2} = \frac{2 + (-4)}{2} = -1$$

$$y_m = \frac{y_1 + y_2}{2} = \frac{6 + 5}{2} = \frac{11}{2}$$

The midpoint is $(-1, \frac{11}{2})$. ■

Example 6   Find the equation of the circle with one diameter having endpoints (2, 6) and (−4, 10).

The center will be the midpoint of the diameter.

$$\text{Center} = \frac{2 + (-4)}{2}, \frac{6 + 10}{2} = (-1, 8)$$

The radius will be the distance from either end to the center.

$$(x - h)^2 + (y - k)^2 = r^2$$
$$(x + 1)^2 + (y - 8)^2 = (\sqrt{(2 - (-1))^2 + (6 - 8)^2})^2$$
$$(x + 1)^2 + (y - 8)^2 = (\sqrt{9 + 4})^2$$
$$(x + 1)^2 + (y - 8)^2 = 13 \quad \blacksquare$$

Example 7   Find the center of the circle through (3, 4) and (0, 7), given that the center lies on the line $x + y = 2$.

If we let the coordinates of the center be (h, k), then the distances of (h, k) from (3, 4) and (0, 7) are equal.

$$\sqrt{(h - 3)^2 + (k - 4)^2} = \sqrt{(h - 0)^2 + (k - 7)^2} \qquad \text{Distance formula}$$
$$(h - 3)^2 + (k - 4)^2 = h^2 + (k - 7)^2 \qquad \text{Square both sides}$$
$$h^2 - 6h + 9 + k^2 - 8k + 16 = h^2 + k^2 - 14k + 49$$

When simplified, this becomes

$$-6h + 6k = 24 \qquad \text{or} \qquad h - k = -4.$$

To solve for h and k, we need a second equation using the same two variables. The center lies on the line $x + y = 2$, so its coordinates must satisfy the equation of this line.

$$h + k = 2$$

Solve the system by elimination to determine the center.

$$
\begin{aligned}
h + k &= 2 \\
\underline{h - k} &= \underline{-4} \\
2h &= -2 \\
h &= -1
\end{aligned}
$$

If $h = -1$ and $h + k = 2$, then $k = 3$. The center is $(-1, 3)$. $\blacksquare$

Exercises 9.3

Find the distances between the given pairs of points.

1. (3, 5), (7, 5)   **2.** (6, 2), (2, 2)   **3.** (4, −3), (4, 7)   **4.** (8, 1), (8, −5)

5. (1, 2), (1, −4)   **6.** (1, −2), (3, 4)   **7.** (9, −7), (6, −4)   **8.** (−2, 3), (−7, 4)

9. (2, 1), (3, −6)   **10.** (13, 2), (5, − 4)

Find the coordinates of the midpoint of the line segment having the given endpoints.

11. (1, 3), (5, 7)   **12.** (2, −3), (−4, 3)   **13.** (4, 9), (−2, 7)

14. (8, −5), (−5, 3)   **15.** (−7, 11), (5, 4)   **16.** (−3, 8), (−2, 5)

Find the coordinates of the center of each circle, find the length of the radius, and graph as required.

17. $x^2 + y^2 = 4$; graph   18. $x^2 + y^2 = 1$; graph

19. $x^2 + y^2 = 9$ 20. $x^2 + y^2 = 16$

21. $x^2 + (y + 4)^2 = 25$; graph   22. $(x - 3)^2 + y^2 = 36$; graph

23. $(x - 3)^2 + (y - 2)^2 = 25$ 24. $(x + 2)^2 + (y + 1)^2 = 1$

25. $(x + 4)^2 + (y - 1)^2 = 16$; graph   26. $(x - 5)^2 + (y + 2)^2 = 25$; graph

27. $(x - 7)^2 + y^2 = 36$ 28. $x^2 + (y - 1)^2 = 9$

Write the equation of each circle in the form $(x - h)^2 + (y - k)^2 = r^2$.

29. Center at (0, 0) and $r = 2$ 30. Center at (0, 0) and $r = 3$

31. Center at (1, 2) and $r = 4$ 32. Center at (2, 4) and $r = 5$

33. Center at (0, −2) and $r = 6$ 34. Center at (5, 0) and $r = 7$

35. Center at (−3, 0) and $r = 3$ 36. Center at (0, −3) and $r = 3$

37. Center at (−1, −3) and $r = 5$ 38. Center at (−4, −1) and $r = 4$

39. Passes through the origin with center at (3, −5)

40. Passes through the origin with center at (−2, 6)

41. Is tangent to the x-axis with center at (−3, −4)

42. Is tangent to the y-axis with center at (3, 2)

43. Consists of the set of all points 5 units from (−3, −5)

44. Consists of the set of all points 6 units from (2, −6)

Determine the coordinates of the center and the length of the radius of each circle by completing the square in each equation.

45. $x^2 + y^2 + 2x - 2y = 0$ 46. $x^2 + y^2 - 6x + 8y = 0$

47. $x^2 + y^2 + 8x - 2y + 8 = 0$ 48. $x^2 + y^2 + 4x + 6y - 12 = 0$

49. $x^2 + y^2 - 5x + 7y + 1 = 0$ 50. $x^2 + y^2 - x - y - 1 = 0$

51. $x^2 + y^2 - x + 3y = \dfrac{3}{2}$ 52. $x^2 + y^2 - 6x - 10y = -33$

53. $x^2 + y^2 - 8x = -4$ 54. $x^2 + y^2 + 18y = -41$

55. $4x^2 + 4y^2 + 20x + 10y = 0$ 56. $2x^2 + 2y^2 + 8x + 4y = 0$

57. $3x^2 + 3y^2 + 30x - 18y = 12$ 58. $5x^2 + 5y^2 - 10x + 15y = 1$

59. $9x^2 + 9y^2 + 24x + 108y + 232 = 0$ 60. $16x^2 + 16y^2 - 128x + 24y - 55 = 0$

Write the equation of each circle with the given conditions.

61. Passes through the point (−1, 5) with center at (5, −2)

62. Passes through the point (3, 7) with center at (6, 11)

63. One diameter is the line segment joining the points (5, −1) and (−3, 7).

64. One diameter is the line segment joining the points (6, 7) and (2, 5).

65. Passes through the points (2, 3) and (−1, 1) and has its center on the line $x - 3y = 11$

66. Passes through the points $(\frac{3}{4}, 0)$ and $(0, -\frac{1}{2})$ and has its center on the line $x + y = 0$

67. Passes through the points (4, 5) and (−2, 3) and has its center on the x-axis

68. Passes through the points (5, 7) and (3, −1) and has its center on the y-axis

9.4   The Parabola

The circle of Section 9.3 and the parabola that we will study in this section are two of the curves that can be obtained by cutting a cone with a plane P. For a circle, the cutting plane is parallel to the base, as in Figure 9.11(a). A parabola results when the cutting plane is parallel to a lateral line, as in Figure 9.11(b). The curve obtained when a plane cuts a cone is called a **conic section.**

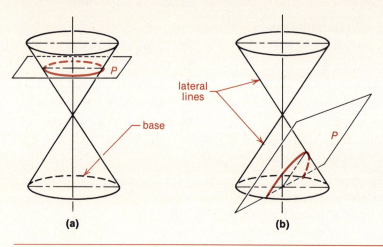

Figure 9.11

Parabolas occur naturally, as in the path that a ball follows when thrown. The reflectors on radar antennae and flashlights are parabolic in shape. Mirrors on telescopes are often parabolic, and the suspension cables on many bridges form parabolas.

> A **parabola** is the set of all points in a plane equidistant from a fixed point F, called the **focus,** and a fixed line D, called the **directrix.**

We will derive the equation of the parabola with focus at $(0, p)$, directrix $y = -p$ and vertex at the origin, as shown in Figure 9.12. (The vertex of a parabola will be discussed later in this section.)

If $P(x, y)$ is any point on the parabola, then by definition

distance PF = distance PD.

The coordinates of a point on D are $(x, -p)$. Thus,

$$\sqrt{(x - 0)^2 + (y - p)^2} = |y - (-p)|$$
$$x^2 + (y - p)^2 = (y + p)^2 \qquad \text{Square each side}$$
$$x^2 + y^2 - 2py + p^2 = y^2 + 2py + p^2$$

or
$$x^2 = 4py.$$

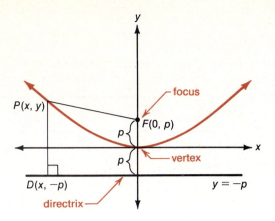

Figure 9.12

Note from Figure 9.12 that p is the distance from the vertex to the focus or the vertex to the directrix.

Example 1   Find the equation of the parabola with vertex at $(0, 0)$ and focus at $(0, 4)$. What is the equation of the directrix?

Here $p = 4$. The equation is

$$x^2 = 4py \qquad \text{or} \qquad x^2 = 16y.$$

The directrix is $y = -4$.   ■

Parabolas can open up, down, to the right, or to the left. The four cases are shown in Figure 9.13(a) through 9.13(d) on the next page. The line through the vertex perpendicular to the directrix is called the **axis of symmetry**.

Parabolas	Equation	Vertex	Focus	Directrix	Axis of Symmetry
Figure 9.13(a)	$x^2 = 4py$	$(0, 0)$	$(0, p)$	$y = -p$	y-axis
Figure 9.13(b)	$x^2 = -4py$	$(0, 0)$	$(0, -p)$	$y = p$	y-axis
Figure 9.13(c)	$y^2 = 4px$	$(0, 0)$	$(p, 0)$	$x = -p$	x-axis
Figure 9.13(d)	$y^2 = -4px$	$(0, 0)$	$(-p, 0)$	$x = p$	x-axis

Example 2   Determine the direction each parabola opens, the vertex, focus, and directrix.

(a) $y^2 = 3x$

The parabola opens to the right, with the vertex at $(0, 0)$. Since $4p = 3$, $p = \frac{3}{4}$. The focus is $(\frac{3}{4}, 0)$, and the directrix is $x = -\frac{3}{4}$.

(b) $x^2 = -6y$

The parabola opens downward, with vertex at $(0, 0)$. Since $4p = 6$, $p = \frac{3}{2}$. The focus is $(0, -\frac{3}{2})$, and the directrix is $y = \frac{3}{2}$.   ■

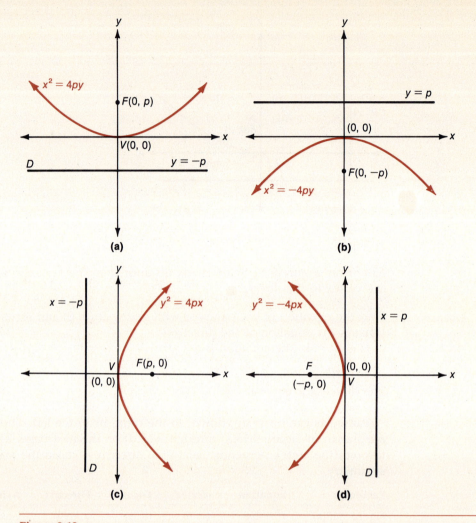

Figure 9.13

The focus is an important consideration on a parabola. As one application, consider the parabolic reflector. In physics it can be demonstrated that rays parallel to the axis of symmetry are reflected to the focus when they strike a parabolic surface. (See Figure 9.14.)

A flashlight's bulb is located at the focus of a parabola. A solar oven concentrates the heat at the focus.

We now consider the graph of the quadratic equation

$$y = ax^2 + bx + c,$$

which is the **standard form** of the equation of a parabola opening upward or downward. If $b = c = 0$, the equation becomes

$$y = ax^2 \quad \text{or} \quad x^2 = \frac{1}{a}y.$$

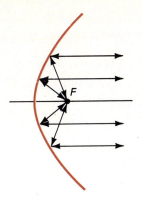

Figure 9.14

Recalling the equation of the parabola discussed earlier, $x^2 = 4py$, we see that $\frac{1}{a} = 4p$. Thus $a > 0$ when $p > 0$, and $a < 0$ when $p < 0$. Thus the parabola opens **upward if $a > 0$** and **downward if $a < 0$.** If a parabola opens upward, its vertex is said to be its **minimum point.** If it opens downward, the vertex is the **maximum point.**

Example 3 Graph each of the following parabolas by constructing a table of values.

(a) $y = x^2$ (b) $y = x^2 + 1$ (c) $y = -(x + 1)^2 + 1$

x	y
-2	4
-1	1
0	0
1	1
2	4

x	y
-2	5
-1	2
0	1
1	2
2	5

x	y
-3	-3
-2	0
-1	1
0	0
1	-3

The graphs are shown as Figures 9.15(a), (b) and (c). ■

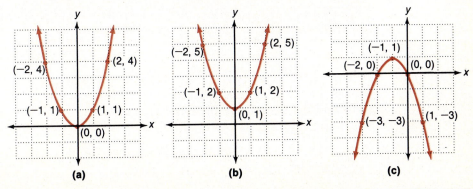

Figure 9.15

Notice that the three parabolas have the same shape. The modifications to the formula shifted the vertex or changed the direction in which the parabola opened. If $y = x^2$ is the basic parabola then $y = x^2 + k$ will shift the vertex k units upward if $k > 0$ or k units downward if $k < 0$. Similarly, $y = (x + k)^2$ shifts the vertex k units to the left if $k > 0$ and k units to the right if $k < 0$.

Example 4   Find the vertex of the parabola $y = (x - 2)^2 - 3$.

With respect to $y = x^2$, the vertex of $y = (x - 2)^2 - 3$ is shifted 2 units to the right and 3 units down. The vertex is $(2, -3)$. ■

It is not difficult to determine the vertex of a parabola with a vertical axis of symmetry. This is accomplished by completing the square in x. Consider

$$y = ax^2 + bx + c,$$

which can be written as

$$y = a(x^2 + \frac{b}{a}x) + c.$$

We now complete the square as follows, adding and subtracting the same amount on the right side of the equation.

$$y = a\left(x^2 + \frac{b}{a}x + \frac{b^2}{4a^2}\right) + c - \frac{b^2}{4a} \qquad a\left(\frac{b^2}{4a^2}\right) = \frac{b^2}{4a}$$

$$y = a\left(x + \frac{b}{2a}\right)^2 + \frac{4ac - b^2}{4a} \qquad \text{LCD} = 4a$$

If we let $\dfrac{-b}{2a} = h$ and $\dfrac{4ac - b^2}{4a} = k$, the equation becomes

$$y = a(x - h)^2 + k.$$

The equation

$$y = a(x - h)^2 + k$$

represents a parabola with vertex (h, k) opening upward if $a > 0$ and downward if $a < 0$. The axis of symmetry is $x = h$.

Example 5   Determine the vertex, axis of symmetry, and the points where the graph crosses the x-axis (x-intercepts) for $y = x^2 - 6x + 5$. Graph.

To find the vertex, complete the square.

$$y = (x^2 - 6x) + 5$$
$$y = (x^2 - 6x + 9) + 5 - 9 \qquad \text{Add and subtract 9}$$
$$y = (x - 3)^2 - 4$$

The vertex is $(3, -4)$. The axis of symmetry is $x = 3$. To determine the x-intercepts, let $y = 0$.

$$0 = x^2 - 6x + 5$$
$$0 = (x - 5)(x - 1)$$
$$x = 5 \quad \text{or} \quad x = 1$$

The intercepts are 5 and 1. The graph is shown in Figure 9.16. ■

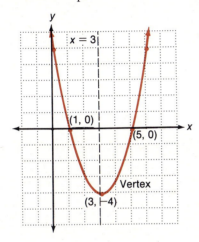

Figure 9.16

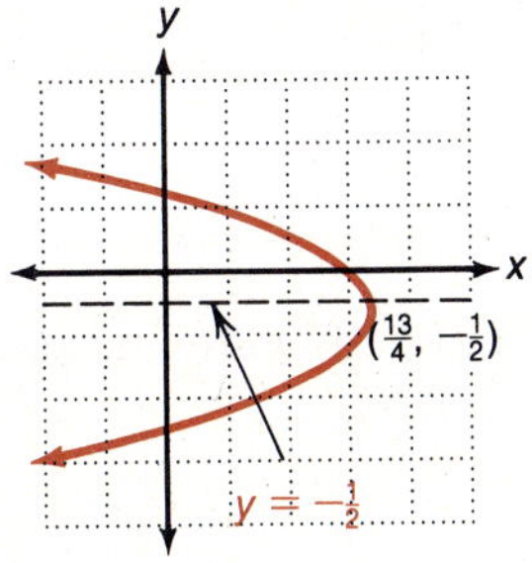

Figure 9.17

Example 6 Determine the vertex and axis of symmetry for $y^2 + y + x - 3 = 0$.

In this example the y term is squared. The axis of symmetry is horizontal. Completing the square on y yields

$$y^2 + y + x - 3 = 0$$
$$x = -y^2 - y + 3 \qquad \text{Solve for } x$$
$$x = -\left(y^2 + y + \frac{1}{4}\right) + 3 + \frac{1}{4} \qquad \text{Subtract and add } \tfrac{1}{4}$$
$$x = -\left(y + \frac{1}{2}\right)^2 + \frac{13}{4}.$$

The vertex is $(\frac{13}{4}, -\frac{1}{2})$. The axis of symmetry is $y = -\frac{1}{2}$. The graph is shown in Figure 9.17. ■

Example 7 Determine the equation of the parabola shown in Figure 9.18 on the next page.

The axis of symmetry is vertical. The general form is $y = a(x - h)^2 + k$. The vertex is $(4, 4)$, so $h = k = 4$. Thus

$$y = a(x - 4)^2 + 4.$$

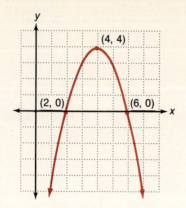

Figure 9.18

Since the parabola passes through (2, 0), its coordinates must satisfy the equation.

$$0 = a(2 - 4)^2 + 4$$
$$-4 = 4a$$
$$-1 = a$$

The equation is $y = -1(x - 4)^2 + 4$. ■

Exercises 9.4

Determine the vertex, axis of symmetry and x-intercepts for each of the following parabolas. Graph where indicated.

1. $y = x^2$; graph

2. $y = x^2 + 1$; graph

3. $y = x^2 - 4$

4. $y = x^2 - 2$

5. $y = -x^2 + 2$

6. $y = -x^2 + 3$

7. $y = x^2 - 3x + 2$; graph

8. $y = x^2 - 5x + 6$; graph

9. $y = -x^2 - 2x - 1$

10. $y = -x^2 - 4x - 4$

11. $y = 2x^2 - 10x - 12$

12. $y = 3x^2 - 3x - 36$

13. $y = -4x^2 - 4x - 2$; graph

14. $y = -2x^2 - 4x - 30$; graph

Determine the vertex and the axis of symmetry for each parabola and state the direction in which it opens.

15. $y = 2(x - 2)^2 - 1$

16. $y = 2x^2 + 3$

17. $x = -2(y - 2)^2 - 1$

18. $x = -(y - 1)^2 + 2$

19. $y = -3(x - 1)^2 - 5$

20. $y = -3(x - 1)^2 + 4$

Rewrite each of the following in the form $x = a(y - k)^2 + h$ or $y = a(x - h)^2 + k$; determine the vertex and the axis of symmetry, and graph as indicated.

21. $x^2 + 4x - 2y - 2 = 0$; graph

22. $x^2 - 2x - 3y + 7 = 0$; graph

23. $y^2 - 10y - 3x + 24 = 0$

24. $y^2 + 2y - 4x - 3 = 0$

25. $x^2 - 3x - 3y + 1 = 0$; graph

26. $x^2 - x + 3y + 1 = 0$; graph

27. $y^2 + 6y + \dfrac{1}{2}x + 7 = 0$

28. $y^2 - 6y + 2x + 17 = 0$

Determine the equation of the parabola from each graph.

29.
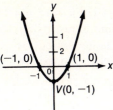

30.
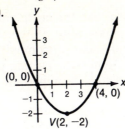

31.
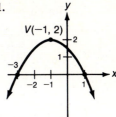

32.
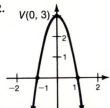

33.
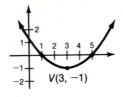

34.
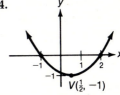

Find the coordinates of the focus and vertex and write the equation of the directrix for each of the following.

35. $y^2 = 6x$

36. $x^2 = 8y$

37. $2x^2 = -8y$

38. $3y^2 = -4x$

39. $x^2 + \left(\dfrac{16}{3}\right)y = 0$

40. $x^2 + \left(\dfrac{25}{4}\right)y = 0$

41. $3y^2 = 8x$

42. $y^2 = -12x$

Solve each of the following as indicated. A calculator may be helpful.

43. A load F applied to a cantilever beam forms a parabolic function according to $F = 32 - 2x^2$ pounds. What is the load when $x = 4$?

44. The velocity V of a rocket forms a parabolic function according to $V = 100 + 75t - 25t^2$, where V is in feet per second when t is in seconds. Sketch the graph for $t \geq 0$. What is the maximum velocity of the rocket? At what time t does the velocity equal zero?

45. The graph of an electric current I forms a parabolic function according to $I = t^2 - 5t + 6$ amperes. For what time t (in seconds) is the current equal to zero?

46. The surface area S of a cube forms a parabolic function according to $S = 6x^2$ where x is the length of an edge of the cube. Graph S as a function of x. For what length of x does S have its minimum value?

47. With a uniformly distributed load, the cable of a suspension bridge hangs in the shape of a parabola according to $1000y = 9x^2 + 10,000$. For what value of x does y have its minimum value (lowest point on the cable) with respect to a level road joining the base of the towers? If the cable is tied to the top of the towers (200 feet apart), how tall are they?

48. If a company produces n power generators per week, its profit P is given by $P = 110n - n^2 - 1000$. How many generators should the company produce each week to have a maximum profit? What minimum and maximum values of n provide a profit greater than or equal to zero? What values of n provide a negative profit (loss)? Assume $n > 0$.

49. Helen wishes to fence a rectangular area adjacent to her house. One side of the rectangle will be the side of the house. What is the largest area she can fence with 100 feet of fencing?

9.5   The Ellipse and Hyperbola

The ellipse and hyperbola, like the circle and parabola, are conic sections obtained when a plane cuts a cone. When the cutting plane is not parallel to a base, a lateral line, or the axis of symmetry, an **ellipse** is generated, as shown in Figure 9.19(a). When the cutting plane is parallel to the axis of symmetry, as in Figure 9.19(b), a **hyperbola** is generated.

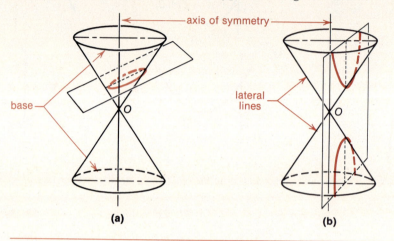

(a)   (b)

Figure 9.19

An **ellipse** is the set of all points, the sum of whose distances from two fixed points, called the **foci,** is a constant. Each of the two foci is called a **focus.**

We will derive the equation for the ellipse with center at the origin and foci at $(c, 0)$ and $(-c, 0)$. To do this we will let (x, y) be any point on the ellipse and $2a$ be the constant referred to in the definition. (See Figure 9.20.)

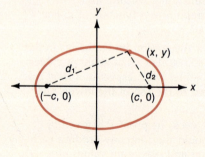

Figure 9.20

If we let d_1 and d_2 be the distances of the point (x, y) from the two foci, then

$$d_1 + d_2 = 2a$$

Both d_1 and d_2 can be found by using the distance formula, as follows.

$$\sqrt{[x - (-c)]^2 + y^2} + \sqrt{(x - c)^2 + y^2} = 2a.$$

To eliminate the radicals, subtract $\sqrt{(x - c)^2 + y^2}$ from each side and square. When the result is simplified, we have

$$cx - a^2 = a\sqrt{(x - c)^2 + y^2}.$$

Square again to get the result

$$(a^2 - c^2)x^2 + b^2y^2 = a^2(a^2 - c^2).$$

Finally, let $a^2 - c^2 = b^2$ and divide both sides by a^2b^2 to get

$$\frac{x^2}{a^2} + \frac{y^2}{b^2} = 1.$$

Since we let $a^2 - c^2 = b^2$ and b^2 must be positive, we know that $a^2 > c^2$ and thus $a > c$. Letting $y = 0$ gives

$$\frac{x^2}{a^2} = 1 \qquad \text{or} \qquad x = \pm a.$$

If $x = 0$, then

$$\frac{y^2}{b^2} = 1 \qquad \text{or} \qquad y = \pm b.$$

Thus a and b are the x- and y-intercepts of the ellipse. Figure 9.21 shows the graph of the ellipse with all points labeled.

The points $(a, 0)$ and $(-a, 0)$ are the **vertices** of the ellipse. The points $(0, b)$ and $(0, -b)$ are **covertices.** For simplicity, we refer to each of these points as a **vertex.**

An ellipse in which the greater elongation is in the x direction, $a > b$, is called an x ellipse. If it is in the y direction, $b > a$, it is called a y ellipse.

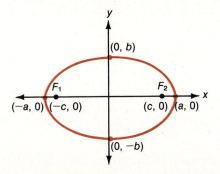

Figure 9.21

Example 1   Sketch the graph of each of the following.

(a) $x^2 + 4y^2 = 16$

We first write the equation in the form

$$\frac{x^2}{a^2} + \frac{y^2}{b^2} = 1$$

by dividing each side by 16 so that

$$\frac{x^2}{16} + \frac{y^2}{4} = 1.$$

The easiest way to sketch an ellipse centered at the origin is to use the x- and y-intercepts.

$$a^2 = 16, \text{ so } a = \pm 4 \qquad \text{and} \qquad b^2 = 4, \text{ so } b = \pm 2$$

The graph of this x ellipse is shown in Figure 9.22(a).

(b) $9x^2 + 4y^2 = 36$

First divide both sides by 36 and then determine the intercepts.

$$9x^2 + 4y^2 = 36$$

$$\frac{x^2}{4} + \frac{y^2}{9} = 1$$

$$a^2 = 4, \text{ so } a = \pm 2 \qquad \text{and} \qquad b^2 = 9, \text{ so } b = \pm 3$$

The graph of this y ellipse is shown in Figure 9.22(b). ■

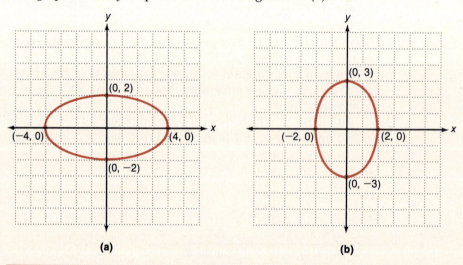

(a)   (b)

Figure 9.22

When the center of an ellipse is shifted from the origin to the point (h, k), its equation becomes

$$\frac{(x - h)^2}{a^2} + \frac{(y - k)^2}{b^2} = 1,$$

which is known as the **standard form.** In this form a and b are the distances from the center (h, k) to the vertices A, B, C, and D, as shown in Figure 9.23.

Example 2   Sketch the graph of $5(x - 1)^2 + 9(y + 1)^2 = 45$.

First put the equation in standard form by dividing each side by 45.

$$\frac{(x - 1)^2}{9} + \frac{(y + 1)^2}{5} = 1$$

The center is at $(h, k) = (1, -1)$; $a^2 = 9$, so $a = \pm 3$; $b^2 = 5$, so $b = \pm\sqrt{5}$. The ellipse is not centered at the origin, so a and b measure the distance from the center to each vertex. (See Figure 9.24.) ∎

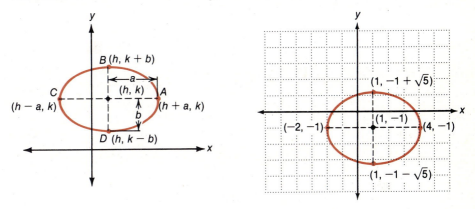

Figure 9.23   **Figure 9.24**

We now turn our attention to the hyperbola.

A **hyperbola** is a set of points, the difference of whose distances from two fixed points, called the foci, is the constant $2a$.

If a hyperbola is centered at the origin and it is symmetric to the x-axis (see Figure 9.25), its equation becomes

$$\frac{x^2}{a^2} - \frac{y^2}{b^2} = 1.$$

The equation can be obtained directly from the definition by letting $d_1 - d_2 = 2a$. The foci are located at $(c, 0)$ and $(-c, 0)$. The points where the graph crosses the axis of symmetry, $(a, 0)$ and $(-a, 0)$, are the **vertices.**

The two broken lines are called the **asymptotes** of the hyperbola. The asymptotes always pass through the center with slope of $\pm\frac{b}{a}$. The graph of the hyperbola will always approach these lines, but it will never reach them.

Hyperbolas are classified as x or y hyperbolas depending on their axis of symmetry. If the axis of symmetry is the y-axis, it is a y hyperbola and its equation becomes

$$\frac{y^2}{b^2} - \frac{x^2}{a^2} = 1.$$

The foci are located at $(0, c)$ $(0, -c)$ and the vertices at $(0, b)$ and $(0, -b)$. The slopes of the asymptotes remain $\pm\frac{b}{a}$. (See Figure 9.26.)

Figure 9.25 is an x hyperbola. Figure 9.26 is a y hyperbola.

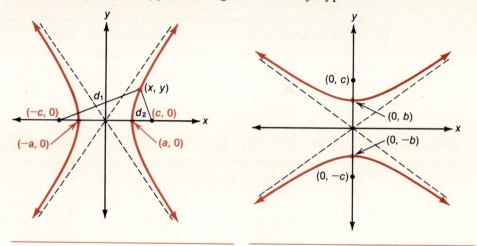

Figure 9.25   **Figure 9.26**

Example 3   Sketch the graph of each of the following hyperbolas.

(a) $\dfrac{x^2}{9} - \dfrac{y^2}{16} = 1$

Since $a^2 = 9$, $a = \pm 3$; since $b^2 = 16$, $b = \pm 4$.

This is an x hyperbola centered at the origin. The asymptotes are $y = \pm\frac{4}{3}x$. The graph is shown in Figure 9.27(a).

(b) $4y^2 - x^2 = 16$

First divide each side by 16.

$$\frac{y^2}{4} - \frac{x^2}{16} = 1$$

This equation represents a y hyperbola centered at the origin. Since $a^2 = 16$, $a = \pm 4$; since $b^2 = 4$, $b = \pm 2$. The asymptotes are $y = \pm\frac{1}{2}x$. The graph is shown in Figure 9.27(b). ■

When the center of the hyperbola is shifted from the origin to (h, k), its **standard form** becomes

$$\frac{(x - h)^2}{a^2} - \frac{(y - k)^2}{b^2} = 1 \quad \text{or} \quad \frac{(y - k)^2}{b^2} - \frac{(x - h)^2}{a^2} = 1.$$

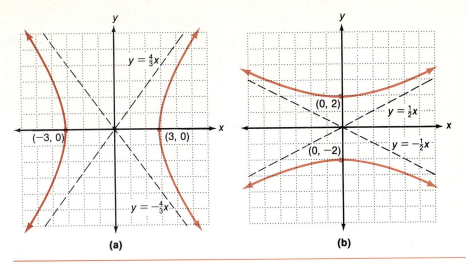

Figure 9.27

Example 4   Write $x^2 + 6x - 2y^2 + 4y - 1 = 0$ in standard form and graph.

Begin by completing the square in both x and y.

$$x^2 + 6x - 2y^2 + 4y - 1 = 0$$
$$x^2 + 6x - 2(y^2 - 2y) = 1$$
$$x^2 + 6x + 9 - 2(y^2 - 2y + 1) = 1 + 9 - 2 \qquad -2 = -2(1)$$
$$(x + 3)^2 - 2(y - 1)^2 = 8$$
$$\frac{(x + 3)^2}{8} - \frac{(y - 1)^2}{4} = 1$$

The center is at $(h, k) = (-3, 1)$. The asymptotes pass through the center with slopes

$$\pm \frac{b}{a} = \pm \frac{2}{2\sqrt{2}} = \pm \frac{\sqrt{2}}{2}.$$

The graph is shown in Figure 9.28. ■

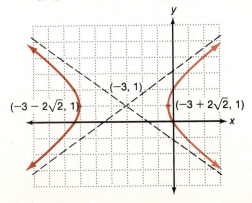

Figure 9.28

Ellipse and Hyperbola

Equation	Location	Identification
$\dfrac{(x-h)^2}{a^2} + \dfrac{(y-k)^2}{b^2} = 1$	Center at (h, k)	x ellipse if $a > b$ y ellipse if $b > a$
$\dfrac{(x-h)^2}{a^2} - \dfrac{(y-k)^2}{b^2} = 1$	Center at (h, k) Slopes of asymptotes $= \pm b/a$	x hyperbola
$\dfrac{(y-k)^2}{b^2} - \dfrac{(x-h)^2}{a^2} = 1$	Center at (h, k) Slopes of asymptotes $= \pm b/a$	y hyperbola

Exercises 9.5

Determine the coordinates of the center and the vertices of each ellipse. Graph where indicated.

1. $x^2 + 4y^2 = 1$; graph
2. $9x^2 + y^2 = 1$; graph
3. $2x^2 + 9y^2 = 18$
4. $3x^2 + y^2 = 6$
5. $3(x-3)^2 + 4(y+1)^2 = 12$
6. $12(x-6)^2 + 8(y-4)^2 = 96$
7. $7(x-2)^2 + 2(y+1)^2 = 14$; graph
8. $9(x-2)^2 + 4(y+1)^2 = 9$; graph

Determine the coordinates of the center and vertices of each hyperbola. When graphing, show the asymptotes.

9. $2x^2 - 4y^2 = 8$
10. $5x^2 - 4y^2 = 20$
11. $16y^2 - 9x^2 = 144$; graph
12. $3y^2 - 6x^2 = 18$; graph
13. $3x^2 - 4y^2 - 4 = 0$
14. $2x^2 - 4y^2 - 1 = 0$
15. $8(x+3)^2 - 4(y-1)^2 = 32$
16. $(x-1)^2 - 3y^2 = 3$
17. $16(y-3)^2 - 9(x-2)^2 = 144$; graph
18. $2(y-1)^2 - (x+2)^2 = 2$; graph

Name each of the following conic sections.

19. $x^2 + y^2 = 4$
20. $x^2 - y = 0$
21. $3x^2 - y^2 = 108$
22. $3x^2 + y^2 = 108$
23. $5x^2 + 5y^2 = 125$
24. $2x^2 - 3y^2 = 5$
25. $y^2 = 16x$
26. $x^2 + 9y^2 = 36$

Rewrite each of the following conic sections, in standard form, as a circle, ellipse, parabola, or hyperbola.

27. $4y^2 = x^2 - 3x + 4$
28. $x^2 + 9y^2 - 2x + 36y = -28$
29. $4y^2 - 24x - 12y = 15$
30. $3x^2 + 3y^2 - 5x + 7y + 4 = 0$
31. $x^2 + 6x = -3y^2 - 6$
32. $3x^2 - 4y^2 + 6x + 6y = 0$
33. $x^2 + 4x = y^2 - 4y + 1$
34. $x^2 + 25 = 3y + 10x$
35. $x^2 + y^2 = 6y - 2x - 5$
36. $y^2 - 6x - 3y = 5$
37. $9x^2 - 8y + 31 = 36x - 4y^2$
38. $x^2 + 2x = 4y^2 - 8y + 7$

Review Exercises

Solve each of the following systems of equations by the method indicated.

39. $3x + 2y = 5$
    $5x - 3y = 7$ by elimination

40. $x + 2y + z = 5$
    $2x + y - z = 3$ by elimination
    $x + y + 2z = 0$

41. $7x - 5y = 8$
$\quad\ \ 2x + 3y = 1$ by substitution

42. $4x - 5y = 1$
$\quad\ \ 2x +\ \ y = 9$ by Cramer's rule

43. $\ \ x + y -\ \ z = 3$
$\quad 3x + y + 2z = 8$ by elimination
$\quad 2x - y - 3z = 1$

44. $5x + 2y - 3z = 0$
$\quad 2x - 3y + 2z = 2$ by Cramer's rule
$\quad\ \ x +\ \ y + 4z = 5$

9.6   Nonlinear Systems of Equations

In Section 3.3 we saw that the solution to a consistent system of linear equations was given by the coordinates of the point of intersection of their graphs. In this section we consider some of the methods used to solve a system of equations in which one or both equations are of degree two. Such a system is called a **nonlinear system** of equations, and a solution will be considered only when it is a point of intersection of their graphs.

Example 1   Find the solution set to the system. Graph the system.

$\quad$ **I.** $x^2 + y^2 = 25$
$\quad$ **II.** $y\ + 2x = 10$

Begin by solving $y + 2x = 10$ for y.

$$y = 10 - 2x$$

Now substitute this value for y into Equation I.

$$x^2 + (10 - 2x)^2 = 25$$
$$x^2 + 100 - 40x + 4x^2 = 25$$
$$5x^2 - 40x + 75 = 0$$
$$x^2 - 8x + 15 = 0 \qquad \text{Divide each side by 5}$$
$$(x - 5)(x - 3) = 0$$
$$x = 5 \quad \text{or} \quad x = 3$$

To find y, substitute $x = 5$ and $x = 3$ into Equation II.

$$y + 2x = 10 \qquad\qquad\qquad y + 2x = 10$$
$$y + 2 \cdot 5 = 10 \quad x = 5 \qquad\qquad y + 2 \cdot 3 = 10 \quad x = 3$$
$$y = 0 \qquad\qquad\qquad\qquad\quad y = 4$$

The solution set consists of two ordered pairs. The solution set is $\{(5, 0), (3, 4)\}$. To sketch the graph, note that Equation I is a circle with center at the origin, radius 5. Equation II is a straight line. The graphs are shown in Figure 9.29 on the next page.   ■

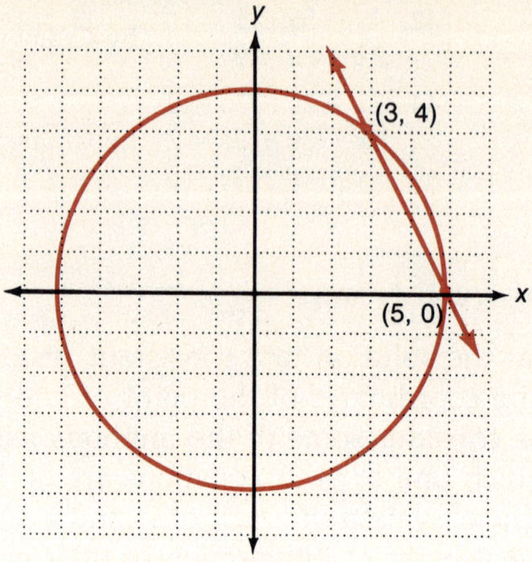

Figure 9.29

From Figure 9.29 it is easy to see that the maximum number of intersections of a line and a circle is 2. In Example 1, if we had substituted $x = 3$ into equation I instead of II, then

$$x^2 + y^2 = 25$$
$$9 + y^2 = 25$$
$$y^2 = 16$$
$$y = \pm 4.$$

It appears that $(3, 4)$ and $(3, -4)$ are both solutions, but $(3, -4)$ does not lie on the line. Thus $(3, -4)$ is **extraneous.**

▶ *Caution*   When solving nonlinear systems of equations, always check the results to determine whether there are any extraneous solutions.

When the form of each equation in a system is recognizable, it is possible to estimate the maximum number of solutions before solving. For example, a circle and an ellipse could intersect in as many as four points or not intersect at all. (See Figure 9.30.)

Solving systems of linear equations by elimination is applicable to some nonlinear systems.

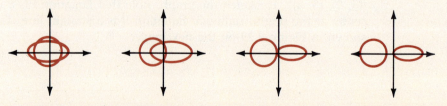

Figure 9.30

Example 2   Solve the system by elimination and graph.

I. $x^2 + y^2 = 7$
II. $3x^2 - 2y^2 = 6$

Begin by multiplying both sides of Equation I by 2 to get III.

III. $2x^2 + 2y^2 = 14$
II. $\underline{3x^2 - 2y^2 = 6}$

$$5x^2 = 20 \qquad \text{Add II to III}$$
$$x = \pm 2$$

Now substitute $x = 2$ or -2 into either of the original equations. Using Equation I, we have

$$x^2 + y^2 = 7 \qquad\qquad\qquad x^2 + y^2 = 7$$
$$2^2 + y^2 = 7 \qquad x = 2 \qquad (-2)^2 + y^2 = 7 \qquad x = -2$$
$$y^2 = 3 \qquad\qquad\qquad\qquad y^2 = 3$$
$$y = \pm\sqrt{3} \qquad\qquad\qquad\qquad y = \pm\sqrt{3}$$

It appears there are four ordered pairs in the solution set:

$$\{(2, \sqrt{3}),\ (2, -\sqrt{3}),\ (-2, \sqrt{3}),\ (-2, -\sqrt{3})\}.$$

The solutions were obtained from Equation I, so we can verify them in Equation II.

$$3x^2 - 2y^2 = 6$$
$$3(\pm 2)^2 - 2y^2 = 6 \qquad x = 2 \text{ or } -2$$
$$3 \cdot 4 - 2y^2 = 6$$
$$-2y^2 = -6$$
$$y^2 = 3$$
$$y = \pm\sqrt{3}$$

Equation I is a circle, and Equation II is a hyperbola. Their graphs are shown in Figure 9.31. ■

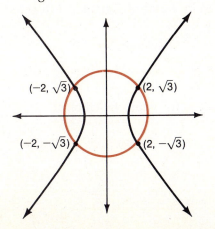

Figure 9.31

Example 3 Solve the system

    I. $2x^2 - y^2 = 1$

    II. $xy = 1.$

Equation II is not a standard form, so we will not estimate the maximum number of possible solutions. Solve Equation II for y and substitute the result into Equation I.

$$xy = 1$$

$$y = \frac{1}{x}$$

$$2x^2 - y^2 = 1$$

$$2x^2 - \left(\frac{1}{x}\right)^2 = 1 \qquad\qquad y = \frac{1}{x}$$

$$2x^4 - 1 = x^2 \qquad\qquad \text{Clear of fractions}$$

$$2x^4 - x^2 - 1 = 0$$

$$(2x^2 + 1)(x^2 - 1) = 0$$

$$2x^2 + 1 = 0 \qquad \text{or} \quad x^2 - 1 = 0 \qquad \text{Factor}$$

$$x^2 = -\frac{1}{2} \qquad \text{or} \qquad x^2 = 1$$

$$x = \pm\frac{\sqrt{2}}{2}i \quad \text{or} \qquad x = \pm 1$$

The only real values for x are 1 and -1. Since $y = \frac{1}{x}$, $y = 1$ or -1. The two solutions are $(1, 1)$ and $(-1, -1)$. The solution set is

    $\{(1, 1), (-1, -1)\}.$ ∎

Example 4 Solve the system

    I. $x^2 - xy - 2y^2 = 0$

    II. $x^2 - \;\; y + \;\; y^2 = 1.$

This system can be solved by factoring Equation I to get x in terms of y.

$$x^2 - xy - 2y^2 = 0$$

$$(x + y)(x - 2y) = 0 \qquad\qquad \text{Factor}$$

$$x + y = 0 \quad \text{or} \quad x - 2y = 0$$

$$x = -y \quad \text{or} \qquad x = 2y$$

We now substitute these values for x into Equation II.

    II. $\qquad x^2 - y + y^2 = 1$

$$(-y)^2 - y + y^2 = 1 \qquad\qquad x = -y$$

$$2y^2 - y - 1 = 0$$

$$(2y + 1)(y - 1) = 0$$

$$2y + 1 = 0 \qquad \text{or} \quad y - 1 = 0$$

$$y = -\frac{1}{2} \quad \text{or} \qquad y = 1$$

Since $x = -y$, $x = \frac{1}{2}$ or $x = -1$. Two solutions are $(\frac{1}{2}, -\frac{1}{2})$ and $(-1, 1)$. When $x = 2y$,

II.
$$x^2 - y + y^2 = 1$$
$$(2y)^2 - y + y^2 = 1 \qquad x = 2y$$
$$4y^2 - y + y^2 = 1$$
$$5y^2 - y + 1 = 0.$$

Since this equation does not factor, we will use the quadratic formula to determine its solution set.

$$y = \frac{1 \pm \sqrt{1 - 20}}{10} = \frac{1 \pm \sqrt{19}i}{10}$$

Since y is not a real number, the solution set is $\{(\frac{1}{2}, -\frac{1}{2}), (-1, 1)\}$. The check is left to the reader.   ■

Exercises 9.6

Solve each of the following systems using the substitution method.

1. $x^2 + y^2 = 9$
   $x = 3$

2. $x^2 + y^2 = 16$
   $y = -4$

3. $y^2 = 4x$
   $x = -y$

4. $x^2 + 4y^2 = 32$
   $x + 2y = 0$

5. $3y^2 - 2x^2 = 25$
   $x + y = 0$

6. $x^2 = 2y$
   $y = x - 3$

7. $x^2 + 3y^2 = 3$
   $x = 2y$

8. $xy = 6$
   $x - y = 1$

9. $x^2 + y^2 + 2x + 4y = 4$
   $x = 4y$

10. $y^2 + x^2 + 2x = 1$
    $x - y = 1$

11. $(x - 6)^2 = 2(y + 2)$
    $x - y = 3$

12. $x^2 + y^2 + 3x - 4y = 10$
    $3x + 2y = -7$

Solve each of the following systems by using the elimination method.

13. $x^2 + y^2 = 9$
    $x^2 - y^2 = 7$

14. $2x^2 + 2y^2 = 3$
    $3x^2 - 2y^2 = 2$

15. $4x^2 + y^2 = 25$
    $x^2 - y^2 = -5$

16. $x^2 + y^2 = 10$
    $16x^2 + y^2 = 25$

17. $x^2 + y^2 = 16$
    $y^2 = 4 - x$

18. $x^2 + y^2 = 5$
    $x^2 + 4y^2 = 8$

19. $x^2 - 2y^2 = 1$
    $x^2 + 4y^2 = 25$

20. $2x^2 + y^2 = 24$
    $-x^2 + y^2 = 12$

21. $2x^2 + 3y^2 = -4$
    $4x^2 + 2y^2 = 8$

22. $x^2 + 4y^2 = -17$
    $-2x^2 + 3y^2 = -10$

23. $x^2 - y^2 = 2$
    $-x + y^2 = 0$

24. $x^2 + y^2 = 20$
    $x^2 - y = 0$

Solve each of the following systems by using any method.

25. $x^2 - 4x + y^2 - 4y = 1$
    $x^2 - 4x + \qquad y = -5$

26. $x^2 + y^2 + 2x - y = 14$
    $x^2 + y^2 + x - 2y = 9$

27. $x^2 - xy - y^2 = 1$
    $x^2 + 3xy - y^2 = 1$

28. $2x^2 + xy - 4y^2 = -12$
    $x^2 - 2y^2 = -4$

29. $x^2 - xy + y^2 = 21$
$x^2 + 2xy - 8y^2 = 0$

30. $2x^2 + 3xy + 3y^2 = 48$
$2x^2 - xy - 3y^2 = 0$

31. $x^2 - xy = 54$
$xy - y^2 = 18$

32. $x^2 + 3xy - 2y^2 = 2$
$- 5xy + 6y^2 = 0$

Write each of the following as a system of equations (in two variables) and solve.

33. Find two numbers whose sum is $\frac{65}{8}$ and whose product is one.

34. Find two numbers whose difference is one and whose product is one.

35. Find the dimensions of a rectangle with an area of 100 square meters and a perimeter of 58 meters.

36. Find the dimensions of a rectangle with an area of 128 sq ft and a perimeter of 132 ft.

37. Find two natural numbers with a product of 63 and a difference of 32.

38. Find two natural numbers with a product of 42 and a sum of squares of 85.

39. Find two numbers whose product is 72 if the sum of their reciprocals is $\frac{1}{4}$.

40. Find two numbers such that the product of their reciprocals is $\frac{1}{9}$ and the sum of their reciprocals is $\frac{5}{6}$.

41. The product of two numbers is 48 and the difference of their reciprocals is $\frac{1}{24}$. Find the numbers.

42. Find two numbers such that the sum of 6 times the smaller and 4 times the larger is 3, and the sum of their squares is $\frac{5}{18}$.

Review Exercises

Solve each system of inequalities by graphing.

43. $x + y \leq 5$
$x - y > 2$

44. $x - 5y \leq 3$
$x + y \leq 7$

45. $x \leq 1$
$y \leq 3$
$y + x \geq 0$

46. $x \geq -2$
$y \geq -2$
$y < x + 1$

9.7   Nonlinear Systems of Inequalities

A **nonlinear inequality** is one in which one or more of the terms is of degree two or greater. For example

$$x^2 + y^2 \leq 4,$$
$$x^2 + 2y^2 - x + y > 3$$

and $$x - y^2 \leq 0$$

are all nonlinear inequalities.

The solution to a nonlinear inequality is a **region of the Cartesian plane.** Consider the four graphs shown in Figures 9.32(a) through 9.32(d).

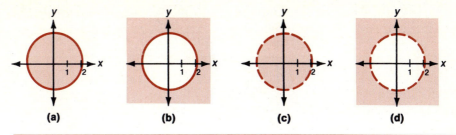

(a)   (b)   (c)   (d)

Figure 9.32

Figure 9.32(a) is the graph of $x^2 + y^2 \le 4$. To determine the region that is the solution set (shaded) a test point is used in the same manner as for linear inequalities. For example, if the test point is (0, 0) then

$$x^2 + y^2 \le 4$$
$$0^2 + 0^2 \overset{?}{\le} 4 \qquad x = 0, \, y = 0$$
$$0 \overset{\checkmark}{\le} 4.$$

Since this is a true statement, the region containing this point (the interior of the circle) and its **boundary,** the circle itself, form the solution set.

Figure 9.32(b) is the graph of $x^2 + y^2 \ge 4$. In this case the test point does not satisfy the inequality, since

$$0^2 + 0^2 \not\ge 4.$$

The solution includes the region outside the circle together with its boundary, the circle.

Figure 9.32(c) and 9.32(d) are the graphs of $x^2 + y^2 < 4$ and $x^2 + y^2 > 4$ respectively. In this case the circle itself is not a part of the solution set. It is shown as a broken line.

The solution to a nonlinear system of inequalities is that portion of the Cartesian plane that satisfies all inequalities.

Example 1   Solve the system $x^2 + y^2 < 16$ and $x + y \le -2$ by graphing.

The equation $x^2 + y^2 = 16$ is a circle with radius 4 and center at the origin. Since the test point (0, 0) satisfies the inequality $x^2 + y^2 < 16$,

$$0^2 + 0^2 < 16,$$

all points on the interior of the circle satisfy it. The inequality $x + y \le -2$ is not satisfied by the test point, (0, 0), since

$$0 + 0 \not\le -2.$$

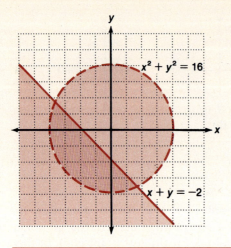

Figure 9.33

Thus, the region that does not contain (0, 0) satisfies $x + y \leq -2$. The solution set is that region that satisfies both inequalities. It is shown with double shading in Figure 9.33. ■

Example 2   Solve $y + x^2 \leq 1$, $x^2 + 4y^2 \geq 16$ by graphing.

The graph of $y + x^2 \leq 1$ or $y \leq -x^2 + 1$ is a parabola opening downward with vertex at (0, 1). If (0, 0) is used as a test point in $y + x^2 \leq 1$ then

$$0 + 0^2 \overset{?}{\leq} 1$$

$$0 \overset{\checkmark}{\leq} 1.$$

Now consider $x^2 + 4y^2 = 16$. This is the equation of an ellipse with center at the origin and vertices at (0, 2), (0, −2), (4, 0), and (−4, 0). Again, if (0, 0) is used as the test point, then $0^2 + 4 \cdot 0^2 \geq 16$. The region outside the ellipse is the solution. The solution to the system is shown with double shading in Figure 9.34 on the facing page. It contains its boundary lines. ■

In Examples 1 and 2 we considered systems involving two inequalities. A system may involve any number of inequalities.

Example 3   Solve $y \geq x^2 + 4x + 4$, $x^2 + y^2 \leq 16$ and $|y| < 4$ by graphing.

The equation $y = x^2 + 4x + 4$, factored as $y = (x + 2)^2$, represents a parabola opening upward with vertex at (−2, 0). The test point (0, 0) does not satisfy the inequality. The inequality $x^2 + y^2 < 16$ is the interior of the circle $x^2 + y^2 = 16$, and $|y| < 4$ is the region between $y = 4$ and $y = -4$. The solution to the system is the region with triple shading in Figure 9.35. ■

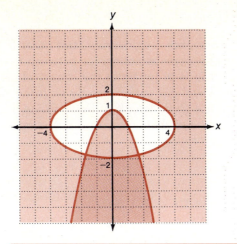

Figure 9.34

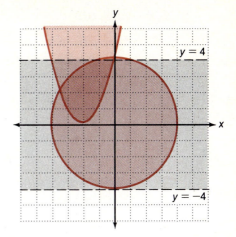

Figure 9.35

Exercises 9.7

Solve each system of inequalities by graphing.

1. $x^2 + y^2 < 25$
   $x + y < 5$

2. $y > x^2 + 2x + 1$
   $y \le x + 3$

3. $x^2 + y^2 \le 4$
   $x + y \ge \left(\dfrac{1}{2}\right)x - 2$

4. $x^2 + y^2 \le 16$
   $x + y \ge 4$

5. $x + y < 4$ (quadrant I only)
   $xy \ge 1$

6. $x^2 + y^2 + 2x - 6y + 1 \le 0$
   $x + y \le 4$

7. $y \ge x^2 - 1$
   $y \le x$

8. $x^2 + y^2 \le 36$
   $x \ge 3$

9. $x^2 - y^2 < 25$
   $x + y \le 4$

10. $x^2 - 4 \le y$
    $x^2 + y^2 \le 25$

11. $9x^2 + 25y^2 \le 225$
    $x^2 + y^2 \ge 16$

12. $16x^2 + 25y^2 \le 400$
    $x^2 - y^2 \ge 4$

13. $x^2 - 6y + y^2 \le -5$
    $9x^2 + 16y^2 < 144$

14. $y > x^2 - 4$
    $y - x < 2$

15. $y - x^2 > -1$
    $y + 2x < -1$

16. $x^2 - y^2 \ge 4$
    $x^2 + y^2 \ge 25$

17. $4x^2 + y^2 \le 16$
    $x^2 + 4y^2 \ge 16$

18. $x^2 + y^2 - 6x - 8y \le 0$
    $xy < 12$

19. $x^2 + y^2 - 10x + 21 \ge 0$
    $x^2 + y^2 - 10x + 2y + 1 \le 0$

20. $x^2 + y^2 \le 9$
    $16x^2 - 4y^2 \le 64$

21. $x^2 - y^2 \ge -4$
    $x^2 + y^2 \le 10$

22. $y < -x^2 + 1$
    $y \ge x^2 - 2$

23. $y \le 4 - x^2$
    $y \ge x^2 - 4$
    $|x| < 1$

24. $x^2 + y^2 < 16$
    $x^2 + y^2 > 4$
    $|x| > 1$

25. $x^2 + y^2 \le 36$
    $x^2 + y^2 \ge 16$
    $|x| > 3$

26. $x^2 + y^2 \le 25$
    $x \ge 1$
    $y < 4$

27. $x^2 + y^2 \le 9$
    $|x| < 2$
    $|y| < 2$

28. $y \le 4 - x^2$
    $y > x^2 - 4$
    $|x| < 1$

29. $x^2 + y \le 5$
    $x^2 + y^2 < 25$
    $|x| < 2$

30. $y \ge x^2 - 4x + 4$
    $x^2 + y^2 < 16$
    $|y| < 3$

9.8 Variation

Many jobs in science, mathematics, engineering, and other fields involve working with formulas that express relationships between two or more quantities. The area of a circle is pi (π) times the square of the radius; the pressure exerted by an enclosed gas varies directly with the absolute temperature; the stiffness of a given length of beam varies directly with the product of the width and cube of the depth. In this section we will discover how to write and use such formulas.

> **Direct Variation**
>
> The statement **y varies directly with x** means
>
> $$y = kx$$
>
> where k is called the **constant of variation.**

Example 1 Translate each equation into words.

(a) $y = kx^2$ means y varies directly with the square of x.

(b) $v = ks^3$ means v varies directly with the cube of s.

(c) $C = \pi d$ means C varies directly with d. ∎

The formula in Example 1(c) gives the circumference of a circle in terms of its diameter. Here the constant of variation is π.

Not all variation is direct. Two additional forms that we will consider are inverse and joint variation.

> **Inverse Variation**
>
> The statement **y varies inversely with x** means
>
> $$y = \frac{k}{x}$$
>
> where k is the constant of variation.

Example 2 Translate each equation into words.

(a) $y = \dfrac{k}{x^2}$ means y varies inversely with the square of x.

(b) $y = \dfrac{k}{\sqrt{x}}$ means y varies inversely with the square root of x. ∎

> ### Joint Variation
> When one variable varies as the product of two or more variables it is said to **vary jointly** and the relation is known as **joint variation.**

Example 3   Translate each equation into words.

(a) $y = kxz$ means y varies jointly with x and z.

(b) $y = \dfrac{kst^2}{r^3}$ means y varies jointly with s and the square of t, and inversely with the cube of r. ■

The constant of variation, as the name implies, retains the same value for all values of the variables satisfying the given condition. For this reason, its value can be readily determined.

Example 4   A quantity y varies directly with x. When $y = 4$, $x = 2$. Find y when $x = 3$.
The variation equation is

$$y = kx.$$

To find k, substitute 4 for y and 2 for x.

$$4 = k \cdot 2$$
$$2 = k$$

The equation becomes

$$y = 2x.$$

When $x = 3$,

$$y = 2 \cdot 3 = 6. \quad ■$$

Example 5 W varies jointly with the cube of x and the square of y and inversely with z. Find the constant of variation if $W = 8$ when $x = 2$, $y = 3$, and $z = 4$.
The variation equation is

$$W = \frac{kx^3y^2}{z}$$

$$8 = \frac{k(2)^3(3)^2}{4} \qquad W = 8,\ x = 2,\ y = 3,\ z = 4$$

$$8 = 18k$$

$$\frac{4}{9} = k \quad ■$$

Example 6   According to Hooke's law, the force necessary to stretch a spring a given distance varies directly with the distance. If a 40-kilogram weight stretches the spring 3 centimeters, how far would a 60-kilogram weight stretch it?

The variation equation is

$$F = kx$$

where F is the weight and x is the vertical displacement.

$$40 = k \cdot 3 \qquad F = 40 \text{ when } x = 3$$

$$\frac{40}{3} = k$$

Thus $F = \dfrac{40}{3}x.$

If $F = 60$, then

$$60 = \frac{40}{3}x$$

$$\frac{180}{40} = x$$

$$4.5 = x.$$

The weight would stretch the spring 4.5 centimeters.   ■

In the following exercises it is unnecessary to be familiar with the units of measurement in order to solve the problems.

Exercises 9.8

Write each indicated variation in the form of an equation.

1. y varies directly with x.
2. R varies directly with s.
3. M varies directly with the square of p.
4. H varies directly with the square of q.
5. V varies directly with the cube of s.
6. R varies directly with the cube of t.
7. y varies inversely with x.
8. A varies inversely with d.
9. S varies directly with the square root of a.
10. H varies directly with the square root of j.
11. V varies jointly with h and the square of r.
12. V varies jointly with L, W and H.
13. P varies directly with c and inversely with the square root of n.
14. y varies directly with the square of x and inversely with z.

Solve each of the following variation problems.

15. If y varies directly with x and $y = 3$ when $x = 9$, find y when $x = 12$.
16. If p varies directly with t and $p = 8$ when $t = 2$, find p when $t = 5$.
17. If A varies directly with the square of s and $A = 8$ when $s = 2$, find A when $s = 7$.
18. If y varies directly with the square of x and $y = 3$ when $x = 4$, find y when $x = 16$.
19. If a varies directly with the cube of b and $a = 6$ when $b = 10$, find a when $b = 2$.

20. If w varies directly with the cube of u and $w = 12$ when $u = 20$, find w when $u = 2$.

21. Z varies directly with the square of x and inversely with y, $Z = 8$ when $x = 4$ and $y = 3$. Find Z when $x = 3$ and $y = 8$.

22. P varies directly with the cube of m and inversely with the square of n. $P = 8$ when $m = 4$ and $n = 6$. Find P when $m = 2$ and $n = 12$.

23. T varies jointly with p and the cube of v and inversely with the square of u. If $T = 24$ when $p = 3$, $v = 2$ and $u = 4$, find T when $p = 4$, $v = 3$ and $u = 2$.

24. A varies jointly with the square of b and the square of c and varies inversely with the cube of d. If $A = 18$ when $b = 4$, $c = 3$ and $d = 2$, find A when $b = 2$, $c = 6$ and $d = 4$.

25. The intensity of illumination I varies inversely with the square of its distance d from the light source. The intensity is 4 foot-candles at 15 feet from the source. Find the intensity 10 feet from the source.

26. The distance d an object falls varies directly with the square of the time t (in seconds) it falls. If the object falls 16 feet in 2 seconds, how far would it fall in 10 seconds?

27. The resistance R of a conductor varies inversely with the area A of its cross section. If $R = 20$ ohms when $A = 8$ square centimeters, find R when $A = 12$ square centimeters.

28. The weight W of an object in space varies inversely with the square of its distance d from the center of the earth. If an object weighs 400 pounds on the surface of the earth, how much does it weigh 1000 miles from the surface of the earth? Assume the radius of the earth to be 4000 miles.

29. The current I in a wire varies directly with the electromotive force E and inversely with the resistance R. In a wire whose resistance is 10 ohms a current of 36 amperes is obtained when the electromotive force is 120 volts. Find the current produced when $E = 220$ volts and $R = 30$ ohms.

30. The distance d required to stop a midsize car varies directly with the square of its speed s. If the car stops in 36 feet at a speed of 24 miles per hour, find the distance required to stop it at a speed of 54 miles per hour.

31. The pressure P of a certain gas in a container varies directly with the temperature T and inversely with the volume V. If 300 cubic feet of this gas exert a pressure of 20 pounds per square foot at a temperature of 500°K, what is the pressure exerted by this gas when the temperature is 400°K and the volume is 500 cubic feet?

32. The surface area S of a hollow cylinder varies jointly with the height H and the radius R of the cylinder. If a cylinder with a radius of 5 inches and a height of 7 inches has a surface area of 213.5 square inches, find the surface area when the radius is 8 inches and the height is 11 inches.

33. The pressure P on a wall varies jointly with the area A of the wall and the square of the velocity V of the wind. If the pressure is 120 pounds when the area is 100 square feet and the velocity of the wind is 20 miles per hour, find the pressure when the area is 200 square feet and the velocity of the wind is 40 miles per hour.

34. Simple interest I earned over a given time varies jointly with the principal p and the rate of interest r. If \$100.00 at 4% earns \$14.00, how much does \$250.00 at 5% earn in the same time period?

35. The amount of pollution P entering the atmosphere in a certain area is found to vary directly with the 2/3 power of the number N of people in that area. If a population of 8000 people produces 900 tons of pollution per year, find a formula for P in terms of N.

36. Kepler's third law states that the time T required for a planet to make one revolution about the sun varies directly with the 3/2 power of the maximum distance of its orbit. If the maximum distance of the earth's orbit is 93 million miles and the maximum distance of Mars' orbit is 142 million miles, what is the time required for Mars to make one orbit?

9.9   Inverse Functions and Relations

In Section 9.1 a relation was defined as a set of ordered pairs. The set of first elements in the ordered pairs formed the domain, and the set of second elements formed the range. In this section we will consider a new relation formed by interchanging the domain and range.

If R is a relation, then R^{-1} is a new relation called the **inverse** of R in which the domain and range elements of each ordered pair have been interchanged. For example,

if $R = \{(1, 2), (3, 4), (5, 6)\}$,   then $R^{-1} = \{(2, 1), (4, 3), (6, 5)\}$.

The domain of R, $\{1, 3, 5\}$, is the range of R^{-1}. The range of R, $\{2, 4, 6\}$, is the domain of R^{-1}.

Relations are often defined in terms of equations. When a relation is defined in this way, its inverse is found by interchanging the variables.

Example 1   Find the inverse of $y = 2x + 3$.

Interchange x and y. The inverse is

$x = 2y + 3$.

Now solve the equation for y.

$$x = 2y + 3$$

$$\frac{x - 3}{2} = y \quad \text{or} \quad y = \frac{1}{2}x - \frac{3}{2} \quad \blacksquare$$

If the inverse of the function f is a function, it is written as f^{-1}. If a function is written as $f(x)$, its inverse is $f^{-1}(x)$.

Example 2   Given $f(x) = 3x + 4$, find $f^{-1}(x)$.

First let $y = f(x)$, then interchange x and y.

$y = 3x + 4$

$x = 3y + 4$ Interchange x and y

$y = \dfrac{x - 4}{3}$ Solve for y

When y in the inverse is replaced with $f^{-1}(x)$, the inverse becomes

$$f^{-1}(x) = \frac{x - 4}{3}.$$ ■

An interesting property that applies to relations and their inverses is that their graphs are mirror images of one another when the mirror is the line $y = x$. For example, $y = x^2$ and $x = y^2$ are inverses. Their graphs together with the graph of $y = x$ are shown in Figure 9.36. As the figure shows, $y = x^2$ is the reflection of $x = y^2$. Notice that the inverse of $y = x^2$, $x = y^2$, is not a function. A vertical line would cut it in more than one point.

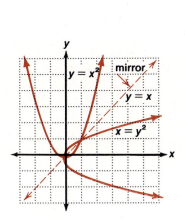

Figure 9.36

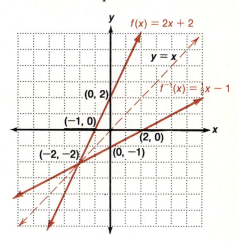

Figure 9.37

Example 3   Graph $f(x) = 2x + 2$ and its inverse on the same axes.
Let $f(x) = y$, giving

$$y = 2x + 2.$$

The inverse, $f^{-1}(x)$, is

$$x = 2y + 2 \quad \text{or} \quad y = \frac{1}{2}x - 1.$$

The graph is shown in Figure 9.37.   ■

In Example 3 both $f(x)$ and $f^{-1}(x)$ are functions. One might wonder what form a function must have in order for its inverse to be a function. The answer is that it must be *one-to-one*.

A function is **one-to-one** if each first element in its ordered pairs corresponds to a unique second element and each second element corresponds to a unique first element.

Example 4   Find the inverse of {(1, 2), (3, 4), (5, 2)}. Is the inverse a function?
       The inverse is {(2, 1), (4, 3), (2, 5)}. The first element 2 corresponds to second elements 1 and 5. The inverse is not a function.   ■

We now consider a theorem that can be used to determine whether a function has an inverse that is a function.

Theorem 9.2   A function has an inverse that is a function if and only if there is a one-to-one correspondence between the domain and range elements:

$$f[f^{-1}(x)] = x \qquad \text{and} \qquad f^{-1}[f(x)] = x.$$

Example 5   Use Theorem 9.2 to show that

$$f(x) = 3x + 4 \qquad \text{and} \qquad f^{-1}(x) = \frac{x - 4}{3}$$

are inverses.
       We must show that $f[f^{-1}(x)] = f^{-1}[f(x)] = x$. Begin by substituting $f^{-1}(x)$ for x in $f(x)$.

$$f(x) = 3x + 4$$

$$f[f^{-1}(x)] = f\left(\frac{x - 4}{3}\right) = 3\left(\frac{x - 4}{3}\right) + 4 = x$$

Now substitute $f(x)$ for x in $f^{-1}(x)$.

$$f^{-1}(x) = \frac{x - 4}{3}$$

$$f^{-1}[f(x)] = f^{-1}(3x + 4) = \frac{3x + 4 - 4}{3} = x$$

The functions are inverses. Each function is one-to-one.   ■

A numerical example of Theorem 9.2, for the functions

$$f(x) = 3x + 4 \qquad \text{and} \qquad f^{-1}(x) = \frac{x - 4}{3},$$

sometimes helps to clarify its meaning. If we let $x = 2$, then

$$f(2) = 3 \cdot 2 + 4 = 10$$

and $\qquad f^{-1}[f(2)] = f^{-1}(10) = \dfrac{10 - 4}{3} = \dfrac{6}{3} = 2.$

Also $\qquad f^{-1}(2) = \dfrac{2 - 4}{3} = \dfrac{-2}{3}$

and $\qquad f[f^{-1}(2)] = 3\left(\dfrac{-2}{3}\right) + 4 = 2.$

Example 6   Find the inverse of $4x^2 + 9y^2 = 36$ and sketch both graphs on the same coordinate axes.

     The inverse is obtained by interchanging x and y:

$$4y^2 + 9x^2 = 36.$$

Both equations are ellipses. The graphs are shown in Figure 9.38.   ■

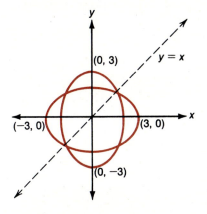

Figure 9.38

     Example 6 illustrates that it is not necessary to solve an equation for y in order to determine its inverse. In fact, when absolute values are involved, it is not even desirable.

Exercises 9.9

For each of the following functions, (a) form the inverse, (b) determine the domain and range of the inverse, and (c) determine whether it is a one-to-one function.

1. {(0, 1), (1, 2), (2, 3), (3, 4)}   **2.** {(−4, 2), (−2, 0), (0, 2), (2, 4)}

3. {(−2, 4), (−1, 1), (0, 0), (1, 1), (2, 4)}   **4.** {(−2, −4), (−1, −1), (0, 0), (1, −1), (2, −4)}

5. {(1, 1), (4, 2), (9, 3), (16, 4)}   **6.** {(0, 0), (1, 1), (8, 2), (27, 3)}

7. {(−3, −5), (−1, 1), (1, 7), (2, 13)}   **8.** {(0, 1), (1, −4), (2, −9), (3, −14)}

Find the inverse of each of the following functions. Sketch the graph of each function as indicated, graphing its inverse and the line $y = x$ on the same coordinate axes. Determine whether it is a one-to-one function.

9. $f(x) = 2x$

10. $f(x) = x$

11. $f(x) = 2x - 1$; graph

12. $f(x) = 2x + 3$; graph

13. $f(x) = -2x + 5$

14. $f(x) = -x + 1$

15. $f(x) = \left(\dfrac{-4}{3}\right)x + \dfrac{1}{2}$

16. $f(x) = \left(\dfrac{-1}{2}\right)x + \dfrac{1}{3}$

17. $f(x) = x^2 + 1$; graph

18. $f(x) = x^2 - 1$; graph

19. $f(x) = 9 - x^2$

20. $f(x) = 16 - x^2$

21. $f(x) = 3 - x^2$

22. $f(x) = 4 - 2x^2$

23. $f(x) = -4 - x^2$

24. $f(x) = -25 - x^2$

25. $f(x) = |x + 2|$; graph

26. $f(x) = |x - 3|$; graph

27. $f(x) = -3x$

28. $f(x) = -6x$

29. $f(x) = x^3 - 1$; graph

30. $f(x) = x^3 + 2$; graph

31. $f(x) = |x| + 1$

32. $f(x) = |x| - 1$

For each of the following one-to-one functions, (a) find f^{-1}, (b) show that $f[f^{-1}(x)] = f^{-1}[f(x)] = x$.

33. $f(x) = x$

34. $f(x) = -x$

35. $f(x) = -x + 2$

36. $f(x) = x + 3$

37. $f(x) = \left(\dfrac{3}{4}\right)x - \dfrac{1}{4}$

38. $f(x) = \left(\dfrac{-3}{5}\right)x + \dfrac{2}{5}$

39. $f(x) = -4x + 8$

40. $f(x) = \left(\dfrac{4}{3}\right)x - 4$

Each of the following defines a relation. Find the inverse, and sketch the graph of the relation and its inverse on the same coordinate axes as indicated. Determine whether the relation or its inverse is a function.

41. $x^2 + y^2 = 9$; graph

42. $x^2 + 4y^2 = 16$; graph

43. $x^2 - y^2 = 4$; graph

44. $y = (x + 1)^2 - 2$; graph

45. $3x^2 + 4y^2 = 12$

46. $x^2 = 1 - y^2$

47. $y^2 = x + 4$; graph

48. $x - 2 = y^2$; graph

49. $y = |x - 1| + 1$

50. $y = |x + 2| - 3$

51. $y = (x - 4)^2 + 1$

52. $x^2 - y^2 = -4$

Chapter 9   Summary

[9.1]   A **relation** is defined as a set of ordered pairs. The set of first elements in the ordered pairs is called the **domain** of the relation, while the set of second elements is called the **range**. When a relation is described in terms of an equation such as

$$\left\{ (x, y) \mid y = \frac{x - 2}{x + 3} \right\},$$

the domain and range are found by eliminating the values for which the equation is undefined.

A **function** is a relation in which each element in the domain corresponds with exactly one element in the range. The notation f or $f(x)$ is used to show that a relation is a function. The notation $f[g(x)] = f \circ g(x)$ is used to indicate the **composition** of functions. When the graph of a relation is known, the

vertical line test can be used to determine whether or not it is a function. If a vertical line intersects the graph in more than one point, it is not the graph of a function.

[9.2]   A **linear function** is one whose graph is a straight line. Vertical lines are not functions. Equations of lines appear in many different forms and can be found by using these forms when enough information is known.

General form	$ax + by + c = 0$
Point-slope form	$y - y_1 = m(x - x_1)$ Point (x_1, y_1); slope $= m$
Slope-intercept form	$y = mx + b$ Slope $= m$; y-intercept $= b$
Horizontal line	$y =$ **constant** Slope is zero
Vertical line	$x =$ **constant** Slope is undefined

[9.3]   When the points (x_1, y_1) and (x_2, y_2) on a plane are known, the distance between them can be found by using the **distance formula:**

$$d = \sqrt{(x_2 - x_1)^2 + (y_2 - y_1)^2}.$$

If the points lie on the same horizontal or vertical line, the distance between them is $|x_2 - x_1|$ or $|y_2 - y_1|$, respectively.

The distance formula is used to find the equation of a circle. A circle with center at (h, k) and radius r has equation

$$(x - h)^2 + (y - k)^2 = r^2.$$

The midpoint of a line segment is found by averaging the coordinates of its endpoints. The midpoint of the line segment joining (x_1, y_1) and (x_2, y_2) is given by the formula

$$(x_m, y_m) = \left(\frac{x_1 + x_2}{2}, \frac{y_1 + y_2}{2} \right).$$

[9.4] [9.5]   A **conic section** is the curve traced when a cone is cut by a plane. The four conic sections studied in this chapter were the circle, parabola, ellipse, and hyperbola. Their standard equations are listed below.

Circle	$(x - h)^2 + (y - k)^2 = r^2$
Parabola	$y = a(x - h)^2 + k$ or $x = a(y - k)^2 + h$
Ellipse	$\dfrac{(x - h)^2}{a^2} + \dfrac{(y - k)^2}{b^2} = 1$
Hyperbola	$\dfrac{(x - h)^2}{a^2} - \dfrac{(y - k)^2}{b^2} = 1$ or $\dfrac{(y - k)^2}{b^2} - \dfrac{(x - h)^2}{a^2} = 1$

Parabolas open upward or downward, to the right or to the left, depending on whether a is positive or negative. An ellipse is called an **x ellipse** if $a > b$ and a **y ellipse** if $a < b$. Hyperbolas symmetric to a horizontal line are called *x* **hyperbolas.** If they are symmetric to a vertical line they are *y* **hyperbolas.** The asymptotes of a hyperbola are the two lines passing through its center with slope $\pm\frac{b}{a}$.

[9.6]   A **nonlinear system of equations** is one in which one or more of the equations is of degree two or greater. Such systems are solved by using a variety of methods, including substitution, elimination, and factoring with substitution. The number of solutions to such a system will vary depending on the degree of the equations.

[9.7]   The solution set to a nonlinear system of inequalities is found by sketching the graph of each member of the system using shading to indicate its individual solution set. The portion of the plane that contains shading from each inequality forms the solution set of the system.

[9.8]   **Variation** is a term used in mathematics to indicate how two or more quantities are related. Three types of variation are **direct variation,** $y = kx$, **inverse variation,** $y = \frac{k}{x}$; and **joint variation,** $y = kxyz$. In each equation, *k* is the **constant of variation.**

[9.9]   The **inverse** of a relation R, written R^{-1}, is that relation obtained when the domain and range elements are interchanged. If a relation is a one-to-one function, its inverse will be a function. If a function and its inverse are both one-to-one, then $f[f^{-1}(x)] = f^{-1}[f(x)] = x$.

Chapter 9   Review

Find f(−2), f(0), and f(3) for each of the following.

1. $f(x) = 3x - 5$ **2.** $f(x) = -5x + 3$ **3.** $f(x) = x^2 - 4$ **4.** $f(x) = 1 - x^3$

Determine the domain and range for each of the following.

5. $f(x) = -6x$ **6.** $f(x) = 3$ **7.** $f(x) = \dfrac{1}{x - 3}$

8. $f(x) = \dfrac{3}{2x + 3}$ **9.** $f(x) = \sqrt{9 - x^2}$ **10.** $f(x) = \sqrt{9 + x^2}$

In exercises 11–17 write the equation of the line having the given characteristics.

11. Passes through (3, 1), (−6, 2)   **12.** $m = 3$, (0, −5)

13. Passes through (−5, 1) and is parallel to $2x - y = 4$

14. Slope $= \frac{-1}{2}$, *y*-intercept $= 3$

15. Passes through the origin, $m = \frac{2}{3}$

16. Passes through the point of intersection of the lines $x + 2y = 6$ and $3x - y = 1$ and through the point (0, 4)

17. Perpendicular to $3x - 5y = 0$ at its point of intersection with $2x + y = 1$

18. Find the distance between the points (5, −6) and (−7, 5).

19. Find the coordinates of the midpoint of the line segment from $(-3, 1)$ to $(2, 4)$.

20. Find the coordinates of the center and the length of the radius for $(x - 2)^2 + (y + 1)^2 = 16$.

21. Determine the coordinates of the center and the length of the radius for $x^2 + y^2 + 2x - 2y = 2$.

22. Write the equation of the circle with center at $(3, -1)$, passing through the point $(5, 2)$.

Identify each of the following conic sections and graph as indicated.

23. $y = -x^2$

24. $y = x^2 + 2x + 3$

25. $3y^2 = 4x$

26. $x^2 - y^2 = 4$; graph

27. $2x^2 + 2y^2 = 32$; graph

28. $5x^2 - 4y^2 = 20$

29. $x^2 - y^2 = -4x - 4y + 1$

30. $x^2 + 2y^2 + 6x - 8y = 0$

31. $3x^2 - 5x + 7y = -3y^2 + 1$

32. $x - 3 = 2y^2 + 4y + 2$

Solve each of the following systems of equations.

33. $x^2 + y^2 = 6$
    $x \quad\;\; = y$

34. $x^2 - y^2 = 2x - 1$
    $\quad\; y = x$

35. $4x^2 + 4y^2 = 6$
    $6x^2 - 4y^2 = 4$

36. $5x^2 + y^2 = 10$
    $x^2 - y^2 = 2$

37. $x^2 + y^2 = 1$
    $x^2 - y^2 = -7$

38. $2x^2 - xy + y^2 = 8$
    $x^2 - \quad\;\; y^2 = 0$

39. $x^2 - 4x = \quad\; y - 3$
    $x^2 \quad\;\; = 3x + y - 2$

40. $x^2 - 2xy + y^2 = 0$
    $x^2 + \; xy + y^2 = 3$

Solve each of the following systems of inequalities by graphing.

41. $x^2 + y^2 < 36$
    $x + y < 4$

42. $x^2 + y^2 \le 24$
    $|x| \quad\; < 4$

43. $x^2 + 36y^2 \le 72$
    $x^2 + \quad y^2 \ge 16$

44. $y \le -x^2 + 4$
    $y \ge \quad x^2 - 1$

45. $y > -x^2 - 2x - 1$
    $y < \quad x + 2$

46. $y \ge \quad |x + 1| - 2$
    $y \le -|x + 1| + 3$

Find the domain and range of each function. Determine whether it is a one-to-one function.

47. $f(x) = -x + 6$

48. $f(x) = x^2$

49. $f(x) = \sqrt{x^2 - 1}$

50. $f(x) = (x - 3)^2 + 1$

51. $f(x) = \sqrt{4 + x^2}$

52. $f(x) = |x - 2| + 1$

Solve each of the following variation problems.

53. The kinetic energy E of an object varies directly with its weight w and the square of its velocity v. An 8-pound object moving at 4 feet per second has 2 foot-pounds of kinetic energy. Find the kinetic energy of a 6000-pound truck moving at 60 miles per hour.

54. The atmospheric pressure P on a given mass of ideal gas varies inversely with the volume v and directly with the absolute temperature t. To what pressure must 100 cubic feet of helium at 1 atmosphere of pressure and 253° temperature be subjected to be compressed to 50 cubic feet when the temperature is 313°?

55. The force of the wind on a sail varies directly with the area of the sail and the square of the velocity of the wind. On one square foot of sail the force is 1 pound when the velocity of the wind is 15 miles per hour. Find the force of a 45-mile-per-hour wind on a sail with an area of 10 square yards.

10

Exponential and Logarithmic Functions

Chapter 4 developed the laws of exponents, and Chapter 7 covered rational exponents. This chapter will examine exponential functions, which have a wide variety of applications. The exponential function is used in physics to describe rates of radioactive decay, in the biological sciences to discuss the growth of bacterial cultures and populations, and in finance to calculate compound interest.

10.1 Exponential Functions

In Chapter 7 we considered exponential expressions in which the base was a real number and the exponent was a rational number. By definition,

$$b^{m/n} = \sqrt[n]{b^m} = (\sqrt[n]{b})^m$$

provided that $\sqrt[n]{b}$ exists. For example,

$$9^{1/2} = \sqrt{9} = 3$$
$$16^{3/4} = (\sqrt[4]{16})^3 = 2^3 = 8$$

and
$$27^{-1/3} = \frac{1}{\sqrt[3]{27}} = \frac{1}{3},$$

but $(-4)^{1/2}$ does not exist in the set of real numbers.

We now ask the question, "What is meant by b^m where b is a real number and m is irrational?" The complete answer requires the concept of the limit from calculus and is beyond the scope of this book. It is possible, however, to gain some insight by considering an expression such as $3^{\sqrt{2}}$. Recall that $\sqrt{2}$ is irrational and as such is a non-terminating, non-repeating decimal. An approximation is

$$\sqrt{2} \approx 1.4142136.$$

The values of $3^1 = 3$, $3^{1.4} = 3^{14/10}$, $3^{1.41} = 3^{141/100}$, $3^{1.414} = 3^{1414/1000}$ . . . can be computed because the exponents are rational numbers. If the process were to continue in the same manner forever, the results would approach $3^{\sqrt{2}}$ so closely that any difference would be insignificant.

Exponential functions can be defined as follows.

A function of the form $f(x) = b^x$, $b > 0$, $b \neq 1$, is an **exponential function** whose domain is the set of real numbers and whose range is the set of positive real numbers.

The behavior of exponential functions is best observed by considering their graphs.

Example 1   Sketch the graph of $f(x) = 2^x$.
First make a table of values for some rational values of x. Assume from the discussion above that a smooth curve will pass through these points as well as all others corresponding to irrational values of x. The graph is shown in Figure 10.1.

x	2^x
-3	$\frac{1}{8}$
-2	$\frac{1}{4}$
-1	$\frac{1}{2}$
0	1
1	2
2	4
3	8

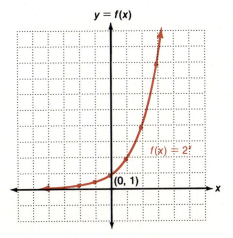

Figure 10.1

Example 2   Sketch the graph of $f(x) = 2^{-x}$.

Begin by constructing a table of values. The graph is shown in Figure 10.2.

x	2^{-x}
−3	8
−2	4
−1	2
0	1
1	$\frac{1}{2}$
2	$\frac{1}{4}$	■

Notice in Examples 1 and 2 that the graph crosses the y-axis where $y = 1$.

Example 3   Graph $f(x) = 2^x + 2$ and $f(x) = 2^{|x|}$ on the same coordinate system.

x	$2^x + 2$		x	$2^{|x|}$
−3	$\frac{17}{8}$		−3	8
−2	$\frac{9}{4}$		−2	4
−1	$\frac{5}{2}$		−1	2
0	3		0	1
1	4		1	2
2	6		2	4
3	10		3	8

The graphs are shown in Figure 10.3. Note that the graph of $f(x) = 2^x + 2$ is the graph of $f(x) = 2^x$ raised 2 units. How does $2^{|x|}$ relate to 2^x? ■

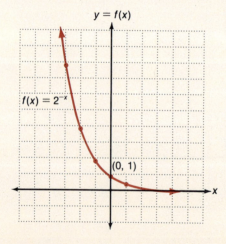

Figure 10.2

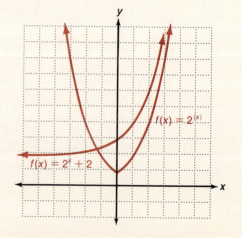

Figure 10.3

As mentioned earlier, the decay of radioactive materials and the growth of bacterial cultures can be represented by exponential functions. The base for the decay or growth is the number

$e = 2.7182818284590 \ldots$

where e is irrational.

Example 4 A particular type of radioactive material decays according to the equation $A(t) = 200e^{-0.05t}$ where $A(t)$ is in grams.

(a) How much material is present initially?
   The initial amount occurs when $t = 0$, so $A(0) = 200e^0 = 200$ grams.

(b) How much is present when $t = 10$?
   When $t = 10$, $A(10) = 200e^{-0.05(10)} = 200e^{-0.5}$. The value of $e^{-0.5}$ can be found by using a calculator with an $\boxed{e^x}$ key.
   Thus $A(10) = 121.3$ grams.

(c) Sketch the decay graph.
   The graph is shown in Figure 10.4.
   From the graph it can be seen that half of the original amount has decayed after 12 to 14 years. In section 10.4 we will see how to calculate the half-life of a radioactive substance. ■

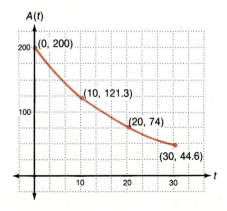

Figure 10.4

Theorem 10.1

If $b > 0$, $b \neq 1$, and m and n are any real numbers, then

$$b^m = b^n$$

if and only if $m = n$.

Example 5 Solve for x if $2^x = 32$.

First express both sides in terms of the same base.

$2^x = 32$

$2^x = 2^5$ $\qquad 2^5 = 32$

$x = 5$ $\qquad$ Theorem 10.1

The solution set is $\{5\}$. ∎

Example 6 Solve $3^{2y} = \dfrac{1}{27}$.

Since $27 = 3^3$, we have

$3^{2y} = \dfrac{1}{3^3}$

$3^{2y} = 3^{-3}$

$2y = -3$ $\qquad$ Theorem 10.1

$y = \dfrac{-3}{2}$.

The solution set is $\left\{\dfrac{-3}{2}\right\}$. ∎

Example 7 Solve $8^{x+3} = 64^x$ for x.

$8^{x+3} = 64^x$ $\qquad$ Both 8 and 64 are powers of 2

$(2^3)^{x+3} = (2^6)^x$

$2^{3x+9} = 2^{6x}$ $\qquad$ $(a^m)^n = a^{m \cdot n}$

$3x + 9 = 6x$

$9 = 3x$

$3 = x$

The solution set is $\{3\}$. ∎

When funds are invested at *compound annual interest* for a given period of years, they grow to an amount given by the formula

$A = P(1 + r)^t,$

where P is the amount originally invested, r is the annual rate of interest, and t is the time in years. For example, an investment of $1000 at 8% interest for two years would amount to

$A = 1000(1 + .08)^2$

$A = 1000(1.08)^2$

$A = \$1166.40.$ $\qquad$ Use a calculator

When interest is compounded more than once a year, the annual interest rate (r) must be divided by the number of times it is compounded (n), and the time period (t) must be multiplied by n.

$$A = P\left(1 + \frac{r}{n}\right)^{nt}$$

Example 8 Find the final amount on deposit if $1300 is invested for 9 years at 9%, compounded monthly.

There are 12 payment periods each year. The formula becomes

$$A = P\left(1 + \frac{r}{12}\right)^{12}$$

$$A = 1300\left(1 + \frac{0.09}{12}\right)^{108} \qquad nt = 12 \cdot 9 = 108$$

$$A = 1300(1 + 0.0075)^{108} = \$2913.46 \quad \blacksquare$$

Exercises 10.1

Complete the ordered pairs or graph as indicated.

1. $f(x) = 2^x$; (0,   ), (1,   ), (2,   )

2. $f(x) = 3^x$; (−2,   ), (0,   ), (2,   )

3. $f(x) = \left(\frac{1}{2}\right)^x$; graph

4. $f(x) = \left(\frac{1}{3}\right)^x$; graph

5. $f(x) = 3^{2x}$; (−2,   ), (0,   ), (2,   )

6. $f(x) = 2^{3x}$; (−1,   ), (0,   ), (1,   )

7. $f(x) = 2^{x-1}$; (−2,   ), (−1,   ), (1,   )

8. $f(x) = 2^{-x+1}$; (−1,   ), (0,   ), (1,   )

9. $f(x) = 5^{-x}$; graph

10. $f(x) = 2^{-3x}$; graph

11. $f(x) = (0.5)^{-x}$; (0,   ), (1,   )

12. $f(x) = (0.25)^{-x}$; (−1,   ), (0,   ), (1,   )

13. $f(x) = 3^x + 1$; graph

14. $f(x) = 2^x - 1$; graph

15. $f(x) = \left(\frac{2}{3}\right)^x$; (−2,   ), (−1,   ), (0,   )

16. $f(x) = \left(\frac{3}{4}\right)^x$; (0,   ), (2,   ), (4,   )

17. $f(x) = 2^{3-2x}$; (0,   ), (1,   ), (2,   )

18. $f(x) = 3^{2-3x}$; (0,   ), (1,   ), (2,   )

19. $f(x) = 1^x$; (0,   ), (1,   ), (2,   )

20. $f(x) = 1^{2x}$; (−2,   ), (0,   ), (2,   )

21. $f(x) = (0.1)^x$; graph

22. $f(x) = (0.1)^{1+x}$; graph

23. $f(x) = \left(\frac{3}{2}\right)\left(\frac{1}{3}\right)^{-x}$; (0,   ), (1,   )

24. $f(x) = 3(2)^{-x}$; (−1,   ), (0,   ), (1,   )

25. $f(x) = 3 - 2(2)^{x-2}$; (2,   ), (4,   )

26. $f(x) = 1 - 2(2)^{x-4}$; (3,   ), (4,   ), (5,   )

27. $f(x) = 3^x$; graph

28. $f(x) = 3^{-x}$; graph

Solve each of the following exponential equations.

29. $2^x = 4$

30. $3^x = 9$

31. $4^x = 64$

32. $5^x = 125$

33. $3^{-x} = \dfrac{1}{27}$

34. $2^{-x} = \dfrac{1}{16}$

35. $2^{3x} = 8$

36. $5^{3x} = 625$

37. $6^{-2x} = \dfrac{1}{216}$

38. $4^{-3x} = \dfrac{1}{16}$

39. $4^x = 8$

40. $9^x = 27$

41. $\left(\frac{1}{2}\right)^{3x+1} = 64$ **42.** $\left(\frac{1}{3}\right)^{2x+3} = 81$ **43.** $(216)^{x+2} = \frac{1}{36}$

44. $(81)^{4x-3} = \frac{1}{27}$ **45.** $\left(\frac{2}{5}\right)^{x} = \frac{125}{8}$ **46.** $\left(\frac{3}{4}\right)^{-x} = \frac{16}{9}$

47. $4^{3x-1} = 256$ **48.** $9^{4x-3} = 243$

Use a calculator if needed to find the amount of money A accumulated for each of the following investments. Daily compounding is based on a 360-day year.

49. $1000 invested for 2 years at 8% compounded annually

50. $1000 invested for 5 years at 7% compounded annually

51. $500 invested for 12 years at 10% compounded semiannually

52. $300 invested for 7 years at 11% compounded semiannually

53. $1250 invested for 5 years at 12% compounded quarterly

54. $2000 invested for 3 years at 8% compounded quarterly

55. $1000 invested for 4 years at 12% compounded monthly

56. $1100 invested for 8 years at 9% compounded monthly

57. $1500 invested for 6 years at 11% compounded daily

58. $1750 invested for 2 years at 13% compounded daily

59. $1000 invested for 11 years at 8.5% compounded daily

60. $2000 invested for 6 years at 7.5% compounded daily

Solve each of the following as indicated. Round to 4 decimal places as needed.

61. The number of bacteria $n(t)$ in a culture after t hours is given by $n(t) = 200e^{0.25t}$.
   (a) How many bacteria are in the culture at $t = 0$ hours?
   (b) How many are in the culture at $t = 5$ hours?

62. A particular substance has a decay rate $D(t)$ that is given by $D(t) = 500e^{-0.05t}$ after t hours.
   (a) Determine the amount of the substance at $t = 0$ hours.
   (b) Determine the amount at $t = 4$ hours.
   (c) Determine the amount at $t = 100$ hours.

63. The number of grams $N(t)$ of potassium present in a solution t hours after a chemical reaction is given by $N(t) = 400e^{-0.055t}$.
   (a) How much potassium remains at $t = 0$ hours?
   (b) How much remains at $t = 3$ hours?
   (c) How much remains at $t = 7$ hours?

64. The world population $P(t)$ is approximately equal to $P(t) = (5 \cdot 10^9)e^{0.02t}$ where t is in years.
   (a) Find the population when $t = 0$ years.
   (b) Find the population when $t = 10$ years.
   (c) Find the population when $t = 30$ years.

10.2   Logarithmic Functions

Closely related to the exponential function is another class of functions called logarithms.

> The **logarithm** of a number x is the power to which the base b must be raised to produce x. In symbols,
>
> $y = \log_b x$ if and only if $b^y = x$, where $b > 0$, $b \neq 1$.

$\text{Log}_b x$ is read, "the logarithm to the base b of x."
The various parts of a logarithm are identified below.

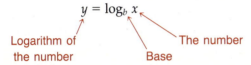

Since y is the logarithm of the number and since $b^y = x$, it can be seen that *a logarithm is an exponent.* Since $y = \log_b x$ and $b^y = x$, substitution for y gives

$$b^{\log_b x} = x.$$

Certain logarithms remain constant for any base: $\log_b b = 1$, since $b^1 = b$; $\log_b 1 = 0$, since $b^0 = 1$.

> 1. The logarithm of any number to the same base is one: $\log_b b = 1.$
> 2. The logarithm of 1 to any base is zero: $\log_b 1 = 0.$

Example 1   Write each expression in exponential form.

(a) $\log_2 8 = 3$
    Recall that $\log_b x = y$ if and only if $b^y = x$.
    $$\log_2 8 = 3 \Longleftrightarrow 2^3 = 8$$

(b) $\log_3 81 = 4$
    $$\log_3 81 = 4 \Longleftrightarrow 3^4 = 81$$

(c) $\log_{10} 1000 = 3$
    $$\log_{10} 1000 = 3 \Longleftrightarrow 10^3 = 1000 \quad \blacksquare$$

Example 2   Write each expression in logarithmic form.

(a) $3^2 = 9$
    Note that $b^y = x$ if and only if $\log_b x = y$.
    $$3^2 = 9 \Longleftrightarrow \log_3 9 = 2$$

(b) $$2^5 = 32 \Longleftrightarrow \log_2 32 = 5$$

(c) $$10^2 = 100 \Longleftrightarrow \log_{10} 100 = 2 \quad \blacksquare$$

Example 3   Solve each of the following logarithmic equations for x, y, or b.

(a) $\log_b 64 = 2$

First write the equation in exponential form.

$\log_b 64 = 2$ if and only if $b^2 = 64$

$$b = 8 \qquad \text{By definition, } b > 0$$

The solution set is $\{8\}$.

(b) $\log_{3/2} \dfrac{27}{8} = y$

$$\log_{3/2} \frac{27}{8} = y \quad \text{if and only if} \quad \left(\frac{3}{2}\right)^y = \frac{27}{8}$$

$$\left(\frac{3}{2}\right)^y = \left(\frac{3}{2}\right)^3$$

$$y = 3$$

The solution set is $\{3\}$.

(c) $\log_{16} x = 1$

$$\log_{16} x = 1 \quad \text{if and only if} \quad 16^1 = x$$

$$16 = x$$

The solution set is $\{16\}$. ■

Example 4   Simplify each of the following.

(a) $\log_3 (\log_8 8)$

$$\log_3 (\log_8 8) = \log_3 1 \qquad\qquad \log_8 8 = 1$$

$$= 0 \qquad\qquad\qquad \log_3 1 = 0$$

(b) $\log_4 [\log_4 (\log_4 16)]$

Begin by noting that $\log_4 16 = 2$.

$$\log_4 [\log_4 (\log_4 16)] = \log_4 [\log_4 2]$$

$$= \log_4 \frac{1}{2} \qquad\qquad \log_4 2 = \frac{1}{2}$$

$$= \frac{-1}{2} \quad ■$$

In Section 10.1 we considered the exponential function $y = f(x) = b^x$. The inverse of this function, which is obtained by interchanging x and y, is $x = b^y$ or $y = f^{-1}(x) = \log_b x$.

Exponential and logarithmic functions are **inverses** of each other.

The graphs of the two functions together with $y = x$ are shown in Figure 10.5.

Example 5 Sketch the graph of $y = \log_2 x$.

First write the equation in exponential form and construct a table of values.

$$y = \log_2 x \quad \text{if and only if} \quad 2^y = x.$$

Assign values to y and compute values for x.

x	y
$\frac{1}{8}$	-3
$\frac{1}{4}$	-2
$\frac{1}{2}$	-1
1	0
2	1
4	2
8	3

The graph is shown in Figure 10.6. ■

Notice in Figures 10.5 and 10.6 that x is always positive. *The logarithm of a negative number or zero does not exist.*

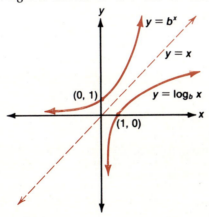

Figure 10.5

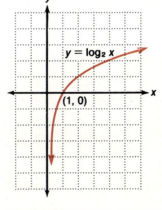

Figure 10.6

Exercises 10.2

Write each of the following in logarithmic form.

1. $3^3 = 27$
2. $4^2 = 16$
3. $2^5 = 32$
4. $5^3 = 125$
5. $10^0 = 1$
6. $5^0 = 1$
7. $9^{1/2} = 3$
8. $36^{1/2} = 6$
9. $10^{-1} = \dfrac{1}{10}$
10. $7^{-1} = \dfrac{1}{7}$
11. $3^{-4} = \dfrac{1}{81}$
12. $2^{-3} = \dfrac{1}{8}$

13. $8^{2/3} = 4$

14. $16^{3/4} = 8$

15. $81^{-3/4} = \dfrac{1}{27}$

16. $125^{-2/3} = \dfrac{1}{25}$

17. $8^1 = 8$

18. $10^1 = 10$

19. $\left(\dfrac{3}{4}\right)^2 = \dfrac{9}{16}$

20. $\left(\dfrac{2}{3}\right)^3 = \dfrac{8}{27}$

21. $(0.3)^3 = 0.027$

22. $(0.15)^2 = 0.0225$

23. $0.001^{1/3} = 0.1$

24. $0.008^{1/3} = 0.2$

Write each of the following in exponential form.

25. $\log_6 36 = 2$

26. $\log_9 81 = 2$

27. $\log_9 9 = 1$

28. $\log_3 3 = 1$

29. $\log_6 1 = 0$

30. $\log_5 1 = 0$

31. $\log_{16} 8 = \dfrac{3}{4}$

32. $\log_{27} 9 = \dfrac{2}{3}$

33. $\log_{10} 0.001 = -3$

34. $\log_{10} 0.01 = -2$

35. $\log_3 \left(\dfrac{1}{9}\right) = -2$

36. $\log_2 \left(\dfrac{1}{8}\right) = -3$

37. $\log_{1/5} 25 = -2$

38. $\log_{1/3} 81 = -4$

39. $\log_{1/2} 32 = -5$

40. $\log_{2/3} \left(\dfrac{27}{8}\right) = -3$

Solve for the variable in each of the following equations.

41. $\log_2 x = 2$

42. $\log_3 x = 2$

43. $\log_y 0.001 = -3$

44. $\log_y 0.01 = -2$

45. $\log_{1/3} 9 = k$

46. $\log_{25} \left(\dfrac{1}{5}\right) = m$

47. $\log_b 1000 = \dfrac{3}{2}$

48. $\log_t 8 = \dfrac{3}{2}$

49. $\log_{2/3} c = -2$

50. $\log_{1/4} w = -3$

51. $\log_2 \left(\dfrac{1}{8}\right) = p$

52. $\log_5 \left(\dfrac{1}{5}\right) = v$

Simplify in Exercises 53–60.

53. $\log_4 (\log_5 5)$

54. $\log_7 (\log_8 8)$

55. $\log_9 [\log_3 (\log_3 27)]$

56. $\log_4 [\log_2 (\log_2 16)]$

57. $\log_{10} (\log_{10} 10)$

58. $\log_2 (\log_7 49)$

59. $\log_b (\log_a a^b)$; $a, b > 0$

60. $\log_a (\log_b b^a)$; $a, b > 0$

61. Determine the domain and range of the function $f = \{(x, y) \mid y = 1^x\}$. Is it a one-to-one function?

62. Determine the domain and range of the function $f = \{(x, y) \mid y = (1/2)^x\}$. Is it a one-to-one function?

10.3   Properties of Logarithms

In the days before calculators and computers were widely used, long calculations such as

$$\sqrt[4]{\dfrac{(3.265)^6(24.28)^7}{(0.036)^4(25.41)^3}}$$

were carried out by using logarithms. Now that the price of a fairly sophisticated calculator is within the reach of most people, using logarithms to make such calculations is no longer necessary. The theorems that were used to make these calculations are still very useful in higher mathematics, and much insight is gained by using them to carry out a few numerical calculations.

Theorem 10.2

Properties of Logarithms

In Symbols	In Words
(a) $\log_b xy = \log_b x + \log_b y$	The logarithm of a product is the sum of the logarithms of its factors.
(b) $\log_b \dfrac{x}{y} = \log_b x - \log_b y$	The logarithm of a quotient is the difference of the logarithm of the numerator and the denominator.
(c) $\log_b x^k = k \log_b x,\ k \in R$	The logarithm of a number raised to a power is the power times the logarithm of the number.

We will prove Theorem 10.2(a) and leave the proofs of 10.2(b) and (c) as exercises. The proofs of the theorems are based upon the relationship between a logarithm and an exponential expression.

Proof:

Let $\log_b x = M$ and $\log_b y = N$. From the definition of a logarithm, we have

$$x = b^M \text{ and } y = b^N.$$

Multiplying the left and right members gives

$$x \cdot y = b^M \cdot b^N$$

or $x \cdot y = b^{M+N}$. $a^m \cdot a^n = a^{m+n}$

This can be written in logarithmic form as

$$\log_b x \cdot y = M + N$$

or $\log_b x \cdot y = \log_b x + \log_b y.$ $\log_b x = M;\ \log_b y = N$

In the examples that follow and throughout the remainder of the text the base will not be indicated when it is 10:

$$\log_{10} x = \log x.$$

$\log_{10} x$ is called a **common logarithm.**

Example 1   Express each of the following as an algebraic sum of logarithms.

(a) $\log_b rst$

$$\log_b rst = \log_b r + \log_b s + \log_b t$$ Theorem 10.2(a)

(b) $\log x^2 y$

$$\log x^2 y = \log x^2 + \log y \qquad \text{Theorem 10.2(a)}$$
$$= 2 \log x + \log y \qquad \text{Theorem 10.2(c)}$$

(c) $\log \dfrac{st}{r}$

$$\log \frac{st}{r} = \log st - \log r \qquad \text{Theorem 10.2(b)}$$
$$= \log s + \log t - \log r \qquad \text{Theorem 10.2(a)} \quad \blacksquare$$

When the logarithm of a number involves a radical, the radical is converted to exponential form before Theorem 10.2 is applied. For example,

$$\log_b \sqrt{x} = \log_b x^{1/2} = \left(\frac{1}{2}\right) \log_b x.$$

Example 2   Write $\log \sqrt[3]{\dfrac{ab}{c^2}}$ as an algebraic sum of logarithms.

$$\log \sqrt[3]{\frac{ab}{c^2}} = \log \left(\frac{ab}{c^2}\right)^{1/3}$$
$$= \frac{1}{3}[\log ab - \log c^2]$$
$$= \frac{1}{3}[\log a + \log b - 2 \log c] \quad \blacksquare$$

Just as we used Theorem 10.2 to write a logarithm as an algebraic sum, we can use it to write the sum or difference of logarithms as a single logarithm with coefficient 1.

Example 3   Write $2 \log_e x + 5 \log_e y - \log_e z$ as a single logarithm with coefficient of 1.

$$2 \log_e x + 5 \log_e y - \log_e z$$
$$= \log_e x^2 + \log_e y^5 - \log_e z \qquad \text{Theorem 10.2(c)}$$
$$= \log_e x^2 y^5 - \log_e z \qquad \text{Theorem 10.2(a)}$$
$$= \log_e \frac{x^2 y^5}{z} \qquad \text{Theorem 10.2(b)} \quad \blacksquare$$

Example 4   Use the properties of logarithms to verify that

$$\frac{1}{2} \log 36 - \frac{1}{3} \log 8 + \frac{3}{5} \log 32 = \log 24.$$

The expression can be written as

$$\log 36^{1/2} - \log 8^{1/3} + \log 32^{3/5} = \log 24$$

Since $36^{1/2} = 6$, $8^{1/3} = 2$, and $32^{3/5} = 8$,

$$\log 6 - \log 2 + \log 8 = \log 24$$

$$\log \frac{6 \cdot 8}{2} = \log 24$$

$$\log 24 = \log 24 \quad \blacksquare$$

▶▶ **Caution** $\log x - \log y + \log z \neq \log x - (\log y + \log z)$

Example 5   Solve for x in $\log_b x = \log_b 2 + \log_b y$, where $x > 0$, $y > 0$.
We first simplify each side

$$\log_b x = \log_b 2 + \log_b y$$
$$\log_b x = \log_b 2y$$

Since the logarithms are equal, the numbers must be equal. Thus

$$x = 2y.$$

The solution set is $\{2y\}$.   ■

Example 6   Solve for x in $\log_3 x + \log_3 (x + 8) = 2$.

$$\log_3 x + \log_3 (x + 8) = 2$$
$$\log_3 x(x + 8) = 2$$
$$x(x + 8) = 3^2 \qquad \text{Definition of a logarithm}$$
$$x^2 + 8x - 9 = 0$$
$$(x + 9)(x - 1) = 0 \qquad \text{Factor}$$
$$x = -9 \quad \text{or} \quad x = 1$$

If $x = -9$ then $\log_3 x$ is the logarithm of a negative number. Thus -9 is not in the solution set. The solution set is $\{1\}$. The check is left to the reader.   ■

Suppose it is known that $\log 2 = 0.3010$, $\log 3 = 0.4771$, and $\log 5 = 0.6990$. These values can be used to find the logarithm of any number that can be expressed in terms of 2, 3, and 5.

Example 7   Find the log $18\sqrt{10}$.

$$\log 18\sqrt{10} = \log 2 \cdot 3^2 \cdot 10^{1/2}$$

$$= \log 2 + 2 \log 3 + \frac{1}{2} \log 10$$

$$= 0.3010 + 0.9542 + 0.5 \qquad \text{log 10 = 1}$$

$$= 1.7552 \quad \blacksquare$$

Example 8   Find $\log_3 2$, given that $\log 2 = 0.3010$.
Since logarithms to base 10 are known, convert base 3 to base 10. To do this, let

$$x = \log_3 2$$
$$3^x = 2. \qquad \text{Definition of a logarithm}$$

Take $\log_{10}$ of both sides.

$$\log 3^x = \log 2$$
$$x \log 3 = \log 2$$

or $\qquad x = \dfrac{\log 2}{\log 3} \qquad\qquad$ Divide by log 3

$$\log_3 2 = \dfrac{\log 2}{\log 3} \qquad\qquad x = \log_3 2$$

$$= \dfrac{0.3010}{0.4771} = 0.6309 \quad\blacksquare$$

Exercises 10.3

Express each of the following as an algebraic sum of logarithms.

1. $\log uvw$

2. $\log rst$

3. $\log_b \dfrac{xy}{z}$

4. $\log_b \dfrac{pq}{n}$

5. $\log_4 \dfrac{a^2}{b^3}$

6. $\log_6 \dfrac{k^3}{m^2}$

7. $\log_e \sqrt{\dfrac{x+y}{x-y}}$

8. $\log_e \sqrt{(x-y)^2(y-z)}$

9. $\log \left(\dfrac{xy^{2/3}}{z^2}\right)^{3/2}$

10. $\log \left(\dfrac{x^2}{x^{1/3}z^{1/2}}\right)^{1/2}$

11. $\log_e y \cdot \sqrt[5]{x}$

12. $\log_e x^3 \cdot \sqrt[4]{y}$

13. $\log_b \sqrt[3]{\dfrac{x^3y}{z^5}}$

14. $\log_b \sqrt[5]{\dfrac{x^4y^2}{z^7}}$

15. $\log_e \dfrac{\sqrt{x^3}}{\sqrt[3]{y^4}}$

16. $\log_e \dfrac{\sqrt[3]{a^2}}{\sqrt[7]{b^5}}$

17. $\log \sqrt[4]{a^2b^{-3/4}c^{1/3}}$

18. $\log \sqrt[3]{xy^{2/3}z^{-1/4}}$

19. $\log 2\pi\sqrt{\dfrac{L}{g}}$

20. $\log \pi\sqrt{\dfrac{2L}{R^2}}$

21. $\log \sqrt{(s-a)(s-b)(s-c)}$

22. $\log \sqrt{x^2(x-a)^3}$

Express each of the following as a single logarithm with a coefficient of 1.

23. $\log a + \log b + \log c$

24. $\log x + \log y + 2 \log z$

25. $\log_e r - \log_e s + \log_e t$

26. $\log_e m + \log_e n - \log_e p$

27. $2 \log_e b + 3 \log_e c$

28. $4 \log_2 s - 2 \log_2 t$

29. $\dfrac{1}{2} \log 12 - \dfrac{1}{2} \log 9 + \dfrac{1}{3} \log 27$

30. $\dfrac{1}{4} \log 8 + \dfrac{1}{4} \log 2 + \dfrac{1}{3} \log 8$

31. $\dfrac{1}{2} (\log a + 4 \log b - 2 \log c)$

32. $\dfrac{1}{3} (\log x + 6 \log y - 3 \log z)$

33. $\dfrac{3}{2} \log_e x - 4 \log_e y$

34. $\dfrac{2}{3} \log_e a - 5 \log_e b$

Use the properties of logarithms to verify that the left side of each equation is equal to the right side.

35. $\log 2 - \log 3 + \log 5 = \log \dfrac{10}{3}$

36. $3 \log 2 - 4 \log 3 = \log \dfrac{8}{81}$

37. $\dfrac{1}{2} \log 25 - \dfrac{1}{2} \log 64 + \dfrac{2}{3} \log 27 = \log \dfrac{45}{8}$

38. $\log 5 - 1 = \log \dfrac{1}{2}$

39. $\log 10 - 2 = \log 0.1$

40. $2 \log 3 + 4 \log 2 - 3 = \log 0.144$

41. $\log 2x^3 - \log 2\sqrt{x^3} - \log \sqrt{x} = \log x$

42. $2 \log b - \log b + \log \sqrt{c} + \log \sqrt{c^5} = \log bc^3$

43. $\dfrac{2}{3} \log x + \dfrac{1}{3} \log y + \dfrac{5}{3} \log y - \dfrac{5}{3} \log x = \log \dfrac{y^2}{x}$

44. $\log \dfrac{9}{12} - \log 38 + 4 \log 2 + \log \dfrac{76}{12} + \log \dfrac{8}{4} = 2 \log 2$

45. $\log \dfrac{16}{27} - \log \dfrac{2}{3} - \log \dfrac{8}{9} = 0$

46. $\log \dfrac{5}{36} - \log \dfrac{2}{27} + \log \dfrac{8}{15} = 0$

47. $\log \dfrac{54}{35} + \log \dfrac{55}{4} - \log \dfrac{18}{70} - \log \dfrac{165}{4} = \log 2$

48. $\log \dfrac{9}{4} + \log \dfrac{6}{45} - \log \dfrac{15}{10} = -\log 5$

Solve for the indicated letter in each of the following.

49. $\log_2 x = y + c$; x

50. $\log a = 2 \log b$; a

51. $\log_e I = \log_e a - T$; I

52. $2 \log x + 3 \log y = 4 \log z - 2$; y

53. $\log (x + 3) = \log x + \log 3$; x

54. $\log_2 8 + \log_2 x = 4$; x

55. $\log_2 x + \log_2 (x - 2) = 3$; x

56. $\log_3 (2x - 5) - 2 = \log_3 (3x - 2)$; x

Given log 2 = 0.3010, log 3 = 0.4771 and log 5 = 0.6990, use the properties of logarithms to evaluate Exercises 57–70. Use a calculator and round answers to 4 decimal places as needed.

57. $\log 8$

58. $\log 30$

59. $\log 12$

60. $\log 100$

61. $\log 16\sqrt{2}$

62. $\log 27\sqrt{3}$

63. $\log 0.25$

64. $\log \sqrt[3]{6}$

65. $\log_3 2$

66. $\log_2 5$

67. $\log_5 8$

68. $\log_5 18$

69. $\log_3 16$

70. $\log_3 125$

71. Prove that $\log_b \dfrac{x}{y} = \log_b x - \log_b y$.

72. Prove that $\log_b x^y = y \log_b x$.

10.4 Solving Logarithmic and Exponential Equations

Many of the problems in this section require the use of a calculator capable of displaying the logarithm of a number and raising a number to a power. It should be noted that results obtained in this manner are approximations.

Sometimes equations are stated in terms of logarithms, and the properties of logarithms must be applied to find the solution set. The next few examples expand on the technique that was introduced in Example 6 of Section 10.3.

Example 1 Solve $\log (x - 2) + \log (x - 3) = \log (x + 13)$.

$$\log (x - 2)(x - 3) = \log (x + 13) \qquad \text{Theorem 10.2(a)}$$
$$(x - 2)(x - 3) = x + 13$$
$$x^2 - 5x + 6 = x + 13$$
$$x^2 - 6x - 7 = 0$$
$$(x - 7)(x + 1) = 0$$
$$x - 7 = 0 \quad \text{or} \quad x + 1 = 0$$
$$x = 7 \quad \text{or} \qquad x = -1$$

We reject $x = -1$ since $\log (x - 2) = \log (-1 - 2) = \log (-3)$ does not exist. The solution set is $\{7\}$. ■

The number e is encountered so frequently in mathematics as the base of logarithms that a special symbol, ln, is used to designate logarithms with a base of e.

$$\log_e a = \ln a$$

The symbol $\ln a$ is read, "the **natural logarithm** of a".

Example 2 Solve $\ln(x - 2) = 5$.

The base of the logarithm is e. By definition,

$$x - 2 = e^5$$
$$x = e^5 + 2.$$

The solution set is $\{e^5 + 2\}$. ■

To get a decimal approximation for the result in Example 2, use a calculator. (Note: calculator designs vary, and the processes outlined below may not work on your calculator.) Many calculators have a key marked $\boxed{e^x}$. To compute e^5, enter 5 and press the $\boxed{e^x}$ key: $e^5 \approx 148.41316$. If your calculator does not have an $\boxed{e^x}$ key, it may have a key marked y^x. To calculate e^5, enter 2.718 ($e \approx 2.718$), press $\boxed{y^x}$, enter 5, and press $\boxed{=}$: $(2.718)^5 \approx 148.33624$. The results differ because 2.718 is an approximation for e. Returning to Example 2,

$$x = e^5 + 2$$
$$x \approx 148.41316 + 2 \qquad \text{Use } \boxed{e^x} \text{ key}$$
$$x \approx 150.41316$$

Example 3   Solve $\log (x^2 - 1) = 1 + \log 2 + \log (x - 1)$.

$$\log (x^2 - 1) = 1 + \log 2(x - 1)$$
$$\log (x^2 - 1) - \log (2x - 2) = 1$$
$$\log \frac{x^2 - 1}{2x - 2} = 1$$

Now change the equation to exponential form:

$$\frac{x^2 - 1}{2x - 2} = 10^1,$$

or

$$x^2 - 20x + 19 = 0$$
$$(x - 19)(x - 1) = 0$$
$$x = 19 \quad \text{or} \quad x = 1.$$

Reject 1 since $\log (x^2 - 1) = \log (1 - 1) = \log 0$, which does not exist. The solution set is {19}. ■

Example 4   Solve $3^x = 8$. Round answer to three decimal places.
Take the logarithm of each side.

$$\log 3^x = \log 8$$
$$x \log 3 = \log 8 \qquad \text{\color{red}{Theorem 10.2(c)}}$$
$$x = \frac{\log 8}{\log 3} \qquad \text{\color{red}{Divide by log 3}}$$

To find log 8 on your calculator, enter 8 and then press the key marked $\boxed{\log}$: $\log 8 \approx 0.9031$. Similarly, $\log 3 \approx 0.4771$. Thus,

$$x \approx \frac{0.9031}{0.4771}$$
$$x \approx 1.893. \qquad \text{\color{red}{Use a calculator}}$$

The solution set is {1.893}. ■

Example 5   Solve $15^{2x+1} = 20^{x+2}$. Round answer to four decimal places.
Take the logarithm of each side.

$$\log 15^{2x+1} = \log 20^{x+2}$$
$$(2x + 1) \log 15 = (x + 2) \log 20$$

Isolate x on one side of the equation.

$$2x \log 15 + \log 15 = x \log 20 + 2 \log 20$$
$$2x \log 15 - x \log 20 = 2 \log 20 - \log 15$$
$$x(2 \log 15 - \log 20) = 2 \log 20 - \log 15$$

Divide both sides by the coefficient of x.

$$x = \frac{2 \log 20 - \log 15}{2 \log 15 - \log 20}$$

Use a calculator to approximate the logarithms and carry out the arithmetic.

$$x \approx \frac{2.6021 - 1.1761}{2.3522 - 1.3010}$$

$$x \approx 1.3527$$

The solution set is $\{1.3566\}$.   ■

Example 6   A radioactive substance decays according to the formula $A = A_0 e^{-0.05t}$, where t is measured in years and A_0 is the initial amount present. What is its half-life?

Half-life is the length of time it takes for one-half of the substance to decay. If A_0 is the original amount, then $\frac{1}{2}A_0$ is the amount that remains at half-life.

$$A = A_0 e^{-0.05t}$$

$$\frac{1}{2}A_0 = A_0 e^{-0.05t} \qquad A = \frac{1}{2}A_0$$

$$\frac{1}{2} = e^{-0.05t} \qquad \text{Divide by } A_0$$

Since we are considering a power of e, take the natural logarithm of each side.

$$\ln 0.5 = \ln e^{-0.05t}$$

$$\ln 0.5 = -0.05t \ln e$$

To find $\ln 0.5$, enter 0.5 and then press $\boxed{\ln}$: $\ln 0.5 \approx -0.6931$.

$$-0.6931 \approx -0.05t \qquad \ln e = 1$$

$$t \approx \frac{-0.6931}{-0.05} \approx 13.9$$

The half-life is approximately 13.9 years.   ■

Exercises 10.4

Solve each of the following logarithmic equations.

1. $\log (x + 2) = 3$

2. $\log (x + 1) = 1$

3. $\log (8y + 40) = \log 1000$

4. $\log (5y - 40) = \log 100$

5. $\ln (y + 3) = 4$

6. $\ln (x - 2) = 5$

7. $\ln (2x - 5) = \ln 3$

8. $\ln (6y + 7) = \ln 8$

9. $\ln t + \ln (2t - 5) = \ln 3$

10. $\ln (k + 2) + \ln (k - 1) = \ln 6$

11. $\log (y + 3) - \log (2y - 5) = -3$

12. $\log (4x + 8) - 2 = -\log 3x$

13. $\log_5 x - \log_5 (3x + 2) = -\log_5 4$

14. $\log_2 (a - 1) + \log_2 3 = \log_2 (a + 5)$

15. $\log y = 1 + \log (y - 1)$

16. $\log (2p^2 + 3) = 1 + \log (p + 1)$

17. $\log_3 (x - 3) + \log_3 x = \log_3 4$

18. $\log 100y + \log (y - 9) = \log 1000$

19. $\log s - \log 8 = \log 10$

20. $\log_8 x + \log_8 5 = \log_8 64$

21. $\log_3 3^x = \log_3 27$

22. $\log_2 2^p = \log_2 16$

23. $\log (u + 4)^2 - \log (18u + 7) = \log 10$

24. $\log (m + 6)^2 - \log (9m + 14) = \log 2$

Solve each of the following exponential equations. Round answers to 4 places if needed. If no real number solution exists, write ∅.

25. $4^x = 7$

26. $3^y = 25$

27. $5^{2d} = 4$

28. $10^{3v-2} = 37$

29. $3^{2+p} = 5^p$

30. $3^{w+1} = 4^{w-1}$

31. $10^{3a} = 57$

32. $10^{2k} = 65$

33. $e^{3x} = 21$

34. $e^{x+2} = 15$

35. $e^{0.32x} = 632$

36. $e^{-1.4x} = 13$

37. $500e^{-0.12x} = 123$

38. $100e^{0.25x} = 48$

39. $9^x = 7^{x+2}$

40. $20^{2x-1} = 30^{3x-2}$

41. $1000^{x^2+2x} = 100^{x^2-2}$

42. $7^{a^2+2a+1} = 1$

43. $7^{\log x} = 12$

44. $3^{\log x} = 5$

45. $10^{3 \log x} = 8$

46. $e^{5 \ln x} = 6$

47. $e^{x^2} = 5$

48. $e^{x^2} = e^{-x^2}$

49. $2^{x^2} = 4^{x-2}$

50. $5^{x^2+x} = 25^{x+1}$

51. $5^{a^2+a} = 25$

52. $7^{t^2+t} = 1$

Solve each of the following. Round answers to 4 decimal places if needed.

53. The amount of a radioactive substance present after t days is given by the decay formula $A = 400e^{-kt}$. If $A = 430$ when $t = 3$, find k.

54. The amount of a radioactive substance after t weeks is given by the formula $A = A_0e^{-0.04t}$, where A_0 is the initial amount. How many weeks does it take for A to decay to one-half its initial amount?

55. A worker can produce P items per day after t days of practice according to the formula $P = 400(1 - e^{-t})$. How many whole days of practice would it take for this worker to be able to produce at least 300 items per day?

56. The population P of a city t years from now is given by the formula $P = 30,000e^{0.05t}$. How many years from now will the population be 70,000?

10.5 More Applications

Using logarithms to solve applied problems requires knowing how to find the logarithm of a number and how to find a number if its logarithm is known. For example, using a calculator,

$\log 18 = 1.2552725$ to seven decimal places.

In exponential form this logarithm is

$10^{1.2552725} = 18.$

If you had been asked to find x where

$\log x = 1.2552725,$

then $x = 10^{1.2552725}.$

Using a calculator with a key marked $\boxed{10^x}$, proceed as follows.

Enter 1.2552725, then press $\boxed{10^x}$.

The result is 17.99999999 ≈ 18. For calculators with a $\boxed{y^x}$ key, the steps are as follows:

Enter 10, press $\boxed{y^x}$, enter 1.2552725, press $\boxed{=}$.

The result is 17.99999999 ≈ 18.

Example 1   Find the value of the variable in each of the following.

(a) log x = −1.235

Write log x = −1.235 in exponential form as $10^{-1.235} = x$. Enter −1.235, then press $\boxed{10^x}$, getting x ≈ 0.05821032.

(b) log x = 3.714

Enter 3.714, then press 10^x, getting x ≈ 5176.0683.

(c) ln x = 0.3041

The base is e. If your calculator has a key marked $\boxed{e^x}$, enter 0.3041 and press $\boxed{e^x}$, getting x ≈ 1.3554045. If an $\boxed{e^x}$ key is not available, a close approximation can be found by using the key marked $\boxed{y^x}$ with $y ≈ e$. ■

The following examples show how to use logarithms to solve some applied problems.

Example 2   An amount P invested at $r\%$ interest compounded annually for t years will have a final value of $A = P(1 + r)^t$. How many years will it take P dollars invested at 8% to double?

When the investment doubles, $A = 2P$. Thus

$$2P = P(1 + 0.08)^t$$
$$2 = (1.08)^t \qquad \text{Divide by } P$$
$$\log 2 = \log (1.08)^t$$
$$\log 2 = t \cdot \log 1.08$$
$$\frac{\log 2}{\log 1.08} = t$$
$$\frac{0.3010}{0.0334} ≈ t \qquad \text{Use a calculator}$$
$$9.01 ≈ t$$

The investment will double in approximately 9 years. ■

Example 3   How much money would have to be invested today at 9% interest compounded annually to have a value of $10,000 in 8 years? Use logarithms.

We wish to find P when $A = 10{,}000$, $r = 0.09$, and $t = 8$.

$$10{,}000 = P(1 + 0.09)^8$$

$$P = \frac{10{,}000}{(1.09)^8}$$

$$\log P = \log \frac{10{,}000}{(1.09)^8} \qquad \text{Take the logarithm of both sides}$$

$$\log P = \log 10{,}000 - 8 \cdot \log 1.09$$

$$\log P \approx 4 - 0.29941 \qquad \text{Use a calculator}$$

$$\log P \approx 3.70059$$

$$P \approx 10^{3.70059}$$

$$P \approx 5018.69$$

If \$5018.69 is invested today at 9% compounded annually, it will be worth approximately \$10,000 in 8 years.   ■

Example 2 could have been calculated without logarithms by using a calculator to divide 10,000 by $(1.09)^8$.

Example 4   Certain quantities involving optics are governed by the equation $L = a + b \log D$. Find D if $L = 14$, $a = 7.3$, and $b = 3.5$.

Substitution yields

$$14 = 7.3 + 3.5 \log D.$$

$$\frac{14 - 7.3}{3.5} = \log D$$

$$82.1 \approx D \qquad \text{Use a calculator} \quad ■$$

Exercises 10.5

Use logarithms to solve each of the following. Round answers to 3 decimal places if needed.

1. A body falling in a vacuum falls s feet in t seconds, where $t = \sqrt{2s/g}$ and $g = 32$ feet per second squared. Find t when $s = 1653$ feet.

2. The period T of a pendulum of length L is given by the formula $T = 2\pi\sqrt{L/g}$, where $g = 981.0$ centimeters per second squared. If $L = 281.3$ centimeters, find T. Use $\pi = 3.1416$.

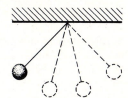

3. A person using a parachute falls with a velocity V feet per second where $V = w(1 - e^{-32t/w})$, w is the weight of the person in pounds, and t is the time falling in seconds. Find the velocity of a 200-pound person who has been falling for 10 seconds.

4. The current I in amperes of an inductive electrical circuit is given by the formula $I = (E/R)(1 - e^{-RT/L})$, where E is the voltage, R is the resistance, T is the time in seconds, and L is the inductance. If $E = 500$, $R = 50$, and $L = 0.5$, how many seconds will it take for the current to equal 4.8 amperes?

5. A radioactive substance decays according to the formula $A = A_0 e^{-0.00041t}$. How long (t in years) will it take for the amount A to be equal to one-half the initial amount A_0?

6. A colony of bacteria with initial size of 1000 increases its population by 10% each day. After x days its population will be approximated by the formula $P = 1000 e^{0.1x}$. How long will it take the colony to double in size?

7. If an initial investment of P dollars earns interest at an annual interest rate r compounded annually, the value after t years is given by $A = P(1 + r)^t$. At an annual rate of 10%, how long will it take an initial investment of $3000 to accumulate to an amount of at least $6000?

8. A radioactive substance has a half-life of 138 days. How long will it take for 90% of a sample of this substance to decay, given that $A = A_0 e^{-0.005t}$. (See Exercise 5.)

9. Radioactive carbon, ^{14}C, is used by geologists to estimate the age of fossils. If an organism contains an amount A_0 of ^{14}C when it dies, then the amount A of ^{14}C remaining t years later is given by $A = A_0 e^{-0.000124t}$. If a fossil contains 60% of the amount of ^{14}C that was present when it died, what is the approximate age of the fossil?

10. The growth equation of a certain city is approximated by $A = A_0 e^{0.0282t}$ where the initial population A_0 in 1970 was 935,000 ($t = 0$). Calculate the population for 1987. Note t is in years.

11. The number of bacteria in a colony grows from 100 to 400 in 16 hours. Using $A = A_0 e^{kt}$, where A_0 is the initial amount and A is the amount after 16 hours, determine k and find the number of bacteria present after 20 hours.

12. An initial amount A_0 of money invested at 8% compounded continuously grows to an amount A according to the formula $A = A_0 e^{0.08t}$, where t is in years. How long would it take for the initial investment to double?

13. The expected population P of a city with initial population P_0 is given by the formula $P = P_0 3^{kt}$ where t is in years. The initial population was 16,000 at $t = 0$, and 20,000 at $t = 10$.
    (a) Determine k.
    (b) Find the expected population for $t = 40$.

14. The atmospheric pressure P varies approximately according to the formula $P = 30(10)^{-0.09d}$ where d is the distance in miles above sea level. Find the atmospheric pressure at the top of Mount Whitney, which is 14,496 feet above sea level.

15. Use the formula in Exercise 14 to determine the atmospheric pressure at the top of Mount Everest, 29,028 feet above sea level.

16. The difference in intensity level of two sounds with intensities I and i is defined by $L = 10 \cdot \log(I/i)$ decibels. Find the intensity level L in decibels of the sound from an electric pump, if L is 175.6 times greater than the intensity of i.

17. Find the intensity level, L (in decibels) of each of the following sounds, if $L = 10 \cdot \log(I/i)$, where $i = 10^{-16}$ watts per centimeter squared.
    (a) A whisper at 10^{-13} watts per centimeter squared

(b) Normal conversation at $3.16 \cdot 10^{-10}$ watts per centimeter squared

(c) A loud motorcycle at 10^{-8} watts per centimeter squared

18. The voltage across a certain capacitor in a circuit is given by $V = 80(1 - e^{-0.04t})$ where t is in seconds. How much time is required for V to equal 50 volts?

19. The light intensity I when light passes through a relatively clear body of water is reduced according to the formula $I = I_0 e^{-kd}$ where I_0 is the light intensity at the surface, d is the distance below the surface in feet, and k is called the coefficient of extinction. Find the depth in Lake Nacimiento, with $k = 0.0485$, at which the light is reduced to 1% of that at the surface.

20. The learning curve in typing is given by the formula $N = 80(1 - e^{-0.07n})$ where N is the number of words typed per minute after n weeks of study. About how many weeks would it take an individual to learn to type 50 words per minute?

21. The limiting magnitude L of an optical telescope with lens diameter D inches is given by $L = 8.8 + 5.1 \log D$. Find the diameter of a telescope with limiting magnitude of 20.6.

Chapter 10 Summary

[10.1]   An **exponential function** takes the form $f(x) = b^x$ where $b > 0$ and $b \neq 1$. The **domain** of exponential functions is the set of real numbers and the **range** is the set of positive real numbers. The graphs of exponential functions of the form $f(x) = b^x$ cross the y-axis at $y = 1$. If both sides of an exponential equation can be expressed in terms of the same base,

$$b^m = b^n,$$

the solution is found by equating the exponents.

[10.2]   The **logarithmic function,** $y = \log_b x$ if and only if $b^y = x$, is the inverse of the exponential function $y = b^x$. From the definition it can be seen that *a logarithm is an exponent.* Some useful observations concerning logarithms, where $b \neq 1$, $b > 0$, are listed below.

1. $b^{\log_b x} = x$

2. The logarithm of any number to the same base is one: $\log_b b = 1$.

3. The logarithm of 1 to any base is zero: $\log_b 1 = 0$.

4. The logarithm of a negative number does not exist.

[10.3]   Historically, logarithms were used to carry out complex numerical computations. Computers and calculators have taken over this role, but the properties involved are still very useful in higher mathematics. The basic properties are listed below.

1. $\log_b xy = \log_b x + \log_b y$

2. $\log_b \dfrac{x}{y} = \log_b x - \log_b y$

3. $\log_b x^k = k \log_b x, \ k \in R$

A logarithm to the base 10 is often written without indicating the base:

$$\log_{10} x = \log x.$$

[10.4]   If the base of a logarithm is $e \approx 2.718$ it is written using the symbol ln:

$$\log_e x = \ln x.$$

The expression $\ln x$ is read, "the **natural logarithm** of x."

To solve a logarithmic equation use the properties of logarithms. Solutions to logarithmic equations should always be checked to ensure that none require taking the logarithm of a negative number.

Chapters 9 and 10   Review

1. Find $f(-1)$, $f(2)$, and $f(0)$ for $f(x) = x^2 - x - 2$.
2. Determine the domain and range of $g(x) = \sqrt{4 - x^2}$.

Write the equation of the line with the given characteristics in Exercises 3–7.

3. Passes through $(4, -1)$ and $(-3, 2)$
4. $m = \frac{1}{2}$; passes through $(2, -5)$
5. Passes through $(0, 6)$ and is perpendicular to $2x - y = 4$
6. $m = \frac{3}{4}$; y-intercept $= 3$
7. Passes through the point of intersection of $x + 2y = 4$ and $2x - y = 2$ and is parallel to $3x + 2y = 5$
8. Find the distance from $(-6, 2)$ to $(5, -1)$.
9. Find the coordinates of the midpoint of the line segment joining $(0, 7)$ to $(2, -3)$.

Identify each of the following conic sections. Graph as indicated.

10. $x^2 + y^2 = 16$; graph
11. $y = x^2 + 7x + 6$
12. $2x^2 + 3y^2 = 12$
13. $x^2 - y^2 = -4$
14. $x^2 + 2x - 2 + y^2 - 2y = 0$
15. $2x^2 + 8x + y^2 - 4y = 4$; graph
16. $(x - 2)^2 + (y + 3)^2 = 16$
17. $2(x - 1)^2 + 4(y + 2)^2 = 24$
18. $y = -(x + 4)^2 + 5$; graph
19. $(x - 1)^2 - (y + 1)^2 = 9$

Solve each system of equations.

20. $x^2 + y^2 = 9$
    $x^2 - y^2 = 1$
21. $x^2 - xy + 2y^2 = 8$
    $x^2 - y^2 = 0$
22. $x^2 + 6y^2 = 18$
    $x^2 - y^2 = -12$
23. $y = x^2 - 5x + 4$
    $y = x^2 + 3x - 4$

Solve each system of inequalities by graphing.

24. $x + y < 6$
    $x^2 + y^2 < 64$
25. $x^2 + y^2 \leq 9$
    $|x| < 2$
26. $y \leq -x^2 + 2$
    $y \geq x^2$
27. $x^2 + 4y^2 \leq 16$
    $x^2 + y^2 \geq 1$

Supply the information requested in Exercises 28–35. Graph as indicated.

28. Given $f(x) = 3x - 5$, find $f^{-1}(x)$.

29. Given $h(x) = x^2 - 6$, find $h^{-1}(x)$.

30. Graph $y = \left(\dfrac{1}{4}\right)^x$.

31. Graph $y = 3^{-x}$.

32. Solve $3^x = 27$.

33. Solve $\left(\dfrac{1}{2}\right)^{2x+3} = \dfrac{1}{64}$.

34. Solve $7^{3x-2} = 343$.

35. Solve $2^{-5x+1} = 128$.

Write each equation in logarithmic form.

36. $2^3 = 8$

37. $3^5 = 243$

38. $\left(\dfrac{2}{3}\right)^2 = \dfrac{4}{9}$

39. $(0.12)^2 = 0.0144$

Write each equation in exponential form.

40. $\log_7 49 = 2$

41. $\log_8 1 = 0$

42. $\log_{1/2} 64 = -6$

43. $\log_{1/6} 36 = -2$

Solve for the variable in each of the following.

44. $\log_5 x = 5$

45. $\log_{1/2} 4 = k$

46. $\log_{3/4} c = -3$

47. $\log_9 \dfrac{1}{9} = w$

Express each of the following as a logarithmic sum or difference.

48. $\log xy\sqrt{z}$

49. $\log \dfrac{x^{1/3} y^{3/4}}{z^{1/2}}$

50. $\log_e \dfrac{a^4}{b^3}$

51. $\log_b \dfrac{a^{1/3}}{b^{-3/4}}$

Express each of the following as a single logarithm with a coefficient of 1.

52. $2 \log 3 - 5 \log 2 + \log 6$

53. $\log_e x - \log_e y + 3 \log_e z$

54. $\dfrac{1}{2} \log 16 + \log 9 - \dfrac{1}{3} \log 125$

55. $\dfrac{1}{2} (\log x - 5 \log y + 2 \log z)$

Solve each of the following equations. Round answers to 4 decimal places, if needed.

56. $\log 6x - \log (2 + x) = 1$

57. $\log_6 x + \log_6 5 = \log_6 30$

58. $7^x = 4$

59. $10^{3x} = 25$

60. $7^x = 5^{x+1}$

61. $e^{x-3} = 17$

Work each of the following applied problems.

62. The amount A of a radioactive substance present after t days is given by $A = 600e^{-kt}$. If $A = 300$ when $t = 8$, find k. Find the amount present after 16 days.

63. The intensity L of sound is given by $L = 10 \log \left(\dfrac{I}{10^{-16}}\right)$, where I is the power of the sound in watts per centimeter squared. Find L for $I = 2.82 \cdot 10^{-8}$.

64. The amount of money A accumulated at compound interest is given by

$$A = P\left(1 + \frac{r}{n}\right)^{nt},$$

where n is the number of times the money is compounded per year, t is the number of years, r is the annual interest rate and P is the initial amount invested. Find A if $P = \$3000$, $r = 9\%$, $n = 12$, and $t = 3$ years.

65. The force of wind on the side of a house trailer varies directly with the area of the side and the square of the velocity of the wind. There is a force of 2 pounds per square foot when the velocity of the wind is 20 miles per hour. Find the force of a wind with a velocity of 40 miles per hour on the side of a trailer with an area of 240 square feet.

11

Sequences, Series, and the Binomial Expansion

The sequences and series to be introduced in this chapter are a small part of an extensive branch of mathematics. They have many applications in the mathematics of finance, engineering, and numerous scientific fields. The values of the logarithms discussed in Chapter 10 are derived by using a series, and many of the other operations that your calculator performs are based on series.

11.1 Sequences and Series, Summation Notation

We begin this section by defining a *sequence*.

> A **sequence** is a function whose domain is the set of positive integers.

For example, the sequence 2, 4, 6, 8, . . . is defined by the function

$f(x) = 2x$,

since $f(1) = 2 \cdot 1 = 2$, $f(2) = 2 \cdot 2 = 4$, $f(3) = 2 \cdot 3 = 6$, and so on. The sequence 1, 4, 9, 16, . . . is defined by the function $g(x) = x^2$. It is not necessary that a sequence follow a pattern; $h(1) = 6$, $h(2) = -56$, $h(3) = 4$ is a sequence.

The numbers that make up a sequence are called its **terms.** When a sequence is written in the form

$a_1, a_2, a_3, a_4, \ldots, a_n$

the **subscript** indicates where a term appears. For example, a_1 is the first term, a_6 is the sixth term, and a_n the *n*th or **general term.**

Sequences are generated by formulas that describe each term corresponding to a value of n.

Example 1   What are the terms for the sequences $a_n = 2n$ and $a_n = 3 + 5n$?

The sequence $a_n = 2n$ has as its terms 2, 4, 6, 8, . . . corresponding to $n = 1, 2, 3, 4, \ldots$. The *n*th term of the sequence is $2n$.

The sequence $a_n = 3 + 5n$ has as its terms 8, 13, 18, 23, . . . corresponding to $n = 1, 2, 3, 4, \ldots$.   ■

Sequences such as those in Example 1 are said to be **infinite** since no last value of n is specified. The sequence $a_n = 2n - 1$ for $n = 1, 2, 3,$ and 4 is a **finite sequence** since its terms are limited to the following:

$$a_1 = 2(1) - 1 = 1 \qquad a_2 = 2(2) - 1 = 3$$
$$a_3 = 2(3) - 1 = 5 \qquad a_4 = 2(4) - 1 = 7$$

Example 2   Find the first four terms of each sequence.

(a) $a_n = \dfrac{n + 1}{n + 2}$

$$a_1 = \frac{1 + 1}{1 + 2} = \frac{2}{3} \qquad n = 1$$

$$a_2 = \frac{2 + 1}{2 + 2} = \frac{3}{4} \qquad n = 2$$

$$a_3 = \frac{3 + 1}{3 + 2} = \frac{4}{5} \qquad n = 3$$

$$a_4 = \frac{4 + 1}{4 + 2} = \frac{5}{6} \qquad n = 4$$

The first four terms are $\frac{2}{3}, \frac{3}{4}, \frac{4}{5},$ and $\frac{5}{6}$.

(b) $a_n = x^{2n}$

$$a_1 = x^{2 \cdot 1} = x^2 \qquad a_2 = x^{2 \cdot 2} = x^4$$
$$a_3 = x^{2 \cdot 3} = x^6 \qquad a_4 = x^{2 \cdot 4} = x^8$$

The first four terms are $x^2, x^4, x^6,$ and x^8.   ■

If the terms of a sequence are known it is often possible to find the general term by inspection. Such a process depends on insight and trial and error.

Example 3   Find the general term of each of the following sequences.

(a) 3, 5, 7, 9, . . .

The sequence is the set of odd integers starting with 3. Since the domain is $n \in \{1, 2, 3, . . .\}$, we must express the sequence in terms of these values of n.

Trial	n	Term	Result	
$n + 2$	1	$1 + 2$	3	*Right*
	2	$2 + 2$	4	*Wrong*

The second term must be 5. Odd integers are 1 greater or less than the even integers, giving the following results.

Trial	n	Term	Result	
$2n + 1$	1	$2 \cdot 1 + 1$	3	*Right*
	2	$2 \cdot 2 + 1$	5	*Right*

An inspection for $n = 3, 4, 5, \ldots$ confirms that $a_n = 2n + 1$.

(b) $\frac{-1}{2}, \frac{3}{4}, \frac{-5}{8}, \frac{7}{16}, \ldots$

The signs alternate. The sign for any term is given by $(-1)^n$. The numerators are the odd numbers beginning with 1. Thus the general term for the numerator is $2n - 1$. The denominators are $2^1, 2^2, 2^3, 2^4, \ldots$ or 2^n. Thus,

$$a_n = \frac{(-1)^n(2n - 1)}{2^n}. \quad \blacksquare$$

Some sequences are described in terms of a **recursion formula** that relates any term a_{k+1} to the preceding term a_k. Such a description is called a **recursive definition.** For example, if $a_{k+1} = a_k + 2$ then

$$a_4 = a_3 + 2, \quad a_7 = a_6 + 2, \quad \text{and} \quad a_{26} = a_{25} + 2.$$

To find the terms of a sequence by using a recursion formula, one term must be known.

Example 4 Find the first five terms of the sequence in which $a_1 = 4$ and $a_{k+1} = a_k + 3$.

$$a_2 = a_1 + 3 = 4 + 3 = 7 \qquad a_1 = 4$$

The formula states that each term is three greater than the preceding term.

$$a_1 = 4$$
$$a_2 = 4 + 3 = 7$$
$$a_3 = 7 + 3 = 10$$
$$a_4 = 10 + 3 = 13$$
$$a_5 = 13 + 3 = 16$$

The first five terms are 4, 7, 10, 13, and 16. $\quad \blacksquare$

A **series** is the indicated sum of a sequence. Such a sum can be written in a compact form called **sigma** or **summation notation.** The Greek letter Σ is used to indicate that a sum is to be found.

Summation Notation

$$\sum_{i=1}^{n} a_i = a_1 + a_2 + a_3 + \cdots + a_n$$

The subscript i, which ranges from 1 to n, is called the **index of summation.**

Example 5 Evaluate the series $\sum_{i=1}^{5} 2^i$.

The series is obtained by replacing i successively with the positive integers 1 through 5, as indicated by the index of summation.

$$\sum_{i=1}^{5} 2^i = 2^1 + 2^2 + 2^3 + 2^4 + 2^5$$

$$= 2 + 4 + 8 + 16 + 32$$

$$= 62 \quad \blacksquare$$

An index of summation can begin with any integer.

Example 6   Evaluate $\sum_{i=3}^{5} (i^2 - 2)$.

$$\sum_{i=3}^{5} (i^2 - 2) = (3^2 - 2) + (4^2 - 2) + (5^2 - 2)$$

$$= 7 + 14 + 23 = 44 \quad \blacksquare$$

Example 7   Evaluate $\sum_{i=0}^{2} (-1)^i(4i - 1)$.

$$\sum_{i=0}^{2} (-1)^i(4i - 1) = (-1)^0(4 \cdot 0 - 1) + (-1)^1(4 \cdot 1 - 1) + (-1)^2(4 \cdot 2 - 1)$$

$$= (1)(-1) + (-1)(3) + (1)(7)$$

$$= 3 \quad \blacksquare$$

Example 8   A motorcycle loses 10% of its value each year. Write a sequence to show its value at the end of each of the first three years if its original cost was $3000.

$$a_1 = 3000 - 0.10(3000) = 2700$$
$$a_2 = 2700 - 0.10(2700) = 2430$$
$$a_3 = 2430 - 0.10(2430) = 2187$$

If a_i is its value at the end of each year then

$$a_1 = a_i - 0.10a_i = 0.90a_i$$
$$a_2 = 0.90a_i - 0.10(0.90)a_i = 0.81a_i$$
and $$a_3 = 0.81a_i - 0.10(0.81)a_i = 0.729a_i$$

Thus $a_n = (0.90)^n a_i$ or $a_n = (0.90)^n(3000)$. $\blacksquare$

Exercises 11.1

Find the first five terms of each sequence with the given general term.

1. $a_n = n$

2. $a_n = n + 1$

3. $a_n = 2n + 3$

4. $a_n = 2n$

5. $a_n = n^3$

6. $a_n = n^2$

7. $a_n = \dfrac{n - 1}{n + 1}$

8. $a_n = \dfrac{n + 3}{n + 2}$

9. $a_n = (-2)^{n+1}$

10. $a_n = (-3)^{n-1}$

11. $a_n = \dfrac{n}{n^2 + 1}$

12. $a_n = \dfrac{n^2 - 1}{n}$

13. $a_n = 3^{-n}$

14. $a_n = 2^{-n}$

15. $a_n = n^2 + n^{-2}$

16. $a_n = n^2 + n^{-1}$

17. $a_n = 3n(-x)^n$

18. $a_n = 2n(-x)^{n+1}$

19. $a_n = (nx)^{n+1}$

20. $a_n = (nx)^n$

Find the general term for each of the following sequences.

21. 2, 3, 4, 5, . . .

22. 3, 4, 5, 6, . . .

23. 1, 4, 9, 16, . . .

24. 1, 8, 27, 64, . . .

25. 3, 9, 27, 81, . . .

26. 2, 4, 8, 16, . . .

27. -1, 1, -1, 1, . . .

28. -10, -20, -30, -40, . . .

29. $\dfrac{1}{2}, \dfrac{2}{3}, \dfrac{3}{4}, \dfrac{4}{5}$, . . .

30. $\dfrac{1}{3}, \dfrac{1}{9}, \dfrac{1}{27}, \dfrac{1}{81}$, . . .

31. 4, 8, 12, 16, . . .

32. 4, 7, 10, 13, . . .

33. $-1, \dfrac{4}{3}, \dfrac{-3}{2}, \dfrac{8}{5}$, . . .

34. $\dfrac{-1}{2}, \dfrac{4}{3}, \dfrac{-9}{4}, \dfrac{16}{5}$, . . .

35. x^2, x^3, x^4, x^5, . . .

36. x, x^4, x^9, x^{16}, . . .

37. $\dfrac{\sqrt{x}}{3}, \dfrac{\sqrt[3]{x}}{9}, \dfrac{\sqrt[4]{x}}{27}, \dfrac{\sqrt[5]{x}}{81}$, . . .

38. $\dfrac{-x^2}{2}, \dfrac{x^3}{4}, \dfrac{-x^4}{8}, \dfrac{x^5}{16}$, . . .

Evaluate each of the following series.

39. $\displaystyle\sum_{i=1}^{4} i$

40. $\displaystyle\sum_{i=1}^{3} (2 + i)$

41. $\displaystyle\sum_{i=2}^{3} (2i - 1)$

42. $\displaystyle\sum_{i=3}^{4} (5i + 2)$

43. $\displaystyle\sum_{i=1}^{7} (-1)^i$

44. $\displaystyle\sum_{i=1}^{4} (i - 2)(i + 3)$

45. $\displaystyle\sum_{i=2}^{6} i(-1)^i$

46. $\displaystyle\sum_{i=3}^{8} (-2)^i$

47. $\displaystyle\sum_{i=2}^{4} \dfrac{i + 1}{i - 1}$

48. $\displaystyle\sum_{i=1}^{5} \dfrac{2i - 1}{i + 2}$

49. $\displaystyle\sum_{i=1}^{7} i^2$

50. $\displaystyle\sum_{i=3}^{7} i^2 - 1$

51. $\displaystyle\sum_{i=2}^{3} (-1)^i (3i - 1)$

52. $\displaystyle\sum_{i=1}^{3} (-1)^{i+1}(i^2 - 1)$

Write each of the following using summation notation.

53. $3 + 6 + 9 + 12$

54. $2 + 4 + 6 + 8 + 10$

55. $1 + 4 + 9 + 16 + 25$

56. $16 + 25 + 36 + 49$

57. $\dfrac{1}{2} + \dfrac{2}{3} + \dfrac{3}{4} + \dfrac{4}{5} + \dfrac{5}{6}$

58. $\dfrac{2}{5} + \dfrac{4}{6} + \dfrac{6}{7} + \dfrac{8}{8}$

59. $-\dfrac{1}{3} + \dfrac{2}{4} - \dfrac{3}{5} + \dfrac{4}{6} - \dfrac{5}{7}$

60. $\dfrac{5}{2} - \dfrac{7}{3} + \dfrac{9}{4}$

61. $7 - 11 + 15 - 19$

62. $2 - 5 + 10 - 17$

63. $x - \dfrac{1}{2}x^3 + \dfrac{1}{3}x^5 - \dfrac{1}{4}x^7$

64. $\dfrac{1}{2} - \dfrac{1}{4}x^2 + \dfrac{1}{6}x^4 - \dfrac{1}{8}x^6$

Find the first 6 terms of each of the following recursive pattern sequences.

65. $a_1 = 2$
$a_n = 3a_{n-1}$ for $n \geq 2$

66. $a_1 = 1$
$a_n = a_{n-1} + 2$ for $n \geq 2$

67. $a_1 = -1$
$a_n = 4a_{n-1} + 5$ for $n \geq 2$

68. $a_1 = -2$
$a_n = 5a_{n-1} + 3$ for $n \geq 2$

11.2   Arithmetic Sequences and Series

In this section we will study a particular type of sequence called an **arithmetic sequence.**

> An **arithmetic sequence** is one in which each term after the first differs from the preceding term by a constant called the **common difference.**

If $a_1, a_2, a_3, a_4, \ldots , a_n$ are terms of an arithmetic sequence and the constant is d, then $a_2 = a_1 + d$, $a_3 = a_2 + d$, and $a_4 = a_3 + d$. When an arithmetic sequence is defined recursively,

$$a_{n+1} = a_n + d$$

where a_1 is the first term.

Example 1   Determine whether or not each of the following sequences is an arithmetic sequence. If it is an arithmetic sequence, identify the common difference.

(a) 3, 5, 7, 9, . . . , $(2n + 1)$
This is an arithmetic sequence with a common difference of 2.

(b) 5, 2, -1, -4, . . . , $(8 - 3n)$
Arithmetic sequence; common difference $= -3$

(c) $1^2, 2^2, 3^2, 4^2, \ldots$
This is not an arithmetic sequence, because there is no common difference. ■

To find an expression for the *n*th term of a given arithmetic sequence, consider the pattern formed when d is added to each term to get the next term.

Term Number	Term	Times d Is Added
1	a_1	0
2	$a_2 = a_1 + d$	1
3	$a_3 = a_1 + 2d$	2
4	$a_4 = a_1 + 3d$	3
⋮	⋮	⋮
n	$a_n = a_1 + (n - 1)d$	$n - 1$

The nth term of an arithmetic sequence is given by

$$a_n = a_1 + (n - 1)d$$

where a_1 is the first term and d is the common difference.

Example 2   Find the ninth term of the sequence 3, 7, 11, ...
In the ninth term a_n becomes a_9.

$a_n = a_1 + (n - 1)d$
$a_9 = 3 + (9 - 1)(4)$ $a_1 = 3,\ d = 4,\ n = 9$
$a_9 = 3 + 32$
$a_9 = 35$ ■

Example 3   The seventh term of an arithmetic sequence is 20, and the eighteenth term is 53. Find the first term and the common difference.
Since the seventh term is 20, we have

$$20 = a_1 + (7 - 1)d. \qquad a_7 = 20,\ n = 7$$

I. $20 = a_1 + 6d$

The eighteenth term is 53, so

$$53 = a_1 + (18 - 1)d. \qquad a_{18} = 53,\ n = 18$$

II. $53 = a_1 + 17d$

We now have two equations, I and II, in two variables, which can be solved by elimination.

I. $a_1 + 6d\ \ = 20$
II. $a_1 + 17d = 53$

$\qquad\qquad 11d = 33$ Subtract *I* from *II*
$\qquad\qquad\ \ d = 3$

Substituting $d = 3$ into I yields

$a_1 + 6 \cdot 3 = 20$
$\qquad a_1 = 2$ ■

Theorem 11.1

The sum S_n of n terms of an arithmetic sequence is given by the formula

$$S_n = \frac{n}{2}(a_1 + a_n).$$

Recall that the indicated sum of a sequence is a series.
Observe that

$$S_n = a_1 + (a_1 + d) + (a_1 + 2d) + (a_1 + 3d) + \cdots + a_n.$$

If we write the series in reverse order each term will be d less than the preceding term:

$$S_n = a_n + (a_n - d) + (a_n - 2d) + (a_n - 3d) + \cdots + a_1$$

Adding both sides of the two equations gives

$$2S_n = (a_1 + a_n) + (a_1 + a_n) + (a_1 + a_n) + \cdots + (a_1 + a_n).$$

Since the expression $(a_1 + a_n)$ occurs once for each of the n terms, the right side of the equation can be written as $n(a_1 + a_n)$. Thus,

$$2S_n = n(a_1 + a_n) \qquad \text{or}$$

$$S_n = \frac{n}{2}(a_1 + a_n).$$

The formula can also be written in an alternate form.

$$S_n = \frac{n}{2}(a_1 + a_n)$$

$$S_n = \frac{n}{2}\,[a_1 + a_1 + (n-1)d] \qquad a_n = a_1 + (n-1)d$$

$$S_n = \frac{n}{2}\,[2a_1 + (n-1)d]$$

Example 4   Find the sum of the sequence 1, 2, 3, . . . , 72.

$$S_n = \frac{n}{2}(a_1 + a_n)$$

$$S_{72} = \frac{72}{2}(1 + 72) \qquad a_1 = 1,\ n = 72,\ a_n = 72$$

$$S_{72} = 2628 \quad \blacksquare$$

Example 5   Given $a_n = 4 + 3n$, find S_{16}.
Begin by finding a_1.

$$a_1 = 4 + 3 \cdot 1 \qquad n = 1$$

$$a_1 = 7$$

Use the alternate formula $S_n = \dfrac{n}{2}\,[2a_1 + (n-1)d]$.

$$S_{16} = \frac{16}{2}[2 \cdot 7 + (16 - 1)3] \qquad a_1 = 7,\ n = 16,\ d = 3$$

$$S_{16} = 8(14 + 45)$$

$$S_{16} = 472 \quad \blacksquare$$

Example 6   Find the number of terms (n) in an arithmetic sequence in which $a_1 = 5$, $a_n = 32$,   and $S_n = 185$.

$$S_n = \frac{n}{2}(a_1 + a_n)$$

$$185 = \frac{n}{2}(5 + 32) \qquad \text{Clear of fractions and simplify}$$

$$370 = 37n$$

$$10 = n \quad \blacksquare$$

Example 7   Evaluate $\sum\limits_{i=1}^{18} (-2i + 6)$.

This expression represents the sum of an arithmetic sequence. The index shows there are 18 terms. We need to find the first and eighteenth terms.

$$a_1 = -2 \cdot 1 + 6 = 4; \quad a_{18} = -2 \cdot 18 + 6 = -30$$

$$S_{18} = \frac{18}{2}[4 + (-30)] = -234 \quad \blacksquare$$

Example 8   Show that the sum of the first n terms of $5 + 9 + 13 + \ldots$ is $2n^2 + 3n$.

$$S_n = \frac{n}{2}[2a_1 + (n - 1)d]$$

Since n is unknown,

$$S_n = \frac{n}{2}[2 \cdot 5 + (n - 1)4] \qquad a_1 = 5,\ d = 4$$

$$= \frac{n}{2}[10 + 4n - 4]$$

$$= 2n^2 + 3n \quad \blacksquare$$

Example 9   Three numbers that form an arithmetic sequence have a sum of 15 and a product of 105. Find the numbers.

Let the numbers be a_1, a_2, and a_3. The sequence is a_1, $(a_1 + d)$, $(a_1 + 2d)$.

The sum is $a_1 + (a_1 + d) + (a_1 + 2d) = 15$.

$$3a_1 + 3d = 15$$

$$a_1 + d = 5$$

$$d = 5 - a_1$$

The product is $a_1(a_1 + d)(a_1 + 2d) = 105$.

Substitute $5 - a_1$ for d.

$$a_1(a_1 + 5 - a_1)[a_1 + 2(5 - a_1)] = 105$$

$$a_1(5)[-a_1 + 10] = 105$$

$$a_1(-a_1 + 10) = 21 \qquad \text{Divide by 5}$$

$$(a_1)^2 - 10a_1 + 21 = 0$$

$$(a_1 - 7)(a_1 - 3) = 0$$

$$a_1 = 7 \qquad \text{or} \qquad a_1 = 3$$

Since $d = 5 - a_1$, d is either -2 or 2. The sequence is 7, 5, 3 or 3, 5, 7. $\quad \blacksquare$

Exercises 11.2

Determine which of the following sequences are arithmetic. Find the common difference for each arithmetic sequence.

1. 1, 2, 3, 4, 5
2. 2, 4, 6, 8, 10
3. 3, 6, 9, 12, 15
4. 4, 8, 12, 16, 20
5. 1, 4, 7, 11
6. 2, 5, 8, 12
7. 20, 15, 10, 5
8. 30, 24, 18, 12
9. 2, 4, 8, 16
10. 1, 3, 6, 9
11. $x + 3y$, $2x + 2y$, $3x + y$
12. $x - 2y$, $2x - y$, $3x$
13. -7, -14, -21, -28
14. 1, 0, -1, -2

Find the indicated unknowns.

15. $a_1 = -3$, $d = 5$; a_{10}; a_8
16. $a_1 = -1$, $d = 3$; a_{18}; a_5
17. $a_1 = 3$, $d = 6$; a_{13}; a_6
18. $a_1 = 4$, $d = 2$; a_7; a_{14}
19. $a_1 = 4$, $d = 3$; a_7; S_{10}
20. $a_1 = 1$, $d = 3$, $a_n = 37$; a_3; S_n
21. $a_{15} = 45$, $S_{15} = 255$; a_1; d
22. $a_{24} = 163$, $S_{24} = 1980$; a_1; d
23. $a_1 = 82$, $d = -13$, $a_n = -9$; n; S_n
24. $a_1 = 46$, $d = -9$, $a_n = -125$; n; S_n
25. $a_9 = -12$, $a_{13} = 3$; a_1; S_6
26. $a_1 = 24$, $a_{24} = -22$; S_{24}; a_4
27. $a_n = 9n - 3$; S_{20}; S_{24}
28. $a_n = 4 + 3n$; S_{15}; S_{20}
29. $a_1 = 36$, $a_3 = 66$, $S_n = 234$; n; d
30. $a_1 = 12$, $a_n = 38$, $S_n = 1325$; n; d
31. $a_1 = 12$, $S_{28} = 1092$; d; a_{28}
32. $a_1 = 71$, $S_{20} = -100$; d; a_{20}
33. $a_1 = 4$, $d = 0.4$, $S_n = 30$; n; a_n
34. $a_1 = 6$, $d = -\dfrac{5}{3}$, $S_n = -15$; n; a_n
35. $d = -4$, $a_n = -56$, $S_n = -420$; a_1, n
36. $d = 3$, $a_n = 1$, $S_n = -39$; a_1; n

Find the sum of each of the following.

37. $\displaystyle\sum_{i=1}^{10} (3i - 1)$
38. $\displaystyle\sum_{i=1}^{7} (2i - 1)$
39. $\displaystyle\sum_{i=5}^{12} (2i - 5)$
40. $\displaystyle\sum_{i=13}^{20} (5i - 1)$
41. $\displaystyle\sum_{i=12}^{18} \dfrac{2i - 1}{3}$
42. $\displaystyle\sum_{i=12}^{16} \dfrac{8 - 2i}{3}$

Solve each of the following.

43. Find the sum of the first 50 natural numbers.
44. Find the sum of the odd integers from 1 through 101.
45. Find the sum of all positive two-digit integers.
46. Find the sum of all the integer multiples of 8 between 9 and 199.
47. Find the sum of all the integer multiples of 7 between 8 and 103.
48. Find three numbers that form an arithmetic sequence, if their sum is 18 and their product is 192.
49. Find three numbers that form an arithmetic sequence, if their sum is 21 and their product is 315.
50. Find three numbers that form an arithmetic sequence, if their sum is 27 and the sum of their squares is 275.

51. Find three numbers that form an arithmetic sequence, if their sum is 24 and the sum of their squares is 200.

52. Show that the sum of the first n natural numbers is $\dfrac{n(n + 1)}{2}$.

53. Show that the sum of the first n terms of $3 + 5 + 7 + 9 + \ldots$ is $2n + n^2$.

54. Show that the sum of the first n terms of $10 + 14 + 18 + 22 + \ldots$ is $8n + 2n^2$.

55. Show that the sum of the first n even natural numbers is $n + n^2$.

56. Marie deposited \$100 in a savings account on her daughter's first birthday and \$200 on her second birthday. If she continues this pattern through her daughter's eighteenth birthday, how much money will she have deposited in the account?

57. Cynthia started a new job with an annual salary of \$16,500 in 1986. If she receives a \$900 raise each year, how much will her annual salary be in 1997?

11.3 Geometric Sequences and Series

The sequence 1, 2, 4, 8, . . . is not an arithmetic sequence because it has no common difference. The pattern of terms shows that any term can be obtained by multiplying the preceding term by 2. Sequences such as the one above are called **geometric sequences.**

> A **geometric sequence** is one in which each term after the first is obtained by multiplying the preceding term by a constant, r, called the **common ratio.**

A geometric sequence is defined recursively by

$$a_{k+1} = ra_k$$

where a_1 is the first term.

When two successive terms of a geometric sequence are known, the common ratio can be found by using this version of the recursion formula:

$$r = \frac{a_{k+1}}{a_k}.$$

For example, in

$$2, \frac{4}{3}, \frac{8}{9}, \frac{16}{27}, \ldots$$

the ratio can be found by taking the ratio of any two consecutive terms. Try using the second and third terms:

$$r = \frac{a_{k+1}}{a_k} = \frac{\dfrac{8}{9}}{\dfrac{4}{3}} = \frac{8}{9} \cdot \frac{3}{4} = \frac{2}{3}.$$

Example 1   Find the common ratio for each geometric sequence.

(a) 1, 3, 9, 27, . . .

Since $3 \div 1 = 3$, $9 \div 3 = 3$, and $27 \div 9 = 3$, the common ratio is 3.

(b) 8, 4, 2, $\frac{1}{2}$, . . .

The common ratio is $\frac{1}{2}$.

(c) $\frac{1}{3}$, $\frac{-2}{15}$, $\frac{4}{75}$, $\frac{-8}{375}$, . . .

The common ratio is $\frac{-2}{5}$. ■

A formula for the nth term of a geometric sequence can be found by observing the pattern of the terms.

Term Number	Term	Number of Times r is Multiplied
1	a_1	0
2	$a_1 r$	1
3	$a_1 r^2$	2
4	$a_1 r^3$	3
⋮	⋮	⋮
n	$a_1 r^{n-1}$	$n - 1$

The nth term of a geometric sequence is given by the formula

$$a_n = a_1 r^{n-1}.$$

Example 2   Find r and a_n for the sequence 3, 6, 12, ...

$$r = \frac{a_{k+1}}{a_k} = \frac{6}{3} = 2$$

$$a_n = a_1 r^{n-1} = 3(2)^{n-1} \qquad r = 2 \quad ■$$

Example 3   The first term of a geometric sequence is 3 and the sixth term is 729. Find r.

$$a_n = a_1 r^{n-1}$$

$729 = 3 \cdot r^{6-1}$ $a_n = a_6 = 729, n = 6$

$3^5 = r^5$ Divide by 3, $729 = 3^6$

$3 = r$ ■

A formula for the sum of the terms in a geometric sequence can be found in a manner analagous to that used to find the sum of the terms of an arithmetic sequence. Let S_n represent the sum.

$$S_n = a_1 + a_1 r + a_1 r^2 + a_1 r^3 + a_1 r^4 + \cdots + a_1 r^{n-1}$$

Multiply both sides of the equation by r to get

$$r S_n = a_1 r + a_1 r^2 + a_1 r^3 + a_1 r^4 + \cdots + a_1 r^n.$$

Now subtract the second equation from the first, noting that the term preceding a_1r^n is a_1r^{n-1}. All terms on the right side except a_1 and a_1r^n are eliminated.

$$S_n - rS_n = a_1 - a_1r^n$$
$$S_n(1-r) = a_1(1-r^n) \qquad \text{Factor}$$
$$S_n = \frac{a_1(1-r^n)}{1-r} \qquad \text{Divide by } 1-r$$

Theorem 11.2

The sum of n terms of a geometric sequence is given by

$$S_n = \frac{a_1(1-r^n)}{1-r}$$

where a_1 is the first term and r is the common ratio ($r \neq 1$). The indicated sum of a geometric sequence is called a **geometric series**.

The sum of a geometric sequence, like the sum of an arithmetic sequence, can be written in an alternate form. To do this, solve $a_n = a_1r^{n-1}$ for a_1r^n.

$$a_n = a_1r^{n-1}$$
$$ra_n = r(a_1r^{n-1}) \qquad \text{Multiply by } r$$
$$ra_n = a_1r^n \qquad r \cdot r^{n-1} = r^n$$

Now substitute this value into the formula for the sum.

$$S_n = \frac{a_1(1-r^n)}{1-r} = \frac{a_1 - a_1r^n}{1-r} = \frac{a_1 - ra_n}{1-r} \qquad ra_n = a_1r^n$$

Alternate Form

$$S_n = \frac{a_1 - ra_n}{1-r}$$

Example 4 Find the sum of a geometric sequence of 5 terms with $a_1 = 2$ and $r = \frac{1}{3}$.

$$S_n = \frac{a_1(1-r^n)}{1-r}.$$

$$S_5 = \frac{2\left(1 - \left(\frac{1}{3}\right)^5\right)}{1 - \frac{1}{3}} = \frac{2\left(1 - \frac{1}{243}\right)}{\frac{2}{3}}$$

$$S_5 = 2 \cdot \frac{3}{2}\left(\frac{242}{243}\right) = \frac{242}{81} \quad \blacksquare$$

Example 5   The sixth term of a geometric sequence is $\frac{1}{2}$, and $r = \frac{1}{2}$. Find a_1 and S_9.
Begin by finding a_1.

$$a_n = a_1 r^{n-1}$$

$$\frac{1}{2} = a_1 \left(\frac{1}{2}\right)^{6-1} \qquad a_6 = \frac{1}{2}, r = \frac{1}{2}, n = 6$$

$$\frac{1}{2} = a_1 \left(\frac{1}{32}\right)$$

$$16 = a_1$$

Now use the value of a_1 in the formula for S_n.

$$S_n = \frac{a_1(1 - r^n)}{1 - r}$$

$$S_9 = \frac{16\left(1 - \left(\frac{1}{2}\right)^9\right)}{1 - \frac{1}{2}} = \frac{16\left(\frac{511}{512}\right)}{\frac{1}{2}}$$

$$S_9 = 32\left(\frac{511}{512}\right) = \frac{511}{16} \quad \blacksquare$$

Example 6   Find n for a geometric sequence in which $a_1 = 2$, $r = -2$, and $S_n = -42$.

$$S_n = \frac{a_1(1 - r^n)}{1 - r}$$

$$-42 = \frac{2(1 - (-2)^n)}{1 - (-2)}$$

$$-42 = \frac{2(1 - (-2)^n)}{3}$$

Multiply each side by $\frac{3}{2}$.

$$-63 = 1 - (-2)^n$$
$$-64 = -(-2)^n$$
$$64 = (-2)^n$$
$$(-2)^6 = (-2)^n \qquad 64 = (-2)^6$$
$$6 = n \quad \blacksquare$$

Example 7   Given the geometric sequence with $r = 3$, $a_n = 486$, $S_n = 728$, find n and a_1.
Begin by using the alternate formula for the sum to find a_1.

$$S_n = \frac{a_1 - ra_n}{1 - r}$$

$$728 = \frac{a_1 - 3 \cdot 486}{1 - 3}$$

$$-1456 = a_1 - 1458$$

$$2 = a_1$$

Now find a_n.

$$a_n = a_1 r^{n-1}$$
$$486 = 2 \cdot 3^{n-1}$$
$$243 = 3^{n-1}$$
$$3^5 = 3^{n-1} \qquad 243 = 3^5$$
$$5 = n - 1$$
$$6 = n \quad \blacksquare$$

If $|r| < 1$, it is possible to find the sum of a geometric sequence containing an infinite number of terms. To see how the formula is modified for n infinitely large and $|r| < 1$, consider $r = 0.01$.

$$(0.01)^2 = 0.0001$$
$$(0.01)^3 = 0.000001$$
$$(0.01)^4 = 0.00000001$$

The value of $(0.01)^n$ can be made as small as desired by letting n become sufficiently large. In the formula

$$S_n = \frac{a_1(1 - r^n)}{1 - r},$$

r^n will approach zero if $|r| < 1$ and n increases without bound.

The sum of an infinite geometric sequence is given by

$$S_\infty = \frac{a_1}{1 - r},$$

where $|r| < |$ and ∞ signifies an infinite number of terms.

If $|r| > 1$, then r^n does not approach zero as n becomes infinite. The infinite sum does not exist.

Example 8   Find the sum of the infinite geometric sequence 4, 2, 1, $\frac{1}{2}$, ...
Here $r = \frac{1}{2}$, $a_1 = 4$. The sum is

$$S_\infty = \frac{4}{1 - \dfrac{1}{2}} = \frac{4}{\dfrac{1}{2}} = 8. \quad \blacksquare$$

Example 9   Write 0.222 . . . as a rational number.
The number 0.222 . . . can be thought of as a sum.

$$0.222 \ldots = 0.2 + 0.02 + 0.002 + 0.0002 + \ldots$$

This is an infinite geometric series with $a_1 = 0.2$ and $r = 0.1$.

$$S_\infty = \frac{a_1}{1 - r} = \frac{0.2}{1 - 0.1} = \frac{0.2}{0.9} = \frac{2}{9} \quad \blacksquare$$

A geometric series, like an arithmetic series, can be expressed using summation notation.

Example 10   Evaluate $\displaystyle\sum_{i=1}^{4} \left(\frac{1}{3}\right)^i$.

$$\sum_{i=1}^{4} \left(\frac{1}{3}\right)^i = \frac{1}{3} + \frac{1}{9} + \frac{1}{27} + \frac{1}{81}$$

Thus $r = \frac{1}{3}$ and $a_1 = \frac{1}{3}$.

$$S_n = \frac{a_1(1 - r^n)}{1 - r}$$

$$S_4 = \frac{\dfrac{1}{3}\left(1 - \left(\dfrac{1}{3}\right)^4\right)}{1 - \dfrac{1}{3}} = \frac{40}{81} \quad \blacksquare$$

Example 11   Evaluate $\displaystyle\sum_{i=1}^{\infty} \left(\frac{1}{3}\right)^i$.

Again $r = \frac{1}{3}$, $a_1 = \frac{1}{3}$.

$$S_\infty = \frac{a_1}{1 - r}$$

$$S_\infty = \frac{\dfrac{1}{3}}{1 - \dfrac{1}{3}} = \frac{\dfrac{1}{3}}{\dfrac{2}{3}} = \frac{1}{2} \quad \blacksquare$$

Exercises 11.3

Decide whether each sequence is geometric, arithmetic, or neither. If the sequence is geometric, find r and a_n; if arithmetic, find d and a_n.

1. 1, 2, 4, 8, 16, . . .

2. 3, 6, 12, 24, . . .

3. 5, 10, 15, 20, 25

4. 6, 10, 14, 18

5. $\dfrac{1}{4}, \dfrac{1}{8}, \dfrac{1}{16}, \ldots$

6. 1, −1, 1, −1, . . .

7. $\dfrac{x}{2}, \dfrac{x+1}{2}, \dfrac{x+2}{2}, \dfrac{x+3}{2}$

8. $\dfrac{x-6}{5}, \dfrac{x-5}{5}, \dfrac{x-4}{5}, \dfrac{x-3}{5}$

9. 5, 10, 30, 120

10. 4, 8, 24, 48

11. 3, 3, 4, 4, 5, 5

12. 2, −1, 3, −2

13. $\dfrac{1}{2}, \dfrac{1}{8}, \dfrac{1}{32}, \ldots$

14. $\dfrac{2}{5}, \dfrac{2}{20}, \dfrac{2}{80}, \ldots$

15. $\dfrac{1}{3}, -\dfrac{1}{15}, \dfrac{1}{75}, \ldots$

16. $-\dfrac{1}{5}, \dfrac{1}{30}, -\dfrac{1}{180}, \ldots$

Find the indicated unknowns for each of the following.

17. $a_1 = 7$, $r = -3$; a_4; S_4

18. $a_1 = 4$, $r = 3$; a_5; S_5

19. $a_1 = 64$, $r = -\dfrac{1}{2}$; a_7; S_7

20. $a_1 = 81$, $r = \dfrac{1}{3}$; a_6; S_6

21. $a_4 = 16$, $r = -\dfrac{1}{4}$; a_1; S_4

22. $a_3 = 100$, $r = \dfrac{1}{10}$; a_1; S_3

23. $a_5 = 81$, $r = -\dfrac{1}{3}$; a_1; S_∞

24. $a_7 = -128$, $r = \dfrac{1}{2}$; a_1; S_∞

25. $a_5 = (1.02)^9$, $r = 1.02$; a_2; a_1

26. $a_4 = (1.01)^8$, $r = 1.01$; a_2; a_1

27. $a_1 = 3$, $r = -2$, $S_n = 129$; n; a_n

28. $a_1 = 5$, $r = -3$, $S_n = -910$; n; a_n

Evaluate each sum, if it exists. If the sum does not exist, state why.

29. $1 - \dfrac{1}{3} + \dfrac{1}{9} - \dfrac{1}{27} + \ldots$

30. $1 - \dfrac{1}{2} + \dfrac{1}{4} - \dfrac{1}{8} + \ldots$

31. $8 + 4 + 2 + 1 + \ldots$

32. $12 + 6 + 3 + \ldots$

33. $\dfrac{1}{10} + \dfrac{1}{5} + 1 + \ldots$

34. $\dfrac{1}{14} + \dfrac{1}{7} + \dfrac{2}{7} + \ldots$

35. $1 - 1 + 1 - 1 + \ldots$

36. $2 - 2 + 2 - 2 + \ldots$

37. $\displaystyle\sum_{i=1}^{8} \left(-\dfrac{3}{4}\right)^i$

38. $\displaystyle\sum_{i=1}^{6} \left(-\dfrac{1}{2}\right)^i$

39. $\displaystyle\sum_{i=1}^{\infty} \left(\dfrac{1}{2}\right)^i$

40. $\displaystyle\sum_{i=1}^{\infty} \left(\dfrac{1}{3}\right)^i$

41. $\displaystyle\sum_{i=1}^{\infty} \left(\dfrac{3}{7}\right)^i$

42. $\displaystyle\sum_{i=1}^{\infty} \left(\dfrac{5}{8}\right)^i$

Write each of the following geometric series as a rational number.

43. $0.333\ldots$

44. $0.111\ldots$

45. $0.666\ldots$

46. $0.999\ldots$

47. $0.212121\ldots$

48. $0.474747\ldots$

49. $0.123123123\ldots$

50. $0.102102102\ldots$

Solve each of the following. Round to 4 decimal places as needed.

51. A ball is dropped from a height of 3 feet. Each time it strikes the ground it bounces upward two-thirds of the distance it fell. Find the total distance traveled by the ball before coming to rest.

52. From what height should the ball in Exercise 51 be dropped so that the total distance traveled is 25 feet?

53. An object moves along a straight line, and each second it moves one-half the distance it moved in the preceding second. If the object moves 8 centimeters the first second, how far will it move in 10 seconds?

54. The tip of a pendulum moves through an arc length of 14 inches its first swing. If it moves through an arc nine-tenths as long as its previous swing each time, find the distance it travels in 6 swings.

55. A point on a pendulum travels 6 centimeters on the first swing and 5 percent less distance on each succeeding swing. Find the distance traveled by the point in 8 swings.

56. The city of Morro Bay, California had a population of 10,000 in 1985. If the population increases by 2 percent per year, what will be the population of Morro Bay after 10 years have passed?

11.4   The Binomial Expansion

Suppose you found it necessary to find the expanded form of $(a + b)^n$ where n is a positive integer. One method would be to find the product using $(a + b)$ as a factor n times. This method is too time-consuming for large values of n and can be replaced by a formula called the **binomial expansion.**

Before we consider the formula we need to consider a new symbol $n!$ called **n factorial**.

$$n! = n(n - 1)(n - 2)(n - 3) \ldots 3 \cdot 2 \cdot 1$$

Thus $n!$ means the product of the first n positive integers.

Example 1   Evaluate each factorial.

(a) $4! = 4 \cdot 3 \cdot 2 \cdot 1 = 24$

(b) $6! = 6 \cdot 5 \cdot 4 \cdot 3 \cdot 2 \cdot 1 = 720$

(c) $5 \cdot 4! = 5(4 \cdot 3 \cdot 2 \cdot 1) = 120$ ∎

Example 1(c) illustrates that factorials can be defined recursively.

$$n! = n(n - 1)!$$

Example 2   Evaluate each of the following.

(a) $\dfrac{7!}{3!} = \dfrac{7 \cdot 6 \cdot 5 \cdot 4 \cdot 3!}{3!}$

$\qquad = 7 \cdot 6 \cdot 5 \cdot 4$ Divide out 3!

$\qquad = 840$

(b) $\dfrac{(n + 1)!}{(n - 1)!} = \dfrac{(n + 1)(n)(n - 1)!}{(n - 1)!}$

$\qquad\qquad = n(n + 1)$ Divide out $(n - 1)!$ ∎

Another new symbol that is used in the binomial expansion is $\binom{n}{r}$.

$$\binom{n}{r} = \frac{n!}{r!(n - r)!}, \text{ where } n \geq r$$

Example 3   Evaluate each of the following.

(a) $\binom{6}{2} = \dfrac{6!}{2!(6-2)!} = \dfrac{6!}{2!4!}$

$= \dfrac{6 \cdot 5 \cdot 4!}{2 \cdot 1 \cdot 4!} = \dfrac{30}{2} = 15$ Write 6! recursively

(b) $\binom{7}{3} = \dfrac{7!}{3!4!} = \dfrac{7 \cdot 6 \cdot 5 \cdot 4!}{3 \cdot 2 \cdot 1 \cdot 4!} = 35$ ■

What is meant by 1! and 0!? Consider

$n! = n(n-1)!$
$1! = 1(1-1)! = 1 \cdot 0!$

Since $1! = 1$, $0!$ must also be 1 for consistency.

$1! = 0! = 1$

Consider the expansion of $(a + b)^n$. The reader should confirm the expansion for $n \in \{1, 2, 3, 4, 5\}$ shown below.

$(a + b)^1 = a + b$
$(a + b)^2 = a^2 + 2ab + b^2$
$(a + b)^3 = a^3 + 3a^2b + 3ab^2 + b^3$
$(a + b)^4 = a^4 + 4a^3b + 6a^2b^2 + 4ab^3 + b^4$
$(a + b)^5 = a^5 + 5a^4b + 10a^3b^2 + 10a^2b^3 + 5ab^4 + b^5$

Several observations can be made about each of the above expansions that will be applied to $(a + b)^n$ in general.

1. The exponent on the first and last term of each expansion is n.

2. The sum of the exponents on any term in the expansion is n.

3. The exponents on a decrease by one in each term. The exponents on b increase by one in each term. For example, for $n = 4$,

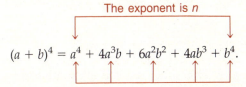

The exponent is n

$(a + b)^4 = a^4 + 4a^3b + 6a^2b^2 + 4ab^3 + b^4.$

The sum of the exponents on each term is n, and the exponents on a decrease by 1 as the exponents on b increase by 1

4. The coefficients of any term can be expressed as $\binom{n}{r}$, where r is one less than the number of the term: $r = 1$ in the second term, $r = 2$ in the third term, and so on.

When these observations are applied to $(a + b)^n$, the binomial expansion is the result.

The Binomial Expansion

$$(a + b)^n = \binom{n}{0} a^n + \binom{n}{1} a^{n-1}b + \binom{n}{2} a^{n-2}b^2 + \binom{n}{3} a^{n-3}b^3 + \cdots + \binom{n}{n} b^n$$

When each coefficient in the binomial expansion is written in factorial notation and simplified, an alternate form results.

The Binomial Expansion, Alternate Form

$$(a + b)^n = a^n + na^{n-1}b$$
$$+ \frac{n(n-1)}{2!} a^{n-2}b^2 + \frac{n(n-1)(n-2)}{3!} a^{n-3}b^3 + \cdots + b^n$$

Example 4   Verify the binomial expansion for $(a + b)^5$ by comparing it to the above results.

$$(a + b)^5 = \binom{5}{0} a^5 + \binom{5}{1} a^4b + \binom{5}{2} a^3b^2 + \binom{5}{3} a^2b^3 + \binom{5}{4} ab^4 + \binom{5}{5} b^5$$

$$= a^5 + \frac{5!}{1!4!} a^4b + \frac{5!}{2!3!} a^3b^2 + \frac{5!}{3!2!} a^2b^3 + \frac{5!}{4!1!} ab^4 + b^5$$

When each factorial is evaluated, we have

$$= a^5 + 5a^4b + 10a^3b^2 + 10a^2b^3 + 5ab^4 + b^5. \quad \blacksquare$$

Example 5   Use the binomial expansion to find $(x - 3y)^4$.

Consider $(x - 3y)^4 = [x + (-3y)]^4$. If $(a + b)^n = [x + (-3y)]^4$, then $a = x$, $b = (-3y)$, and $n = 4$. This time, use the alternate form of the binomial expansion.

$$(x - 3y)^4 = x^4 + 4x^3(-3y) + \frac{4 \cdot 3}{2!} x^2(-3y)^2 + \frac{4 \cdot 3 \cdot 2}{3!} x(-3y)^3 + (-3y)^4$$

$$= x^4 - 12x^3y + 54x^2y^2 - 108xy^3 + 81y^4 \quad \blacksquare$$

The binomial expansion makes it possible to find a particular term without carrying out the entire expansion. Notice that

$$\binom{n}{3}a^{n-3}b^3$$

represents the fourth term. In the same manner,

$$\binom{n}{r-1}a^{n-(r-1)}b^{r-1}$$

represents the rth term of the binomial expansion.

Example 6   Find the fifth term of the expansion $(c^2 + 2d)^8$.
Here $r = 5$, $a = c^2$, $b = 2d$, and $n = 8$.

$$\binom{n}{r-1}a^{n-(r-1)}b^{r-1} = \binom{8}{5-1}(c^2)^{8-(5-1)}(2d)^{5-1}$$

$$= \binom{8}{4}(c^2)^4(2d)^4$$

$$= 1120c^8d^4 \quad \blacksquare$$

Example 7   Find the term containing x^8 in the expansion of $(x^2 + y^3)^7$.
Here $a = x^2$, $b = y^3$, and $n = 7$. Note that
$x^8 = (x^2)^4 = a^4$.

Use the formula:

$$\binom{n}{r-1}a^{n-(r-1)}b^{r-1} = \binom{7}{3}(x^2)^4(y^3)^3 \qquad \text{The sum of the exponents} = 7 = n$$

$$= 35x^8y^9. \quad \blacksquare$$

Exercises 11.4

Evaluate each of the following.

1. $3!$

2. $5!$

3. $10!$

4. $8!$

5. $\dfrac{20!}{19!}$

6. $\dfrac{18!}{17!}$

7. $\dfrac{6!}{4!2!}$

8. $\dfrac{7!}{5!2!}$

9. $\dfrac{5!}{0!(5-0)!}$

10. $\dfrac{9!}{0!(9-0)!}$

11. $\dfrac{11!}{8!(11-8)!}$

12. $\dfrac{12!}{7!(12-7)!}$

13. $\binom{6}{5}$

14. $\binom{4}{3}$

15. $\binom{5}{1}$

16. $\binom{3}{1}$

17. $\binom{7}{7}$

18. $\binom{2}{2}$

19. $\binom{5}{0}$

20. $\binom{4}{0}$

Write each of the following as the quotient of two factorials.

21. 8

22. 6

23. $5 \cdot 6 \cdot 7$

24. $7 \cdot 8 \cdot 9$

25. $12 \cdot 11 \cdot 10 \cdot 9$

26. $8 \cdot 7 \cdot 6 \cdot 5$

27. $3 \cdot 2 \cdot 1$

28. $4 \cdot 3 \cdot 2 \cdot 1$

Write each of the following without factorial notation and simplify.

29. $\dfrac{n!}{(n-1)!}$

30. $\dfrac{(n+1)!}{n!}$

31. $\dfrac{(n+3)!}{(n-1)!}$

32. $\dfrac{n!}{(n-3)!}$

Use the binomial expansion for each.

33. $(x+y)^3$

34. $(u+v)^4$

35. $(a+2b)^5$

36. $(h+3k)^6$

37. $(u-v)^7$

38. $(x-y)^5$

39. $(2x-y)^4$

40. $(3u-2v)^6$

41. $(3h^2-5k)^4$

42. $(3r^2-4s)^5$

Find the indicated term in each expansion and simplify.

43. $(x^2-y)^8$; third

44. $(t+3)^7$; fourth

45. $(r+2s)^9$; sixth

46. $(3x-2y)^7$; fifth

47. $(2p^2-q^2)^7$; fifth

48. $(u^3-3v^2)^6$; third

49. $\left(\dfrac{x}{2}-3\right)^{12}$; fourth

50. $\left(\dfrac{p}{2}-\dfrac{1}{3}\right)^{14}$; seventh

Approximate the following by using the binomial expansion. Use the first four terms. A calculator may be helpful. Do not round off the results.

51. $(1.01)^8$ [*Hint:* $(1.01)^8 = (1+0.1)^8$.]

52. $(0.99)^7$

53. $(0.98)^{12}$

54. $(2.01)^5$

Chapter 11   Summary

[11.1]   A **sequence** is a function whose domain is the set of positive integers. The numbers that make up a sequence are called its **terms.** When a sequence is written in the form

$$a_1, a_2, a_3, a_4, \ldots, a_n$$

the subscripts indicate the number of the term. If no last value of n is specified, a sequence is **infinite.** When a last term is specified, the sequence is **finite.**

Sequences can be defined by a formula for the nth term that describes every term in the sequence based upon the value of n, or they can be defined by a **recursion formula.** A recursion formula relates the value of any term a_{k+1} to the term before it, a_k.

The indicated sum of a sequence is called a **series.** The series $a_1 + a_2 + a_3 + \ldots + a_n$ can be written using **summation notation** as

$$\sum_{i=1}^{n} a_i \quad \text{Index of summation}$$

where the number of terms is indicated in the index.

[11.2] [11.3]   Two types of sequences are **arithmetic sequences** and **geometric sequences.** Any two successive terms of an arithmetic sequence differ by a constant called the **common difference.** In a geometric sequence any two successive terms are in the same ratio

$$r = \frac{a_{k+1}}{a_k},$$

where r is a constant called the **common ratio.**

The following **recursion formulas** describe arithmetic and geometric sequences respectively:

$$a_{k+1} = a_k + d \text{ and } a_{k+1} = a_k r.$$

The nth term and sum of a sequence can be found by using the formulas listed below.

Type of Sequence	nth Term	Sum of n Terms	Abbreviations Used
Arithmetic	$a_n = a_1 + (n-1)d$	$S_n = \dfrac{n}{2}(a_1 + a_n)$	a_1 = first term
		$S_n = \dfrac{n}{2}[2a_1 + (n-1)d]$	n = term number d = common difference
Geometric	$a_n = a_1 r^{n-1}$	$S_n = \dfrac{a_1(1 - r^n)}{1 - r}$	a_1 = first term
		$S_n = \dfrac{a_1 - r a_n}{1 - r}$	n = term number r = common ratio
Infinite Geometric	$a_n = a_1 r^{n-1}$	$S_n = \dfrac{a_1}{1 - r}$ $\|r\| < 1$	a_1 = first term n = term number r = common ratio

[11.4]   **Factorial notation** is used to indicate the product of the first n positive integers:

$$n! = n(n-1)(n-2)(n-3) \cdots 3 \cdot 2 \cdot 1.$$

When defined recursively, $n! = n(n-1)!$.

The **binomial expansion** is a formula for evaluating binomials raised to positive integer powers. Two common forms exist,

$$(a + b)^n = \binom{n}{0} a^n + \binom{n}{1} a^{n-1}b + \binom{n}{2} a^{n-2}b^2 + \cdots + \binom{n}{n} b^n$$

and $$(a + b)^n = a^n + na^{n-1}b + \frac{n(n-1)}{2!}a^{n-2}b^2 + \cdots + b^n,$$

where $\binom{n}{r} = \dfrac{n!}{r!(n-r)!}$ and $0! = 1! = 1$.

The rth term in the binomial expansion is found by using the formula

$$\binom{n}{r-1} a^{n-(r-1)}b^{r-1}.$$

Chapter 11   Review

Write the first four terms of each sequence having the given value for a_n.

1. $a_n = n^2$

2. $a_n = (-2)^{n+1}$

3. $a_n = (-1)^n(2n + 1)$

4. $a_n = (-1)^{n+1}(x)^n$

Find the indicated unknowns for each of the following.

5. $a_1 = 1$, $d = 4$; a_8; a_{11}

6. $a_1 = 2$, $d = -3$; a_6; a_9

7. $a_9 = 18$, $S_9 = 80$; a_1; d

8. $a_n = 2 + 5n$; S_{12}; S_{18}

9. $a_1 = 5$, $r = 1$; a_6; S_6

10. $a_1 = 64$, $r = \dfrac{1}{2}$; a_6; S_6

11. $r = -3$, $a_1 = 5$; a_6; S_6

12. $a_3 = 125$, $r = \dfrac{1}{5}$; a_1; S_4

Write in expanded form and find the sum.

13. $\displaystyle\sum_{i=1}^{10} (i - 1)$

14. $\displaystyle\sum_{i=1}^{4} (2)^i$

15. $\displaystyle\sum_{i=1}^{5} (-1)^i$

16. $\displaystyle\sum_{i=1}^{8} (3i - 2)$

17. Find the sum of the even integers between 5 and 59.

18. Find three numbers that form an arithmetic sequence, if their sum is 33 and their product is 1320.

Evaluate each of the following.

19. $7!$

20. $\dfrac{35!}{33!}$

21. $\dbinom{8}{6}$

22. $\dfrac{n!}{(n - 2)!}$

23. $\dfrac{13!}{9!(13 - 9)!}$

24. $\dbinom{5}{0}$

Use the binomial expansion for each.

25. $(a - b)^5$

26. $(3x - 2y)^6$

27. $(2p^2 - 3q^2)^4$

28. $(u - v)^8$

Find the indicated term in each expansion.

29. $(x - 2y)^{10}$; fourth

30. $(x - y)^{14}$; middle term

31. $(2a - b)^{11}$; fifth

32. $\left(\dfrac{p}{2} + \dfrac{1}{3}\right)^{15}$; seventh

Solve each of the following. Round answers to 4 decimal places as needed.

33. At the end of each year a certain motorcycle is worth 91% of its value at the beginning of that year. If the motorcycle cost $3800 new, what is its value at the end of three years?

34. Twyeffort Plastering had an income of $56,000 during the first year of operation. If the income increases by $11,000 each succeeding year, find the total income during the first six years of operation.

35. A ball is dropped from a height of 6 feet. Each time the ball strikes the ground it bounces upwards seven-eighths of the distance it fell. Find the total distance the ball travels before coming to rest.

36. The tip of a pendulum travels 11 inches on the first swing and 8% less distance on each succeeding swing. Find the distance traveled by the tip of the pendulum during 12 swings.

37. Use a calculator to approximate the value of $(1.03)^{12}$ using the first four terms of the binomial expansion. Do not round off.

Appendix:
Geometric Formulas; Tables

Right triangle

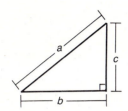

Pythagorean theorem

$$a^2 + b^2 = c^2$$

Triangle

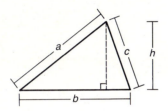

Perimeter	Area
$a + b + c$	$\frac{1}{2}bh$

Rectangle

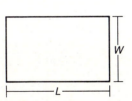

$2L + 2W$	LW

Square

$4S$	S^2

Circle

$2\pi r$	πr^2

Cube

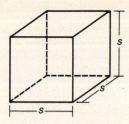

Surface area

$6s^2$

Volume

s^3

Rectangular prism

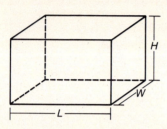

$2WH + 2LH + 2LW$

HLW

Cylinder

$2\pi r^2 + 2\pi rh$

$\pi r^2 h$

Cone

$\pi rs + \pi r^2$

$\frac{1}{3}\pi r^2 h$

Sphere

$4\pi r^2$

$\frac{4}{3}\pi r^3$

Table 1   Squares and Square Roots

n	n^2	$\sqrt{n}$	$\sqrt{10n}$	n	n^2	$\sqrt{n}$	$\sqrt{10n}$
1	1	1.000	3.162	51	2601	7.141	22.583
2	4	1.414	4.472	52	2704	7.211	22.804
3	9	1.732	5.477	53	2809	7.280	23.022
4	16	2.000	6.325	54	2916	7.348	23.238
5	25	2.236	7.071	55	3025	7.416	23.452
6	36	2.449	7.746	56	3136	7.483	23.664
7	49	2.646	8.367	57	3249	7.550	23.875
8	64	2.828	8.944	58	3364	7.616	24.083
9	81	3.000	9.487	59	3481	7.681	24.290
10	100	3.162	10.000	60	3600	7.746	24.495
11	121	3.317	10.488	61	3721	7.810	24.698
12	144	3.464	10.954	62	3844	7.874	24.900
13	169	3.606	11.402	63	3969	7.937	25.100
14	196	3.742	11.832	64	4096	8.000	25.298
15	225	3.873	12.247	65	4225	8.062	25.495
16	256	4.000	12.649	66	4356	8.124	25.690
17	289	4.123	13.038	67	4489	8.185	25.884
18	324	4.243	13.416	68	4624	8.246	26.077
19	361	4.359	13.784	69	4761	8.307	26.268
20	400	4.472	14.142	70	4900	8.367	26.458
21	441	4.583	14.491	71	5041	8.426	26.646
22	484	4.690	14.832	72	5184	8.485	26.833
23	529	4.796	15.166	73	5329	8.544	27.019
24	576	4.899	15.492	74	5476	8.602	27.203
25	625	5.000	15.811	75	5625	8.660	27.386
26	676	5.099	16.125	76	5776	8.718	27.568
27	729	5.196	16.432	77	5929	8.775	27.749
28	784	5.292	16.733	78	6084	8.832	27.928
29	841	5.385	17.029	79	6241	8.888	28.107
30	900	5.477	17.321	80	6400	8.944	28.284
31	961	5.568	17.607	81	6561	9.000	28.460
32	1024	5.657	17.889	82	6724	9.055	28.636
33	1089	5.745	18.166	83	6889	9.110	28.810
34	1156	5.831	18.439	84	7056	9.165	28.983
35	1225	5.916	18.708	85	7225	9.220	29.155
36	1296	6.000	18.974	86	7396	9.274	29.326
37	1369	6.083	19.235	87	7569	9.327	29.496
38	1444	6.164	19.494	88	7744	9.381	29.665
39	1521	6.245	19.748	89	7921	9.434	29.833
40	1600	6.325	20.000	90	8100	9.487	30.000
41	1681	6.403	20.248	91	8281	9.539	30.166
42	1764	6.481	20.494	92	8464	9.592	30.332
43	1849	6.557	20.736	93	8649	9.644	30.496
44	1936	6.633	20.976	94	8836	9.695	30.659
45	2025	6.708	21.213	95	9025	9.747	30.822
46	2116	6.782	21.448	96	9216	9.798	30.984
47	2209	6.856	21.679	97	9409	9.849	31.145
48	2304	6.928	21.909	98	9604	9.899	31.305
49	2401	7.000	22.136	99	9801	9.950	31.464
50	2500	7.071	22.361	100	10000	10.000	31.623

Table 2 Common Logarithms

n	0	1	2	3	4	5	6	7	8	9
1.0	.0000	.0043	.0086	.0128	.0170	.0212	.0253	.0294	.0334	.0374
1.1	.0414	.0453	.0492	.0531	.0569	.0607	.0645	.0682	.0719	.0755
1.2	.0792	.0828	.0864	.0899	.0934	.0969	.1004	.1038	.1072	.1106
1.3	.1139	.1173	.1206	.1239	.1271	.1303	.1335	.1367	.1399	.1430
1.4	.1461	.1492	.1523	.1553	.1584	.1614	.1644	.1673	.1703	.1732
1.5	.1761	.1790	.1818	.1847	.1875	.1903	.1931	.1959	.1987	.2014
1.6	.2041	.2068	.2095	.2122	.2148	.2175	.2201	.2227	.2253	.2279
1.7	.2304	.2330	.2355	.2380	.2405	.2430	.2455	.2480	.2504	.2529
1.8	.2553	.2577	.2601	.2625	.2648	.2672	.2695	.2718	.2742	.2765
1.9	.2788	.2810	.2833	.2856	.2878	.2900	.2923	.2945	.2967	.2989
2.0	.3010	.3032	.3054	.3075	.3096	.3118	.3139	.3160	.3181	.3201
2.1	.3222	.3243	.3263	.3284	.3304	.3324	.3345	.3365	.3385	.3404
2.2	.3424	.3444	.3464	.3483	.3502	.3522	.3541	.3560	.3579	.3598
2.3	.3617	.3636	.3655	.3674	.3692	.3711	.3729	.3747	.3766	.3784
2.4	.3802	.3820	.3838	.3856	.3874	.3892	.3909	.3927	.3945	.3962
2.5	.3979	.3997	.4014	.4031	.4048	.4065	.4082	.4099	.4116	.4133
2.6	.4150	.4166	.4183	.4200	.4216	.4232	.4249	.4265	.4281	.4298
2.7	.4314	.4330	.4346	.4362	.4378	.4393	.4409	.4425	.4440	.4456
2.8	.4472	.4487	.4502	.4518	.4533	.4548	.4564	.4579	.4594	.4609
2.9	.4624	.4639	.4654	.4669	.4683	.4698	.4713	.4728	.4742	.4757
3.0	.4771	.4786	.4800	.4814	.4829	.4843	.4857	.4871	.4886	.4900
3.1	.4914	.4928	.4942	.4955	.4969	.4983	.4997	.5011	.5024	.5038
3.2	.5051	.5065	.5079	.5092	.5105	.5119	.5132	.5145	.5159	.5172
3.3	.5185	.5198	.5211	.5224	.5237	.5250	.5263	.5276	.5289	.5302
3.4	.5315	.5328	.5340	.5353	.5366	.5378	.5391	.5403	.5416	.5428
3.5	.5441	.5453	.5465	.5478	.5490	.5502	.5514	.5527	.5539	.5551
3.6	.5563	.5575	.5587	.5599	.5611	.5623	.5635	.5647	.5658	.5670
3.7	.5682	.5694	.5705	.5717	.5729	.5740	.5752	.5763	.5775	.5786
3.8	.5798	.5809	.5821	.5832	.5843	.5855	.5866	.5877	.5888	.5899
3.9	.5911	.5922	.5933	.5944	.5955	.5966	.5977	.5988	.5999	.6010
4.0	.6021	.6031	.6042	.6053	.6064	.6075	.6085	.6096	.6107	.6117
4.1	.6128	.6138	.6149	.6160	.6170	.6180	.6191	.6201	.6212	.6222
4.2	.6232	.6243	.6253	.6263	.6274	.6284	.6294	.6304	.6314	.6325
4.3	.6335	.6345	.6355	.6365	.6375	.6385	.6395	.6405	.6415	.6425
4.4	.6435	.6444	.6454	.6464	.6474	.6484	.6493	.6503	.6513	.6522
4.5	.6532	.6542	.6551	.6561	.6571	.6580	.6590	.6599	.6609	.6618
4.6	.6628	.6637	.6646	.6656	.6665	.6675	.6684	.6693	.6702	.6712
4.7	.6721	.6730	.6739	.6749	.6758	.6767	.6776	.6785	.6794	.6803
4.8	.6812	.6821	.6830	.6839	.6848	.6857	.6866	.6875	.6884	.6893
4.9	.6902	.6911	.6920	.6928	.6937	.6946	.6955	.6964	.6972	.6981
5.0	.6990	.6998	.7007	.7016	.7024	.7033	.7042	.7050	.7059	.7067
5.1	.7076	.7084	.7093	.7101	.7110	.7118	.7126	.7135	.7143	.7152
5.2	.7160	.7168	.7177	.7185	.7193	.7202	.7210	.7218	.7226	.7235
5.3	.7243	.7251	.7259	.7267	.7275	.7284	.7292	.7300	.7308	.7316
5.4	.7324	.7332	.7340	.7348	.7356	.7364	.7372	.7380	.7388	.7396
n	0	1	2	3	4	5	6	7	8	9

Table 2 (continued)

n	0	1	2	3	4	5	6	7	8	9
5.5	.7404	.7412	.7419	.7427	.7435	.7443	.7451	.7459	.7466	.7474
5.6	.7482	.7490	.7497	.7505	.7513	.7520	.7528	.7536	.7543	.7551
5.7	.7559	.7566	.7574	.7582	.7589	.7597	.7604	.7612	.7619	.7627
5.8	.7634	.7642	.7649	.7657	.7664	.7672	.7679	.7686	.7694	.7701
5.9	.7709	.7716	.7723	.7731	.7738	.7745	.7752	.7760	.7767	.7774
6.0	.7782	.7789	.7796	.7803	.7810	.7818	.7825	.7832	.7839	.7846
6.1	.7853	.7860	.7868	.7875	.7882	.7889	.7896	.7903	.7910	.7917
6.2	.7924	.7931	.7938	.7945	.7952	.7959	.7966	.7973	.7980	.7987
6.3	.7993	.8000	.8007	.8014	.8021	.8028	.8035	.8041	.8048	.8055
6.4	.8062	.8069	.8075	.8082	.8089	.8096	.8102	.8109	.8116	.8122
6.5	.8129	.8136	.8142	.8149	.8156	.8162	.8169	.8176	.8182	.8189
6.6	.8195	.8202	.8209	.8215	.8222	.8228	.8235	.8241	.8248	.8254
6.7	.8261	.8267	.8274	.8280	.8287	.8293	.8299	.8306	.8312	.8319
6.8	.8325	.8331	.8338	.8344	.8351	.8357	.8363	.8370	.8376	.8382
6.9	.8388	.8395	.8401	.8407	.8414	.8420	.8426	.8432	.8439	.8445
7.0	.8451	.8457	.8463	.8470	.8476	.8482	.8488	.8494	.8500	.8506
7.1	.8513	.8519	.8525	.8531	.8537	.8543	.8549	.8555	.8561	.8567
7.2	.8573	.8579	.8585	.8591	.8597	.8603	.8609	.8615	.8621	.8627
7.3	.8633	.8639	.8645	.8651	.8657	.8663	.8669	.8675	.8681	.8686
7.4	.8692	.8698	.8704	.8710	.8716	.8722	.8727	.8733	.8739	.8745
7.5	.8751	.8756	.8762	.8768	.8774	.8779	.8785	.8791	.8797	.8802
7.6	.8808	.8814	.8820	.8825	.8831	.8837	.8842	.8848	.8854	.8859
7.7	.8865	.8871	.8876	.8882	.8887	.8893	.8899	.8904	.8910	.8915
7.8	.8921	.8927	.8932	.8938	.8943	.8949	.8954	.8960	.8965	.8971
7.9	.8976	.8982	.8987	.8993	.8998	.9004	.9009	.9015	.9020	.9025
8.0	.9031	.9036	.9042	.9047	.9053	.9058	.9063	.9069	.9074	.9079
8.1	.9085	.9090	.9096	.9101	.9106	.9112	.9117	.9122	.9128	.9133
8.2	.9138	.9143	.9149	.9154	.9159	.9165	.9170	.9175	.9180	.9186
8.3	.9191	.9196	.9201	.9206	.9212	.9217	.9222	.9227	.9232	.9238
8.4	.9243	.9248	.9253	.9258	.9263	.9269	.9274	.9279	.9284	.9289
8.5	.9294	.9299	.9304	.9309	.9315	.9320	.9325	.9330	.9335	.9340
8.6	.9345	.9350	.9355	.9360	.9365	.9370	.9375	.9380	.9385	.9390
8.7	.9395	.9400	.9405	.9410	.9415	.9420	.9425	.9430	.9435	.9440
8.8	.9445	.9450	.9455	.9460	.9465	.9469	.9474	.9479	.9484	.9489
8.9	.9494	.9499	.9504	.9509	.9513	.9518	.9523	.9528	.9533	.9538
9.0	.9542	.9547	.9552	.9557	.9562	.9566	.9571	.9576	.9581	.9586
9.1	.9590	.9595	.9600	.9605	.9609	.9614	.9619	.9624	.9628	.9633
9.2	.9638	.9643	.9647	.9652	.9657	.9661	.9666	.9671	.9675	.9680
9.3	.9685	.9689	.9694	.9699	.9703	.9708	.9713	.9717	.9722	.9727
9.4	.9731	.9736	.9741	.9745	.9750	.9754	.9759	.9763	.9768	.9773
9.5	.9777	.9782	.9786	.9791	.9795	.9800	.9805	.9809	.9814	.9818
9.6	.9823	.9827	.9832	.9836	.9841	.9845	.9850	.9854	.9859	.9863
9.7	.9868	.9872	.9877	.9881	.9886	.9890	.9894	.9899	.9903	.9908
9.8	.9912	.9917	.9921	.9926	.9930	.9934	.9939	.9943	.9948	.9952
9.9	.9956	.9961	.9965	.9969	.9974	.9978	.9983	.9987	.9991	.9996
n	0	1	2	3	4	5	6	7	8	9

Table 3   Natural Logarithms

x	ln x	x	ln x	x	ln x
0.0		4.5	1.5041	9.0	2.1972
0.1	−2.3026	4.6	1.5261	9.1	2.2083
0.2	−1.6094	4.7	1.5476	9.2	2.2192
0.3	−1.2040	4.8	1.5686	9.3	2.2300
0.4	−0.9163	4.9	1.5892	9.4	2.2407
0.5	−0.6931	5.0	1.6094	9.5	2.2513
0.6	−0.5108	5.1	1.6292	9.6	2.2618
0.7	−0.3567	5.2	1.6487	9.7	2.2721
0.8	−0.2231	5.3	1.6677	9.8	2.2824
0.9	−0.1054	5.4	1.6864	9.9	2.2925
1.0	0.0000	5.5	1.7047	10	2.3026
1.1	0.0953	5.6	1.7228	11	2.3979
1.2	0.1823	5.7	1.7405	12	2.4849
1.3	0.2624	5.8	1.7579	13	2.5649
1.4	0.3365	5.9	1.7750	14	2.6391
1.5	0.4055	6.0	1.7918	15	2.7081
1.6	0.4700	6.1	1.8083	16	2.7726
1.7	0.5306	6.2	1.8245	17	2.8332
1.8	0.5878	6.3	1.8405	18	2.8904
1.9	0.6419	6.4	1.8563	19	2.9444
2.0	0.6931	6.5	1.8718	20	2.9957
2.1	0.7419	6.6	1.8871	25	3.2189
2.2	0.7885	6.7	1.9021	30	3.4012
2.3	0.8329	6.8	1.9169	35	3.5553
2.4	0.8755	6.9	1.9315	40	3.6889
2.5	0.9163	7.0	1.9459	45	3.8067
2.6	0.9555	7.1	1.9601	50	3.9120
2.7	0.9933	7.2	1.9741	55	4.0073
2.8	1.0296	7.3	1.9879	60	4.0943
2.9	1.0647	7.4	2.0015	65	4.1744
3.0	1.0986	7.5	2.0149	70	4.2485
3.1	1.1314	7.6	2.0281	75	4.3175
3.2	1.1632	7.7	2.0412	80	4.3820
3.3	1.1939	7.8	2.0541	85	4.4427
3.4	1.2238	7.9	2.0669	90	4.4998
3.5	1.2528	8.0	2.0794	95	4.5539
3.6	1.2809	8.1	2.0919	100	4.6052
3.7	1.3083	8.2	2.1041		
3.8	1.3350	8.3	2.1163		
3.9	1.3610	8.4	2.1281		
4.0	1.3863	8.5	2.1401		
4.1	1.4110	8.6	2.1518		
4.2	1.4351	8.7	2.1633		
4.3	1.4586	8.8	2.1748		
4.4	1.4816	8.9	2.1861		

Table 4   Powers of *e*

x	e^x	e^{-x}	x	e^x	e^{-x}
0.00	1.00000	1.00000	1.60	4.95302	0.20189
0.01	1.01005	0.99004	1.70	5.47394	0.18268
0.02	1.02020	0.98019	1.80	6.04964	0.16529
0.03	1.03045	0.97044	1.90	6.68589	0.14956
0.04	1.04081	0.96078	2.00	7.38905	0.13533
0.05	1.05127	0.95122			
0.06	1.06183	0.94176	2.10	8.16616	0.12245
0.07	1.07250	0.93239	2.20	9.02500	0.11080
0.08	1.08328	0.92311	2.30	9.97417	0.10025
0.09	1.09417	0.91393	2.40	11.02316	0.09071
0.10	1.10517	0.90483	2.50	12.18248	0.08208
			2.60	13.46372	0.07427
0.11	1.11628	0.89583	2.70	14.87971	0.06720
0.12	1.12750	0.88692	2.80	16.44463	0.06081
0.13	1.13883	0.87810	2.90	18.17412	0.05502
0.14	1.15027	0.86936	3.00	20.08551	0.04978
0.15	1.16183	0.86071			
0.16	1.17351	0.85214	3.50	33.11545	0.03020
0.17	1.18530	0.84366	4.00	54.59815	0.01832
0.18	1.19722	0.83527	4.50	90.01713	0.01111
0.19	1.20925	0.82696			
			5.00	148.41316	0.00674
0.20	1.22140	0.81873	5.50	224.69193	0.00409
0.30	1.34985	0.74081			
0.40	1.49182	0.67032	6.00	403.42879	0.00248
0.50	1.64872	0.60653	6.50	665.14163	0.00150
0.60	1.82211	0.54881			
0.70	2.01375	0.49658	7.00	1096.63316	0.00091
0.80	2.22554	0.44932	7.50	1808.04241	0.00055
0.90	2.45960	0.40656	8.00	2980.95799	0.00034
1.00	2.71828	0.36787	8.50	4914.76884	0.00020
1.10	3.00416	0.33287			
1.20	3.32011	0.30119	9.00	8130.08392	0.00012
1.30	3.66929	0.27253	9.50	13359.72683	0.00007
1.40	4.05519	0.24659			
1.50	4.48168	0.22313	10.00	22026.46579	0.00005

Answers to Selected Exercises

Exercises 1.1 (page 5)

1. {A, D}   **3.** Ø   **5.** {4}   **7.** {1, 2, 3, 4}   **9.** {1, 2, 3, 4, 5}   **11.** T   **13.** F   **15.** F   **17.** T
19. F   **21.** {1, 2, 3, 4, 5, 6, 8, 9}   **23.** {2, 6}   **25.** {1, 2, 3, 6}   **27.** {1, 2, 3, 4, 5, 6, 8, 9}
29. Ø   **31.** Ø   **33.** {1, 2, 3, 6}

35. 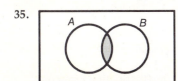   **37.** 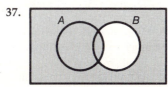   **39.**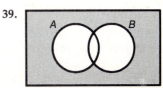

41. 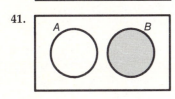   **43.** 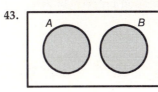   **45.**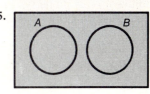

47. Infinite   **49.** Empty   **51.** {3}, Ø   **53.** {r}, {t}, {r, t}, Ø   **55.** {0}, {1}, {2}, {0, 1}, {0, 2}, {1, 2},
{0, 1, 2}, Ø   **57.** {x|x is a positive even number less than 10}   **59.** {x|x is a vowel}
61. {x|x is a natural number greater than 6}

63. 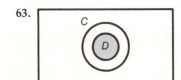   **65.** 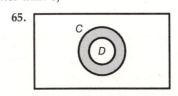   **67.** 4

Exercises 1.2 (page 15)

1. <   **3.** >   **5.** <   **7.** >   **9.** ∈   **11.** ∉   **13.** ∈   **15.** ∉   **17.** −6   **19.** 17   **21.** |x|
23. m   **25.** ⊆   **27.** ⊆   **29.** ⊄   **31.** ⊆   **33.** 0.5, rational   **35.** 0.75, rational
37. $0.\overline{428571}$, rational   **39.** $0.6\overline{6}$, rational   **41.** 7 < 11   **43.** x + 5 > 0   **45.** 3m + 1 ≥ 0
47. 4 < x < 7   **49.** 0 ≤ 3p < 9

51.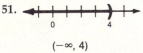

(−∞, 4)

53.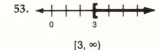

[3, ∞)

55.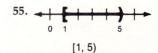

[1, 5)

57.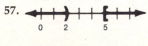

(−∞, 2) ∪ [5, ∞)

59. Ø

61.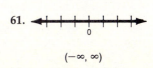

(−∞, ∞)

63. $\{x \mid 3 < x < 6\}$ **65.** $\{x \mid -2 \le x < 8\}$ **67.** $\{x \mid -1 \le x \le 0\}$ **69.** $\{x \mid x \ge -3\}$ **71.** $\{x \mid x < -2\}$

73.   **75.**   **77.**

79.

Exercises 1.3   (page 20)

1. 9   **3.** 0   **5.** $3x - 3t$ **7.** $(7 \cdot 8)y$ **9.** a **11.** $t \cdot r$ **13.** $5 + b$ **15.** p **17.** $3 + 5$
19. $x + 4$ **21.** Yes; $-3 + 4 = 1$ **23.** No; $1 + 1 = 2$ **25.** Yes; $2 + 4 = 6$
27. No; $-1 + (-1) = -2$ **29.** $6x + 6$ **31.** $5(m + 3)$ **33.** $9(a + b)$ **35.** $13r + 52t$
37. Associative property of addition   **39.** Commutative property of multiplication
41. Multiplicative inverse property   **43.** Multiplicative identity property   **45.** Reflexive property
47. Distributive property   **49.** Closure property   **51.** Additive inverse property
53. $(5 + 6)t = 11t$ **55.** $(13 - 8)b = 5b$ **57.** $(9 + 1)y = 10y$ **59.** $6 - 4 \ne 4 - 6$
61. $(8 \div 4) \div 2 \ne 8 \div (4 \div 2)$ **63.** (a) Identity property of multiplication (b) Distributive property
(c) Additive inverse property (d) Multiplication property of zero (e) Additive inverse property
(f) Substitution property

Exercises 1.4   (page 26)

1. 3   **3.** 1   **5.** $\dfrac{8}{3}$ **7.** $\dfrac{15}{11}$ **9.** $\dfrac{7}{13}$ **11.** $\dfrac{3}{2}$ **13.** $\dfrac{11a}{4}$ **15.** $\dfrac{y}{2}$ **17.** $\dfrac{59s}{60}$ **19.** $\dfrac{9}{8}$

21. $\dfrac{97}{28}$ **23.** $\dfrac{19}{4}$ **25.** $\dfrac{25}{8}$ **27.** $\dfrac{163}{66}$ **29.** $\dfrac{33}{8}$ **31.** $\dfrac{3}{8}$ **33.** $\dfrac{1}{6}$ **35.** $\dfrac{3}{2}$ **37.** 1

39. 1   **41.** $\dfrac{15}{56}$ **43.** $\dfrac{7}{24}$ **45.** $\dfrac{7}{80}$ **47.** $\dfrac{377}{100}$ **49.** $\dfrac{1}{12}$ **51.** $\dfrac{357}{400}$ **53.** $\dfrac{127}{84}$ **55.** $\dfrac{2}{3}$

57. $\dfrac{1}{2}$ **59.** $\dfrac{7}{9}$ **61.** 4   **63.** 3   **65.** 3   **67.** 8   **69.** 105   **71.** 195   **73.** 20 inches

Exercises 1.5   (page 31)

1. 10   **3.** 6   **5.** -32 **7.** 18   **9.** -8 **11.** -15 **13.** 6   **15.** -18 **17.** -23 **19.** -6
21. -3 **23.** 18   **25.** -10 **27.** 9   **29.** 3   **31.** 28   **33.** -3 **35.** $11x$ **37.** $5p$
39. $-24m$ **41.** $-9p$ **43.** $19r + 5$ **45.** $-5a + 20b$ **47.** $-13m - 68p$ **49.** -12 **51.** 18
53. 36   **55.** 14,729 feet   **57.** $-2°$ **59.** -594.669 **61.** -95.896 **63.** (a) Additive inverse
property (b) Additive property of equality (c) Commutative property of addition (d) Definition of
subtraction (e) Associative property of addition (f) Additive inverse property

Exercises 1.6   (page 37)

1. 33   **3.** -78 **5.** 72   **7.** 48   **9.** -440 **11.** 84   **13.** 2   **15.** -3 **17.** 32   **19.** 0
21. Undefined   **23.** -3 **25.** 0   **27.** 0   **29.** 10   **31.** -43 **33.** -13 **35.** -2 **37.** -5
39. -100 **41.** Undefined   **43.** $-\dfrac{1}{8}$ **45.** 560   **47.** Undefined   **49.** $\dfrac{1}{2}$ **51.** $12x + 17y$
53. $-9r - 38t$ **55.** $29a + 21b$ **57.** -0.3 **59.** 62.8   **61.** -61.7 **63.** Negative   **65.** Yes
67. Yes   **69.** 0

Exercises 1.7   (page 42)

In 1–19, let n = the number.

1. $2n$ **3.** $\frac{1}{2}n$ **5.** $n + 7$ **7.** $8 + n$ **9.** $n + 1$ **11.** $n - 30$ **13.** $6n$ **15.** $\frac{n}{6}$ **17.** $80 - n$

19. $19n$ **21.** $19x$ **23.** $\frac{6}{2n}$ **25.** $3h - 26$ **27.** $\frac{y}{11}$ **29.** $\frac{1}{4}a + 8$ **31.** $6r - 51$ **33.** $36 - x$

35. Let w = wages for one hour; $8w$ **37.** Let c = cost of a gallon of gas; $18c$ **39.** Let x = cost of a thirty-pound bag of fertilizer; $6x$ **41.** Let n be an even natural number; $n + (n + 2)$ **43.** Let r be a real number; $r + (r + 1) + 6$ **45.** Let n be an even integer, $n + 2$ be the next consecutive even integer, and $n + 4$ be the next consecutive even integer; $n + (n + 2) + (n + 4)$.   **47.** Let x be an even natural number, $x + 2$ be the next consecutive even natural number, and $x + 4$ be the next consecutive even natural number, $x + (x + 2) + (x + 4) - 30$.   **49.** Let x = a whole number, $83 - x$ = the other whole number; $x + (83 - x)$ **51.** Let x be an integer, $2x + 1$ be the other integer; $x + (2x + 1)$ **53.** Let x = the number; $x + 8 = 10$ **55.** Let n = the number; $11 + 3n = 24$ **57.** Let x centimeters = a side of the square; $x \cdot x = 64$ **59.** Let n inches = a side; $n + n + n = 81$ **61.** $I = (0.08)(1000)(1)$ **63.** $y - 6$ = the number of dimes; $0.05(6) + 0.10(y - 6) = 2.75$ **65.** Let n = number of dresses purchased; $n = \frac{134.85}{44.95}$ **67.** $\frac{x + (x + 10)}{2} < 90$ **69.** Let n = the number; $n + 10 = 3n - 6$

Chapter 1   Review   (page 46)

1. Set   **2.** Infinite   **3.** {a, i, l, n, o, r, t}   **4.** {1, 3}, {1}, {3}, ∅   **5.** {−2, 2, 4, 6, 8}   **6.** Intersection   **7.** No   **8.** Subset   **9.** Less than   **10.** To the left   **11.** Additive inverses   **12.** Integers   **13.** Rational   **14.** Irrational   **15.** 0.2727   **16.** Closure   **17.** Commutative   **18.** Distributive   **19.** Symmetric   **20.** $-\frac{57}{70}$ **21.** $-\frac{5}{2}$ **22.** $-\frac{416}{153}$ **23.** $-\frac{21}{4}$ **24.** $\frac{361}{45}$ **25.** $-\frac{61}{36}$ **26.** $-\frac{52}{45}$ **27.** -15 **28.** -4 **29.** -45 **30.** -13 **31.** 4   **32.** Negative   **33.** 4   **34.** Undefined   **35.** 3   **36.** $\frac{1}{3}$ **37.** $-19x + 13y$ **38.** $x + (x + 1) = -33$ **39.** $x - 5 = 4$ **40.** $2n + 3 = 18$ **41.** $d = 55(3)$ **42.** Let x = number of dimes; $18 - x$ = number of nickels; $0.10x + 0.05(18 - x) = 1.40$ **43.** Let L = length; $L - 6$ = width; $2L + 2(L - 6) = 48$ **44.** Let w = width; $w + 1$ = length; $w(w + 1) = 132$ **45.** $88 < \frac{x + (x + y)}{2} < 93$

Exercises 2.1   (page 55)

1. {8}   **3.** {−5}   **5.** {3}   **7.** $\left\{\frac{3}{2}\right\}$ **9.** $\left\{\frac{-11}{3}\right\}$ **11.** {−4}   **13.** {−4}   **15.** $\left\{\frac{1}{3}\right\}$ **17.** $\left\{\frac{3}{2}\right\}$

19. {0}   **21.** {10}   **23.** Conditional; {−1}   **25.** Identity; $\{a \mid a \in R\}$ **27.** Conditional; {6}

29. Identity; $\{y \mid y \in R\}$ **31.** {10}   **33.** $\left\{\frac{-3}{5}\right\}$ **35.** $\left\{\frac{3}{4}\right\}$ **37.** {−1}   **39.** $\left\{\frac{-1}{3}\right\}$ **41.** $\left\{\frac{17}{9}\right\}$

43. {1}   **45.** {8}   **47.** $\left\{\frac{-15}{2}\right\}$ **49.** $\left\{\frac{15}{14}\right\}$ **51.** $\left\{\frac{-45}{14}\right\}$ **53.** $\left\{\frac{3}{8}\right\}$ **55.** $\left\{\frac{7}{16}\right\}$ **57.** {−2}

59. {15}   **61.** $\left\{\frac{10}{3}\right\}$ **63.** $\left\{\frac{1}{4}\right\}$, $x \neq 0$ **65.** {−3}, $k \neq 0$ **67.** $\left\{\frac{15}{2}\right\}$, $r \neq 0$

69. $\left\{-\frac{1}{2}\right\}$, $k \neq -1$ **71.** $\left\{-\frac{22}{3}\right\}$ **73.** {−6}, $m \neq -5$ **75.** {2}, $p \neq -2$ **77.** ∅; $a \neq -1$

79. {4.8}   **81.** {3.1}   **83.** 53, 54   **85.** 51, 57   **87.** L = 40, W = 20   **89.** 5 oz

Exercises 2.2 (page 59)

1. {4, −4}   **3.** Ø   **5.** {169, −169}   **7.** $\left\{\dfrac{1}{4}, \dfrac{-1}{4}\right\}$ **9.** $\left\{\dfrac{-7}{36}, \dfrac{7}{36}\right\}$ **11.** Ø   **13.** {7, −11}

15. Ø   **17.** {−2, 8}   **19.** {6, 0}   **21.** $\left\{\dfrac{-1}{2}\right\}$ **23.** Ø   **25.** {14, −10}   **27.** $\left\{\dfrac{-27}{2}, \dfrac{33}{2}\right\}$

29. $\left\{\dfrac{8}{5}, \dfrac{-2}{5}\right\}$ **31.** {−12, 3}   **33.** $\left\{\dfrac{-3}{2}, \dfrac{3}{4}\right\}$ **35.** $\left\{\dfrac{-7}{6}\right\}$ **37.** $\{y \mid y \in R\}$ **39.** $\left\{\dfrac{3}{2}\right\}$

41. $\left\{\dfrac{-8}{3}, \dfrac{10}{9}\right\}$ **43.** Ø   **45.** Ø   **47.** $\{p \mid p > 4\}$ **49.** $\{y \mid y \in R\}$ **51.** $\left\{-3, \dfrac{-1}{5}\right\}$ **53.** Ø

55. $\left\{\dfrac{11}{2}\right\}$ **57.** $\left\{\dfrac{6}{5}, 4\right\}$ **59.** False   **61.** True   **63.** True   **65.** True   **67.** {−3, 3}

69. {−0.0, 1.0}   **71.** {0.9, 1.3}   **73.** $|x| = 8; \{-8, 8\}$ **75.** $\left|\dfrac{2x - 3}{4}\right| = 0; \left\{\dfrac{3}{2}\right\}$

77. $\left|\dfrac{3x}{4} + 1\right| = -10;$ Ø   **79.** $|x| = 4$ **81.** $|x| = 3$

Exercises 2.3 (page 67)

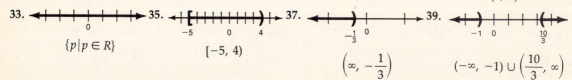

1. $(-\infty, 4)$ **3.** $\{x \mid x \le 4, x \in N\}$ **5.** $\left[\dfrac{8}{3}, \infty\right)$ **7.** $\left(-\dfrac{1}{2}, \infty\right)$

9. $\{x \mid x \ge 1, x \in J\}$ **11.** $\left[-\dfrac{5}{3}, \infty\right)$ **13.** $(-3, 4)$ **15.** $\{x \mid -3 \le x \le 3, x \in J\}$

17. $\{x \mid -2 < x < 3, x \in W\}$ **19.** $\left[\dfrac{13}{5}, \infty\right)$ **21.** $\left(-\infty, \dfrac{-1}{5}\right)$ **23.** $\{x \mid x \ge 12, x \in J\}$

25. $\left\{p \mid p \ge \dfrac{13}{8}, p \in W\right\}$ **27.** $\left\{x \mid x < \dfrac{21}{4}, x \in J\right\}$ **29.** $\left(-\infty, \dfrac{38}{27}\right)$ **31.** $(2, 5)$

33. $\{p \mid p \in R\}$ **35.** $[-5, 4)$ **37.** $\left(\infty, -\dfrac{1}{3}\right)$ **39.** $(-\infty, -1) \cup \left(\dfrac{10}{3}, \infty\right)$

41. $(-\infty, 21.7)$ **43.** $[163.9, \infty)$ **45.** $[-9.4, \infty)$ **47.** $(-\infty, 27]$ The number is less than or equal to
27.   **49.** $(-15, 21)$ The number lies between −15 and 21.   **51.** $(0, 40)$ A side must be less than
40 ft in length.   **53.** $n = 5$ A profit is earned on 5 or more units.   **55.** $(72, 92)$ A score between
72 and 92 would yield an average between 80 and 90.

Exercises 2.4   (page 72)

1. $-3 < x < 3$ **3.** $-4 \le p \le 4$ **5.** $m < -3$ or $m > 3$ **7.** $-3 \le x \le 13$

9. $m \le \dfrac{-3}{2}$ or $m \ge \dfrac{5}{2}$ **11.** $|p| < 2$ **13.** $|m| > 3$ **15.** $|x + 1| \le 6$ **17.** $|5q - 1| > 3$

19. $|p| < 3$ **21.** $|y - 3| \le 5$ **23.** $|m + 3| > 5$ **25.** $|6q - 2| > 4$

27.   **29.**   **31.** No solution

33. No solution   **35.**   **37.**

39.   **41.**   **43.**

45. No solution   **47.**   **49.** No solution

51.   **53.**   **55.**

57. $[-2.00, 1.58]$ **59.** $(-\infty, -3.66] \cup [2.65, \infty)$ **61.** No solution   **63.** $\{1, 2, 3, 4, 5, 6, 7\}$
65. Price is less than or equal to \$3.50.   **67.** Price is at least \$14.00.
69. $|-3 + (-7)| = |-3| + |-7|$
$\qquad\qquad 10 = 3 + 7$
$\qquad\qquad 10 \overset{\checkmark}{=} 10$

Exercises 2.5   (page 77)

1. $\dfrac{I}{pt}$ **3.** $\dfrac{c - b}{x}$ **5.** $\dfrac{P}{2} - W$ or $\dfrac{P - 2W}{2}$ **7.** $\dfrac{S}{r}$ **9.** $\dfrac{y - b}{x}$ **11.** $\dfrac{2A}{h} - b$ or $\dfrac{2A - bh}{h}$

13. $\dfrac{E}{C^2}$ **15.** $\dfrac{2D}{t^2}$ **17.** $\dfrac{A - P}{Pr}$ **19.** $\dfrac{ab}{a + b}$ **21.** $\dfrac{T_2 P_1 V_1}{P_2 V_2}$ **23.** $\dfrac{100B}{C}$ **25.** $\dfrac{R_1 R_2}{R_1 + R_2}$

27. $\dfrac{5}{11} D + 15$ **29.** $\dfrac{f_0}{M}$ **31.** $\dfrac{5}{9} (F - 32)$ **33.** $\dfrac{S - A}{DS}$ **35.** $\dfrac{2A}{al}$ **37.** $\dfrac{3V}{\pi h}$ **39.** $\dfrac{3A}{y_0 + 4y_1 + y_2}$

41. $\dfrac{4 - y}{3}$ **43.** $\dfrac{y - b}{m}$ **45.** $\dfrac{y - y_1 + mx_1}{m}$ **47.** $\dfrac{7 + 4y}{5 - 2y}$ **49.** 0.12   **51.** 95   **53.** $\dfrac{40}{3}$

55. 15 centimeters   **57.** 6.2 square meters   **59.** 24   **61.** 119.5°   **63.** 65.7   **65.** 56 feet
67. 37.7 cubic feet

Exercises 2.6   (page 83)

1. 55, 56, 57   **3.** 14, 15   **5.** 80°   **7.** \$150   **9.** 24   **11.** $L = 21$m, $W = 12$m   **13.** 12
15. 78 at \$3.00, 22 at \$4.50   **17.** \$16,250, \$8750   **19.** 24 minutes at 10 miles per hour, 12 minutes at
15 miles per hour   **21.** 25 pounds at \$5.00 each; 50 pounds at \$6.50 each   **23.** 15 pounds
25. $2\frac{1}{4}$ feet from 540 pounds   **27.** 32 liters   **29.** 2 quarts

Chapters 1 and 2   Review   (page 87)

1. Set   **2.** Element   **3.** Subset   **4.** A set   **5.** Set builder   **6.** Union   **7.** Empty   **8.** $\subseteq$
9. $\subseteq$ **10.** 8   **11.** Whole   **12.** a **13.** 6   **14.** Integers   **15.** Rational   **16.** 0.285714
17. $\{x \,|\, x > 1, x \in R\}$ **18.** $\{x \,|\, -2 \le x \le 3, x \in J\}$ **19.** Associative   **20.** Inverses   **21.** Inverses
22. 1   **23.** $ac = bc$ **24.** $a + (-b)$ **25.** $\dfrac{13}{3}$ **26.** $\{-4, 4\}$ **27.** $[0, 6]$ **28.** $\dfrac{16}{3}$

29. Negative   **30.** $\dfrac{x-21}{7}$ **31.** $2(x-6)$ **32.** $x \neq -4, 0, 2$ **33.** $\emptyset$ **34.** $|x+2| < 4$ **35.** 27

36. 308 square inches   **37.** $\dfrac{2A}{h} - b$ **38.** Positive   **39.** 34, 38   **40.** 41, 42, 43   **41.** {12}

42. $\left\{\dfrac{7}{3}\right\}$ **43.** $\left\{\dfrac{14}{3}\right\}$ **44.** $\emptyset$ **45.** $\{-6, 6\}$ **46.** $\left\{-1, \dfrac{2}{3}\right\}$ **47.** $\{-1, 5\}$

48.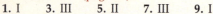
$\quad$ −3 $\quad$ 0 $\quad$ 3

49.
$\quad$ 0 $\quad$ 1

50.
$\quad$ $-\dfrac{8}{3}$ $\quad$ 0

51.
$\quad$ −1 $\quad$ 0 $\quad$ 1 $\quad$ 2

52.
$\quad$ 0 $\quad$ 1 $\quad$ 2 $\quad$ 3

53.
$\quad$ 0 $\quad$ $\dfrac{4}{3}$ $\quad$ 2

54.
$\quad$ 0 $\quad$ $\dfrac{11}{9}$
55. $x = \dfrac{y-b}{m}$ **56.** $t = \dfrac{I}{pr}$ **57.** $r = \dfrac{A-p}{pt}$ **58.** $a = f$

59. $x = \dfrac{5+7y}{4-y}$ **60.** $r = \dfrac{s-a}{s}$ **61.** $|x| \leq 5$ **62.** $|y| \geq 5$ **63.** $x + x + 2 + x + 4 = 114$; $x = 36$, $x + 2 = 38$, $x + 4 = 40$ **64.** $5x + 25(30 - x) = 450$; 15 nickels, 15 quarters
65. $3.75x + 4.00(40 - x) = 153.00$; $x = 28$ pounds at \$3.75 per pound; 12 pounds at \$4.00 per pound

Exercises 3.1 (page 97)
1. I   **3.** III   **5.** II   **7.** III   **9.** I

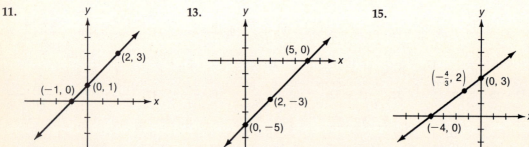

11. $(2, 3)$, $(-1, 0)$, $(0, 1)$ **13.** $(5, 0)$, $(2, -3)$, $(0, -5)$ **15.** $\left(-\dfrac{4}{3}, 2\right)$, $(0, 3)$, $(-4, 0)$

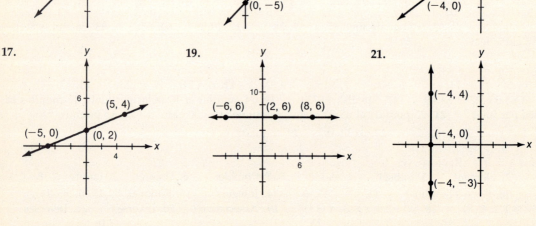

17. $(5, 4)$, $(-5, 0)$, $(0, 2)$ **19.** $(-6, 6)$, $(2, 6)$, $(8, 6)$ **21.** $(-4, 4)$, $(-4, 0)$, $(-4, -3)$

23.

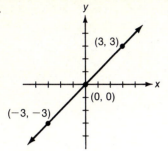

25.

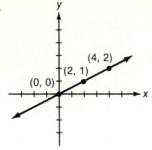

27. (a), (b), (c)

29. (a), (c), (d)

31.

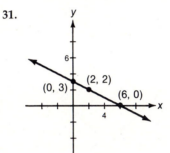

33.

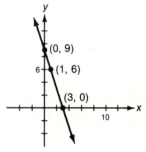

35.

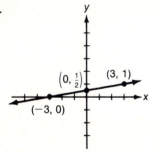

37.

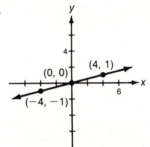

39.

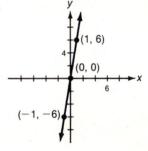

41.

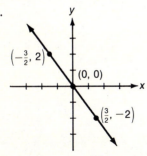

43.

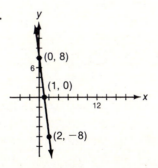

45.

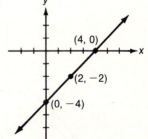

47. (1, 3), (2, 6), (3, 9)

49. $\left(0, \frac{3}{4}\right)$, $\left(\frac{-3}{2}, 0\right)$, $\left(1, \frac{5}{4}\right)$

51. $\left(0, \frac{2}{5}\right)$, $\left(\frac{-1}{2}, 0\right)$, (2, 2)

53.

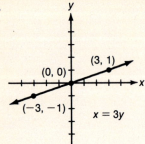

55.

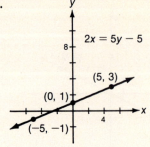

57.

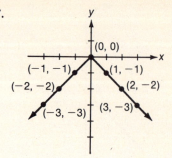

59.

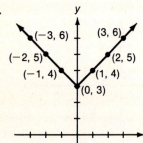

61.

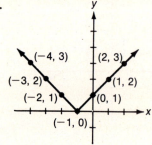

63. $(-2.31, 0.52)$ **65.** No   **67.** (a) \$225 loss (b) 30 (c) 40 (d) \$90

Exercises 3.2 (page 104)

1. $\dfrac{1}{3}$ **3.** -1 **5.** $\dfrac{-1}{4}$ **7.** 0   **9.** Undefined   **11.** $\dfrac{3}{4}$ **13.** -1 **15.** $\dfrac{d-b}{c-a}$ **17.** $\dfrac{b_2 - b_1}{a_2 - a_1}$

19. 0, horizontal   **21.** Undefined, vertical   **23.** 1, rises   **25.** -1, falls   **27.** 1, rises   **29.** No

31. No   **33.** Yes   **35.**

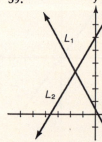

Parallel

37.

y

Perpendicular

39.

y

Neither

41.

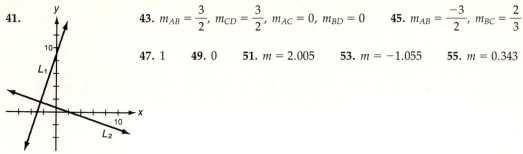

Perpendicular

43. $m_{AB} = \dfrac{3}{2}$, $m_{CD} = \dfrac{3}{2}$, $m_{AC} = 0$, $m_{BD} = 0$ **45.** $m_{AB} = \dfrac{-3}{2}$, $m_{BC} = \dfrac{2}{3}$

47. 1   **49.** 0   **51.** $m = 2.005$ **53.** $m = -1.055$ **55.** $m = 0.343$

Exercises 3.3 (page 110)

1.

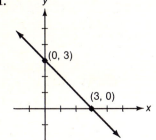

Dependent

3.

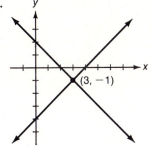

Independent

5.

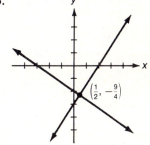

Independent

7.

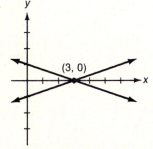

Independent

9.

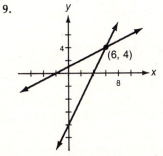

Independent

11.

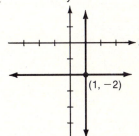

Independent

13.

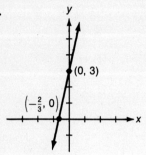

(0, 3)

$\left(-\frac{2}{3}, 0\right)$

Dependent

15.

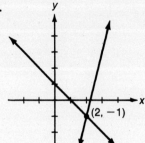

(2, −1)

Independent

17.

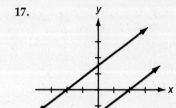

Inconsistent

19.

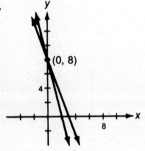

(0, 8)

4

8

Independent

21.

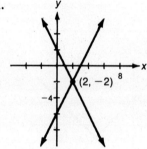

(2, −2)   8

−4

Independent

23.

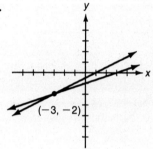

(−3, −2)

25.

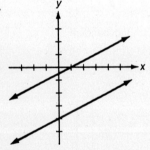

27.

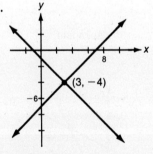

8

(3, −4)

−6

29. Yes

31. No

33.

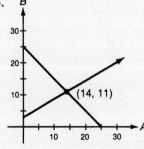

B

30

20

10

(14, 11)

10   20   30   A

35.

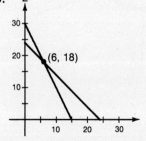

L

30

20

10

(6, 18)

10   20   30

Exercises 3.4 (page 118)

1. $\{(2, -1)\}$ **3.** $\{(3, 0)\}$ **5.** $\left\{\left(\frac{-3}{4}, 2\right)\right\}$ **7.** $\{(3, 2)\}$ **9.** Inconsistent **11.** $\{(6, 4)\}$

13. $\left\{\left(\frac{54}{11}, \frac{-27}{11}\right)\right\}$ **15.** $\left\{\left(\frac{23}{34}, \frac{-5}{34}\right)\right\}$ **17.** $\left\{\left(\frac{-7}{19}, \frac{27}{19}\right)\right\}$ **19.** $\{(3, 2)\}$ **21.** Dependent

23. $\left\{\left(\frac{16}{9}, \frac{8}{9}\right)\right\}$ **25.** $\left\{\left(\frac{-13}{3}, \frac{11}{3}\right)\right\}$ **27.** Inconsistent **29.** $\left\{\left(\frac{22}{3}, 4\right)\right\}$ **31.** $\left\{\left(\frac{2}{5}, -2\right)\right\}$

33. $\left\{\left(\frac{3}{2}, -3\right)\right\}$ **35.** $\left\{\left(\frac{516}{7}, \frac{-516}{41}\right)\right\}$ **37.** $a = 0, b = 4$ **39.** $a = 4, b = \frac{-2}{3}$ **41.** $a = 3, b = 2$

43. $\left\{\left(\frac{-25}{2}, \frac{-33}{2}\right)\right\}$ **45.** $\left\{\left(0, \frac{-20}{3}\right)\right\}$ **47.** $\left\{\left(\frac{-32}{13}, \frac{69}{13}\right)\right\}$ **49.** $\left\{\left(\frac{48}{5}, \frac{16}{5}\right)\right\}$ **51.** $\{(3, 1)\}$

53. $\{(1, 1)\}$ **55.** $x + y = 74; x - y = 10; x = 42; y = 32$

Exercises 3.5 (page 126)

1. $\{(3, -1, 5)\}$ **3.** $\{(1, 2, -1)\}$ **5.** $\{(1, 2, 3)\}$ **7.** $\{(2, 3, 4)\}$ **9.** $\left\{\left(\frac{-1}{3}, \frac{4}{3}, \frac{3}{2}\right)\right\}$

11. $\{(-3, 0, -9)\}$ **13.** $\{(-5, 2, 1)\}$ **15.** $\emptyset$ **17.** $\emptyset$ **19.** $\{(1, 1, 1)\}$ **21.** $\left\{\left(1, \frac{1}{2}, \frac{-1}{3}\right)\right\}$

23. $\{(1, 1, -1)\}$ **25.** 17 nickels, 32 dimes, 6 quarters
27. \$10,000 at 5%, \$5,000 at 6%, \$10,000 at 8% **29.** A is fastest; 120 minutes

Exercises 3.6 (page 132)

1. 1 **3.** 0 **5.** 0 **7.** 5 **9.** 0 **11.** -7 **13.** $\frac{-8}{15}$ **15.** 2 **17.** 312 **19.** 42

21. -42 **23.** 4 **25.** -6 **27.** 1 **29.** $\frac{9}{2}$ **31.** $0 = 0$ **33.** $adk - bck = adk - bck$

37. $3x - 2y - 1 = 0$ is the equation; $(1, 1)$ satisfies; $(-3, -5)$ satisfies **39.** $\left\{\left(\frac{-3}{7}, \frac{1}{14}\right)\right\}$

41. $\left\{\left(\frac{-10}{11}, \frac{18}{11}, \frac{38}{11}\right)\right\}$

Exercises 3.7 (page 138)

1. $\{(4, 3)\}$ **3.** $\left\{\left(\frac{9}{7}, \frac{13}{7}\right)\right\}$ **5.** $\left\{\left(\frac{-4}{5}, \frac{-22}{5}\right)\right\}$ **7.** $\{(6, 4)\}$ **9.** $\left\{\left(\frac{4}{3}, -4\right)\right\}$

11. $\left\{\left(\frac{64}{33}, \frac{-160}{9}\right)\right\}$ **13.** $\{(-1, 2, 1)\}$ **15.** $\{(1, 1, 1)\}$ **17.** Inconsistent **19.** $\{(2, -3, 2)\}$

21. Does not exist **23.** $\{(-1, -1, -1)\}$ **25.** $\left\{\left(\frac{1}{2}, 0, 1\right)\right\}$ **27.** $\left\{\left(\frac{2}{5}, -2, -1\right)\right\}$

29. $\left\{\left(\frac{347}{2548}, \frac{347}{176}, \frac{-347}{254}\right)\right\}$ **31.** Crawl stroke—10 minutes; breast stroke—5 minutes

33. Sherrie had \$4.50, Jason had \$1.50, and Lisa had \$6.00.

Exercises 3.8 (page 146)

1.
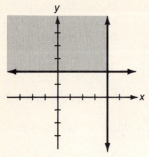

3.
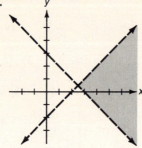

5.
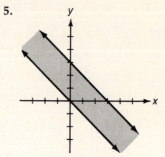

7.
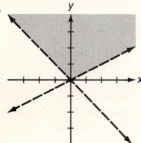

9.
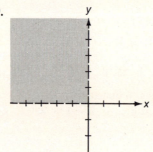

11.
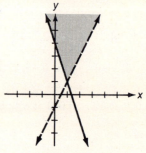

13.
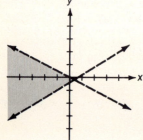

15.
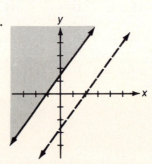

17.
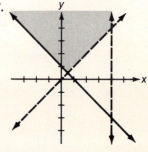

19.
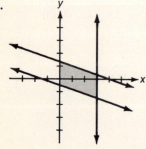

21.
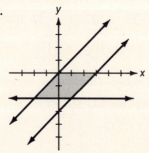

23.
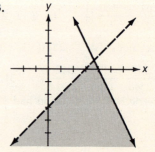

25. 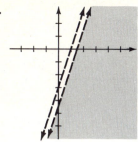   **27.** Yes   **29.** Yes

31. No   **33.**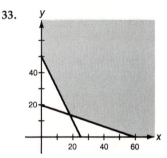

Exercises 3.9   (page 151)

1. $\dfrac{3}{2}$, $\dfrac{-17}{2}$ **3.** First = 17; second = 19   **5.** Speed of boat = $17\frac{1}{2}$ miles per hour; speed of

current = $2\frac{1}{2}$ miles per hour   **7.** Faster = 55 kilometers per hour; slower = 45 kilometers per hour
9. Large cola = 90¢; small cola = 60¢   **11.** $2000   **13.** $\frac{3}{4}$ liter of 40%; $1\frac{1}{4}$ liter of 80%

15. 4 pounds of peanuts; 6 pounds of pecans   **17.** Dictionary is $\dfrac{45}{26}$ inches thick; encyclopedia is

$\dfrac{7}{13}$ inches thick   **19.** $2500 to the first company; $4500 to the second company   **21.** $10,000 at 5%;

$5,000 at 6%; $10,000 at 8%   **23.**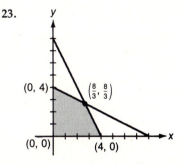

Chapter 3   Review   (page 154)

1. Zero   **2.** $\dfrac{y_2 - y_1}{x_2 - x_1}$ **3.** Parallel   **4.** Undefined or no slope   **5.** Perpendicular   **6.** $\dfrac{-1}{5}$

7. $\dfrac{1}{2}$ **8.** Parallel   **9.** (3, 3)   **10.** -4 **11.** $\left\{ \left(\dfrac{21}{11}, \dfrac{8}{11} \right) \right\}$ **12.** Dependent

13. Inconsistent   **14.** Independent

15.

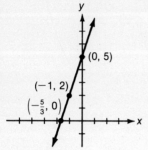

16.

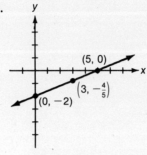

17.

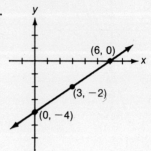

18.

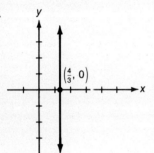

19.

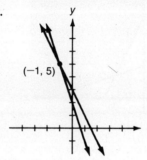

20. $\left\{\left(\dfrac{5}{6}, \dfrac{7}{3}\right)\right\}$

21. $\{(2, 1)\}$

22. $\left\{\left(\dfrac{13}{6}, \dfrac{-17}{6}\right)\right\}$

23.

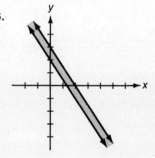

24. $\left\{\left(\dfrac{14}{11}, \dfrac{26}{11}\right)\right\}$

25.

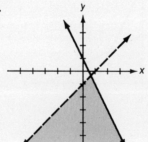

26.

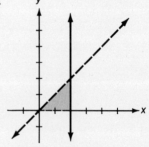

27. $\{(2, -1, -2)\}$

28. $\left\{\left(4, \dfrac{4}{3}\right)\right\}$

29.

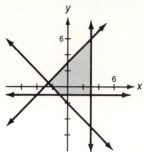

30. $\left\{\left(\dfrac{21}{2}, -84\right)\right\}$ **31.** (a) -10 (b) 5 (c) 2 (d) -10 (e) $\dfrac{22}{5}$ **32.** Oats at \$5.50 per bushel; rye at \$4.00 per bushel

Exercises 4.1 (page 165)

1. 8   **3.** $\dfrac{1}{10}$ **5.** 1   **7.** -32 **9.** 100   **11.** $\dfrac{-1}{5}$ **13.** $\dfrac{1}{9}$ **15.** 0.04   **17.** $\dfrac{4}{9}$ **19.** $\dfrac{25}{81}$

21. 16   **23.** 10   **25.** 1   **27.** $\dfrac{9}{125}$ **29.** $\dfrac{1}{10}$ **31.** 1   **33.** 5^5 **35.** -8^5 **37.** $(x - y)^2(-7)^3$

39. $1 \cdot 10^2 + 2 \cdot 10^1 + 9 \cdot 10^0 + 0 \cdot 10^{-1} + 2 \cdot 10^{-2}$

41. $8 \cdot 10^3 + 7 \cdot 10^2 + 0 \cdot 10^1 + 1 \cdot 10^0 + 9 \cdot 10^{-1} + 0 \cdot 10^{-2} + 6 \cdot 10^{-3}$ **43.** a^9 **45.** t **47.** p^8

49. s^4 **51.** $\dfrac{1}{c^{15}}$ **53.** a^4b^4 **55.** s^2t **57.** $\dfrac{8x^4}{y}$ **59.** $\dfrac{8}{x^2}$ **61.** 1   **63.** $\dfrac{1}{p^2}$ **65.** $\dfrac{4b^3}{3a^5}$

67. $\dfrac{m^9n^2}{8}$ **69.** $4p^8q^2$ **71.** x^{n+m} **73.** a^{5n-1} **75.** p^2 **77.** c^{n^2+n} **79.** $\dfrac{b^4}{a}$ **81.** $x^{3n}y^{2n}$

83. $a + b$ **85.** 1   **87.** $(x + y)^7$ **89.** $a - 3b$ **91.** $\dfrac{a^6}{b^2(a + b)^2}$ **93.** $\left(\dfrac{x}{y}\right)^{-m} = \dfrac{1}{\left(\dfrac{x}{y}\right)^m} = \left(\dfrac{y}{x}\right)^m$

95. $(3 + 4)^2 \neq 3^2 + 4^2$ **97.** 0.2963   **99.** 28.5499   **101.** 0.5186
$49 \neq 25$

103.

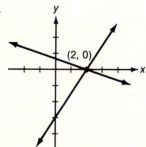

105.

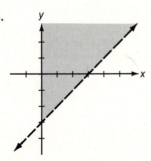

Exercises 4.2 (page 172)

1. Yes   **3.** No   **5.** Yes   **7.** No   **9.** Yes   **11.** No   **13.** (a) $5p^2 + p$ (b) 2nd degree (c) 5
(d) binomial   **15.** (a) $23r$ (b) 1st degree (c) 23 (d) monomial   **17.** (a) $-q^5 - 9q^3 + 5q^0$
(b) 5th degree (c) -1 (d) trinomial   **19.** (a) $2x^3 + 5x^2 - 9x$ (b) 3rd degree (c) 2 (d) trinomial

21. $7a$ **23.** $3y$ **25.** $-6m^2 - 3m$ **27.** $12ab^2$ **29.** $\dfrac{-1}{4}t$ **31.** $8x^2$ **33.** $-b$ **35.** $-3t + 3$

37. $-b^2 + 3b - 1$ **39.** $3a + 5$ **41.** $-4k - 17$ **43.** $-2x^2y - 3xy^2 - 4$ **45.** $-6r^2 + 16r - 5$

47. $-14x^2y + 6xy^2 + 18xy$ **49.** $2x + 4y$ **51.** $-7x + 5y + 8$ **53.** $6m + 1$ **55.** $21s - 14t$

57. $24r - 30s$ **59.** $9x^2 - 15xy + 6y^2 + 3$ **61.** $-1, 3, 8$ **63.** $x^2 + 2x + 2$ **65.** $-6x^2 + 13x - 6$

67. -2 **69.** $-6.06x^2 + 1.142x - 17.849$ **71.** -0.4392 **73.** \$5450   **75.** $\left\{\left(\dfrac{11}{18}, \dfrac{19}{18}\right)\right\}$

77.

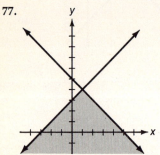

79. $-3xy^5$ **81.** $\dfrac{1}{x^2y^{16}}$

Exercises 4.3   (page 180)

1. $6x^5$ **3.** $56m^6$ **5.** $-p^9q^{12}$ **7.** t^7 **9.** $49a^9$ **11.** $2y^3 - 6y^2$ **13.** $15t^4 + 30t^3 + 45t^2$

15. $-3s^4 + 27s$ **17.** $a^4b + 7a^3b^2 - a^2b^3$ **19.** $-3x^5 + 19x^4 - 57x^3$ **21.** $a^{3n} + 3a^{2n} - 4a^n$

23. $u^3 + 2u^2v + 2uv^2 + v^3$ **25.** $6x^5 + 3x^4y - 2x^3y^2 + 2x^2y^3 - y^5$ **27.** $x^{3n} - y^{3n}$ **29.** $x^2 + 2x + 1$

31. $4p^2 - 12p + 9$ **33.** $r^2 - s^2$ **35.** $9k^2 - 25$ **37.** $4h^4 - 36h^2 + 81$ **39.** $16y^4 - 9$

41. $a^{2m} - 2a^mb^m + b^{2m}$ **43.** $x^{4m} - y^{4m}$ **45.** $a^2 + 2ab + b^2 - 16$

47. $25p^2 + 10pq + q^2 - 50p - 10q + 25$ **49.** $x^2 + 2x - 3$ **51.** $k^2 + 11k + 28$ **53.** $6t^2 + 13t + 7$

55. $28y^2 + 36y - 40$ **57.** $56r^2 + 2rs - 30s^2$ **59.** $21x^4 + 32x^2y^2 - 5y^4$ **61.** $66m^4 + 23m^2n - 63n^2$

63. $10p^3 - 41p^2 + 31p - 6$ **65.** $24a^3 - 71a^2b + 51ab^2 - 10b^3$

67. $q^5 - 3q^4r + 2q^3r^2 + 4q^2r^3 - 3qr^4 - r^5$ **69.** $a^{3n} + b^{3n}$ **71.** $2a^4 - 5a^3 + 5a^2 + 11a - 10$

73. $6x^4 - 13x^3y - 41x^2y^2 - 27xy^3 - 5y^4$ **75.** $-12x$ **77.** $-12x^ny^n$ **79.** $4u^3 + 4u^2v - 5uv^2 - 3v^3$

81. $p^{3m+5} - 2p^{3m+6}q^{m+1} + p^{2m+5}q^3 - p^mq^{m+3} + 2p^{m+1}q^{2m+4} - q^{m+6}$ **83.** $8x^2 + 10x + 3$

85. $9x^2 + 15x + \dfrac{13}{2} + \pi\left(4x^2 + 6x + \dfrac{9}{4}\right)$ **87.** $14x^2 + 20x + 7 + \pi\left(4x^2 + 6x + \dfrac{9}{4}\right)$

89. 39.48 square yards

Exercises 4.4   (page 186)

1. $a + 2$ **3.** $-3t + 4$ **5.** $2t^2 - 3t + 1$ **7.** $8x^2 + 4x - 1$ **9.** $-x - 3 + \dfrac{4}{x}$ **11.** $-9a + 11 - \dfrac{3}{2a^2}$

13. $6ab^2 + 4b$ **15.** $a^{4m} - a^{2m}$ **17.** $-14t^{12n} + 9t^{8n}$ **19.** $-7a^7b^4c^5 + 4a^4b^5c^{12}$

21. $\dfrac{15b^8c^7d}{11} - \dfrac{5b^5c^7d^5}{11} - \dfrac{5}{22}$ **23.** $x + 1$ **25.** $a + 3$ **27.** $t + 3$ **29.** $x^2 + 5x - 6$ **31.** $v - 2$

33. $5a^2 + 9a - \dfrac{1}{7} + \dfrac{\frac{-9}{7}}{7a - 2}$ **35.** $4x^2 + 2x + 1$ **37.** $3s^3 + \dfrac{8}{5}s - \dfrac{8}{25} + \dfrac{\frac{-17}{25}}{5s + 1}$ **39.** $x^n + y^n$

41. $x^n - y^n$ **43.** $x^{2n} - x^ny^n + y^{2n}$ **45.** $Q(x) = x + 4, R(x) = 1$ **47.** $Q(x) = \dfrac{5}{4}x + \dfrac{33}{16}, R(x) = \dfrac{-111}{16}$

49. $5x - 1$ **51.** $5x^2 + 4x - 1$ **53.** $0.3x^2 + 0.1x + \dfrac{0.01}{0.2x - 0.3}$

55. $0.0016s^4 + 0.008s^3 + 0.04s^2 + 0.2s + 1$ **57.** $5x^3 + 15x^2 - x - 1$ **59.** $9m^2 + 12mn + 4n^2$

61. $121a^2 - 36b^2$ **63.** $y^4 - 10y^3 + 25y^2 - 49$ **65.** $21x^2 + xy - 36y^2$

67.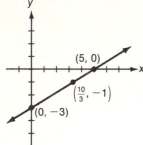

69.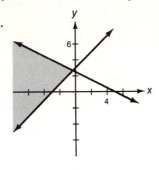

71. $\left\{\left(\dfrac{17}{11}, \dfrac{7}{11}\right)\right\}$

Exercises 4.5 (page 192)

1. $x - 5$ **3.** $a^2 + 6a + 5 + \dfrac{15}{a - 2}$ **5.** $3t^2 - 4t + 4 + \dfrac{4}{t + 1}$ **7.** $x^4 + 2x^3 + 4x^2 + 8x + 16$

9. $2y^3 + y^2 + 3y + 3$ **11.** $r^3 + r^2 + 5r + 3$ **13.** $a^5 + a^4 + a^3 + a^2 + a + 1 + \dfrac{-2}{a - 1}$

15. $2t^3 + 3t^2 + 14t + 70 + \dfrac{357}{t - 5}$ **17.** $5y^3 - 10y^2 + 2y - 10 + \dfrac{23}{y + 2}$

19. $2m^3 - 6m^2 + 15m - 40 + \dfrac{113}{m + 3}$ **21.** 0 **23.** 51 **25.** 0 **27.** -33 **29.** -36 **31.** $x + \dfrac{1}{3}$

33. $x^4 + \dfrac{1}{2}x^2 + \dfrac{5}{4}$ **35.** $7x^2 + \dfrac{21}{2}x + \dfrac{65}{4} + \dfrac{\frac{155}{8}}{x - \frac{3}{2}}$ **37.** $\dfrac{1}{2}x + \dfrac{5}{2}$ **39.** $x + 8 + \dfrac{6}{6x - 5}$

41. $-20x^3y - 7x^2y + 20xy$ **43.** $-29x^2 - 57x + 330$ **45.** $9a^2 + 12ab + 4b^2 - c^2$

47. $9x^4 - 12x^2y + 4y^2$ **49.** $x^{2m+4} + 4x^{m+5}y^{3m-2} - x^{m-1}y^{2m-1} - 4y^{5m-3}$ **51.** $9x^4 - 12x^3y^2 + 4x^2y^4$

53.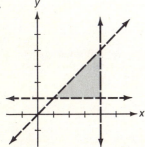

Exercises 4.6 (page 196)

1. 1.8×10^4 **3.** 8.0×10^{-2} **5.** 4.3×10^7 **7.** 8.1×10^{-9} **9.** 8,000   **11.** 0.011   **13.** 1,000
15. 0.01   **17.** 7,800   **19.** 0.000192   **21.** 120   **23.** 770   **25.** 100,000   **27.** 0.07   **29.** 0.12
31. 50   **33.** $10,000,000 = 10^7$ **35.** $128 = 1.28 \times 10^2$ **37.** $8,000 = 8.0 \times 10^3$ **39.** 6.6×10^{21} tons
41. 1150.7 worms/gallon   **43.** \$2367   **45.** 1.0×10^6 **47.** 200   **49.** 29,688.12 cubic feet

Chapters 3 and 4   Review (page 198)

1.

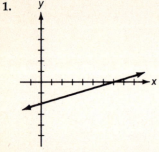

2. $\{(-2, -2)\}$

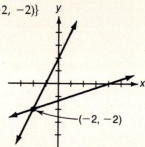

$(-2, -2)$

3. $\{(1, 0)\}$ **4.** $\{(1, 2)\}$

5.

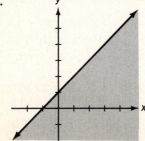

6.

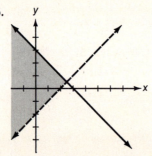

7.

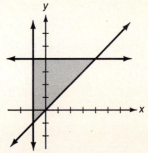

8. $\left\{\left(\dfrac{73}{43}, \dfrac{42}{43}\right)\right\}$ **9.** 64   **10.** $\dfrac{1}{81a^4b^5}$ **11.** (a) $4x^3 + x^2 - x$ (b) 3rd (c) trinomial (d) 4

12. (a) $-7y^2 + 5y + 6$ (b) 2nd (c) trinomial (d) -7 **13.** (a) $k^4 - k^1$ (b) 4th (c) binomial (d) 1

14. Not a polynomial   **15.** Not a polynomial   **16.** Not a polynomial   **17.** $\dfrac{-18}{x^{11}}$

18. $4a^2 + 20ab + 25b^2$ **19.** $3m^4n - 15m^3n^2 + 18m^2n^4$ **20.** $\dfrac{3}{2500}$ **21.** $26r^2 - 137rs - 33s^2$

22. $a^3 + a^2b - ab^2 - b^3$ **23.** $-3p^2 + 4q - 7$ **24.** $m^2 - 1 + \dfrac{7}{m - 1}$ **25.** $10x^2 + 7x + 5$

26. $6 + 12m$ **27.** $3a + 5 + \dfrac{-1}{6a - 7}$ **28.** $\{(3, -1, -2)\}$ **29.** 0.035   **30.** $6y^3 - 7y^2 - 9y - 2$

31. $-4s^2t - st + t^2$ **32.** 5 nickels, 3 quarters   **33.** 40 milliliters of 50%, 60 milliliters of 80%

34. $\dfrac{1}{2}$ kilometer per minute **35.** **36.** 7 **37.** -9 **38.** \$27,900

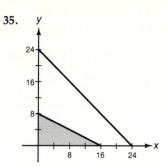

39. 298,400 centimeters **40.** 651,822.563 cubic centimeters

Exercises 5.1 (page 202)

1. $x + 2$ **3.** $2m - 3$ **5.** $7s + 1$ **7.** $y - 3$ **9.** $x^2 + 2x - 3$ **11.** $3t^2 - 4t + 2$

13. $2a^2 - 10a + 1$ **15.** $p^2 + \dfrac{16}{3}p - \dfrac{10}{3}$ **17.** $4s^2 - 6s + 1$ **19.** $8r^2 - 21r + 14$ **21.** $2(x + 5)$

23. $11(y - 5)$ **25.** $5a(3a - 10)$ **27.** $8(1 - 2t)$ **29.** $3(a + 3b - 2)$ **31.** $xy(x - y)$

33. $6mp(2m - p + 3)$ **35.** $t(a + b - c)$ **37.** Prime **39.** $3r^2s^3(5r^3 - 8s)$ **41.** $a(5 + 7a - 9a^2)$

43. $7x^2(4x^2 - 8x - 7)$ **45.** $2r^4s^3(4r - 3s^2 - 10)$ **47.** $16xyz(3 - 4xz - 5y)$ **49.** $9a(7a^2c - 9ad + 8cd)$

51. $(x + 1)(x + 2)$ **53.** $(2a + 1)(a - 1)$ **55.** $(a + c)(m + n)$ **57.** $(x - y)^2$ **59.** $(t - 1)[t(t - 1) - 1]$

61. $(m - 5)(4 - m)$ **63.** $\left(\dfrac{1}{2} - a\right)\left(\dfrac{1}{3} + a\right)$ **65.** $(x - 1)(x - 2)$ **67.** $(2a + 3)(2a + 5)$

69. $(a^2 + b)^2(a^3 + ab - b)$ **71.** $a^m(a + 1)$ **73.** $x^{2m}(x - 1)$ **75.** $y(y^{m-1} + 1)$ **77.** $r^{2m+2}(r - 1)$

79. $x^2 - y^2$ **81.** $-3x - y + 4$ **83.** $9m^2 + 12mn + 4n^2$ **85.** $a^3 - b^3$ **87.** $x^4 + 2x^3 + 4x^2 + 8x + 16$

89. $a^{2m} + 2a^{m+n} + a^{2n}$

Exercises 5.2 (page 206)

1. $(x + 2)(x + 1)$ **3.** $(m + 6)(m + 1)$ **5.** $(y - 1)^2$ **7.** $(p - 3)(p - 2)$ **9.** $(x - 3)(x + 2)$

11. $(t + 3)(t - 1)$ **13.** $(y - 10)(y + 2)$ **15.** $(m + 6)(m - 5)$ **17.** $(r + 8a)(r - 2a)$

19. $(a - 12b)(a + 5b)$ **21.** $2(x + 6)(x + 4)$ **23.** Prime **25.** $5(p - 1)(p - 8)$

27. $(5 + s)(2 - s)$ or $-(s - 2)(s + 5)$ **29.** $x(x - 8)(x + 4)$ **31.** $4a(a - 5b)(a + b)$

33. $\left(r + \dfrac{1}{5}\right)\left(r + \dfrac{1}{3}\right)$ **35.** $\left(k + \dfrac{1}{3}\right)\left(k - \dfrac{1}{2}\right)$ **37.** Prime **39.** $(ab - 2)^2$ **41.** $-3(m + 4)(m - 3)$

43. $-7a\left(a^2 + \dfrac{1}{2}a - \dfrac{1}{16}\right)$ **45.** $-(x - 6)(x - 2)$ **47.** $3st(t + 9s)(t + 2s)$ **49.** $5pq(p + q)^2$

51. $(x^2 + 4)(x^2 + 2)$ **53.** $(a^2 + 4)^2$ **55.** $2(m^2 + 5)(m^2 + 4)$ **57.** $-5(t^2 + 6)(t^2 - 5)$

59. $-k\left(k - \dfrac{1}{6}\right)\left(k - \dfrac{1}{5}\right)$ **61.** $(x + 3)(x - 2)(x + 1)$ **63.** $(a - 3)^2(a + 2)$ **65.** $(t + 7)^3$

67. $\left(k - \dfrac{1}{3}\right)^3$ **69.** $(x^m + 6)(x^m + 2)$ **71.** $(y^m - 0.2)(y^m - 0.1)$ **73.** $c^m(c + 5)(c + 6)$

75. $s^2(s^m - 2)(s^m + 1)$ **77.** $k^3(k^m - 0.7)(k^m - 0.3)$ **79.** $0.65y(0.9y - 0.11)$ **81.** $-0.93t(0.11t + 0.9)$

Exercises 5.3 (page 211)

1. $(x + 1)(x - 1)$ **3.** $2(a + 5)(a - 5)$ **5.** $(xy + 4z)(xy - 4z)$ **7.** $(11p + 12q)(11p - 12q)$

9. $(a^2 + b^2)(a + b)(a - b)$ **11.** $3(r^m + s^m)(r^m - s^m)$ **13.** $b^3\left(\dfrac{1}{4}a + \dfrac{1}{3}b\right)\left(\dfrac{1}{4}a - \dfrac{1}{3}b\right)$

15. $ab(0.8a + 0.7b)(0.8a - 0.7b)$ **17.** $4ab$ **19.** $-24mn$ **21.** $(a^4 + b^4)(a^2 + b^2)(a + b)(a - b)$

23. $(x - y)(x^2 + xy + y^2)$ **25.** $(p + q)(p^2 - pq + q^2)$ **27.** $(3a - 2b)(9a^2 + 6ab + 4b^2)$

29. $(xy + wz)(x^2y^2 - wxyz + w^2z^2)$ **31.** $(a^m - b^m)(a^{2m} + a^m b^m + b^{2m})$ **33.** $(2x - 1)(x^2 - x + 1)$

35. $2(3y^2 + 6y + 4)$ **37.** $(a^2 + b^2)(a^4 - a^2b^2 + b^4)$ **39.** $(x - y)(x^2 + xy + y^2)(x^6 + x^3y^3 + y^6)$

41. $(2b)(3a^2 + b^2)$ **43.** $\left(\dfrac{1}{3} s - \dfrac{1}{2} t\right)\left(\dfrac{1}{9} s^2 + \dfrac{1}{6} st + \dfrac{1}{4} t^2\right)$ **45.** $(a + 2)^2$ **47.** $(x - 1)^2$

49. $(2x + 3y)^2$ **51.** $3p(p - q)^2$ **53.** $(c^m + 7d^n)^2$ **55.** $(y + 3)^2(y - 3)^2$ **57.** $(a + 3)^2$

59. $3m(m^2 - 5)^2$ **61.** $[x - y - 2a + 2b]^2$ **63.** $\left(\dfrac{1}{4} w + \dfrac{1}{5} z\right)^2$ **65.** $(c + d)(x + y)$

67. $(a + b)(a - 2)$ **69.** $(r + 2s)(m + 2n)$ **71.** $(x - 1)(x + 1)^2$ **73.** $(s + t)(s - t)(x + y)(x - y)$

75. $(a - b + 1)(a + b + 1)$ **77.** $(y + 3x + 3)(y - 3x + 3)$ **79.** $(s - 2t - 4)(s + 2t + 4)$

81. $(3b - 7)(2a - c)$ **83.** $(2m + 3n)(4m^2 - 6mn + 9n^2 + 1)$ **85.** $2(4x - y)(16x^2 + 4xy + y^2 - 1)$

87. $(x + y + 5)(x - y + 1)$ **89.** $(a + b - c)(x - y)$

Exercises 5.4   (page 216)

1. $(2a + 1)(a + 1)$ **3.** $(3x + 1)(x + 2)$ **5.** $(2s - 1)(s + 4)$ **7.** $2(3r + 4)(2r - 1)$

9. $(4m - 5)(2m + 1)$ **11.** $2(4x + 17)(x - 1)$ **13.** $(2a + 5)^2$ **15.** $4y^3(2y - 5)(y + 5)$

17. $(3c + 2d)(3c - 10d)$ **19.** $(4 + 3k)(3 - 2k)$ **21.** $(2 + 3t)(1 - 4t)$ **23.** $(3ab - 10)(2ab + 1)$

25. $-2x^2(k + 3a)(k + 4a)$ **27.** $(5y + 8)(3y + 5)$ **29.** $-2(8x - 3y)(2x + y)$ **31.** $(ab + 3cd)(ab + 2cd)$

33. No   **35.** No   **37.** Yes   **39.** Yes   **41.** Yes   **43.** $2(x + 2)^2(x - 2)^2$

45. $3(a + 3b)(a - 3b)(a + b)(a - b)$ **47.** $(c + 2d)(c - d)(c^2 - 2cd + 4d^2)(c^2 + cd + d^2)$

49. $(3x + 2y)(3x - 2y)(x + 3y)(x - 3y)$ **51.** $2(7c + d)(7c - d)(c + d)(c - d)$

53. $\left(\dfrac{2}{3} y + \dfrac{1}{2}\right)\left(\dfrac{2}{3} y + \dfrac{1}{3}\right)$ **55.** $\left(\dfrac{3}{5} m + \dfrac{1}{3} n\right)\left(\dfrac{1}{2} m - \dfrac{1}{4} n\right)$ **57.** $(3x^m - 2y^n)(x^m + 4y^n)$

59. $(4c^m + 9d^n)(3c^m - d^n)$ **61.** $(x + 5)(x + 2)$ **63.** $[6(m + n) + 5][4(m + n) + 3]$

65. $2(3x + 2y)(4a + 13b)$ **67.** $(2x - 1)(2x - y)$ **69.** $(a - b + 6d)(a - b - 6d)$ **71.** $(y - x)(72x - 1)$

73. $(5a + 3b + 4c)(5a + 3b - 4c)$ **75.** $[3(x - y) + 5(a + b)][2(x - y) - 7(a + b)]$ **77.** $63(2y - 1)$

Exercises 5.5   (page 220)

1. $\{-3, 0\}$ **3.** $\{1, 2\}$ **5.** $\{-3, 3\}$ **7.** $\{3, 7\}$ **9.** $\{-4, 12\}$ **11.** $\{9\}$ **13.** $\left\{\dfrac{-3}{4}, 7\right\}$

15. $\left\{\dfrac{-8}{5}, \dfrac{3}{2}\right\}$ **17.** $\{-2, 3\}$ **19.** $\{-18, -2, 0\}$ **21.** $\left\{\dfrac{-1}{8}, 0, \dfrac{1}{2}\right\}$ **23.** $\left\{\dfrac{-1}{4}, 0, \dfrac{2}{3}\right\}$

25. $\{1, 2, 5\}$ **27.** $\left\{-1, 1, \dfrac{4}{3}\right\}$ **29.** $\left\{\dfrac{-1}{4}, \dfrac{1}{3}, \dfrac{1}{2}\right\}$ **31.** $\left\{\dfrac{1}{3}\right\}$ **33.** $\left\{\dfrac{-1}{4}, \dfrac{1}{2}\right\}$ **35.** $\left\{\dfrac{2}{3}, \dfrac{3}{2}\right\}$

37. $x^2 - 5x + 6 = 0$ **39.** $x^2 + 3x = 0$ **41.** $x^3 - x^2 - 2x = 0$ **43.** $x^2 - \dfrac{9}{2}x - \dfrac{5}{2} = 0$

45. $x^3 + \dfrac{1}{21}x^2 - \dfrac{2}{21}x = 0$ **47.** $\left\{\dfrac{-1}{2}\right\}$ **49.** $\left\{-1, \dfrac{13}{15}\right\}$ **51.** $\left\{\dfrac{-2}{3}, \dfrac{5}{2}\right\}$ **53.** $\{-5, 3\}$

55. $\{-1\}$ **57.** $\{-1, 2\}$ **59.** $\{4, 6\}$ **61.** 11, 13   **63.** 30, 70   **65.** 9, 12

Exercises 5.6   (page 224)

1. $\dfrac{1}{3}$ **3.** 3   **5.** W = 5 inches; L = 12 inches   **7.** 5 yards   **9.** 500 miles   **11.** W = 10 feet;

L = 20 feet   **13.** W = 20 feet; L = 50 feet   **15.** 16 inches   **17.** W = 18 inches; L = 36 inches

19. 2 seconds   **21.** 3, 4, 5   **23.** 22   **25.** 0 or 12 hours

Chapter 5 Review (page 227)

1. $3(9a + 17)$ **2.** $2l(2m + 3t - 9z)$ **3.** $2x(2y - 3)(8 - 5x)$ **4.** $(a - 3)(a - 4)$

5. $13xy^2z(7xy - 5z - 3xz^2)$ **6.** $y^{2m}(y^{m+1} - 1)$ **7.** $(t - 8)(t + 1)$ **8.** $(b - 15)(b + 2)$

9. $(s + 8)(s - 8)$ **10.** $2(a - 2)(a^2 + 2a + 4)$ **11.** $6(x - 5)(x - 1)$ **12.** $(4m - 3)^2$

13. $3rs^2(r - 4s)(r + 3s)$ **14.** $(x^2 + 4y^2)(x + 2y)(x - 2y)$ **15.** $(b - 3)^3$ **16.** $(k^m - 4r^m)(k^m - r^m)$

17. $3\left(y + \dfrac{1}{2}\right)\left(y + \dfrac{1}{4}\right)$ **18.** $(s^m - 2t^m)(s^{2m} + 2s^m t^m + 4t^{2m})$ **19.** $(5a + 6b)^2$ **20.** $(x + 3)(x + y)$

21. $(a + 2b + c)(a + 2b - c)$ **22.** $(k + 2)(3k - 2m)$ **23.** $(6 - x)(1 - 2x)$ **24.** $-7(y - 6)(y + 1)$

25. $(3k + m + 3)(3k - m - 3)$ **26.** $(2a - b)(2a + b)(4a^2 + 2ab + b^2)(4a^2 - 2ab + b^2)$

27. $(x - 1)^2(6a + b)$ **28.** $\left(\dfrac{1}{3}t - \dfrac{1}{4}\right)^2$ **29.** $(5s - 1)(7s^2 - s + 1)$ **30.** $-4ab$

31. $n^2(0.7m - 1.1n)(0.7m + 1.1n)$ **32.** $(x^4 + y^4)(x^2 + y^2)(x + y)(x - y)$

33. $\left(\dfrac{1}{9}a^2 + \dfrac{1}{4}b^2\right)\left(\dfrac{1}{3}a + \dfrac{1}{2}b\right)\left(\dfrac{1}{3}a - \dfrac{1}{2}b\right)$ **34.** $-(2x - 5)(10x - 19)$ **35.** $2(3y - 2w)(x + z)$

36. $10a(a - 2b)^2$ **37.** $(s^2 - 6)(s^2 - 3)$ **38.** $2m(m^2 - 5)^2$ **39.** $(a - b + c)(z - t)$

40. $(a - 2)(a^2 + 2a + 3)$ **41.** $xy(y + 3x)(4y - 3x)$ **42.** $2(3k - 8m)(k + 2m)$

43. $(a + b)(a - 3b)(a^2 - ab + b^2)(a^2 + 3ab + 9b^2)$ **44.** $(6x^m + y^n)(3x^m - 4y^n)$ **45.** $\{-2, -1\}$

46. $\left\{-2, 0, \dfrac{4}{3}\right\}$ **47.** $\{-2, 8\}$ **48.** $\left\{\dfrac{-5}{2}, \dfrac{1}{3}\right\}$ **49.** $\left\{\dfrac{-1}{2}, \dfrac{7}{2}\right\}$ **50.** $\{-4, -3, 0\}$

51. $x^2 + x - 2 = 0$ **52.** $x^3 + 2x^2 - 3x = 0$ **53.** $x^2 - \dfrac{1}{6}x - \dfrac{1}{6} = 0$ **54.** $x^3 - \dfrac{34}{5}x^2 - \dfrac{7}{5}x = 0$

55. 8, 9 or $-9, -8$ **56.** 25 feet **57.** 2 feet **58.** 75 miles

Exercises 6.1 (page 234)

1. $\dfrac{1}{3}$ **3.** $\dfrac{3}{4z}$ **5.** -1 **7.** $m + n$ **9.** $\dfrac{a}{4}$ **11.** $\dfrac{a}{a + 4}$ **13.** $\dfrac{3r}{13a^2}$ **15.** $\dfrac{xy}{x - y}$ **17.** 1

19. $\dfrac{m - 4}{m + 2}$ **21.** Equivalent **23.** Equivalent **25.** Equivalent **27.** $x \neq 0$ **29.** $b \neq 0, \dfrac{-3}{2}$

31. $p \neq -1, 4$ **33.** $x \neq 3, y \neq 1$ **35.** $a \neq b$ **37.** $x \neq 3, y \neq 1$ **39.** $30a^3b^2; a, b \neq 0$

41. $3(a + 1); a \neq -3, -1$ **43.** $-5p(p^2 + q^2); p = \pm q$ **45.** $m^2 - 16; m \neq 4$ **47.** $(y + 1)(3x + 2);$

$y \neq 1, x \neq -\dfrac{2}{3}$ **49.** $p - q; p \neq q$ **51.** $\dfrac{6}{13}$ **53.** Undefined **55.** 0 **57.** $\dfrac{1 - y}{3y^2(x - y)}$

59. $x - y$ **61.** $\dfrac{2 - a}{a - b}$ **63.** $\dfrac{a + 3}{a - 3}$ **65.** $\dfrac{x^2 + y^2}{x^2 - y^2}$ **67.** $\dfrac{x^2 - xy + y^2}{x - y}$ **69.** $\dfrac{2a - 5}{a}$

Exercises 6.2 (page 238)

1. $\dfrac{x}{2}$ **3.** $6ap$ **5.** $\dfrac{x - y}{2}$ **7.** $4p$ **9.** $4; b \neq 0, a \neq 0$ **11.** $3n^2; n \neq 0, m \neq 0$

13. $1; m \neq n, q \neq p$ **15.** $\dfrac{y + 2}{2y}; y \neq 0, 2$ **17.** $\dfrac{a + 1}{a + 2}$ **19.** $\dfrac{t - 1}{t - 3}$ **21.** $\dfrac{a + 1}{a + 2}$ **23.** 1

25. $\dfrac{3nx^3}{2y^2}$ **27.** $\dfrac{w(y - 1)}{w - x}$ **29.** $\dfrac{-(r^2 + rs + s^2)(r + s)^2}{r^2 + s^2}$ **31.** $\dfrac{2a + b}{a + b}$ **33.** 3 **35.** $\dfrac{(m + 2)(x + y)}{(2x + y)^2}$

37. $\dfrac{5q}{2}$ **39.** $\dfrac{-7(x - 2)}{3(x - 3)(x + 2)(x - 8)}$ **41.** $\dfrac{k - 1}{k^2 - k + 1}$ **43.** $(x - 3)(x + 2)$ **45.** $-b(a + b)$

47. $\dfrac{w^3z^3}{5a^4b^5}$ **49.** $\dfrac{3x + 5}{x - 1}$ **51.** $\dfrac{x - 1}{x + 2}$ **53.** $\dfrac{p - 3}{p - 1}$ **55.** $\dfrac{a^4}{(a + 3)^3}$ **57.** $\dfrac{1}{a - 3}$

59. $\dfrac{a^6}{(a - 3)(a + 3)^4}$ **61.** 0.809 **63.** -0.092 **65.** $\dfrac{12}{x + 3}$

Exercises 6.3   (page 245)

1. $\dfrac{x}{y}$ **3.** $\dfrac{8}{a+b}$ **5.** $\dfrac{1}{x-2}$ **7.** $\dfrac{1}{r+t}$ **9.** $\dfrac{1}{m^2+m+1}$ **11.** $18a$ **13.** x^3y^2

15. $(p-2)(p-6)$ **17.** $q(q+1)(q-1)$ **19.** $(x+2)(x+1)(x-1)$ **21.** $(3b+1)^2(b+3)$

23. $-(2w-1)(w+1)(w-6)$ **25.** $\dfrac{-1}{10a}$ **27.** $\dfrac{12+7x}{12x^2}$ **29.** $\dfrac{3a-7}{30a}$ **31.** $\dfrac{-6m+3}{(m+3)(m-4)}$

33. 0 **35.** $\dfrac{5}{m+2}$ **37.** $\dfrac{2}{x^2-1}$ **39.** $\dfrac{4k^2+5k+6}{k(2k+3)}$ **41.** $\dfrac{-2x}{x^2-4}$ **43.** $\dfrac{c^2-21c}{(c-4)(c-5)(c+3)}$

45. $\dfrac{6s+5}{(4s-1)(3s-4)}$ **47.** $\dfrac{ab^2-a}{b}$ **49.** $\dfrac{m^2}{m-1}$ **51.** $\dfrac{-k^2}{k^2+2k+1}$ **53.** $\dfrac{a^2+ab+a-v}{u+v}$

55. $5m^2+5m$ **57.** $\dfrac{2x}{(x+1)^2}$ **59.** $\dfrac{(3a+1)(a+3)}{12a^2(a+1)}$ **61.** $\dfrac{11k+5}{3k(k+1)}$ **63.** $\dfrac{y+w}{(x-w)(x-y)}$

65. $\dfrac{2b}{(a-b)(c-b)}$ **67.** $\dfrac{x^n+x^ny^n+y^{2n}}{x^{2n}-y^{2n}}$

Exercises 6.4   (page 251)

1. 2 **3.** $\dfrac{c}{a}$ **5.** $\dfrac{14}{9}$ **7.** $\dfrac{2}{x}$ **9.** $\dfrac{3}{4}$ **11.** $\dfrac{k}{k+1}$ **13.** $\dfrac{ab}{b-a}$ **15.** $\dfrac{a-1}{a^3}$ **17.** $12b$

19. $\dfrac{x}{x-1}$ **21.** $\dfrac{v}{2}$ **23.** 1 **25.** 1 **27.** $1-2q$ **29.** 2 **31.** $\dfrac{7x+1}{x+1}$ **33.** $(2t+3)(t-2)$

35. $\dfrac{3b(2b-3)}{(b-3)(b-1)}$ **37.** $3-2p$ **39.** $2m-9$ **41.** $\dfrac{s^2-17}{s^2-6s-27}$ **43.** $\dfrac{a+1}{a}$ **45.** $\dfrac{1-2p}{2p+1}$

47. $\dfrac{w+5}{w+4}$ **49.** $\dfrac{-(x+y)}{2(x-y)}$ **51.** 12 **53.** $\dfrac{x+y}{y-x}$ **55.** $\dfrac{(rs)^3}{s-r}$ **57.** $\dfrac{1}{p+1}$ **59.** $\dfrac{r-2s}{s+5r}$

61. $\dfrac{1}{2}$ **63.** $\dfrac{(a+3)(a-4)}{3(a^2+a-1)}$ **65.** $\left\{\dfrac{11}{2}\right\}$ **67.** $\left\{\left(\dfrac{14}{5},\dfrac{-3}{5}\right)\right\}$ **69.** $\left[\dfrac{-2}{3},2\right]$

Exercises 6.5   (page 256)

1. $\{-8\}$ **3.** $\{56\}$ **5.** $\{6\}$ **7.** $\{4\}$ **9.** $\{2\}$ **11.** $\{9\}$, $r\neq 0$ **13.** $\left\{\dfrac{1}{7}\right\}$, $y\neq 0$

15. $\{-4\}$, $x\neq -3$ **17.** $\left\{\dfrac{8}{3}\right\}$, $s\neq\dfrac{-2}{3}$ **19.** $\emptyset$, $m\neq 2, 4$ **21.** $\left\{\dfrac{9}{11}\right\}$, $x\neq 0$ **23.** $\emptyset$, $x\neq \pm 4$

25. $\{-7\}$, $p\neq 1, -3$ **27.** $\left\{\dfrac{3}{4}\right\}$, $s\neq 0, 1$ **29.** $\left\{\dfrac{-35}{3}\right\}$, $a\neq 0, \dfrac{7}{3}$ **31.** No   **33.** Yes   **35.** No

37. No   **39.** $p(1+rt)$ **41.** $\dfrac{-a}{5}$ **43.** $\dfrac{16t^2+s}{t}$ **45.** $\dfrac{pF}{F-p}$ **47.** $\dfrac{ab}{5-3a}$ **49.** $\dfrac{x^3-3}{3x^3-2}$

51. $\dfrac{-x^2+2x-2}{x^2-2x-1}$ **53.** $\left\{\dfrac{-1}{4}\right\}$, $a\neq -4, \dfrac{1}{2}, 3$ **55.** $\{-13\}$, $r\neq -5, 1, 9$ **57.** $\{2\}$, $k\neq\dfrac{-3}{5}, \dfrac{5}{4}, 6$

59. $x+\dfrac{1}{x}$ **61.** $\dfrac{600}{x}$ miles   **63.** $\dfrac{126}{N}$

Exercises 6.6   (page 261)

1. -7 or -1 **3.** 9 **5.** $\dfrac{13}{19}$ **7.** $\dfrac{7}{2}$ **9.** $\dfrac{-1}{2}$ **11.** $5\frac{1}{2}$ hours   **13.** 2 hours, 90 miles

15. Penny travels at 50 miles per hour; Bill travels at 70 miles per hour.

17. Janet rows at 10 miles per hour.   **19.** 3 miles per hour   **21.** $6\frac{3}{7}$ hours   **23.** 12 hours

25. 30 hours   **27.** 36 minutes   **29.** $11\frac{1}{4}$ hours

Chapters 5 and 6 Review (page 264)

1. $(a + 9)(a - 9)$ **2.** $6y(x + y)^2$ **3.** $(5t + 3)(4t - 7)$ **4.** $2(m + 2)(m^2 - 2m + 4)$

5. $st\left(\frac{1}{4}s - \frac{1}{5}t\right)^2$ **6.** $(p^2 + 1)(p + 5)(p - 5)$ **7.** $\{1, 4\}$ **8.** $\left\{\frac{-3}{2}, \frac{1}{3}\right\}$ **9.** $\{-2\sqrt{3}, 2\sqrt{3}, -1, 1\}$

10. $\{2, 3\}$ **11.** $\frac{5}{3}$ **12.** $\frac{(a + 3)(b + 3)}{a - 3}$ **13.** $\frac{(m - 3)(n - 2)}{n + 2}$ **14.** $-9q - 7$ **15.** 0 **16.** $\frac{y^3z}{2x^2}$

17. $\frac{4s^2 - 2r^2}{r + s}$ **18.** $\frac{p(p - 4)}{p^2 - 2p + 1}$ **19.** $\frac{2v + 1}{u^2 - v^2}$ **20.** $\frac{5x^2 - 16x + 4}{x - 3}$ **21.** $\frac{x(2x + 1)}{2(x + 1)}$

22. $\frac{3(p^2 - 1)}{5}$ **23.** $1 - 2y$ **24.** $\frac{-1}{2}$ **25.** $\frac{t - 1}{5(2t - 1)}$ **26.** $\frac{y - x}{y + x}$ **27.** $\frac{-a}{b}$ **28.** $\frac{x + y}{a - b}$

29. $\{-30\}$ **30.** $\{-1\}$ **31.** $\left\{\frac{11}{3}\right\}$ **32.** $\left\{\frac{1}{5}\right\}$ **33.** $\left\{\frac{192}{125}\right\}$ **34.** $\emptyset$ **35.** $\{-6, 3\}$ **36.** $\{-1, 9\}$

37. $\{1\}$ **38.** $\left\{\frac{-9}{2}\right\}$ **39.** $\left\{\frac{12}{7}\right\}$ **40.** $\{-7\}$ **41.** -1 **42.** $\frac{9}{13}$ **43.** $3\frac{1}{2}$ hours, 224 miles

44. $2\frac{11}{12}$ minutes **45.** 60 miles per hour **46.** $10\frac{10}{13}$ hours

Exercises 7.1 (page 268)

1. 5 **3.** 3 **5.** $\frac{1}{2}$ **7.** 8 **9.** -3 **11.** 9 **13.** 8 **15.** 3 **17.** $\frac{4}{9}$ **19.** 2 **21.** $\frac{8}{343}$

23. $\frac{5}{2}$ **25.** x **27.** $p^{4/5}$ **29.** $10t$ **31.** a^2 **33.** $18m^{5/6}$ **35.** $\frac{3}{k^{13/6}}$ **37.** $a^{1/2}$ **39.** $2r$

41. x **43.** $\frac{y}{2^{1/3}}$ **45.** $\frac{xy^2}{3}$ **47.** $\frac{2^{2/3}p^{53/63}}{q^{5/3}}$ **49.** $x + x^{3/4}$ **51.** $k - 1$ **53.** $\frac{15}{t^{5/12}} - 6t^{5/6}$

55. $x - y$ **57.** $a + 2a^{1/2}b^{1/4} + b^{1/2}$ **59.** $x + y$ **61.** $a - 2$ **63.** p **65.** $x + 1$ **67.** $1 - \frac{1}{a^{2/3}}$

69. $x^{1/2} - y^{1/2}$ **71.** $2p^{5/4} - 1$ **73.** $x^{2/3} + x^{1/3}y^{1/3} + y^{2/3}$ **75.** $\frac{x^{14/3}}{y^{1/2}z^{5/12}}$ **77.** $-5r^{7/20}s^{5/3}t^{1/12}$

79. $\frac{y^{3/16}z^{3/20}}{x^{1/2}}$ **81.** $\frac{b^2}{162a^{2/3}}$ **83.** $\frac{r^2}{s^4t^9}$ **85.** $\frac{y^{11/3}}{x^{11/4}}$ **87.** y^{2n+2} **89.** t^{3m} **91.** $x^{n+1/2}$

93. $\frac{y^{5n/6}}{x^{5m/6}}$ **95.** 2.2361 **97.** 1.4422 **99.** 0.5774 **101.** 5.1962 **103.** 0.2315 **105.** -1.0384

Exercises 7.2 (page 274)

1. $\sqrt{6}$ **3.** $\sqrt[3]{11}$ **5.** $\sqrt[3]{x^2}$ **7.** $\frac{1}{\sqrt[3]{p}}$ **9.** $\frac{6}{\sqrt[7]{t^3}}$ **11.** $\sqrt[3]{5y}$ **13.** $7^{1/2}$ **15.** $2^{1/3}$ **17.** $x^{2/3}$

19. $y^{3/7}$ **21.** $\frac{1}{5^{2/3}}$ **23.** $5^{1/3}x^{2/3}$ **25.** 7 **27.** $\frac{m^{3/5}}{3^{2/5}}$ **29.** $(x + y)^{1/2}$ **31.** $(p^3 + q^3)^{1/3}$ **33.** 2

35. 4 **37.** x **39.** -2 **41.** $2p$ **43.** $4t^2$ **45.** r^2t^3 **47.** $\frac{x}{3y}$ **49.** $x + y$ **51.** $(p + q)^2$

53. $2\sqrt{2}$ **55.** $2\sqrt{5}$ **57.** $\frac{3\sqrt{3}}{4}$ **59.** $\frac{8}{5}$ **61.** $x\sqrt{x}$ **63.** $p\sqrt{pq}$ **65.** $3\sqrt{2}$ **67.** $\sqrt{10}$

69. $\frac{3}{4}$ **71.** $\frac{x\sqrt{x}}{y}$ **73.** 0.2 **75.** 3×10^2 **77.** $3b\sqrt{7a}$ **79.** $2a^2\sqrt[3]{3a}$ **81.** $2p^2\sqrt[4]{2p}$

83. $rst^2\sqrt[6]{st}$ **85.** $\frac{\sqrt{3}}{9}$ **87.** $\frac{\sqrt[3]{6}}{9}$ **89.** $\frac{\sqrt{3x}}{6}$ **91.** $\frac{\sqrt[3]{4}}{4}$ **93.** $\frac{\sqrt[3]{8}}{4}$ **95.** $\frac{\sqrt[3]{20}}{10}$ **97.** $\frac{\sqrt{3x}}{x^2}$

99. $\dfrac{\sqrt[3]{a^2x}}{a^3}$ **101.** $x + 1$ **103.** $p + 2$ **105.** $x + y$ **107.** $\dfrac{a + 6}{b + 7}$ **109.** $|x|$ **111.** $5k^2$

113. $-3y\sqrt[3]{4y^2}$ **115.** $\dfrac{1}{s^2}$ **117.** $\dfrac{7\sqrt{2}}{|r|}$ **119.** $\dfrac{-2\sqrt{2}}{x}$ **121.** 1.897 **123.** 4.041 **125.** 1.105

Exercises 7.3   (page 278)

1. $7\sqrt{3}$ **3.** $-5\sqrt[3]{5}$ **5.** $4x\sqrt{10}$ **7.** $2\sqrt[4]{2}$ **9.** $x\sqrt{y}$ **11.** $u\sqrt[3]{3v} + u\sqrt[3]{v}$ **13.** $3\sqrt{2}$
15. $13\sqrt[3]{2}$ **17.** $6\sqrt{5}$ **19.** $-7\sqrt{3} + 3\sqrt{5}$ **21.** $-15\sqrt[4]{2}$ **23.** $17\sqrt{2} + 6\sqrt{6}$ **25.** 0
27. $v^2\sqrt[3]{2v}$ **29.** $\sqrt{5}$ **31.** $6\sqrt{7}$ **33.** $\dfrac{8y}{5}\sqrt{2xyz}$ **35.** $50u^2\sqrt{6u}$ **37.** $-7yz^3\sqrt{3y^2z^2}$

39. $6x\sqrt{6xy} - 7y\sqrt{3x} + 33x^2\sqrt{6y}$ **41.** $\dfrac{c}{a}$ **43.** $7 + 5\sqrt{5}$ **45.** $6 - 7\sqrt{7}$ **47.** $13 - 9\sqrt{11}$
49. $15 - 3\sqrt{6}$ **51.** $3\sqrt{11} - 11$ **53.** $12\sqrt{6} - 18$ **55.** $\sqrt{uv} - u$ **57.** -1 **59.** 6 **61.** $28x$
63. $7 + 2\sqrt{10}$ **65.** $a + 3 - 2\sqrt{a + 2}$ **67.** 37 **69.** $\dfrac{c}{a}$ **71.** $a + b$ **73.** $\sqrt{5} + 1$ **75.** $3 - x$
77. $\sqrt[3]{y} - 1$ **79.** $\sqrt[5]{y} - \sqrt[5]{z}$ **81.** $(\sqrt{a} + \sqrt{b})(\sqrt{a} - \sqrt{b})$ **83.** $(\sqrt[3]{r} - \sqrt[3]{s})(\sqrt[3]{r^2} + \sqrt[3]{rs} + \sqrt[3]{s^2})$
85. Yes   **87.** No   **89.** Yes   **91.** No

Exercises 7.4   (page 283)

1. $\dfrac{\sqrt{2}}{2}$ **3.** $2\sqrt{3}$ **5.** $3\sqrt{5}$ **7.** $-4\sqrt{7}$ **9.** $\dfrac{\sqrt{3}}{3}$ **11.** $\dfrac{-\sqrt{30}}{6}$ **13.** $\sqrt[3]{45}$ **15.** $\dfrac{-\sqrt[3]{252}}{5}$
17. $\sqrt[4]{x^2}$ **19.** $2\sqrt[3]{a^2}$ **21.** $\sqrt{x + y}$ **23.** $\dfrac{\sqrt[3]{3r^2}}{r}$ **25.** $\dfrac{\sqrt[4]{mn^3}}{n}$ **27.** $\sqrt[5]{u^4}$ **29.** $\dfrac{\sqrt[3]{4}}{2}$
31. $\dfrac{\sqrt[4]{9x^2}}{3x}$ **33.** $\dfrac{8\sqrt[5]{16w^2}}{w}$ **35.** $6\sqrt{3}$ **37.** $\dfrac{4}{5}\sqrt{5}$ **39.** $\dfrac{4}{3}\sqrt{3p}$ **41.** $\dfrac{6\sqrt{2} + 6 - 8\sqrt{3}}{3}$
43. $\dfrac{33\sqrt{2w} - 60\sqrt{6}}{10}$ **45.** $\dfrac{-\sqrt{x^2+1}}{x(x^2 + 1)}$ **47.** $10\sqrt{2} - 4$ **49.** $3\sqrt{10} + 8$ **51.** $\dfrac{3\sqrt{x} + x}{x}$
53. $\dfrac{5w\sqrt{6z} + 6z\sqrt{5w}}{30wz}$ **55.** $\dfrac{4\sqrt{3} + \sqrt{21}}{9}$ **57.** $\dfrac{-(4\sqrt{3} + \sqrt{5})}{43}$ **59.** $24 + 14\sqrt{3}$ **61.** $\dfrac{11 - 6\sqrt{2}}{7}$
63. $21 + 7\sqrt{5} + 15\sqrt{2} + 5\sqrt{10}$ **65.** $\dfrac{3\sqrt{2} - 4}{2}$ **67.** $\dfrac{x\sqrt{x} - x\sqrt{y} + y\sqrt{x} - y\sqrt{y}}{x - y}$
69. $\dfrac{6u - 5\sqrt{uv} + v}{4u - v}$ **71.** $\dfrac{81xy - 18\sqrt{abxy} + ab}{81xy - ab}$ **73.** $\dfrac{\sqrt[3]{a^2} + \sqrt[3]{ab} + \sqrt[3]{b^2}}{a - b}$ **75.** $\dfrac{\sqrt{m + n} + \sqrt{p}}{m + n - p}$
77. $\dfrac{\sqrt[3]{s} - \sqrt[3]{t}}{s - t}$ **79.** $a^{1/6} + a^{1/3}$ **81.** $\dfrac{(xy)^{14/15} - (xy)^{4/5}}{xy}$ **83.** $1 + (x - y)^{2/3}$ **85.** $\dfrac{3}{3\sqrt{6} + 27\sqrt{2}}$
87. $\dfrac{1}{5 + 2\sqrt{6}}$ **89.** $\dfrac{-1}{5 + 2\sqrt{6}}$ **91.** $\{-1, 5\}$ **93.** $\left\{\dfrac{-3}{2}\right\}$ **95.** $\{-1, 3\}$ **97.** $\{-3, 2\}$
99. $\left\{\dfrac{5}{13}, 3\right\}$

Exercises 7.5   (page 287)

1. $\{4\}$ **3.** $\varnothing$ **5.** $\{24\}$ **7.** $\{2\}$ **9.** $\varnothing$ **11.** $\{3\}$ **13.** $\{0\}$ **15.** $\left\{\dfrac{-35}{3}\right\}$ **17.** $\left\{\dfrac{25}{3}\right\}$

19. $\{8\}$ **21.** $\left\{\dfrac{-1}{2}\right\}$ **23.** $\{-32\}$ **25.** $\varnothing$ **27.** $\{-81\}$ **29.** $\left\{\dfrac{9}{2}\right\}$ **31.** $\{2\}$ **33.** $\left\{\dfrac{1}{2}\right\}$

35. $\left\{\dfrac{2}{5}\right\}$ **37.** $\{1\}$ **39.** $\varnothing$ **41.** $\left\{\dfrac{5}{2}\right\}$ **43.** $\left\{\dfrac{3}{8}\right\}$ **45.** $\varnothing$ **47.** $\varnothing$ **49.** $\varnothing$ **51.** $\{0\}$ **53.** $\{4\}$

55. {5}   **57.** {9}   **59.** {8}   **61.** $\left\{\dfrac{5}{2}\right\}$ **63.** {7}   **65.** {9}   **67.** {−3, 3}   **69.** {−3, 3}

Exercises 7.6 (page 292)

1. −1   **3.** i **5.** 1   **7.** −i **9.** 1   **11.** i **13.** 1   **15.** $5 − i$ **17.** $4 − 10i$ **19.** −12
21. $−12 − i$ **23.** $−4 + 3i$ **25.** 2   **27.** $−47 − 97i$ **29.** $11 − 60i$ **31.** $2 − 11i$ **33.** $−5i$

35. $2 − 2i$ **37.** $\dfrac{1}{2} + \dfrac{i}{2}$ **39.** $\dfrac{7}{29} + \dfrac{26i}{29}$ **41.** $−11 + 17i$ **43.** $2i$ **45.** $5i$ **47.** $6i$

49. $4\sqrt{2}i$ **51.** $7\sqrt{2}i$ **53.** $−30\sqrt{2}i$ **55.** $−3 + 15\sqrt{3}i$ **57.** $4\sqrt{5}i$ **59.** $−\sqrt{3}i$

61. $−4\sqrt{2}i + 16\sqrt{3}i$ **63.** $\dfrac{−\sqrt{3}i}{3}$ **65.** $\dfrac{−\sqrt{3}i}{3}$ **67.** $\dfrac{6}{7} − \dfrac{3\sqrt{3}i}{7}$ **69.** $\dfrac{4 + \sqrt{5}}{2} + \dfrac{(4 − \sqrt{5})i}{2}$

71. $\dfrac{3}{2} − \dfrac{29\sqrt{2}i}{44}$ **73.** $\dfrac{\sqrt{14} − \sqrt{35}}{3} + (2\sqrt{2} − 2\sqrt{5})i$ **75.** $x = 3,\ y = 2$ **77.** $x = \dfrac{4}{5},\ y = \dfrac{12}{5}$

79. $x = 5,\ y = 4$ **81.** $x = 1,\ y = 3$ **83.** $x = \dfrac{2}{3},\ y = \dfrac{3}{4}$ **85.** 1   **87.** −1   **89.** Yes   **91.** Yes

93. No

Exercises 7.7 (page 296)

1. 9   **3.** $\dfrac{E\sqrt{n}}{D}$ **5.** 77.25   **7.** 6.43   **9.** 1.63 inches   **11.** 7.30 inches   **15.** 200nt

17. $\dfrac{−x^2 q}{2v^2}$ **19.** $S = 6V^{2/3}$

Chapter 7 Review (page 299)

1. $\dfrac{1}{(0.3)^3}$ or $\dfrac{1}{0.027}$ **2.** $\dfrac{3^{1/2} + 2^{1/2}}{6^{1/2}}$ **3.** $5^{43/12}$ **4.** $\dfrac{1}{x^{31/12}}$ **5.** $\dfrac{8b}{a^{2/3}}$ **6.** $\dfrac{1}{(a + b)^{1/2}}$

7. $\dfrac{2x + 1}{(x − 2)^{1/2}}$ **8.** $\dfrac{2b^{1/2} + 3a^{1/2}}{a^{1/2}b^{1/2}}$ **9.** $m − n$ **10.** $\dfrac{(s + 1)^2}{s}$ **11.** $2t\sqrt[4]{4}$ **12.** 20   **13.** $−2\sqrt[3]{4}$

14. $10i$ **15.** $n + 4$ **16.** $−4k\sqrt[4]{k^3}$ **17.** 1   **18.** $43\sqrt{2}$ **19.** $x^2\sqrt{x^2+1}$ **20.** $−9 − 5\sqrt{10}$

21. $198 + 120\sqrt{2}$ **22.** $10x + 2\sqrt{2}x$ **23.** $8i$ **24.** $\left(8x − \dfrac{3x^2}{y}\right)\sqrt[3]{x^2 y}$ **25.** 1

26. $x + 2 − 2\sqrt{x + 1}$ **27.** $\dfrac{\sqrt{15}}{5}$ **28.** $\dfrac{1}{3y}\sqrt[3]{12xy^2}$ **29.** $\dfrac{3a\sqrt{7b} − 7b\sqrt{3a}}{21ab}$ **30.** $\dfrac{3\sqrt{6} − 8}{10}$

31. $\dfrac{3}{2}\sqrt[3]{4k}$ **32.** $2\sqrt[3]{4a^2}$ **33.** $\dfrac{25 + 30\sqrt{x} + 9x}{25 − 9x}$ **34.** $\dfrac{2a + b + 2\sqrt{a(a + b)}}{b}$ **35.** {8}   **36.** $\left\{\dfrac{23}{3}\right\}$

37. {−9}   **38.** Ø   **39.** {4}   **40.** Ø   **41.** {5}   **42.** $\left\{\dfrac{−7}{11}\right\}$ **43.** {8}   **44.** $\left\{\dfrac{−3}{2}, \dfrac{1}{3}\right\}$ **45.** Ø

46. $\left\{\dfrac{−1}{2}\right\}$ **47.** $x = 2,\ y = 7$ **48.** $x = 3,\ y = −2$ **49.** 0   **50.** $28\sqrt{2}$ **51.** $\dfrac{4 + 7i}{5}$

52. $\dfrac{10 − 7\sqrt{2}i}{11}$ **53.** $23 − 10\sqrt{2}i$ **54.** $15 − 5i$ **55.** $\dfrac{7 + 6\sqrt{2}i}{121}$ **56.** $−8i$

Exercises 8.1 (page 304)

1. {2, 4}   **3.** $\left\{\dfrac{1}{2}, 2\right\}$ **5.** $\left\{\dfrac{−1}{2}, 3\right\}$ **7.** $\left\{\dfrac{5}{2}, 3\right\}$ **9.** $\left\{\dfrac{1}{3}, \dfrac{3}{2}\right\}$ **11.** $\left\{\dfrac{−7}{5}, \dfrac{−5}{3}\right\}$

13. $\left\{\dfrac{3b}{5}, \dfrac{2b}{3}\right\}$ **15.** $\{\pm 4\}$ **17.** $\{\pm 2i\}$ **19.** $\{\pm 9\}$ **21.** $\left\{\pm\dfrac{1}{1000}\right\}$ **23.** $\left\{\pm\dfrac{1}{2}\right\}$ **25.** $\{\pm\sqrt{7}\}$

27. $\left\{\pm\dfrac{\sqrt{6}}{2}\right\}$ **29.** $\left\{\pm\dfrac{4}{5}\right\}$ **31.** $\left\{\dfrac{1\pm\sqrt{5}}{2}\right\}$ **33.** $\left\{-1, \dfrac{-1}{2}\right\}$ **35.** $\left\{\dfrac{-3\pm\sqrt{3}i}{2}\right\}$ **37.** $\left\{\dfrac{5\pm 2i}{7}\right\}$

39. $\{a \pm |b|\}$ **41.** $\left\{\pm\dfrac{5}{9}i\right\}$ **43.** $x^2 - 5x + 6 = 0$ **45.** $x^3 - 4x^2 - 5x = 0$ **47.** $x^2 - \dfrac{5}{6}x + \dfrac{1}{6} = 0$

49. $x^2 - 8 = 0$ **51.** $x^2 + 9 = 0$ **53.** $x^2 - 4x - 1 = 0$ **55.** $x^2 - 2x + 5 = 0$ **57.** $x^2 - 2x + 4 = 0$

59. $x^2 - (2a + 3b)x + 6ab = 0$ **61.** $ax^2 + bx + c = 0$ **63.** $\{-1 \pm \sqrt{5}\}$ **65.** $\{3 \pm \sqrt{7}\}$

67. $\left\{\dfrac{-3\pm 2\sqrt{2}}{2}\right\}$ **69.** $\left\{\dfrac{3\pm 4\sqrt{2}}{7}\right\}$ **71.** $\left\{\dfrac{3\pm\sqrt{2}i}{2}\right\}$ **73.** $\left\{\dfrac{7\pm 5\sqrt{2}i}{8}\right\}$ **75.** $\{\pm 3, \pm 3i\}$

77. $\{\pm\sqrt{2}, \pm\sqrt{2}i\}$ **79.** $\{\pm b, \pm bi\}$ **81.** $\{\pm\sqrt{2}, \pm\sqrt{3}\}$ **83.** $\dfrac{\sqrt{A\pi}}{\pi}$ **85.** $\dfrac{\sqrt{2sg}}{g}$ **87.** $\dfrac{\sqrt{AP}}{P} - 1$

89. $\pm\sqrt{c}$ **91.** $\dfrac{\pm\sqrt{rst}}{r}$ **93.** $\dfrac{b\pm\sqrt{c}}{a}$ **95.** $\dfrac{-n\pm|n|}{m}$

Exercises 8.2 (page 307)

1. $1, (a + 1)^2$ **3.** $9, (x - 3)^2$ **5.** $25, (r + 5)^2$ **7.** $49, (v - 7)^2$ **9.** $\{-1, 3\}$ **11.** $\{-6, -2\}$

13. $\{-1, 2\}$ **15.** $\{1 \pm \sqrt{2}i\}$ **17.** $\left\{\dfrac{-3\pm\sqrt{17}}{4}\right\}$ **19.** $\left\{\dfrac{5\pm\sqrt{11}i}{6}\right\}$ **21.** $\left\{\dfrac{-3\pm\sqrt{17}}{4}\right\}$

23. $\left\{\dfrac{-2\pm\sqrt{3}}{2}\right\}$ **25.** $\left\{-2, \dfrac{-3}{2}\right\}$ **27.** $\left\{\dfrac{-7\pm\sqrt{13}}{6}\right\}$ **29.** $\left\{\dfrac{-7}{5}, 0\right\}$ **31.** $\left\{\dfrac{2\pm\sqrt{3}}{3}\right\}$

33. $\{-6, -5\}$ **35.** $\left\{\dfrac{3\pm\sqrt{11}i}{2}\right\}$ **37.** $\left\{\dfrac{-1\pm\sqrt{3}i}{2}\right\}$ **39.** $\{\pm 5, \pm 5i\}$ **41.** $\left\{\dfrac{-4}{3}, \dfrac{3}{2}\right\}$

43. $\left\{\dfrac{-1}{3}, 2\right\}$ **45.** $\{4\}$ **47.** $\{5\}$ **49.** $\left\{\dfrac{-5\pm\sqrt{61}}{6}\right\}, x \neq -2$ **51.** $\left\{\dfrac{1\pm\sqrt{15}i}{4}\right\}, v \neq -1, \dfrac{1}{2}$

53. -3 **55.** 7 and 5 or -5 and -7 **57.** 3 and 4 **59.** Width $= 4$, length $= 12$

61. Altitude of 3 meters, base of 8 meters **63.** $(x + 1)^2 + (y + 2)^2 = 9$ **65.** $(x - 2)^2 + (y - 6)^2 = 47$

Exercises 8.3 (page 314)

1. $\{2, 4\}$ **3.** $\{-5, -3\}$ **5.** $\left\{\dfrac{-1}{2}, 4\right\}$ **7.** $\{-2, 1\}$ **9.** $\left\{-4, \dfrac{3}{2}\right\}$ **11.** $\left\{\dfrac{-3\pm\sqrt{7}i}{2}\right\}$

13. $\left\{\dfrac{-1\pm\sqrt{5}i}{3}\right\}$ **15.** $\left\{-6, \dfrac{-1}{3}\right\}$ **17.** $\left\{-2, \dfrac{1}{2}\right\}$ **19.** $\left\{\dfrac{-1\pm\sqrt{13}}{4}\right\}$ **21.** $\left\{\dfrac{1\pm\sqrt{21}}{2}\right\}$

23. $\left\{\dfrac{-3\pm\sqrt{13}}{2}\right\}$ **25.** $\left\{\dfrac{-7\pm\sqrt{1743}i}{8}\right\}$ **27.** $\{-6 \pm 2\sqrt{14}\}$ **29.** $\{\pm 2\sqrt{3}i\}$ **31.** $\{\pm 3\sqrt{2}\}$

33. $\{-5, 0\}$ **35.** $\left\{\dfrac{-3}{5}, 0\right\}$ **37.** $\{2 \pm 2\sqrt{2}i\}$ **39.** $\{2 \pm \sqrt{15}\}$ **41.** $\left\{\dfrac{-7i}{2}, \dfrac{-i}{2}\right\}$ **43.** $\left\{\dfrac{2i\pm\sqrt{21}}{5}\right\}$

45. Two real solutions **47.** Complex conjugates **49.** Complex conjugates

51. Two real solutions **53.** One real solution of multiplicity two **55.** Complex conjugates

57. $k < 9$ **59.** $k = 16$ **61.** $k < \dfrac{-1}{4}$ **63.** $k > 1$ **65.** $k < \dfrac{9}{32}$ **67.** $x^2 - 2x - 15 = 0$

69. $x^2 - \dfrac{4}{3}x + \dfrac{5}{9} = 0$ **71.** $x^2 - 7 = 0$ **73.** $\left\{\dfrac{-5\pm\sqrt{11}i}{2}\right\}$ **75.** $\left\{\dfrac{2}{5}\right\}$ **77.** $\left\{-3, \dfrac{5}{4}\right\}$

79. $\left\{\dfrac{-27}{8}, 1\right\}$ **81.** $\left\{\dfrac{-5}{6}\right\}$ **83.** $\{-5, 21\}$ **85.** $\left\{-3, \dfrac{-1}{4}\right\}, x \neq -4, 1$ **87.** $\left\{\dfrac{-9}{4}\right\}$ **89.** $\{0, 3\}$

Exercises 8.4 (page 318)

1. $\{\pm 1, \pm\sqrt{2}\}$ **3.** $\{\pm\sqrt{3}\}$ **5.** $\{\pm 3, \pm\sqrt{6}i\}$ **7.** $\{0, \pm\sqrt{3}i, \pm\sqrt{2}i\}$ **9.** $\{\pm 2\sqrt{2}, \pm\sqrt{7}\}$

11. $\{2, -1 \pm \sqrt{3}i\}$ **13.** $\left\{-1, \dfrac{1 \pm \sqrt{3}i}{2}\right\}$ **15.** $\left\{0, -5, \dfrac{5 \pm 5\sqrt{3}i}{2}\right\}$ **17.** $\left\{0, 1, \dfrac{-1 \pm \sqrt{3}i}{2}\right\}$

19. $\left\{\pm 1, \dfrac{-1 \pm \sqrt{3}i}{2}, \dfrac{1 \pm \sqrt{3}i}{2}\right\}$ **21.** $\left\{\pm 1, \dfrac{-1 \pm \sqrt{3}i}{2}, \dfrac{1 \pm \sqrt{3}i}{2}\right\}$ **23.** $\left\{\pm\dfrac{3}{2}, \dfrac{-3 \pm 3\sqrt{3}i}{4}, \dfrac{3 \pm 3\sqrt{3}i}{4}\right\}$

25. $\{1, 729\}$ **27.** $\{-27, 64\}$ **29.** $\left\{-32, \dfrac{-1}{32}\right\}$ **31.** $\{-2, -1, 2, 3\}$ **33.** $\left\{\pm\dfrac{\sqrt{5}}{2}, \pm 1\right\}$

35. $\left\{\pm\sqrt{2}i, \pm\dfrac{\sqrt{10}i}{2}\right\}$ **37.** $\left\{\dfrac{-3}{2}, \dfrac{2}{3}\right\}$ **39.** $\left\{\dfrac{16 \pm \sqrt{409}}{17}\right\}; x \neq 1, 7, \dfrac{1}{3}$ **41.** $\left\{\dfrac{1 \pm \sqrt{33}}{2}\right\}; p \neq \pm 1$

43. $\left\{\dfrac{53}{14}\right\}; a \neq 1, 3, 4$ **45.** $\left\{\dfrac{1 \pm \sqrt{57}}{2}\right\}$ **47.** $\{3 \pm 3\sqrt{2}\}$ **49.** $\{4\}$ **51.** 6 hours, 12 hours

53. 3 hours, 6 hours   **55.** 200 miles per hour   **57.** 6 kilometers per hour

Exercises 8.5 (page 324)

1. $(-1, 0)$ **3.**    **5.** $[-3, 2]$

7. (number line graph, 0 and 3)   **9.** $(-\infty, -5) \cup (3, \infty)$ **11.** (number line graph, $-\frac{3}{2}$, 0, $\frac{1}{3}$)

13. $[-2, 1]$ **15.** $(-3, 4)$ **17.** $[-6, -1]$ **19.** $(-\infty, -4) \cup (-4, \infty)$ **21.** $(-\infty, -2] \cup [11, \infty)$

23. $\{2\}$ **25.** $(-3, 3)$ **27.** $\left(-\infty, \dfrac{-7}{6}\right) \cup (0, \infty)$ **29.** $(-2, 0)$ **31.** $(-\infty, -4) \cup [-3, \infty)$

33. $(-\infty, -2)$ **35.** $[-5, -4)$ **37.** $(-\infty, -5) \cup [9, \infty)$ **39.** $\left(-11, \dfrac{-19}{3}\right) \cup (3, \infty)$

41. $(-2, 0] \cup \left[\dfrac{3}{2}, 5\right)$ **43.** $(-\infty, -2) \cup [0, 1]$

45. (number line graph, 0, $\frac{1}{3}$, $\frac{3}{2}$)   **47.** $(-\infty, -1) \cup \left(0, \dfrac{4}{3}\right)$ **49.** $(-4, -1) \cup (1, 4)$ **51.** (number line graph, $-3, -1, 0, 1, 3$)

53. $(-3, -2) \cup (2, 3)$ **55.** (a) $(0, 5)$ (b) $(5, 8]$ (c) $[1, 4]$ **57.** (a) $\left[\dfrac{1}{2}, 2\right]$ (b) $\dfrac{5}{2}$ second

(c) $\left(0, \dfrac{1}{2}\right) \cup \left(2, \dfrac{5}{2}\right]$ **59.** (a) $(10, 20)$ (b) $(0, 10] \cup [20, 50]$ (c) $31°$ **61.** $(0, 6)$

63. $(-\infty, 0) \cup \left(\dfrac{1}{2}, \infty\right)$ **65.** $\left[\dfrac{-7}{3}, 3\right]$ **67.** $(-\infty, \infty)$ **69.** $(-\infty, -7) \cup (17, \infty)$ **71.** $\left(\dfrac{-2}{3}, 2\right)$

Exercises 8.6 (page 327)

1. Base is 10 meters; height is 4 meters.   **3.** Width is 25 inches; length is 75 inches.   **5.** 5 and 12

7. $2\sqrt{3}$ hours   **9.** 2 miles per hour   **11.** $(2 + 2\sqrt{5})$ hours   **13.** 7 members now

15. $[60, 110]$ **17.** 219 miles   **19.** $\dfrac{x\sqrt{2}}{2}$ **21.** \$0.70   **23.** 15 hours and 10 hours

25. 5 miles per hour

Chapters 7 and 8 Review (page 331)

1. $x - y$ **2.** $x + y$ **3.** $\dfrac{5a^{1/2}b - 6ab^{1/2}}{ab}$ **4.** $\dfrac{27}{a^2b}$ **5.** $\dfrac{b^{18}}{a^9}$ **6.** $\dfrac{x^{12-16/n}}{y^{4-8/n}}$ **7.** $-2x$ **8.** $|y^3|$

9. $(a + b)^3$ **10.** 5   **11.** $|x|$ **12.** $\dfrac{u^2}{p^2}\sqrt{u}$ **13.** $\dfrac{\sqrt[4]{6}}{3}$ **14.** $\dfrac{\sqrt[4]{3xy^3}}{|y|}$ **15.** $\dfrac{\sqrt[3]{18}}{3}$ **16.** 0

17. $3y\sqrt{yz} - 4y\sqrt{2xyz}$ **18.** -2 **19.** $8 - 4\sqrt{3}$ **20.** $\dfrac{\sqrt{6} + 2\sqrt{2}}{2}$ **21.** $4 - \sqrt{15}$ **22.** $\{22\}$

23. $\varnothing$ **24.** $\left\{\dfrac{4}{5}\right\}$ **25.** $\{\pm\sqrt{10}\}$ **26.** $\{2\}$ **27.** $\left\{\dfrac{7}{9}\right\}$ **28.** i **29.** $2i$ **30.** $-2 + 5i$

31. $12 - 4\sqrt{2} + (6 - 9\sqrt{2})i$ **32.** $2 - 11i$ **33.** $7 + 24i$ **34.** $\dfrac{-5 + 12i}{13}$ **35.** $\dfrac{9 + 53i}{34}$

36. $\dfrac{-50 - 35\sqrt{2}i}{44}$ **37.** $\dfrac{1 - 5\sqrt{5}i}{9}$ **38.** $x = \dfrac{-14}{5}; y = \dfrac{-21}{5}$ **39.** $x = \dfrac{-1}{5}; y = \dfrac{16}{5}$

40. $\left\{\dfrac{1}{3}, 3\right\}$ **41.** $\{7 \pm \sqrt{6}\}$ **42.** $\left\{\dfrac{5 \pm 2\sqrt{2}i}{7}\right\}$ **43.** $\left\{\dfrac{-11 \pm 3\sqrt{3}i}{2}\right\}$ **44.** $\{-2\}$

45. $\left\{\dfrac{-7 \pm \sqrt{145}}{6}\right\}$ **46.** $\left\{\dfrac{1 \pm \sqrt{23}i}{12}\right\}$ **47.** $\left\{\dfrac{-1 \pm \sqrt{31}i}{4}\right\}$ **48.** $\{1, 4\}$ **49.** $\left\{\dfrac{-5}{2}, 1\right\}$

50. $\left\{0, \dfrac{-1 \pm \sqrt{35}i}{6}\right\}$ **51.** $\left\{\dfrac{3 \pm \sqrt{15}}{3}\right\}$ **52.** $\{2, -1 \pm \sqrt{3}i\}$ **53.** $\{\pm\sqrt{7}, \pm i\}$ **54.** $\{6^6, 1\}$

55. $\left\{\dfrac{1}{5}, \dfrac{1}{6}\right\}$ **56.** $\{1, 3, 1 \pm \sqrt{2}i\}$ **57.** $\left\{\pm\dfrac{\sqrt{2}}{2}, \pm\sqrt{2}\right\}$ **58.** $(-1, 5)$ **59.** $(-\infty, -4] \cup [1, \infty)$

60. $\left(\dfrac{-4}{3}, \dfrac{-1}{2}\right) \cup \left(\dfrac{-1}{3}, \infty\right)$

Exercises 9.1 (page 338)

1. Yes; $D = \{-1, 1, 2, 3\}$; $R = \{2, 3, 7, 8\}$ **3.** Yes; $D = \{-4, 0, 2, 6\}$; $R = \{-8, 16\}$
5. No; $D = \{2, 3, 6\}$; $R = \{-7, -5, 7\}$ **7.** Yes; $D = \{-3, -1, 0, 2\}$; $R = \{-3, -1, 5, 9\}$
9. No; $D = \{2, 3, 4\}$; $R = \{0, 1, 3, 4\}$ **11.** Yes; $D = \{x \,|\, x \in N\}$; $R = \{y \,|\, y \ge -1, y \in J\}$
13. Yes; $D = \{x \,|\, -3 < x < 3\}$; $R = \{y \,|\, 0 \le y < 3\}$ **15.** Yes; $D = \{x \,|\, x \ge -2\}$; $R = \{y \,|\, y \ge 0\}$
17. Yes; $D = \{x \,|\, x \ne 1\}$; $R = \{y \,|\, y \ne 0\}$ **19.** Yes; $D = \{x \,|\, 2 \le x < 3\} \cup \{x \,|\, x \le -2\}$; $R = \{y \,|\, y \ge 0\}$
21. Yes; $D = \{x \,|\, x \ne -3\}$; $R = \{y \,|\, y \ne 1\}$ **23.** Yes; $D = \{x \,|\, x \ge -3\}$; $R = \{y \,|\, y \le 0\}$
25. Yes; $D = \{x \,|\, -5 \le x \le 5\}$; $R = \{y \,|\, 0 \le y \le 5\}$ **27.** $-2, -4, 0$ **29.** $-1, 0, 0$ **31.** $0, 0, 4$

33. $-8, 6, 22, -6$ **35.** $3a - 2, 9b - 2, -6c - 2, 3t^2 - 2$ **37.** $\dfrac{-b^2 + 4ac}{4a}, \dfrac{3b^2 + 4ac}{4a}, 0$

39. $r^2 + 2, s^2 + 2, r^2 + 2rs + s^2 + 2, r^2 + s^2 + 4$ **41.** 1   **43.** 2   **45.** -5 **47.** $2x + h$

49. $-4x - 2h$ **51.** Yes   **53.** No   **55.** Yes   **57.** No   **59.** (a) \$0 (b) \$15,000 (c) $7\frac{1}{2}$ years

61. (a) $p(n) = -n^2 + 52n - 100$ (b) $n = 2$ or 50 (c) $2 < n < 50$ (d) $n < 2$ or $n > 50$ (e) 26 units

Exercises 9.2 (page 343)

1. $x - y + 2 = 0$ **3.** $3x + y - 11 = 0$ **5.** $x - 2y - 3 = 0$ **7.** $2x + 5y - 10 = 0$ **9.** $y - 4 = 0$
11. $x - 2 = 0$ **13.** $3x - y + 13 = 0$ **15.** $x - y - 4 = 0$ **17.** $y - 2 = 0$ **19.** $y - 2 = 0$
21. $2x + y - 8 = 0$ **23.** $5x + 6y + 8 = 0$ **25.** $y - 4 = 0$ **27.** $x + 5 = 0$ **29.** $x - 18y + 3 = 0$

31. $5x - 6y + 1 = 0$ **33.** $y = 2x + 5$ **35.** $y = -\dfrac{1}{3}x + \dfrac{1}{2}$ **37.** $y = -5$ **39.** $x - 3 = 0$

41. $x - 4y + 4 = 0$ **43.** $2x - y = 0$ **45.** $x - 2y + 5 = 0$ **47.** $3x + 5y - 10 = 0$
49. $4x + 3y - 9 = 0$ **51.** $x - 2y - 3 = 0$ **53.** $2x + 5y + 5 = 0$ **55.** $x = 0$ **57.** Yes
59. $BC: x - 3 = 0$; $AB: x + 2y - 5 = 0$; $AC: 3x - y - 1 = 0$
61. $AD: x - y + 5 = 0$; $BC: x - y - 3 = 0$; $AB: x + 7y - 3 = 0$; $CD: x + 3y - 7 = 0$
63. $22x + 25y - 69 = 0$ **65.** $175x - 50y - 448 = 0$

Exercises 9.3 (page 350)

1. 4   **3.** 10   **5.** 6   **7.** $3\sqrt{2}$ **9.** $5\sqrt{2}$ **11.** (3, 5)   **13.** (1, 8)   **15.** $\left(-1, \dfrac{15}{2}\right)$

17. 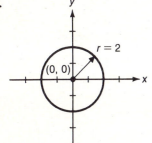   **19.** (0, 0), 3   **21.**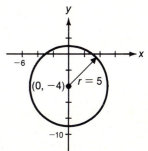

23. (3, 2), 5   **25.** 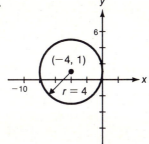   **27.** (7, 0), 6

29. $x^2 + y^2 = 4$ **31.** $(x - 1)^2 + (y - 2)^2 = 16$ **33.** $x^2 + (y + 2)^2 = 36$ **35.** $(x + 3)^2 + y^2 = 9$
37. $(x + 1)^2 + (y + 3)^2 = 25$ **39.** $(x - 3)^2 + (y + 5)^2 = 34$ **41.** $(x + 3)^2 + (y + 4)^2 = 16$
43. $(x + 3)^2 + (y + 5)^2 = 25$ **45.** $(-1, 1)$, $\sqrt{2}$ **47.** $(-4, 1)$, 3   **49.** $\left(\dfrac{5}{2}, \dfrac{-7}{2}\right)$, $\dfrac{\sqrt{70}}{2}$
51. $\left(\dfrac{1}{2}, \dfrac{-3}{2}\right)$, 2   **53.** $(4, 0)$, $2\sqrt{3}$ **55.** $\left(\dfrac{-5}{2}, \dfrac{-5}{4}\right)$, $\dfrac{5\sqrt{5}}{4}$ **57.** $(-5, 3)$, $\sqrt{38}$
59. $\left(\dfrac{-4}{3}, -6\right)$, $2\sqrt{3}$ **61.** $(x - 5)^2 + (y + 2)^2 = 85$ **63.** $(x - 1)^2 + (y - 3)^2 = 32$
65. $\left(x - \dfrac{7}{2}\right)^2 + \left(y + \dfrac{5}{2}\right)^2 = \dfrac{65}{2}$ **67.** $\left(x - \dfrac{7}{3}\right)^2 + y^2 = \dfrac{250}{9}$

Exercises 9.4 (page 358)

1.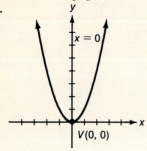

3. $V(0, -4)$; $x = 0$; 0
5. $V(0, 2)$; $x = 0$; $\pm\sqrt{2}$

7.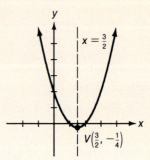

9. $V(-1, 0)$; $x = -1$; -1

11. $V\left(\dfrac{5}{2}, \dfrac{-49}{2}\right)$; $x = \dfrac{5}{2}$; $-1, 6$

13.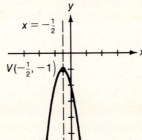

15. $V(2, -1)$; $x = 2$; up
17. $V(-1, 2)$; $y = 2$; left
19. $V(1, -5)$; $x = 1$; down

21.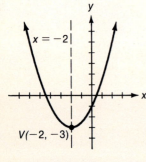

$y = \frac{1}{2}(x + 2)^2 - 3$

23. $x = \dfrac{1}{3}(y - 5)^2 - \dfrac{1}{3}$;

$V\left(\dfrac{-1}{3}, 5\right)$; $y = 5$

25.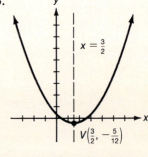

$y = \frac{1}{3}\left(x - \frac{3}{2}\right)^2 - \frac{5}{12}$

27. $x = -2(y + 3)^2 + 4$; $V(4, -3)$; $y = -3$ **29.** $y = x^2 - 1$ **31.** $y = -\dfrac{1}{2}(x + 1)^2 + 2$

33. $y = \dfrac{1}{4}(x - 3)^2 - 1$ **35.** $V(0, 0)$; $f\left(\dfrac{3}{2}, 0\right)$; $d\left(x = \dfrac{-3}{2}\right)$ **37.** $V(0, 0)$; $f(0, -1)$; $d(y = 1)$

39. $V(0, 0)$; $f\left(0, \dfrac{-4}{3}\right)$; $d\left(y = \dfrac{4}{3}\right)$ **41.** $V(0, 0)$; $f\left(\dfrac{2}{3}, 0\right)$; $d\left(x = \dfrac{-2}{3}\right)$ **43.** 0

45. $t = 2$ or 3 seconds   **47.** $x = 0$ for minimum; 100 feet high   **49.** 1250 square feet

Exercises 9.5   (page 366)

1.

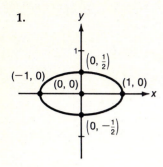

3. Center $(0, 0)$; $V(\pm 3, 0)$, $(0, \pm \sqrt{2})$

5. Center $(3, -1)$; $V(5, -1)$, $(1, -1)$, $(3, -1 \pm \sqrt{3})$

7.

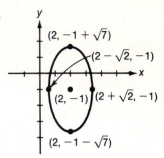

9. Center $(0, 0)$; $V(\pm 2, 0)$ **11.**

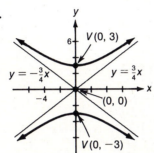

13. Center $(0, 0)$; $V\left(\pm \dfrac{2\sqrt{3}}{3}, 0\right)$

15. Center $(-3, 1)$; $V(-1, 1)$, $(-5, 1)$

17.

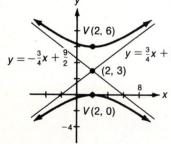

19. Circle   **21.** Hyperbola   **23.** Circle   **25.** Parabola

27. $\dfrac{y^2}{\frac{7}{16}} - \dfrac{\left(x - \frac{3}{2}\right)^2}{\frac{7}{4}} = 1$ **29.** $x = \dfrac{1}{6}\left(y - \dfrac{3}{2}\right)^2 - 1$

31. $\dfrac{(x + 3)^2}{3} + \dfrac{y^2}{1} = 1$ **33.** $(x + 2)^2 - (y - 2)^2 = 1$

35. $(x + 1)^2 + (y - 3)^2 = 5$ **37.** $\dfrac{(x - 2)^2}{1} + \dfrac{(y - 1)^2}{\frac{9}{4}} = 1$ **39.** $\left\{\left(\dfrac{29}{19}, \dfrac{4}{19}\right)\right\}$ **41.** $\left\{\left(\dfrac{29}{31}, \dfrac{-9}{31}\right)\right\}$

43. $\left\{\left(\dfrac{32}{17}, \dfrac{26}{17}, \dfrac{7}{17}\right)\right\}$

Exercises 9.6   (page 371)

1. $\{(3, 0)\}$ **3.** $\{(0, 0), (4, -4)\}$ **5.** $\{(5, -5), (-5, 5)\}$ **7.** $\left\{\left(\dfrac{2\sqrt{21}}{7}, \dfrac{\sqrt{21}}{7}\right), \left(\dfrac{-2\sqrt{21}}{7}, \dfrac{-\sqrt{21}}{7}\right)\right\}$

9. $\left\{\left(\dfrac{-24 + 8\sqrt{26}}{17}, \dfrac{-6 + 2\sqrt{26}}{17}\right), \left(\dfrac{-24 - 8\sqrt{26}}{17}, \dfrac{-6 - 2\sqrt{26}}{17}\right)\right\}$

11. $\{(7 + \sqrt{11}, 4 + \sqrt{11}), (7 - \sqrt{11}, 4 - \sqrt{11})\}$ **13.** $\{(\pm 2\sqrt{2}, 1), (\pm 2\sqrt{2}, -1)\}$

15. $\{(\pm 2, 3), (\pm 2, -3)\}$ **17.** $\{(4, 0), (-3, \pm \sqrt{7})\}$ **19.** $\{(\pm 3, 2), (\pm 3, -2)\}$ **21.** $\emptyset$

23. $\{(2, \pm \sqrt{2})\}$ **25.** $\{(2, -1)\}$ **27.** $\{(\pm 1, 0)\}$ **29.** $\{(2\sqrt{7}, \sqrt{7}), (-2\sqrt{7}, -\sqrt{7}), (4, -1), (-4, 1)\}$

31. $\{(9, 3), (-9, -3)\}$ **33.** $\dfrac{1}{8}$ and 8   **35.** 25 by 4   **37.** 9 and 7   **39.** 6 and 12   **41.** 6 and 8

43.

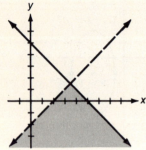

45.

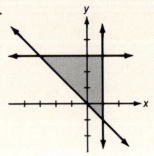

Exercises 9.7 (page 375)

1.

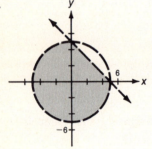

3.

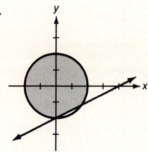

5.

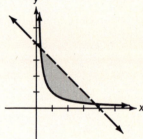

7.

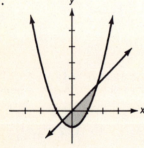

9.

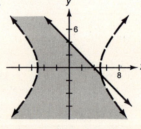

11.

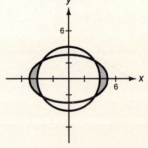

13.

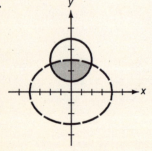

15.

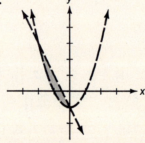

17.

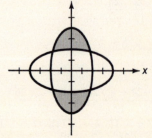

19.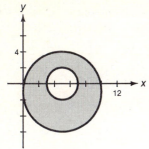

21.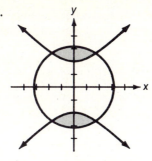

23.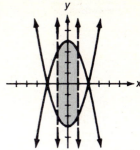

25.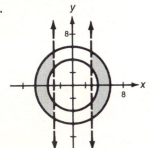

27.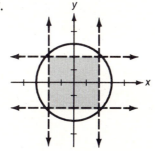

29.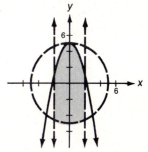

Exercises 9.8 (page 378)

1. $y = kx$ **3.** $M = kp^2$ **5.** $V = ks^3$ **7.** $y = \dfrac{k}{x}$ **9.** $S = k\sqrt{a}$ **11.** $V = khr^2$ **13.** $P = \dfrac{kc}{\sqrt{n}}$

15. 4   **17.** 98   **19.** $\dfrac{6}{125}$ **21.** $\dfrac{27}{16}$ **23.** 432   **25.** 9 foot-candles   **27.** $\dfrac{40}{3}$

29. 22 amperes   **31.** $\dfrac{48}{5}$ pounds per square foot   **33.** 960 pounds   **35.** $P = \dfrac{9}{4}N^{2/3}$

Exercises 9.9 (page 383)

1. (a) {(1, 0), (2, 1), (3, 2), (4, 3)}; (b) D: {1, 2, 3, 4}; R: {0, 1, 2, 3}; (c) one-to-one
3. (a) {(4, −2), (1, −1), (0, 0), (1, 1), (4, 2)}; (b) D: {0, 1, 4}; R: {−2, −1, 0, 1, 2}; (c) not one-to-one
5. (a) {(1, 1), (2, 4), (3, 9), (4, 16)}; (b) D: {1, 2, 3, 4}; R: {1, 4, 9, 16}; (c) one-to-one
7. (a) {(−5, −3), (1, −1), (7, 1), (13, 2)}; (b) D: {−5, 1, 7, 13}; R: {−3, −1, 1, 2}; (c) one-to-one

9. $y = \dfrac{1}{2}x$; one-to-one   **11.**

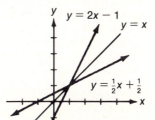

Inverse is a
one-to-one
function.

13. $y = \dfrac{5 - x}{2}$; one-to-one

15. $y = -\dfrac{3}{4}x + \dfrac{3}{8}$; one-to-one

17. Not one-to-one

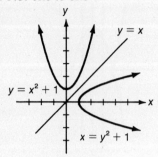

Inverse is not
a one-to-one
function.

19. $y = \pm\sqrt{9 - x}$; not one-to-one
21. $y = \pm\sqrt{3 - x}$; not one-to-one
23. $y = \pm\sqrt{-4 - x}$; not one-to-one

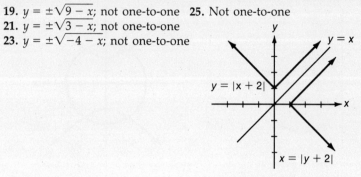

25. Not one-to-one

Inverse is not
a one-to-one
function.

27. $y = -\dfrac{1}{3}x$; one-to-one

29. One-to-one

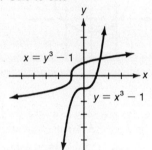

Both function
and inverse are
one-to-one.

31. $|y| = x - 1$, not one-to-one

33. (a) $f^{-1}(x) = x$

35. (a) $f^{-1}(x) = -x + 2$

37. (a) $f^{-1}(x) = \dfrac{4x + 1}{3}$

39. (a) $f^{-1}(x) = \dfrac{-x + 8}{4}$

41.

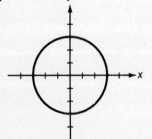

Neither relation
is a function.

43.

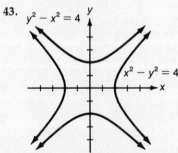

Neither relation
is a function.

45. $3y^2 + 4x^2 = 12$;
neither is a function

47.

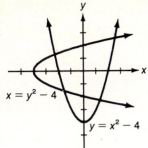

$x = y^2 - 4$

$y = x^2 - 4$

The relation
$y = x^2 - 4$ is
a function.

49. $|y - 1| = x - 1$; inverse is not a function
51. $y = \pm\sqrt{x - 1} + 4$; inverse is not a function

Chapter 9   Review (page 386)
1. $f(-2) = -11$; $f(0) = -5$; $f(3) = 4$ **2.** $f(-2) = 13$; $f(0) = 3$; $f(3) = -12$ **3.** $f(-2) = 0$; $f(0) = -4$;
$f(3) = 5$ **4.** $f(-2) = 9$; $f(0) = 1$; $f(3) = -26$ **5.** D: $\{x \mid x \in R\}$; R: $\{f(x) \mid f(x) \in R\}$ **6.** D: $\{x \mid x \in R\}$;
R: $\{f(x) \mid f(x) = 3\}$ **7.** D: $\{x \mid x \neq 3\}$; R: $\{f(x) \mid f(x) \neq 0\}$ **8.** D: $\left\{x \mid x \neq \dfrac{-3}{2}\right\}$; R: $\{f(x) \mid f(x) \neq 0\}$

9. D: $\{x \mid -3 \leq x \leq 3\}$; R: $\{f(x) \mid f(x) \geq 0\}$ **10.** D: $\{x \mid x \in R\}$; R: $\{f(x) \mid f(x) \geq 3\}$ **11.** $x + 9y - 12 = 0$

12. $y = 3x - 5$ **13.** $2x - y + 11 = 0$ **14.** $y = -\dfrac{1}{2}x + 3$ **15.** $y = \dfrac{2}{3}x$ **16.** $11x + 8y - 32 = 0$

17. $65x + 39y - 34 = 0$ **18.** $\sqrt{265}$ **19.** $\left(\dfrac{-1}{2}, \dfrac{5}{2}\right)$ **20.** Center $(2, -1)$; radius $= 4$

21. Center $(-1, 1)$; radius $= 2$ **22.** $(x - 3)^2 + (y + 1)^2 = 13$ **23.** Parabola   **24.** Parabola
25. Parabola

26.

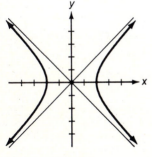

Hyperbola

27.

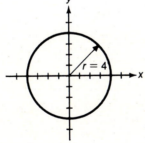

$r = 4$

Circle

28. Hyperbola

29. Hyperbola   **30.** Ellipse   **31.** Circle   **32.** Parabola   **33.** $\{(\sqrt{3}, \sqrt{3}), (-\sqrt{3}, -\sqrt{3})\}$

34. $\left\{\left(\dfrac{1}{2}, \dfrac{1}{2}\right)\right\}$ **35.** $\left\{\left(1, \pm\dfrac{\sqrt{2}}{2}\right), \left(-1, \pm\dfrac{\sqrt{2}}{2}\right)\right\}$ **36.** $\{(\pm\sqrt{2}, 0)\}$ **37.** $\emptyset$

38. $\{(2, 2), (-2, -2), (\sqrt{2}, -\sqrt{2}), (-\sqrt{2}, \sqrt{2})\}$ **39.** $\{(1, 0)\}$ **40.** $\{(1, 1), (-1, -1)\}$

41.

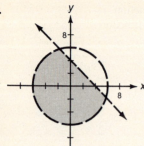

42.

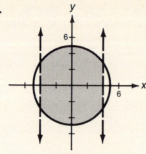

43.

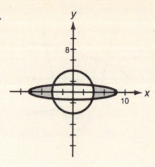

44.

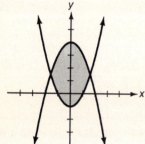

45.

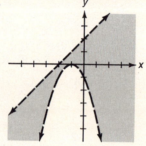

46.

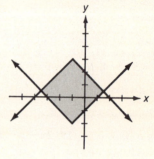

47. D: $\{x \mid x \in R\}$; R: $\{f(x) \mid f(x) \in R\}$; one-to-one   **48.** D: $\{x \mid x \in R\}$; R: $\{f(x) \mid f(x) \geq 0\}$; not one-to-one
49. D: $\{x \mid \; |x| \geq 1\}$; R: $\{f(x) \mid f(x) \geq 0\}$; not one-to-one   **50.** D: $\{x \mid x \in R\}$; R: $\{f(x) \mid f(x) \geq 1\}$;
not one-to-one   **51.** D: $\{x \mid x \in R\}$; R: $\{f(x) \mid f(x) \geq 2\}$; not one-to-one   **52.** D: $\{x \mid x \in R\}$;

R: $\{f(x) \mid f(x) \geq 1\}$; not one-to-one   **53.** 726,000 foot-pounds   **54.** $\dfrac{626}{253}$ atmospheres of pressure

55. 810 pounds

Exercises 10.1   (page 393)

  1. 1, 2, 4

3.

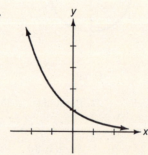

5. $\dfrac{1}{81}$, 1, 81

7. $\dfrac{1}{8}$, $\dfrac{1}{4}$, 1

9.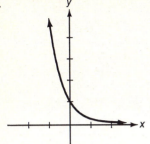

11. 1, 2

13.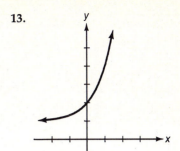

15. $\dfrac{9}{4}, \dfrac{3}{2}, 1$

17. 8, 2, $\dfrac{1}{2}$

19. 1, 1, 1

21.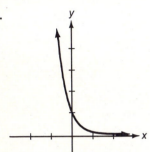

23. $\dfrac{3}{2}, \dfrac{9}{2}$

25. 1, −5

27.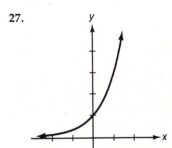

29. {2}   **31.** {3}   **33.** {3}   **35.** {1}   **37.** $\left\{\dfrac{3}{2}\right\}$ **39.** $\left\{\dfrac{3}{2}\right\}$ **41.** $\left\{\dfrac{-7}{3}\right\}$ **43.** $\left\{\dfrac{-8}{3}\right\}$

45. {−3}   **47.** $\left\{\dfrac{5}{3}\right\}$ **49.** $1166.40   **51.** $1612.55   **53.** $2257.64   **55.** $1612.23   **57.** $2901.90

59. $2546.93   **61.** (a) 200 (b) 698   **63.** (a) 400 grams (b) 339.1575 grams (c) 272.1803 grams

Exercises 10.2   (page 397)

1. $\log_3 27 = 3$ **3.** $\log_2 32 = 5$ **5.** $\log_{10} 1 = 0$ **7.** $\log_9 3 = \dfrac{1}{2}$ **9.** $\log_{10} \dfrac{1}{10} = -1$

11. $\log_3 \dfrac{1}{81} = -4$ **13.** $\log_8 4 = \dfrac{2}{3}$ **15.** $\log_{81} \dfrac{1}{27} = \dfrac{-3}{4}$ **17.** $\log_8 8 = 1$ **19.** $\log_{3/4} \dfrac{9}{16} = 2$

21. $\log_{0.3} 0.027 = 3$ **23.** $\log_{0.001} 0.1 = \dfrac{1}{3}$ **25.** $6^2 = 36$ **27.** $9^1 = 9$ **29.** $6^0 = 1$ **31.** $16^{3/4} = 8$

33. $10^{-3} = 0.001$ **35.** $3^{-2} = \dfrac{1}{9}$ **37.** $\left(\dfrac{1}{5}\right)^{-2} = 25$ **39.** $\left(\dfrac{1}{2}\right)^{-5} = 32$ **41.** {4}   **43.** {10}

45. {−2}   **47.** {100}   **49.** $\left\{\dfrac{9}{4}\right\}$ **51.** {−3}   **53.** 0   **55.** 0   **57.** 0   **59.** 1

61. D: $\{x \mid x \in R\}$; R: $\{y \mid y = 1\}$; not one-to-one

Exercises 10.3   (page 402)

1. $\log u + \log v + \log w$ **3.** $\log_b x + \log_b y - \log_b z$ **5.** $2 \log_4 a - 3 \log_4 b$

7. $\dfrac{1}{2}[\log_e (x + y) - \log_e (x - y)]$ **9.** $\dfrac{3}{2} \log x + \log y - \dfrac{3}{4} \log z$ **11.** $\log_e y + \dfrac{1}{5} \log_e x$

13. $\log_b x + \dfrac{1}{3} \log_b y - \dfrac{5}{3} \log_b z$ **15.** $\dfrac{3}{2} \log_e x - \dfrac{4}{3} \log_e y$ **17.** $\dfrac{1}{2} \log a - \dfrac{3}{16} \log b + \dfrac{1}{12} \log c$

19. $\log 2 + \log \pi + \dfrac{1}{2} \log L - \dfrac{1}{2} \log g$ **21.** $\dfrac{1}{2}[\log (s - a) + \log (s - b) + \log (s - c)]$ **23.** $\log abc$

25. $\log_e \dfrac{rt}{s}$ **27.** $\log_e b^2 c^3$ **29.** $\log 2\sqrt{3}$ **31.** $\log \dfrac{b^2}{c} \sqrt{a}$ **33.** $\log_e \dfrac{x^{3/2}}{y^4}$ **49.** $\{2^{y+c}\}$

51. $\{ae^{-T}\}$ **53.** $\left\{\dfrac{3}{2}\right\}$ **55.** $\{4\}$ **57.** $\{0.9030\}$ **59.** $\{1.0791\}$ **61.** $\{1.3545\}$ **63.** $\{-0.6020\}$

65. $\{0.6309\}$ **67.** $\{1.2918\}$ **69.** $\{2.5236\}$

Exercises 10.4   (page 406)

1. $\{998\}$ **3.** $\{120\}$ **5.** $\{e^4 - 3\}$ **7.** $\{4\}$ **9.** $\{3\}$ **11.** $\emptyset$ **13.** $\{2\}$ **15.** $\left\{\dfrac{10}{9}\right\}$ **17.** $\{4\}$

19. $\{80\}$ **21.** $\{3\}$ **23.** $\{1, 9\}$ **25.** $\{1.4037\}$ **27.** $\{0.4307\}$ **29.** $\{4.3013\}$ **31.** $\{0.5853\}$

33. $\{1.0148\}$ **35.** $\{20.1528\}$ **37.** $\{11.6869\}$ **39.** $\{15.4859\}$ **41.** $\{-0.7639, -5.2361\}$

43. $\{18.9230\}$ **45.** $\{2\}$ **47.** $\{\pm 1.2686\}$ **49.** $\emptyset$ **51.** $\{-2, 1\}$ **53.** -0.0241 **55.** 2 days

Exercises 10.5   (page 409)

1. 10.164 seconds   **3.** 159.621 feet per second   **5.** 1690.603 years   **7.** 7.273 years
9. 4119.561 years   **11.** 0.087, 567   **13.** (a) $k = 0.020$ (b) 38,531   **15.** 9.601
17. (a) 30 (b) 64.997 (c) 80   **19.** 94.952 feet   **21.** 205.933 inches

Chapters 9 and 10   Review   (page 412)

1. 0, 0, -2 **2.** $D: \{x \mid -2 \le x \le 2\}$; $R: \{g(x) \mid 0 \le g(x) \le 2\}$ **3.** $3x + 7y - 5 = 0$

4. $x - 2y - 12 = 0$ **5.** $x + 2y - 12 = 0$ **6.** $y = \dfrac{3}{4}x + 3$

7. $15x + 10y - 36 = 0$ **8.** $\sqrt{130}$ **9.** $(1, 2)$ **10.**

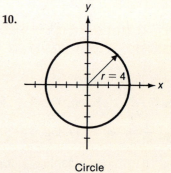

Circle

11. Parabola
12. Ellipse
13. Hyperbola
14. Circle

15.

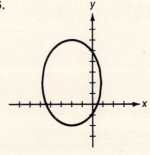

Ellipse

16. Circle
17. Ellipse

18.

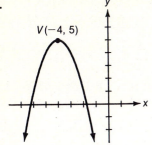

$V(-4, 5)$

Parabola

19. Hyperbola

20. $\{(\pm\sqrt{5}, 2), (\pm\sqrt{5}, -2)\}$

21. $\{(2, 2), (-2, -2), (\sqrt{2}, -\sqrt{2}), (-\sqrt{2}, \sqrt{2})\}$

22. $\emptyset$

23. $\{(1, 0)\}$

24.

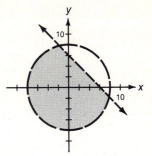

25.

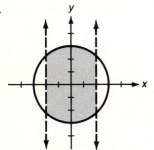

26.

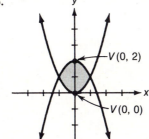

$V(0, 2)$

$V(0, 0)$

27.

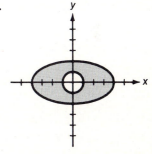

28. $f^{-1}(x) = \dfrac{x+5}{3}$

29. $h^{-1}(x) = \pm\sqrt{x+6}$

30.

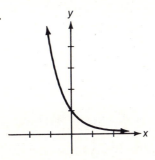

31.

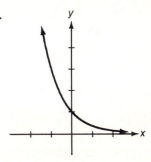

32. $\{3\}$ **33.** $\left\{\dfrac{3}{2}\right\}$ **34.** $\left\{\dfrac{5}{3}\right\}$ **35.** $\left\{\dfrac{-6}{5}\right\}$ **36.** $\log_2 8 = 3$ **37.** $\log_3 243 = 5$

38. $\log_{2/3} \dfrac{4}{9} = 2$ **39.** $\log_{0.12} 0.0144 = 2$ **40.** $7^2 = 49$ **41.** $8^0 = 1$ **42.** $\left(\dfrac{1}{2}\right)^{-6} = 64$

43. $\left(\dfrac{1}{6}\right)^{-2} = 36$ **44.** $\{3125\}$ **45.** $\{-2\}$ **46.** $\left\{\dfrac{64}{27}\right\}$ **47.** $\{-1\}$ **48.** $\log x + \log y + \dfrac{1}{2}\log z$

49. $\dfrac{1}{3}\log x + \dfrac{3}{4}\log y - \dfrac{1}{2}\log z$ **50.** $4\log_e a - 3\log_e b$ **51.** $\dfrac{1}{3}\log_b a + \dfrac{3}{4}$ **52.** $\log \dfrac{27}{16}$

53. $\log_e \dfrac{xz^3}{y}$ **54.** $\log \dfrac{36}{5}$ **55.** $\log\sqrt{\dfrac{xz^2}{y^5}}$ **56.** $\emptyset$ **57.** $\{6\}$ **58.** $\{0.7124\}$ **59.** $\{0.4660\}$

60. $\{4.7833\}$ **61.** $\{5.8332\}$ **62.** $k = 0.0866, 150.1042$ **63.** 84.5025 **64.** \$3925.94

65. 1920 pounds

Exercises 11.1   (page 418)

1. 1, 2, 3, 4, 5   **3.** 5, 7, 9, 11, 13   **5.** 1, 8, 27, 64, 125   **7.** $0, \dfrac{1}{3}, \dfrac{1}{2}, \dfrac{3}{5}, \dfrac{2}{3}$

9. 4, −8, 16, −32, 64   **11.** $\dfrac{1}{2}, \dfrac{2}{5}, \dfrac{3}{10}, \dfrac{4}{17}, \dfrac{5}{26}$ **13.** $\dfrac{1}{3}, \dfrac{1}{9}, \dfrac{1}{27}, \dfrac{1}{81}, \dfrac{1}{243}$

15. $2, \dfrac{17}{4}, \dfrac{82}{9}, \dfrac{257}{16}, \dfrac{626}{25}$ **17.** $-3x, 6x^2, -9x^3, 12x^4, -15x^5$ **19.** $x^2, 8x^3, 81x^4, 1024x^5, 15625x^6$

21. $a_n = n + 1$ **23.** $a_n = n^2$ **25.** $a_n = 3^n$ **27.** $a_n = (-1)^n$ **29.** $a_n = \dfrac{n}{n+1}$ **31.** $a_n = 4n$

33. $a_n = (-1)^n \dfrac{2n}{n+1}$ **35.** $a_n = x^{n+1}$ **37.** $a_n = \dfrac{\sqrt[n+1]{x}}{3^n}$ **39.** 10   **41.** 24   **43.** −1   **45.** 4

47. $\dfrac{20}{3}$ **49.** 140   **51.** −6   **53.** $\displaystyle\sum_{i=1}^{4} 3i$ **55.** $\displaystyle\sum_{i=1}^{5} i^2$ **57.** $\displaystyle\sum_{i=1}^{5} \dfrac{i}{i+1}$ **59.** $\displaystyle\sum_{i=1}^{5} (-1)^i \dfrac{i}{i+2}$

61. $\displaystyle\sum_{i=1}^{4} (-1)^{i+1}(4i + 3)$ **63.** $\displaystyle\sum_{i=1}^{4} (-1)^{i+1} \dfrac{x^{2i-1}}{i}$ **65.** 2, 6, 18, 54, 162, 486   **67.** −1, 1, 9, 41, 169, 681

Exercises 11.2   (page 424)

1. Yes; 1   **3.** Yes; 3   **5.** No   **7.** Yes; −5   **9.** No   **11.** Yes; $x - y$ **13.** Yes; −7

15. $a_{10} = 42; a_8 = 32$ **17.** $a_{13} = 75; a_6 = 33$ **19.** $a_7 = 22; S_{10} = 175$ **21.** $a_1 = -11; d = 4$

23. $n = 8; S_n = S_8 = 284$ **25.** $a_1 = -42; S_6 = \dfrac{-783}{4}$ **27.** $S_{20} = 1830; S_{24} = 2628$

29. $n = 4; d = 15$ **31.** $d = 2; a_{28} = 66$ **33.** $n = 6; a_n = 6$ **35.** $n = 14; a_1 = -4$ **37.** 155

39. 96   **41.** $\dfrac{203}{3}$ **43.** 1275   **45.** 4905   **47.** 728   **49.** 5, 7, 9   **51.** 6, 8, 10   **57.** \$26,400

Exercises 11.3   (page 430)

1. Geometric; $r = 2; a_n = 2^n$ **3.** Arithmetic; $d = 5; a_n = 5^n$ **5.** Geometric; $r = \dfrac{1}{2}; a_n = \left(\dfrac{1}{2}\right)^{n+1}$

7. Arithmetic; $d = \dfrac{1}{2}; a_n = \dfrac{x+n-1}{2}$ **9.** Neither   **11.** Neither   **13.** Geometric; $r = \dfrac{1}{4};$

$a_n = \dfrac{1}{2^{2n-1}}$ **15.** Geometric; $r = \dfrac{-1}{5}; a_n = \dfrac{1}{3}\left(\dfrac{-1}{5}\right)^{n-1}$ **17.** $a_4 = -189; S_4 = -140$ **19.** $a_7 = 1;$

$S_7 = \dfrac{129}{3}$ **21.** $a_1 = -1024; S_4 = -816$ **23.** $a_1 = 3^8 = 6561; S_\infty = \dfrac{19683}{4}$ **25.** $a_1 = (1.02)^5;$

$a_2 = (1.02)^6$ **27.** $n = 7; a_7 = 192$ **29.** $\dfrac{3}{4}$ **31.** 16   **33.** Does not exist; $|r| = 2$

35. Does not exist; $|r| = -1$ **37.** $\dfrac{-55713}{114688}$ **39.** 1   **41.** $\dfrac{3}{4}$ **43.** $\dfrac{1}{3}$ **45.** $\dfrac{2}{3}$ **47.** $\dfrac{7}{33}$

49. $\dfrac{41}{333}$ **51.** 15 feet   **53.** $\dfrac{1023}{64}$ centimeters = 15.9844 centimeters   **55.** 40.3895 centimeters

Exercises 11.4   (page 435)

1. 6   **3.** 3,628,800   **5.** 20   **7.** 15   **9.** 1   **11.** 165   **13.** 6   **15.** 5   **17.** 1   **19.** 1

21. $\dfrac{8!}{7!}$ **23.** $\dfrac{7!}{4!}$ **25.** $\dfrac{12!}{8!}$ **27.** $\dfrac{3!}{1!}$ or $\dfrac{3!}{0!}$ **29.** n **31.** $n(n + 1)(n + 2)(n + 3)$

33. $x^3 + 3x^2y + 3xy^2 + y^3$ **35.** $a^5 + 10a^4b + 40a^3b^2 + 80a^2b^3 + 80ab^4 + 32b^5$

37. $u^7 - 7u^6v + 21u^5v^2 - 35u^4v^3 + 35u^3v^4 - 21u^2v^5 + 7uv^6 - v^7$ **39.** $16x^4 - 32x^3y + 24x^2y^2 - 8xy^3 + y^4$

41. $81h^8 - 540h^6k + 1350h^4k^2 - 1500h^2k^3 + 625k^4$ **43.** $28x^{12}y^2$ **45.** $4032r^4s^5$ **47.** $280p^6q^8$

49. $\dfrac{-1485}{128}x^9$ **51.** 1.082856 **53.** 0.78464

Chapter 11   Review   (page 438)

1. $1, 4, 9, 16$ **2.** $4, -8, 16, -32$ **3.** $-3, 5, -7, 9$ **4.** $x, -x^2, x^3, -x^4$ **5.** $a_8 = 29; a_{11} = 41$

6. $a_6 = -13; a_9 = -22$ **7.** $a_1 = \dfrac{-2}{9}; d = \dfrac{41}{18}$ **8.** $S_{12} = 414; S_{18} = 891$ **9.** $a_6 = 5; S_6 = 30$

10. $a_6 = 2; S_6 = 126$ **11.** $S_6 = -910; a_6 = -1215$ **12.** $a_1 = 5^5 = 3125; S_4 = 3900$ **13.** 45 **14.** 30

15. -1 **16.** 92 **17.** 864 **18.** $10, 11, 12$ **19.** 5040 **20.** 1190 **21.** 28 **22.** $n(n - 1)$

23. 715 **24.** 1 **25.** $a^5 - 5a^4b + 10a^3b^2 - 10a^2b^3 + 5ab^4 + b^5$

26. $729x^6 - 2916x^5y + 4860x^4y^2 - 4320x^3y^3 + 2160x^2y^4 - 576xy^5 + 64y^6$

27. $16p^8 - 96p^6q^2 + 216p^4q^4 - 216p^2q^6 + 81q^8$

28. $u^8 - 8u^7v + 28u^6v^2 - 56u^5v^3 + 70u^4v^4 - 56u^3v^5 + 28u^2v^6 - 8uv^7 + v^8$ **29.** $-960x^7y^3$

30. $3432x^7y^7$ **31.** $42240a^7b^4$ **32.** $\dfrac{5005}{373,248}p^9$ **33.** $\$2863.57$ **34.** $\$501,000$ **35.** 90 feet

36. 86.9459 inches   **37.** 1.42534

Index

PROPERTIES OF RADICALS

$\sqrt[n]{a \cdot b} = \sqrt[n]{a} \cdot \sqrt[n]{b}$ or $(a \cdot b)^{1/n} = a^{1/n}b^{1/n}$, a and b not both 0 for n even

$\sqrt[n]{\dfrac{a}{b}} = \dfrac{\sqrt[n]{a}}{\sqrt[n]{b}}$ or $\left(\dfrac{a}{b}\right)^{1/n} = \dfrac{a^{1/n}}{b^{1/n}}$, a and b not both < 0

$b^{1/n} = \sqrt[n]{b}$ if $x \geq 0$ and n is even

$\sqrt[n]{b^n} = b$ if n is odd; $\qquad \sqrt[n]{b^n} = |b|$ if n is even

COMPLEX NUMBERS

$i = \sqrt{-1}$, $i^2 = -1$

If $a, b \in R$, then

$\qquad a + bi = c + di$ if and only if $a = c$ and $b = d$

$\qquad (a + bi) + (c + di) = (a + c) + (b + d)i$

$\qquad (a + bi)(c + di) = (ac - bd) + (ad + bc)i$

$\qquad a + bi$ and $a - bi$ are conjugates

$\qquad (a + bi)(a - bi) = a^2 + b^2$.

QUADRATIC FORMULA

The solution set of the equation $ax^2 + bx + c = 0$, $a \neq 0$, is given by the formula

$$x = \frac{-b \pm \sqrt{b^2 - 4ac}}{2a}$$

If $b^2 - 4ac > 0$, the equation has two distinct real solutions.

If $b^2 - 4ac = 0$, there is one real solution of multiplicity two.

If $b^2 - 4ac < 0$, the solutions are complex conjugates.

If x_1 and x_2 are solutions of $ax^2 + bx + c = 0$, then

$$x_1 + x_2 = \frac{-b}{a} \qquad \text{and} \qquad x_1 x_2 = \frac{c}{a}.$$

SLOPE OF A LINE

$\text{Slope} = m = \dfrac{y_2 - y_1}{x_2 - x_1}$

MIDPOINT OF A LINE

$(x_m, y_m) = \left(\dfrac{x_1 + x_2}{2}, \dfrac{y_1 + y_2}{2}\right)$

LINEAR EQUATIONS

General form $\qquad ax + by + c = 0$

Point-slope form $\qquad y - y_1 = m(x - x_1)$

Slope-intercept form $\qquad y = mx + b$

Horizontal line $\qquad y = \text{constant}$

Vertical line $\qquad x = \text{constant}$

DISTANCE FORMULA

$d = \sqrt{(x_2 - x_1)^2 + (y_2 - y_1)^2}$